D1276171

# HANDBOOK OF
# DRINKING WATER QUALITY

# HANDBOOK OF
# DRINKING
# WATER
# QUALITY

## Second Edition

### John De Zuane, P.E.
*Consulting Engineer and Former
Director of the Bureau of
Public Health Engineering
New York City Department of Health*

## VAN NOSTRAND REINHOLD

I⟨T⟩P®   A Division of International Thomson Publishing Inc.

New York • Albany • Bonn • Boston • Detroit • London • Madrid • Melbourne
Mexico City • Paris • San Francisco • Singapore • Tokyo • Toronto

Copyright © 1997 by Van Nostrand Reinhold

I(T)P ® An International Thomson Publishing Company
The ITP logo is a registered trademark used herein under license

Printed in the United States of America

For more information, contact:

Van Nostrand Reinhold
115 Fifth Avenue
New York, NY 10003

Chapman & Hall
2–6 Boundary Row
London
SE1 8HN
United Kingdom

Thomas Nelson Australia
102 Dodds Street
South Melbourne, 3205
Victoria, Australia

Nelson Canada
1120 Birchmount Road
Scarborough, Ontario
Canada M1K 5G4

Chapman & Hall GmbH
Pappelallee 3
69469 Weinheim
Germany

International Thomson Publishing Asia
221 Henderson Road #05–10
Henderson Building
Singapore 0315

International Thomson Publishing Japan
Hirakawacho Kyowa Building, 3F
2-2-1 Hirakawacho
Chiyoda-ku, 102 Tokyo
Japan

International Thomson Editores
Seneca 53
Col. Polanco
11560 Mexico D.F. Mexico

1  2  3  4  5  6  7  8  9  10  BKR  01  00  99  98  97  96

Library of Congress Cataloging-in-Publication Data

De Zuane, John
    Handbook of drinking water quality / John De Zuane.—2nd ed.
       p.    cm.
    Rev. ed. of: Handbook of drinking water quality. 1990.
    Includes bibliographical references and index.
    ISBN 0–442–02344–8
    1. Water quality management—Handbooks, manuals, etc.  2. Drinking
water—Standards—Handbooks, manuals, etc.  I. De Zuane, John
Handbook of drinking water quality.  II. Title.
TD365.D4897   1996
363.6′1—dc20                                          96-35504
                                                         CIP

 http://www.vnr.com
product discounts • free email newsletters
software demos • online resources

email: info@vnr.com

A service of  I(T)P ®

# Contents

# Preface to the Second Edition

The "technology years" of the 1990s saw an unexpected development in the field of quality control of drinking water. With all previous research efforts concentrated on carcinogens, pesticides, and herbicides, suddenly the Milwaukee waterborne outbreak caused by the infamous *Cryptosporidium* came with earthquake-like effect. Microbiological parameters and controls returned to glory, after having been demoted in the 1970s and 1980s; disinfection, perhaps with more sophisticated water treatment, was back in favor. *Cryptosporidium* and *Giardia* became the urgent targets of concern, and fear of the carcinogens was no longer at the top of the regulatory agenda—except for radon, lead, and arsenic.

It must be recognized that in the 1970s and 1980s public fear of cancer (a terrible disease, showing a statistically serious increase in the industrialized nations) had sustained the hope that the identification and elimination of some specific organic contaminant might suddenly produce great success in the battle against cancer. A reader of the voluminous "final" or "proposed" rules issued by the United States Environmental Protection Agency (USEPA) in the '90s could also reach the same conclusions (see Chapter 11). Another viewpoint may be reached by looking at the recent monitoring requirements, whereby certain organic chemicals may now be accepted by the states in intervals of sampling every *three or five years*, while other "old-fashioned" parameters, like turbidity, pH, chlorine residuals, temperature, are now subjected to *daily* recording.

A second edition has been necessary to emphasize the new world of water quality control, with more emphasis on the new pathogenic protozoa and their deleterious effects on the "sensitive population" (the infant, chronically ill, elderly, and AIDS population). In the second edition, the important parameters treated are: *Cryptosporidium*; the coliform rule; the enhanced surface water treatment requirements; new disinfectant by-products; the control of watersheds' pollutional causes; also, *radon*, perhaps a more serious contaminant than originally considered; the European drinking water standards; the new World Health Organization Guidelines of 1993; and other new problems in water treatment operation like "Zebra Mussels."

It has also been the intention of this revision to maintain the format of a handbook or manual with the same order: Chapter 1 as an introduction to water consumption and potability; Chapters 2, 3, and 4 as the core of the handbook, listing parameters as contaminants or indicators of contamination; the new emphasis on microbiological parameters and the updating of radionuclides in Chapters 5 and 6, respectively.

Sampling and monitoring required the reengineering of Chapter 8. Chapter 10 is dedicated to understanding the water treatment problems affecting water quality. Chapter 11, entirely new, presents the current and proposed federal regulations for surface water, disinfection and disinfectant by-products, the lead and copper rule, the present and future of groundwater regulations, and the information collection rules.

The preparation of the second edition was helped by comments and recommendations of competent book reviewers and users of this manual in their daily professional activities.

# Preface to the First Edition

This book is intended for use as a handbook for quick reference and technical support on the desks of sanitary engineers, public health administrators, public works engineers, water treatment plant operators, and graduate students in environmental health or public health engineering.

The purpose of this book is to evaluate and emphasize water quality control, from the source to the treatment plant, from the distribution system to the consumer. Recent detection of contamination of potable supplies—through industrial pollution and improper sanitary landfill design and operation—has demanded the quick evaluation of many chemicals, particularly synthetic organic chemicals. Most of these contaminants were never examined before in raw or potable water, and their potential ill effects were not known until the last 15 years.

Professional and public concern has demanded investigation through research, resulting in over 10,000 pages of reports, summarizing epidemiological and animal studies with related, progressive improvement of research criteria. Chapter 1 (Potable Water) and Chapter 9 (Public Health Regulations) give an introduction to major problems and solutions. Chapters 2 through 7 (Physical, Chemical, Microbiological, and Radionuclide Parameters) give detailed and specific information on how standards are evaluated and issued. (This is the "Handbook" part of the book.) Chapter 8 (Water Analyses) provides details of sampling, monitoring, and interpretation of laboratory results. Chapter 10 examines water works (intakes, treatment plants, and distribution systems) following the water path from the source to the consumer, with advice on preventive measures to maintain potability from plant effluent on through to the problems intrinsic of the distribution system.

Throughout the book an attempt has been made to present the content as simply as possible so that readers, without scientific background, can utilize the information without consulting other related books. For example, Chapter 6 (Radionuclides) contains the lengthy but necessary definitions to understand the brief but complicated standards related to radioactivity in water.

Beside explanations and listings of current national and international standards, the book provides all data necessary for the reader to write new standards, particularly where standards are limited or incomplete; for example, in developing countries where only limited information on World Health Organization (WHO) guidelines may be sometimes available.

This book was written by a civil/sanitary/public health engineer, who, after designing sanitary structures, spent 15 years in evaluating environmental problems from a Health Authority's viewpoint, mostly in the water quality control field, in the New York City urban environment, and an additional 5 years in supervising the operation of a conventional water treatment plant and its distribution system.

The fascination of a rational and reasonable evaluation of environmental contamination, based on scientific knowledge and engineering experience, has compelled me to prepare a tool to judge and comprehend the most important compound in our life: *Drinking Water*. If this Handbook will be sufficient to quickly answer the questions of the hour, my mission will be accomplished; if not, I hope it will generate a desire to expand the reader's knowledge and consultation with books on sanitary chemistry, microbiology, engineering water works design, and/or the patient review of the eight volumes of the National Academy of Science (NAS). If this desire of continuing education is stimulated, my mission will be equally accomplished.

I am pleased to thank the encouragement of many scientists in the chemical, microbiological, and virological laboratories of the New York City Health Department and the enthusiasm of the operators at the Poughkeepsie, NY, Water Purification Plant.

Thanks should also be extended to the numerous professional engineers of the technical societies (AWWA, ASCE, WPCF) and of the National Society of Professional Engineers, who, in their daily, silent, serious, unrecognized, and thankless efforts, have done so much for our technological progress; they taught me humility, discipline, and professionalism.

New York                                                    John De Zuane, P.E.

# Acknowledgments

I would like to acknowledge the productive field interviews with Mr. Douglas Fairbanks, Jr., Chief Operator of the Poughkeepsie Water Purification Plant. He is known to me for his experience, and ability to successfully manage a conventional filtration plant in which the raw water is pumped from a critical point of the Hudson River.

I am grateful to Ms. Gloria A. De Zuane for her patience in proofreading all the new material and for establishing the structure that integrated the new with the revised old material.

I would like to acknowledge the support, advice, accurate review, and environmental interest of the Van Nostrand Reinhold staff, particularly Ms. Nancy Olsen, Publisher, Ms. Sharon Gibbons, Executive Assistant, Environmental Sciences, and Ms. Barbara Mathieu, Production Editor, who personally reviewed the manuscript with professional dedication and experience and in record time.

# CHAPTER 1

# Potable Water

## RESPONSIBILITIES TO THE CONSUMER

The intention of this book is to analyze drinking water quality (also defined as potable water) from the viewpoint of sanitary or public health engineering.

In order for water delivered to the "ultimate consumer" (at the kitchen faucet) to be considered safe or potable, it must be scrutinized with a multi-science approach involving bacteriology, chemistry, physics, engineering and public health, preventive medicine, and control and evaluation management. This book provides the professional engineer, as well as the operator of water treatment plants and the student of sanitary or public health engineering or science, with a theoretical and practical guide and consultation manual to be used in the design/construction/operation of water supply facilities guided by the fundamental concepts of water quality control.

This field, which was practically developed only during the last 100 years, with initial emphasis limited to disinfection and filtration, has been revolutionized during the last 30 years when, in particular, chemical examination became more sophisticated under public pressure and sometimes public furor caused by health related environmental problems of the 20th century.

The professional or student must look at all the details, without losing the overall view of the professional responsibility of providing the best water quality to the community. This involves examining trihalomethanes, viruses, polychlorinated biphenols, carcinogens, etc. in addition to the traditional parameters—physical, chemical, and bacteriological.

This responsibility can best be summarized as follows:

- Water must be distributed in sufficient quantity and pressure at all times.
- Storage capacity at the source, as well as at intermediate points of the distribution system, should maintain the water pressure and flow (quantity) within the conventional limits, particularly during water main breaks, major fires, or rehabilitation of the existing system.
- Maintenance of the distribution system must be planned, implemented, and controlled at the same optimum level contemplated for the design, construction, and operation of the treatment facilities and the protection of the water source.
- Community leaders and the general public must receive realistic information on a continuing basis about the functioning of this vital service. This is essential in order to ensure that necessary financing is forthcoming not only for construction, upgrading, and expansion but also for the safe operation, maintenance, and control of the entire water supply system.
- Operational experience has recorded that the cost per liter, gallon, cubic foot, or cubic meter of water delivered to the consumer has increased

progressively due to manpower, automation, laboratory, upgrading, and treatment costs. Therefore, steps toward eliminating wastes, leakages, and unauthorized consumption must be undertaken. This can be accomplished by metering consumers, measuring the water supply flow, and evaluating at least annually the "unaccounted for" as a diagnosis and cure of the entire water supply system.

- Contaminants must be eliminated or reduced to a safe level to minimize menacing waterborne diseases and formation of long-term or chronic injurious health effects.
- Water quality cannot be monitored without adequate laboratory facilities. Some laboratory work is routinely performed at the water treatment plant: but a sophisticated Health Authority's approved laboratory, be it private or part of the community administration, must be available to the public health engineer and water manager so that physical, chemical, microbiological, and radionuclide parameters can be analyzed routinely, and in surveys or emergency investigations.
- Since routine analyses are reduced to the necessary minimum, special surveys must be instituted to evaluate specific problems; for example, lead content in the distribution system and at the consumer's tap, or detection of suspected contaminants at the source (industrial pollution), or suspected contamination due to cross-connection potentials.
- Health Authorities are demanding, or should request, water suppliers to routinely perform a cross-connection control program to make sure that the distribution system, in particular, is protected from plumbing errors and illegal connections that may lead to injection of non-potable water into public or private supplies of drinking water.
- A perfectly designed and built water supply (system or plant) will not survive long before incompetent personnel might undermine operation and maintenance or bypass safety devices. Health Authorities should regulate operators and laboratory personnel in terms of requisites for education, experience, licensing, and continuing education also involving ethical or professional behavior.
- Professionals who supervise operators, maintenance and laboratory personnel are the managers of water plants and related operations. Managers must be professionally responsible to the consumers and the Health Authorities. They must efficiently perform constant supervision of personnel and systems, with water quality and adequacy of flow and pressure among their top priorities. With economics in mind and technical updating in this field of revolutionary progress, continuing education and rigorous professionalism are required.
- Another responsibility of water managers is to submit an accurate annual budget with technical documentation for each budgetary item. Managers must be able to document their request for capital construction funds through a program of at least five years of upgrading and complementing existing facilities and water quality control financing.

To summarize, the following are the main professional responsibilities in the field of water control—which should not be overlooked or forgotten in the intricacy of the many parameters that complicate the task:

- Adequate flow and water pressure.
- Sufficient storage, strategically located.
- Accurate and constant control and maintenance of the distribution system and the protection of watershed and water intake facilities.
- Encouragement of public participation.
- Minimizing of per-capita consumption.
- Elimination or reduction of contaminants to a safe level.
- Adequate laboratory facilities.
- Scheduling and performing surveys for water quality evaluation.
- Implementation of a cross-connection prevention program.
- Continuing education of operators and safety training of maintenance personnel.
- Self-evaluation management performance with sensitivity to economics and technological updating.
- Accurate budgeting and availability of capital construction funds in order to provide quantity and quality to water consumers through long-range planning.

## DEFINITION OF POTABLE WATER

Potable or "drinking" water can be defined as the water delivered to the consumer that can be safely used for drinking, cooking, and washing. A certification by a licensed professional engineer specialized in the field is no longer sufficient. The public health aspects are of such importance and complexity that the Health Authority having jurisdiction in the community now reviews, inspects, samples, monitors, and evaluates on a continuing basis the water supplied to the community, using constantly updated drinking water standards. Such public health control helps to guarantee a continuous supply of water maintained within safe limits.

Water analysis alone is not sufficient to maintain quality but must be combined with the periodic review and acceptance of the facilities involved. This approval consists of the evaluation and maintenance of proper protection of the water source, qualifications of personnel, water supplier's (purveyor's) adequate monitoring work, and also evaluation of the quality and performance of laboratory work.

Hence, it could be summarized that potable water must meet the physical, chemical, bacteriological, and radionuclide parameters when supplied by an approved source, delivered to a treatment and disinfection facility of proper design, construction, and operation, and in turn delivered to the consumer through a protected distribution system in sufficient quantity and pressure.

In addition, water should have palatability, be within reasonable limits of temperature, and possibly gain the confidence of the consumer.

## WATER CONSUMPTION

Water consumption is primarily influenced by the following factors:

- the presence of high water consumption industries
- the economic level of consumers (their use of dishwashers, washing machines, pools, air conditioning, etc.)
- age of the water supply system (leakages)
- local climate and precipitations (lawn and garden watering)
- the cost of water
- the number of persons per unit of surface (concentration of population—multistory dwellings use less water per capita than one-family homes)
- maintenance of the distribution system (frequency of water main breaks)
- month of the year (peak in summer months)
- hour of the day
- presence or absence of transient population (commuters)
- size of the population served

The following data for water consumption apply to the industrialized nations:

Average daily consumption varies from 350 to 850 liters (92 to 225 gallons) per capita per day. Statistical data reporting lower or higher consumption than indicated above should be analyzed carefully and skeptically. This average is computed on an annual basis. It takes into consideration normal commercial and public use. Residential consumption may vary from 270 to 770 liters (71 to 203 gallons) per capita per day. Assuming that an average expected consumption is 600 liters (156 gallons) per capita per day, then deducting commercial/industrial use of 80 liters (21 gallons) gives the residential per capita consumption of 520 liters (135 gallons). Of this amount, only 2 liters (0.52 gallons) is the expected annual base daily consumption through drinking (basis used by toxicologists to evaluate trace metal intake) and a maximum of 8 liters (2.08 gallons) for drinking and cooking. Therefore, less than 2% of the water consumed is the portion requiring first-class quality; another 25% of the total daily consumption requiring a "second-class" quality (not directly ingested but still in contact with skin or tools used in cooking).

Since it is not possible to furnish water to the consumer in first-class, second-class, and third-class quality (while all water must be of potable quality), an effort to avoid losses or excessive consumption of "third-class" quality water appears advisable. Leakages, lawn watering, air conditioning, floor and street cleaning, etc. are considered third-class quality.

## THE MAKING OF POTABILITY

Throughout human history, the search for clean, fresh, and palatable water has been man's priority. During the last 20 centuries, serious attempts have been made to serve communities with a sufficient amount of drinking water; but only during the last two centuries have criteria for acceptability been developed in any complexity, with numbers instead of adjectives, with chemical and bacteriological examination to form the base for standards. When the relationship between waterborne diseases and drinking water was established, then the technology for treatment and disinfection developed rapidly. Standards were developed at the same time, mostly originated by the Health Authorities, and by dedicated sanitary engineers and scientists.

Prior to the Industrial Revolution, farmers and artisans obtained drinking water from shallow wells or from water drawn from streams, rivers, and lakes. The Industrial Revolution caused concentrations of population with resultant localized demand both in quantity and quality (e.g., deeper wells; pipelines from rivers; sedimentation of water in storage; and the onset of understanding the need for better sanitary conditions).

The first filtration plant in the United States was built in Poughkeepsie, New York, in 1870 using the Hudson River as a raw water source and slow sand filtration for treatment. Chlorination for microbiological disinfection was introduced at the beginning of the 20th century.

From 1925 onward, most public water supplies were equipped with some form of treatment or were in the process of planning or building water works improvements.

Construction of sewage treatment plants also developed more rapidly, particularly after World War II, in many cases eliminating the major source of pollution (raw water contamination at the water intake, typical of downstream plants). The conversion of many primary wastewater treatment plants into secondary plants (biological treatment and disinfection of the effluent) has slowed the development of new water plants where no plants exist for lack of competing funds. In order to provide a technology acceptable to the Health Authorities, when the receiving waters were to be used as raw water for water treatment plants downstream or for recreational use, priority was given to sewage treatment plants, slowing down upgrading of water treatment plants. In other words, the huge public spending required by the construction of wastewater pollution control facilities competed with the need for upgrading water works or for development of new regional water supply resources.

In the opinion of public health engineers and health department officials, environmental control priorization programs should have listed drinking water quality control as the first priority. Water quality control involves the use of the best available technology within economic limits from the protected raw water source to the required treatment and to the distribution system leading to the ultimate consumer's tap. The consumer is entitled to receive safe drinking water and be informed on water quality and planning for an adequate supply.

The economics of water supply may be briefly summarized here. The cost of providing safe and palatable water is still easily affordable for the consumer, despite increased construction costs, increased manpower costs, and the introduction of sophisticated operational instrumentation and water control monitoring. The cost of one ton of water is still normally under 50 cents (U.S.). One reason is that water is of primary importance and its need is well understood by the taxpayer. Financial bonds for water utilities are highly rated in the investment market for their safety (defaults are extremely rare). As a result, interest rates are low, and capital construction costs can be translated into very small increases in annual rates. To discourage waste, it is a wise move for communities to provide, maintain, and upgrade water meters, so that consumers pay for the water they use, rather than a flat rate.

The construction of efficient secondary wastewater treatments to prevent pollution is certainly not sufficient. An industrial pollution pretreatment program must be planned, enforced, and constantly evaluated. Similar control must be implemented at community and regional levels for control of hazardous waste from storage to ultimate disposal. A solid waste control program should be implemented by a community and its regional authority, under the overall supervision of the Health Authorities.

While the complete environmental control program is of paramount importance to the citizen, it is the most sensitive aspect of public health and preventive medicine *to protect the water supply* in this quest for better health (see "Public Health Engineering," Chapter 9).

## THE ROLE OF HEALTH AUTHORITIES

Throughout this book, ample reference to the "Health Authorities" is made (see Figure 1-1). Practically no mention has yet been made of the role and responsibility of public works organizations or sanitary engineering scientific organizations, who are certainly involved in developing projects related to water works and its engineering for maintenance and monitoring instrumentation. Such an apparent oversight should be examined and justified.

The Health Authority of this present revolutionary era in environmental control is a new institution no longer limited to contagious and preventable diseases with its biostatistics and clinics. Now, under the same original umbrella of preventive medicine, environmental health is or should be managed by sanitary/public health engineers working in collaboration with other

Figure 1–1. The New York City Health Department Laboratory is located in this building. Teamwork in financing (Public Health Service) and sharing with the New York University Hospital Research Center allow space for the Health Department Laboratory. Its subdivisions include toxicology, microbiology, retrovirology, immunology, cytology, food and water chemistry, and a Poison Control Center that provides information on an around-the-clock basis for handling acute poisoning episodes.

scientific professionals, particularly with chemists, biologists, virologists, non-medical sanitarians, epidemiologists, and toxicologists, who all are part of the Health Authority's inside team. The outside assistance of other scientists and engineers connected with environmental control facilities, but part of environmental control and enforcement governmental agencies, is, of course, essential.

In smaller communities, such a scientific, operational, superstructured agency becomes too expensive. In those cases, the "Health Authorities" referred to in this book are the regional, state, or federal (national) organizations that can provide the necessary team or teams for programming, standard setting, and evaluation of the local, general, and specific environmental conditions. **The drinking water supply and its quality is priority one in public health** but not necessarily so in public works or civil engineering. Potability is not reached with simple setting of standards or collecting samples or evaluating laboratory results.

Also, maintenance of standards is not sufficient to guarantee the safety of drinking water. Standards promulgated even with the best regulations and under the best scientific wisdom are not sufficient to assure that all chemical and bacteriological parameters are met. Moreover, it is impossible to run a "complete" analysis of potable water; this impossibility is due to lack of scientific knowledge, and unlimited funds are never available to complete a laboratory analysis. Even when funds are no obstacle, an elaborate analysis will take weeks, perhaps months, to be completed, while the results will only tell the water quality at the time of sampling. So, at present there is only one solution: a continuous, rigid control from the collection of the raw water at the intake to the treatment and distribution to the "ultimate consumer" (as the U.S. Public Health Service defined the individual to be protected at the end of the line). Hence, the Health Authority team, working in close cooperation with the project engineer, resident engineer, the chief of operation, chief of maintenance, administrative community leaders, public work directors, municipality laboratories, and under well prepared standards and regulations related to water quality, can evaluate, supervise, monitor, inform the consumer, and provide control—in its entirety. If done properly and professionally, this is the only way to guarantee that the best available technology will be used to protect the consumer in a continuously changing scientific world.

It is evident that success is reachable only with the cooperation of competent and dedicated operators and public works professionals. For all this to be accomplished, there is also need of cooperation with intelligent management and dedicated, responsible community leaders.

Health Authorities promulgate laws and regulations that must be enforced: and these must be, therefore, scientifically valid in order to stand the challenge of the courts and public scrutiny. This responsible team of officials in their duties and responsibilities is the "Health Authority" referred to throughout the book (see Figure 1-2).

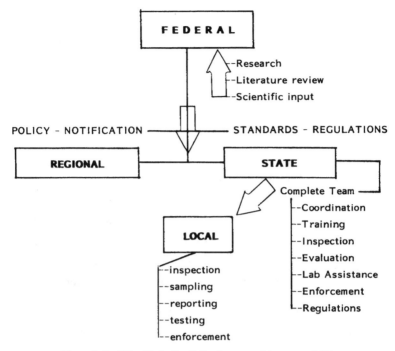

Figure 1–2. "Health Authority"—A geographic responsibility.

The World Health Organization (WHO) should be considered a "Health Authority" for the present and historical prestige of this organization, but it does not directly issue enforceable standards. The WHO is a specialized agency of the United Nations dedicated to public health and international matters related to health. In issuing the 1984 *Guidelines for Drinking-Water Quality* in *Volume 1* ("Recommendations"), WHO stated:

The *Guidelines for Drinking-Water Quality* are intended for use by countries as a basis for the development of standards which, if properly implemented, will ensure the safety of drinking-water supplies. It must be emphasized that the levels for water constituents and contaminants that are recommended in the guidelines are not standards in themselves. In order to define standards, it is necessary to consider these recommendations in the context of prevailing environmental, social, economic, and cultural conditions.

These guidelines are intended to supersede both the European (1970) and International (1971) Standards for Drinking-Water, which have been in existence for over a decade. The main reason for departing from the previous practice of prescribing international standards for drinking-water quality is the desirability of adopting a risk-benefit approach (qualitative or quantitative) to national standards and regulations. Standards and regulations achieve nothing unless they can be implemented and enforced, and this requires relatively expensive facilities and expertise.

Although the main purpose of these guidelines is to provide a basis for the development of standards, the information given may also be of assistance in developing alternative control procedures where the implementation of drinking-water standards is not feasible. For example, the existence of adequate codes of practice for the installation and operation of water-treatment plants and water supply and storage systems, and for household plumbing may promote safer drinking-water supplies by increasing the reliability of the service, avoiding the use of undesirable materials (e.g. lead pipes exposed to plumbo-corrosive water), and by simplifying repair and maintenance.

In characterizing the "Health Criteria and Other Supporting Information" of Volume 2, WHO also summarized the publication stating:

The *Guidelines for Drinking-Water Quality* are intended for use by countries as a basis for the development of standards which, if properly implemented, will ensure the safety of drinking-water supplies. This volume contains a review of the toxicological, epidemiological, and clinical evidence that was available and used in deriving the recommended guideline values that are given in Volume 1. In addition, it provides information regarding the detection of contaminants in water and measures for their control.

Although the main purpose of the guidelines is to provide a basis for the development of standards, the information given may also be of assistance to countries in developing alternative control procedures where the implementation of drinking-water standards is not feasible.

The WHO naturally has unique problems in terms of evaluating water consumption in very different climates, technological development, financial possibilities, and populations that vary from cities of over 8 million to very rural, desolate villages in the desert.

For the European Economic Community (EEC or European Union) the "Health Authority" is the European Agency that issues drinking water standards applicable and enforceable *in each nation* of the 15 countries of the European Community. The standards for potable water were issued in 1980 and became effective and enforceable in each country in 1985. (See specific standards in Appendix A-III.)

In the United States, the Health Authority may be defined as follows:

The Environmental Protection Agency (USEPA), mandated by Congress to protect the environment, acts in water programs under the Public Health Service Act as amended by the Safe Drinking Water Act (SDWA). It is the responsibility of this Agency, at the national level, to issue rules and regulations containing specific standards for drinking water distributed by public water systems.

The USEPA, after evaluation of research work and specific advisory boards, issues an Advance Notice of Proposed Rulemaking, inviting comments and scheduling public hearings. After review of comments received and additional data received in the meantime from the scientific community, the USEPA issues proposed regulations containing recommended maximum contaminant levels, with the intention to publish after a period of comments and review the final Maximum Contaminant Level or the drinking water standards that are official and enforceable.

The USEPA could not enforce these standards without the help and contribution of the States. Normally, an agreement is reached with a State as to the delegation of authority and jurisdiction involving also a federal financing of this program. The States then are expected to issue standards that are at least as restrictive as the federal standards. So the State Department of Health and/or Department of Environmental Conservation becomes the Health Authority with jurisdiction within the State. Such jurisdiction may be partially or totally delegated to County Health Departments/Environmental Conservation Departments that become the Health Authority with local jurisdiction. Normally when large cities are involved in the State, absorbing more than one county, the jurisdiction is passed to the City Health Department, which becomes the local Health Authority, with the right to issue drinking water standards that are at least as restrictive as the State standards.

In some cases, where states are in proximity, an interstate Health Authority is needed to coordinate joint efforts of environmental control, with limited power to issue standards.

## THE SAFE DRINKING WATER ACT

Until 1974, in the United States the federal standards were issued by the United States Public Health Service (USPHS), directed by the Surgeon General in Washington, DC. Public Health was the guiding force and the field of application was limited to the interstate carriers (such as trains, ships, planes, and so forth).

The USPHS issued standards in 1914, then revised them in 1925, 1942, 1946, and 1962. The 1962 standards were more specific in physical, chemical, and microbiological parameters. The 50 states accepted those standards with minor modifications. Organic chemicals were neglected for another 10 years; subsequently, the public concern about drinking water, potentially contaminated by volatile or synthetic organic chemicals, stimulated Congress to issue federal rules, the Safe Drinking Water Act of 1974, (SDWA). Contamination of drinking water by pesticides and industrial discharges is in the field of environmental control, therefore the United States Environmental Protection Agency (USEPA) became the federal agency responsible for issuing new standards, mainly the maximum contaminant levels and the rules for enforcement of nationwide improvements in drinking water quality.

The SDWA of 1974 stated specific responsibilities, dictating time frames and defining parameters. Included were provisions for procedures for a state to assume primacy (delegation of enforcement with financial help) and provisions for public notice and development of a program intended to protect the underground sources of drinking water.

The SDWA was subjected to amendments in 1977, 1979, 1980, and a major amendment in 1986, with additional rules for lead contamination control in 1988. A new amendment was scheduled for 1996.

## HOW TO USE THIS BOOK

All manuals provide a large quantity of data subdivided into an orderly form to be usefully at hand for quick reference. Consequently, these data appear uncoordinated when compared with the typical bond that textbooks provide. In consideration that such data, as presented here cannot be memorized, even by the most dedicated individual, this book should be considered a manual by the user who intends to gather information (for example: standards for "lead," health effects of "aluminum" in water, significance of "Radium-226," or the threat of "viruses").

However, to the individual interested in drinking water evaluation, be it theory or application, this handbook should be used as a textbook from the first definition of potable water and responsibilities of water works professionals (Chapter 1) to the public health regulations (Chapter 9).

The following summarizes the progressive assimilation of the public health engineering principles enumerated and described in this book.

*Chapter 1.*   Briefly describes the field of water quality control and the responsibilities to the consumer.

*Chapter 2.*   Covers the traditional analysis of physical parameters, their significance in public health, and usefulness in interpretation of related laboratory reports.

*Chapter 3.*   Introduces the inorganic chemicals, their occurrence in our environment, their toxicity, and how standards are derived and issued by the regulatory agencies in comparison with international standards and historical development. Concepts on safety factors are presented to enable the reader to judge the standard in its reality and significance of each parameter.

*Chapter 4.*   Deals with the organic chemicals and their recent introduction among the parameters fundamental for water quality control. The detergents, phenols, and trihalomethanes (THMs) determination precede the presentation of the volatile chemicals (VOCs) restricted to the most probable potential contaminants of drinking water. Also the synthetic chemicals (SOCs) are examined in detail where contamination potential and health threats are considered high enough to generate drinking water standards.

The occurrence in the environment, the health effects from toxicity to carcinogenicity, the existing standards and methods of determination in laboratories are listed for each regulated parameter. VOCs are also synthetic chemicals, but they are treated separately in public health engineering for standard setting.

*Chapter 5.*   Examines the bacteriological contamination of water, its significance in preparation of standards after brief review of waterborne diseases and the subsequent basis for water disinfection, understanding of laboratory reports, and laboratory limitation in testing (particularly routine testing).

*Chapter 6.*   Introduces the radiological standards and their verbal complexity. To understand radioactivity and its terminology, an effort is made to define all related terms and the public health significance of low-level radioactivity. Additional material is presented in Appendix A-VI, reproducing an intelligent effort on the part of USEPA (the Federal agency struggling to provide valid standards) to meet congressional deadlines for notification and enforcement.

*Chapter 7.*   Provides basic concepts in reviewing carcinogenic substances. When organic chemicals or radionuclides are reviewed, their potential carcinogenicity is listed: but in this chapter carcinogenesis is examined with concepts for research work. Chapter 7 is the last of the six chapters dedicated to significant parameters of water quality or factors to be used in judgment of potability.

*Chapter 8.*   Examines the application of standards (when, where, and how). To analyze water for evaluation of quality and routine monitoring, a proper program for sampling must be developed. The public health engineer selects the frequency and specificity of sampling and must understand the related laboratory activity. In addition, this chapter provides a guideline for laboratory data interpretation.

*Chapter 9.*   Provides the core of public health engineering activities. A list of environmental programs is followed by concepts for issuing regulations. The special significance of turbidity, a physical parameter of bacteriological importance, is reviewed here to measure what can be called the "bacteriological temperature" of the patient—water. The program of fluoridation is described here and this potentially controversial issue is analyzed.

*Chapter 10.*   Offers the final review of water quality control and structures from the source to the consumers. This section presents to the water treatment operator, public health official, or scientist involved in drinking water the examination of basic structures of water works and their relationship to water quality. Hopefully, this brief description of water works components is sufficient to follow the path of water from the state of natural condition at the source to the distribution system, where problems of water quality reappear after potability is achieved at the treatment plant effluent. Cross-connection and corrosion problems are reviewed in this important section of water quality control (the relatively large and invisible water system component: the distribution system).

*Chapter 11.*   Presents the federal regulations of top priority importance, particularly for the surface water supplies with or without conventional treatment and the introduction of the latest technology as the possible Best Available Technology (BAT). Both, the Surface Water Treatment Rule and the Enhance Surface Water Treatment Rule are included as well as the federal jurisdiction on lead and copper control. Groundwater disinfection has been proposed by the federal agency with dominant regulatory power in the environmental field, USEPA, particularly with its proposals contained in the Information Collection Rule (ICR) that will change many existing regulations in the early years of the third millenium.

*Appendixes.*   Contain a glossary of commonly used acronyms and abbreviations; list world (WHO) and European guidelines; provide an example of raw and treated water quality analyses; give the New York State standard for monitorings; radionuclide data; fluoridation, and the New York City experience; list meaningful general references for essential "company" in our library; and include a table of conversion factors for convenient reference to interpret the numerical values used in this manual.

## REFERENCES

American Water Works Association. 1995. **The S.D.W.A. Advisor**. Denver, CO: A.W.W.A.
American Water Works Association. 1991. **Journal**. Vol. 83 (6). Denver, CO: A.W.W.A.
Babbit, Dolan, and Cleasby. 1969. **Water Supply Engineering**. New York: McGraw-Hill.
California Institute of Technology. 1963. **Water Quality Criteria**. Pasadena, CA: NTIS, Springfield, VA.
Degler, S. E. and S. C. Bloom. 1969. **Federal Pollution Control Programs**. BNA's Environmental Management Series.
Hilleboe, H. E. and G. W. Larimore. 1965. **Preventive Medicine**. Philadelphia and London: W. B. Saunders Co.
Laas, W. and S. S. Beicos. 1967. "The Water In Your Life," (#75-8050). NY: Popular Library.
National Academy of Sciences. 1975. **Principles for Evaluating Chemicals in the Environment**. Washington, DC.
New York State Department of Health, Division of Environmental Health Services. **Water Supply Control**, Bulletin #22. Albany, NY.
Salvato, J. A. 1992. **Environmental Engineering and Sanitation**. New York: J. Wiley & Sons.
Skeat, W. O. 1969. **Manual of British Water Engineering Practice**, Vol. 3. Water Quality and Treatment. London: The Institution of Water Engineers.
**The Safe Drinking Water Act of 1974**. P.L. 93-523, 93 U.S.C. § (1974).

USEPA. 1986. Safe Drinking Water Act. 1986 Amendments, EPA 570/9-86-002. Washington, DC.

U.S. Public Health Service, Bureau of Water Hygiene. 1969. **Manual For Evaluating Public Drinking Water Supplies**. Cincinnati, OH.

Wolman, A. 1969. **Water, Health and Society**. Scarborough, Ontario, Canada (Fitzhenry & Whiteside, Ltd.): Indiana University Press.

# General or Physical Parameters

TASTE AND ODOR

COLOR

TEMPERATURE

pH Value

ALKALINITY

HARDNESS

SOLIDS

TURBIDITY

SOLUBILITY

WATER: CHEMICAL AND PHYSICAL PARAMETERS

SECONDARY DRINKING WATER STANDARDS

PARTICLE SIZE COUNT

ADDITIONAL GENERAL OR PHYSICAL PARAMETERS

# Physical Characteristics

The physical characteristics examined in this chapter apply both to principles of drinking water evaluation and to analysis of water found in nature as potential or actual sources of water supplies.

The typical, meaningful, or traditional parameters described here are: *Temperature, pH, Color, Taste, Odor, Turbidity, Electrical Conductivity*, and *Solids*. Turbidity is also a parameter used in the bacteriological evaluation. Solids may be classified also under chemical examination. In evaluating solids, it is useful to compare, at the same time, *Alkalinity* and *Hardness* (they also can be considered as chemical parameters). Because of the importance of pH in water treatment, a review of the basic concepts of water *Solubility* is summarized here.

## TASTE AND ODOR

Objectionable taste and odor can more likely be found at the source (raw water) than at the consumer's tap. Earthy-musty odors are normally derived from natural biological processes. They can be classified as organic compounds, but there are also synthetic compounds of agricultural and/or industrial origin.

In analyzing the odor thresholds of various substances in water, it can be stated that human abilities indicate the presence of concentrations as low as less than 1 µg/L for substances such as geosmin or methylisoborneol, or as high as 20 mg/L for chloroform.

Testing in laboratories is done according to *Standard Methods for the Examination of Water and Wastewater* (see Appendix A-VIII, "General References"). Qualitative terms used to describe taste and odor are often cataloged as swampy, grassy, medicinal, septic, phenolic, musty, fishy, and sweet. Problems with tastes and odors are normally associated with surface water vs. groundwater (at least in the ratio of 2:1).

Large surface supplies are detrimentally affected by the biological degradation of algae and their waste products; algae are usually bloomed by nutrients produced by human habitation (failing septic tanks, sewage treatment plant effluents, landfills leachate, and so forth).

Groundwaters also may be influenced by human habitation (particularly landfill leachate) in addition to possible dissolution of cells and minerals and/or bacterial activity in the aquifers. In groundwaters, one of the most common offensive chemicals is hydrogen sulfide ($H_2S$), which imparts a rotten egg odor at high concentration or simply a swampy, musty odor at low concentration. The origin of hydrogen sulfide is attributed to anaerobic bacterial action on organic sulfur, elemental sulfur, sulfates, or sulfites. Minor problems are caused in underground waters by dissolved iron and manganese.

Taste and odor problems in surface waters are usually caused by algae, actinomycetes, and other microorganisms. Actinomycetes, defined as moldlike bacteria, decomposers of microbiota, are usually associated with algae. They are most commonly detected following blue-green algae bloom. Streptomyces are commonly found, with nocardia and micromonospora also detected.

One of the most often reported complaints by drinking water consumers is "chlorine taste," where the odor threshold is sometimes as low as 0.2–0.4 mg/L at the typical pH value. Hence, disinfection itself becomes one of the major sources of complaint.

To control taste and odor problems, the best approach is watershed supervision, use of algicides, aeration, destratification of reservoirs, and pretreatment. Water filtration treatment and good management of the distribution system will help to minimize the taste and odor problems.

When preventive actions and conventional treatment are not sufficiently effective, activated carbon treatment, particularly in the powdered form, has been used extensively and successfully for taste and odor control (see Chapter 10, "Water Treatment," in the section entitled *Activated Carbon (PAC and GAC) Treatment* on page 449).

## Taste

Taste is used to determine the acceptability of drinking water from a judgment based on sensory evaluation (nerve endings in papillae located on the tongue and stimulated by chemicals). Here the major concern is with the *Taste Rating Test* and not the *Taste Threshold Test*, which is used for quantitative measurement of minimum detectable taste.

The temperature of the sampled water should be around 15°C. Usually panel members selected report the results by comparing their reactions against a rating scale (see *Standard Methods*). As this is an organoleptic test, it has no particular scientific value. Nevertheless, this parameter is important because it attempts to measure the acceptability of drinking water.

## Odor

It is assumed that odor in water is created by chemicals—particularly organic chemicals—or natural processes of decomposition of vegetable matter or microorganism activity. Industrial pollution or landfill leachate may be potential sources of taste- and odor-causing substances. So it can be stated that odor stimuli are of a chemical nature (distilled water has no taste or odor).

The odor test is an organoleptic determination and, as such, has the same limitations stated for the taste test. The instrument used for odor testing is the nose; therefore, only a qualitative test is available for evaluation of water. The *Odor Threshold Test* is, within its limitation, a quantitative test because the scope of the analysis is to determine the minimum quantity of a chemical that produces an odor. This is accomplished using dilution water and possibly a large number of persons to make up the panel, in consideration that olfactory capability varies not only from person to person but also in a given person from day to day, or even during the same day. Odor should be tested immediately after collection or preparation with odor-free water. Temperature should be controlled and recorded during the test. Odor from laboratory material and equipment may influence the determination. Smoking and eating before the test should be avoided. Unquestionably, psychological factors do influence responses. Quantitative determinations may be also influenced by the freshness of the samples, sizeable losses by evaporation, storage temperature, and improper glassware, rubber, cork, or plastic stoppers. Odor threshold tests are performed at 60°C to obtain a better evaporation (hot threshold test), but 40°C is also used.

Water tested from the distribution system may contain chlorine residual. Of course, the chlorine odor should be noted, but the test may continue with the dechlorination of the sample, normally with thiosulfate.

In 1989, the USEPA issued a "Secondary Maximum Contaminant Level" (SMCL) of 3 Threshold Odor Number (TON) for odor (Part 143—National Secondary Drinking Water Regulations—*Fed. Reg.* Vol. 54, No. 97, May 22, 1989). Secondary standards are parameters not related to health.

When a dilution is used, a number can be devised in clarifying odor. Sometimes laboratories report such units in TON:

$$\text{Threshold Odor Number} = \frac{V_T + V_D}{V_T}$$

where:

$V_T$ = volume tested

$V_D$ = volume of dilution with odor-free distilled water

for $V_D = 0$,   TON = 1 (lowest obtainable value)

for $V_D = V_T$, TON = 2

for $V_D = 2V_T$, TON = 3, etc.

"Taste and odor" and "color" determinations, in spite of their aesthetic and organoleptic limitations and the fact that they are seldom connected to toxicologic effects, are nevertheless important in potability consideration. This is because they may be a first alarm signal for a potential health hazard, and they play an important role in the consumer's evaluation of drinking water.

## COLOR

Color, when noticed in drinking water by the consumer, is an objectionable characteristic that would make the water supply psychologically unacceptable. Unusually high intensity of color at the source is caused by the presence of iron and manganese, humus and peat materials, plankton and weeds. If these causes do not exist at present or historically, industrial wastes may be a possible cause.

Color measured as a "true color" (turbidity removed) is classified by comparison of the sample with a known concentration of colored solutions. Color in raw water must be removed by pretreatment (sedimentation or dilution) and conventional treatment (coagulation, sedimentation, filtration, chlorination, or ozonation). It is mostly an aesthetic parameter evaluated at the consumer's faucet.

When turbidity is not removed, "apparent color" is noted. It is possible that removing turbidity (filtration or centrifugation) may remove some "true color."

*Standard Methods* recognizes the "Visual Comparison Method" as a reliable method for analyzing water from the distribution system and normally from raw water color determination. Other methods are listed as acceptable, but they are more applicable to highly colored water as found in industrial wastes or in some domestic waste. The "Spectrophotometric Method" may be used for some raw waters.* The "Visual Comparison Method" consists of comparing the sample with known concentrations of colored solutions or with properly calibrated glass-colored disks. The pH influences the color of water; therefore, the color value should also report the pH value.

In the laboratory, standardization of color is reached with the "Platinum-Cobalt Method"; in the field, the color determination is made with glass disks. The results are acceptable if the disks have been calibrated with the colors of the platinum–cobalt scale.

## Standards

It is difficult to connect color to health concern. Color is a physical parameter that is not necessarily related to toxicity or pathogenic contamination of water. Nevertheless, harmful color can create psychological rejection and fears, leading to limitation of water intake with consequent effects on personal health. Moreover, there have been cases in which sudden changes in color have stopped people from drinking contaminated water. Such contamination cases were caused by defective plumbing connections in buildings leading to cross-connection. Several cases were recorded by the New York City Department of Health with contamination due to chromates (air conditioning, cooling towers treatment for anticorrosion) or high concentration of copper brought into solution by low pH and long stay in stand-by tanks. Particularly, chromates impart a yellow-greenish color already at 6 mg/L when toxicity is still not significant.

---

*Approved by the Standard Methods Committee in 1993, an additional method for color was added, the *Tristimulus Filter Method* (N. 2120 of the 19th edition of 1995).

Health Authorities have regulated color in drinking water. The U.S. Public Health Service Drinking Water Standards of 1962 set a maximum of 15 units for color. The World Health Organization (WHO) set a recommendation of 5 units, and a maximum permissible level of 50 units.* The European Community (1980) set a guide number of 1 and a maximum limit of 20 units. USEPA recommends 15 units as a guide.

In 1989, USEPA issued a Secondary MCL of 15 color units for color (Part 143—National Secondary Drinking Water Regulations—*Fed. Reg.* Vol. 54, No. 97, May 22, 1989). Secondary standards are parameters not related to health.

All the above-mentioned units are in *color units* or CU; i.e., the color produced by 1 mg platinum/L (in the form of the chloroplatinate ion); the same unit is sometimes reported as mg/L scale Pt/Co (in reference to the Platinum–Cobalt Method of color analysis). To compare these units, it is possible not to notice a color of 10–15 units, while at 100 units water may have the appearance of tea.

## Notes

- The standard of 15 TCU (Total Color Unit) is based on acceptance of the fact that at 15 TCU no color is detected visually and, therefore, consumers' acceptance is expected: therefore, standards are not based on health effects.
- Water showing high color values may require treatment for reduction to an acceptable level (15 TCU). The process used is chemical oxidation to supplement the coagulation and filtration process. A study undertaken in 1976 in Sweden on 200 plants showed that colors (TCU) ranging from 5 to 150 TCU of the raw water were reduced to 5–25 TCU in treated water. (Reported by the World Health Organization (WHO) *Guidelines for Drinking-Water Quality* Vol. 2, Geneva, 1984).
- The presence of organic color (humic substances) in treated water may create difficulties in maintaining a free available chlorine residual in the distribution system. This can be confirmed with the determination of trihalomethanes, which are the reaction products of chlorine with humic substances.

## TEMPERATURE

An ideal water supply should have, at all times, an almost constant temperature or one with minimum variation; for example, 8°C to 14°C (46°F to 57°F). This small variation of temperature would indicate a physically protected water supply, with closed underground reservoirs, connected with a source of

---

*The recommended standard by WHO is 15 TCU (1984).

underground origin or deep surface reservoirs. Surface supplies, such as those in almost all cities, may not be able to avoid temperatures that seasonably vary between 4°C and 23°C (39°F and 73°F), with variation strongly influenced by climate and latitude. To maintain temperatures as constant as possible, water works, from intake structures for raw water to the mains supplying all consumers, should be designed and built in a protected underground environment. Temperature also has an influence on water treatment and in limnological evaluations. Water intakes in artificial or natural reservoirs should be designed to provide withdrawal at various depths with considerable distance between water surface and bottom, to avoid surface floating solids and bottom deposits. Portholes at different levels provide better water quality with the possibility of controlling temperatures within limits.

Accuracy in temperature measurements is not a critical factor. Temperature values are not normally standardized by public health criteria, because of the insignificant health effects. The U.S. Public Health Service and the USEPA have not issued guide numbers or limits for temperature. The WHO has not issued limitations on temperature, but the European Community (1980) uses a guide number of 12°C and a maximum of 25°C (46°F and 75°F).

In general, it can be stated that temperature has an influence on the treatment of water supplies; on the aquatic life of water reservoirs (biochemical reactions may double the reaction rate for a 10°C increase in temperature); on the taste of drinking water; on the oxygen dissolved; on the activity of organisms producing bad taste and odor; on solubility of solids in water; and on the rate of corrosion in the distribution system.

Temperature measurement of the water analyzed can be done with a mercury-filled Celsius (centigrade) thermometer, which must be periodically checked against a precision thermometer. Thermometers in metal cases are used in field determination to avoid breakage. *Standard Methods* has no specific regulations, with the exception of information on the use of reversing thermometers used to measure temperature in reservoirs for limnological determinations.

In sampling, temperature readings must be done immediately due to changes caused by air temperature and manipulation of the sample. The thermometer must be left in water long enough to get a final, constant reading.

## Notes

- Consumers agree that cool water is more palatable (psychologically associating with running water; also due to lower detectability of taste and odor).
- Chemical and bacteriological reactions increase with increasing temperature (in the raw and treated water temperature ranges).
- Trihalomethane (THM) formation and corrosion increase with increased temperature.

- Health authorities have not issued recommended standards for present inability to control temperature in the distribution system from water intake to the ultimate consumer.

## pH VALUE

### Definition

The pH is the logarithm in base 10 of the reciprocal of the hydrogen ion activity given in moles per liter.

The pH scale of values extends from 0 (very acidic) to 14 (very alkaline), with the middle value (pH = 7) corresponding to exact neutrality at 25°C (46°F). The pH value represents the instantaneous hydrogen ion activity, while values of "alkalinity" and "acidity" represent the buffering capacity of the sampled water.

$$pH = -\log[H^+] \ or \ pH = \log 1/[H^+]$$

The logarithmic function is selected not only for the simplicity of value representation but also because electrochemical potentials vary with the logarithm of the concentration, and the wide range of ionic activities is better included in the logarithmic graphical representation.

### pH in Water

Raw water examined for potential use as drinking water has an expected pH value between 4 and 9 but, more than likely, encountered values will be between 5.5 and 8.6.

Values higher than 7 are normally expected in raw water due to the presence of carbonates and bicarbonates (from contact with rocks and stones), but acid rain may lower the pH to values under 7. The pH is altered in the water treatment process by the addition of chemicals required by treatment. Before leaving the treatment plant, the pH may be again corrected to assure a slightly basic value—7.0 to 7.8—to avoid corrosion problems in the distribution system.

A general drinking water survey may encounter variation of pH at the plant effluents or in the distribution system between 6.5 and 8.0. The large majority of public water supplies will register a pH range between 6.9–7.4.

### pH Determination

pH can be determined through a colorimetric or an electrometric method. The colorimetric method is often used in field determination to measure variation in pH or to have general reference values since the interference caused by color, turbidity, colloidal matter, free chlorine, and chemicals makes readings questionable. There is an additional problem with slow deterioration of the test kit color standards.

The electrometric method known as the "Glass Electrode Method" is the standard method described by the *Standard Methods for the Examination of Water and Wastewater*. (See Appendix A-VIII, "General References").

There are several commercially available makes and models of pH meters. In addition to the necessity of following manufacturer's instructions, the expected accuracy of the instrument can be determined per *Standard Methods* calibration.

It must be noted that pH comparators have to use indicator solutions selected to cover the anticipated pH range; such indicators are not reliable in determining pH values of water for less than 20 ppm in alkalinity.

Some physical (and chemical) parameters can and should be determined first in a water treatment plant laboratory. For operational control and record, such samples can be run by the plant operator-on-duty around the clock and reviewed at least twice daily by the chief operator.

Also, in sampling, care must be taken to avoid agitation of the sample because the loss of carbon dioxide may change the actual pH value.

## pH Interpretation

pH values less than 7.0 indicate an acid reaction. Because pH is expressed in a logarithmic scale, a drop of 1.0 in pH value (pH = 6.0) is equivalent to an increase of 10 in acidic intensity.

Even mildly acidic waters influence the corrosive action of water in the distribution system and house plumbing metals. Particularly important for health effects are conditions when lead pipes exist in the distribution system or house plumbing (see under *Lead* in Chapter 3, page 80.), and the pH is lower than 7.0.

It is important to understand the difference of acidity, alkalinity, and pH in water chemistry. For example, the following acids are equal by definition in their equally "normal" solution, for example 0.1, but the pH varies considerably (as a reminder, pH is a measure of the equilibrium concentration of the dissociated hydrogen ion present in solution as $H^+$).

|  |  | At 25°C (46°F) | pH* |
|---|---|---|---|
| *ACIDS*: | hydrochloric | 0.1N | 1.1 |
|  | sulfuric | " | 1.2 |
|  | acetic | " | 2.9 |
|  | boric | " | 5.2 |
| *BASES*: | sodium hydroxide | 0.1N | 13.0 |
|  | ammonia | " | 11.1 |
|  | sodium bicarbonate | " | 8.4 |

*Handbook of Chemistry and Physics*, 75th edition, 1995.

## Health Effects

In the range of pH expected in raw and drinking water, there is no immediate direct effect of pH in health of humans. Carbonated beverages commonly consumed even by the more sensitive population[†] have pH values well below the above-mentioned acidic pH value of raw or drinking water (for example, the expected pH of soft drinks is between 2.0–4.0). The same can be said for foods (apples 2.9–3.3) generally on the acid side of the pH scale. Human urine has an expected pH range between 4.8–8.4, indicating the metabolic changes and the human tolerance of pH variations.

The minimum and maximum allowable pH range for potability is 6.5–8.5 (WHO—1984) and 6.5–8.5 (Safe Drinking Water Act—SDWA).[‡] The European Community lists a guide number to be within 6.5–8.5 (not applicable to bottled water) with a maximum of 9.5. Low pH has a more effective bactericidal, virucidal, and cysticidal action in disinfection provided by chlorination. However, the formation of THMs is favored by high pH (significantly if above 8.5).

The role of pH in water chemistry is also associated with corrosivity, alkalinity, hardness, acidity, chlorination, coagulation, and carbon dioxide stability (particularly related to groundwaters and algae flourishing). For a better understanding of pH, see also "Water Solutions—Solubility," on pages 42–44.

## ALKALINITY

### Definition

In water, alkalinity is caused by bicarbonates, carbonates, and hydroxide components in a raw or treated water supply. Bicarbonates are the major components because of carbon dioxide action on "basic" materials of soil; borates, silicates, and phosphates may be minor components. Also, alkalinity of raw water may contain salts formed from organic acids, such as humic acid.

Alkalinity is defined as the capacity of water to accept protons (this concept is used in water chemistry only). Alkalinity can also be defined as the capacity to neutralize acid.

### Alkalinity in Water

Because alkalinity is due to the presence of salts of weak acids, the major influencing components as far as pH variation is concerned are hydroxides, carbonates, and bicarbonates.

---

[†]The ill, the elderly, the very young, and the hypersensitive.
[‡]In 1989, USEPA issued a "Secondary Maximum Contaminant Level" (SMCL) of 6.5 to 8.5 for pH (Part 143—National Secondary Drinking Water Regulations—*Fed. Reg.* Vol. 54, No. 97, May 22, 1989). Secondary standards are parameters not related to health.

From the potability viewpoint, alkalinity is not a significant parameter. Variations of concentration from 5 to 125 mg/L are expected, and the extremes of these values are tolerated in water supplies. While raw waters normally contain alkalinity in the form of carbonate of soda or bicarbonate of calcium and magnesium, treated water alkalinity is mainly influenced by the chemical treatment for coagulation, water softening, and corrosion control.

## Determination

**Total alkalinity** is determined by titration with sulfuric acid or other strong acids of known strength to the end point of indicators (see *Standard Methods* in Appendix A-VIII, "General References") with the result expressed in mg/l of calcium carbonate equivalent to the determined alkalinity.

Interference is expected from free chlorine residual in some water supplies due to the bleaching action on the indicator. When the lime-soda softening process produces finely divided calcium carbonate and magnesium hydroxide, interference is expected and filtering of the sample is necessary for accuracy. Polyethylene or pyrex bottles are preferred for sample collection. Alkalinity determination should be performed as soon as possible after sample collection (certainly within one day) when considerable causticity or carbon dioxide is expected. Three types of alkalinity can be determined in water analysis: hydroxide, carbonate, and bicarbonate alkalinity. This subdivision is useful in softening processes and in boiler water analysis.

## Interpretation

Raw water alkalinity is examined with special interest when the coagulation process is utilized in treatment. A minimum of approximately 15 ppm in raw water is necessary for coagulation; a maximum of 100 ppm is acceptable for domestic use. Therefore, the in-between readings normally found for alkalinity have scarce significance from a sanitary engineering or public health viewpoint.

The value of alkalinity examined actually gives an estimate of nonacid constituents of water. When the "basic" constituents in water are limited to salts of calcium and magnesium, alkalinity values are equal to hardness values. When alkalinity is greater than hardness, basic salts, such as sodium and potassium salts, may be present in addition to calcium and magnesium.

When alkalinity is less than hardness, the presence of salts of calcium and magnesium are more likely to be sulfates instead of carbonates. For a better understanding of alkalinity, see also *Hardness*, below, *pH*, above, page 27, and "Solubility," page 40.

## Standards

There is no correlation between alkalinity and health in evaluation of drinking water quality parameters. Alkalinity in water is related to pH, hardness, calcium, magnesium, sodium, potassium and sulfates, total solids—parameters usually examined in water analyses. Alkalinity is also measured routinely in water, particularly when conventional treatment is considered. By definition, alkalinity represents a "status" of water; "capacity to accept protons" or "capacity to neutralize acid."

## HARDNESS

### Definition

Hardness may be considered a physical or chemical parameter of water. It represents the total concentration of calcium and magnesium ions, reported as calcium carbonate. Other metallic ions may be present in specific cases in such a considerable amount as to require inclusion in hardness reporting quantity. Therefore, it can also be defined as the sum of polyvalent cations present in the analyzed water. Originally, hardness was examined and evaluated in raw water sampling as an indicator of water quality in terms of precipitating soap. In this measurement, calcium and magnesium are the major precipitating ions, followed by iron and aluminum, with manganese, strontium, zinc (negligible solubility yields negligible contribution to hardness), and hydrogen ions as additional chemicals that may precipitate soap.

Since calcium and magnesium are normally the only significant ions, hardness is reported as calcium and magnesium in the calcium carbonate form. In other words, "hard" water requires more soap to produce foam or lather. The other negative aspect of "hard" water versus "soft" water is the natural capacity of hard water to produce scale in hot water pipes, boilers, and heaters. A positive aspect of hard water is less danger of corrosivity and, within certain limits, a better taste.

Surface raw water is softer than groundwater (more rain, less contact with soil minerals).

From a practical viewpoint, the degree of hardness can be interpreted as:

$$0–50 \ mg/L = soft$$
$$50–150 \ mg/L = moderately \ hard$$
$$150–300 \ mg/L = hard$$
$$300 \ up \ mg/L = very \ hard$$

Since excessive hardness can be removed by softening processes, its determination is necessary for treatment requirements and to judge the efficiency of treatment when required for industrial or commercial and, rarely, community water supply treatment. Since hardness may considerably affect industrial/commercial processes, treatment is better left to the individual consumer rather than having the supplier undertake an expensive treatment of all raw water supplies. Softening treatment, nevertheless, may be required for public water supplies when hardness is greater than 150 mg/L.

**Carbonate hardness**, also approximately the same as **temporary hardness** (hardness that can be removed by boiling), is represented by the carbonate and bicarbonate salts of calcium and magnesium.

Dissolved calcium and magnesium bicarbonates can be removed by boiling the water (temporary hardness). Precipitation of these carbonates is represented as

$$Ca(HCO_3)_2 + heat \rightarrow CaCO_3 + H_2O + CO_2$$
$$Mg(HCO_3)_2 + heat \rightarrow MgCO_3 + H_2O + CO_2$$

**Noncarbonate hardness**, also approximately the same as **permanent hardness** (hardness that cannot be removed by boiling), is represented by any salts of calcium and magnesium (such as sulfates and chlorides) except carbonates and bicarbonates.

Salt water, such as that encountered in the sea, brackish water, and waters with high sodium concentration, also interfere with soap effectiveness; but are not classified as a contribution to hardness. These are neutral salts. Fresh water contains dissolved minerals that are associated with hardness and alkalinity. Potassium bicarbonate ($KHCO_3$), potassium carbonate ($K_2CO_3$), and sodium bicarbonate—soda ($NaHCO_3$), and sodium carbonate ($Na_2CO_3$) are alkalines causing sodium and potassium alkalinity; while calcium bicarbonate ($Ca(HCO_3)_2$) and magnesium carbonate ($MgCO_3$) cause carbonate hardness. Sometimes in the past the term **negative hardness** was used to represent the difference between alkalinity and the total calcium and magnesium compounds as calcium carbonate when alkalinity in the water is reported higher than hardness. As stated above, in the United States hardness is reported as a calcium carbonate equivalent in mg/L. Other reporting systems are or have been used to report hardness. The French, German, and British Degrees are units found in water literature; but the tendency now is to report hardness as calcium. The relationship of various reporting systems is represented in Tables 2-1, 2-2, and 2-3 and the following examples.

### Examples

Water having 0.20 grams of CaO and 0.10 grams of MgO per liter contains a total hardness of

$$MgO : CaO = 0.1 : x \quad x = \frac{0.1 \times 56}{40} = 0.14 \text{ grams}$$
$$(40) \quad (56)$$

where:

(MgO hardness equivalent) 0.14 + 0.20 (CaO) = 0.34 grams/L
or 34°—German Degrees
or 34° × 1.79 = 61° French Degrees

Table 2-1.  Comparison of Hardness Units of Measure.

|  | French Degrees | British Degrees | German Degrees | Calcium Milligrams | Calcium Millimoles |
|---|---|---|---|---|---|
| French Degrees | 1 | 0.70 | 0.56 | 4.008 | 0.1 |
| British Degrees | 1.43 | 1 | 0.80 | 5.73 | 0.143 |
| German Degrees | 1.79 | 1.25 | 1 | 7.17 | 0.179 |
| Calcium Milligrams | 0.25 | 0.175 | 0.140 | 1 | 0.025 |
| Calcium Millimoles | 10 | 7 | 5.6 | 40.08 | 1 |

Table 2-2.  Comparison of Hardness Units with
Comparison of ppm of $CaCO_3$.

| Parts per Million $CaCO_3$ | Grains per U.K. gal. $CaCO_3$ (Clark Scale) (British Degrees) | Grains per U.S. Gal. $CaCO_3$ (American Degrees) | Parts per 100,000 $CaCO_3$ (French Degrees) | Parts per 100,000 CaO (German Degrees) | Parts per Million Ca (Russian Degrees) |
|---|---|---|---|---|---|
| 1.00 | 0.07 | 0.058 | 0.10 | 0.056 | 0.40 |
| 14.29 | 1.00 | 0.83 | 1.43 | 0.80 | 5.72 |
| 17.15 | 1.20 | 1.00 | 1.72 | 0.96 | 6.86 |
| 10.00 | 0.70 | 0.58 | 1.00 | 0.56 | 4.00 |
| 17.86 | 1.25 | 1.04 | 1.79 | 1.00 | 7.14 |
| 2.57 | 0.18 | 0.15 | 0.26 | 0.14 | 1.03 |

*Notes*: French Degree    1 part of $CaCO_3$ per 100,000 of water
British Degree    1 grain of $CaCO_3$ per gallon of water
German Degree    1 part of CaO per 100,000 of water

Table 2-3.  Comparison of Hardness Units.

|  | Parts per Million | Parts per 100,000 | Grains per British Gallon | Grains per U.S. Gallon |
|---|---|---|---|---|
| 1 part per million | 1.0 | 0.10 | 0.07 | 0.058 |
| 1 part per 100.000 | 10.0 | 1.00 | 0.70 | 0.580 |
| 1 grain per British Gal. | 14.3 | 1.43 | 1.00 | 0.830 |
| 1 grain per U.S. Gallon | 17.1 | 1.71 | 1.20 | 1.000 |

or 34° × 1.25 = 42.5° British Degrees
or 61° (French) × 10 = 610 mg/L as $CaCO_3$

$$CaO : CaCO_3 = 0.34 : x \quad x = \frac{(0.34)(100)}{56} = 0.61 \text{ grams}$$
$$(56) \quad (100)$$

or $0.61 \times 1{,}000 = 610$ mg/L

or French Degrees : German Degrees $= CaCO_3:CaO$
$$(100) \quad (56)$$

or 100 mg/L of hardness $(CaCO_3) = 10°$ French
$$= 5.6° \text{ German } (\times 0.56)$$
$$= 7° \quad \text{British } (\times 0.70)$$

## Public Health Aspects of Hardness

Already twenty years ago, some incomplete or preliminary studies relating to soft water and a higher incidence of cardiovascular diseases as compared with the population served by nonsoft water (much higher hardness) suggested the need to relate the risk factors involved, such as protective effects from constituents of hard water or trace elements in hard water or harmful elements in soft water, and other related potential health effects. Many studies in the United States, Canada, and Europe in the 1970s sought to define the causes and numerical incidence of this inverse correlation between hardness and cardiovascular diseases. The National Academy of Sciences in its 1977 *Drinking Water and Health—Volume 1* (see Appendix A-VIII, "General References") summarized water hardness and health as follows:

There is a large body of scientific information that indicates certain inorganic or mineral constituents of drinking water are correlated with increased morbidity and mortality rates. These constituents by usual definition are not considered to be "contaminants," as they often are associated with the level of "hardness" of drinking water and occur naturally or are picked up from water-treatment or distribution systems.

A voluminous body of literature suggests that in the United States and other developed nations, the incidence of many chronic diseases, but particularly cardiovascular diseases (heart disease, hypertension, and stroke), is associated with various water characteristics related to hardness. Most of these reports indicate an *inverse* correlation between the incidence of cardiovascular disease and the amount of hardness. A few reports also indicate a similar inverse correlation between the hardness of water and the risk from several noncardiovascular causes of death as well.

Several hypotheses are reported on how water factor(s) may effect health; these mostly involve either a protective action attributed to some elements found in hard water or harmful effects attributed to certain metals often found in soft water.

The theorized protective agents include calcium, magnesium, vanadium, lithium, chromium, and manganese. The suspect harmful agents include the metals cadmium, lead, copper, and zinc, all of which tend to be found in higher concentrations in soft water as a result of the relative corrosiveness of soft water.

It is evident from the review of the literature that there is considerable disagreement concerning the magnitude or even the existence of a "water factor" risk, the identity of the specific causal factor(s), the mode of action, and the specific pathologic effects.

Nevertheless, the preponderance of reported evidence reflects a consistent trend of statistically significant inverse correlations between the hardness of water and the incidence of cardiovascular diseases. As a result, there is a general impression that harmful elements in soft water and/or protective elements in hard water are causally implicated in the pathogenesis of cardiovascular and possibly other chronic diseases.

The wide spectrum of alleged associated effects, the lack of consistency in theorized or reported etiologic factors, the very small quantities of suspect elements in water relative to other sources, and the discrepancies between studies raise serious questions as to whether drinking water really serves as a vehicle of causal agents, is an indicator of something broader within the environment, or represents some unexplained spurious associations. Despite these uncertainties, the body of evidence is sufficiently compelling to treat the "water story" as plausible, particularly when the number of potentially preventable deaths from cardiovascular diseases is considered. In the United States, cardiovascular diseases account for more than one-half of the approximate 2 million deaths occurring each year. On the assumption that water factor(s) are causally implicated, it is estimated that optimal conditioning of drinking water could reduce this annual cardiovascular disease mortality rate by as much as 15% in the United States.

In view of this potential health significance, it is essential to ascertain whether water factors are causally linked to the induction of cardiovascular or other diseases and, if so, to identify the specific factors that are involved. Much more definitive information is needed in order to identify what remedial water treatment actions, if any, can be considered.

## Methods of Determination of Hardness

*Standard Methods* gives preference for accuracy to the method of measuring hardness by calculation,* applicable to all waters. With this method, hardness is computed from the calcium and magnesium determination; and, if present in significant amounts, other cations, determined by a mineral analysis or specifically in search for hardness producing cations, must be taken into consideration.

Another standard method is the *EDTA titration method* that can be applied with modifications to all waters. *EDTA* is the abbreviation for ethylenediaminetetraacetic acid, used with its salts in the EDTA titrimetric method. For interference, precision, and accuracy, consult the latest edition of *Standard Methods*.

## SOLIDS

In drinking water total dissolved solids are made up primarily of inorganic salts with small concentrations of organic matter. Contributory ions are mainly carbonate, bicarbonate, chloride, sulfate, nitrate, sodium, potassium,

---

*Hardness—mg equivalent $CaCO_3$ per liter = 2.497 (Ca, mg/L) + 4.118 (Mg, mg/L).

calcium, and magnesium. Major contribution to total dissolved solids in water is the natural contact with rocks and soil with minor contribution from pollution in general, including urban runoff. In some cases, however, considerable impact occurs from snow and ice control on roads in winters.

The purpose of this parameter is to evaluate and measure all suspended and dissolved matters in water. In spite of the chemical composition, solids are classified among the physical or general parameters of water quality. Total solids content of 500 mg/L is a desirable upper limit. Values of up to 1,000 mg/L of total solids have been encountered in drinking waters. Several different tests may be performed on raw and treated waters in relation to solids: dissolved solids, settleable solids, suspended solids, total solids, and specific conductance. The following definitions are applicable to these tests:

- **Total dissolved solids (TDS)**. TDS in water sampled are limited to the solids in solution.
- **Settleable solids**. Solids in suspension that can be expected to settle by gravity only in a quiescent status like an oversized settling tank. Period of time must be defined. Commonly used in analysis of sewage, this test may indicate data useful to evaluate the sedimentation process but only when very high turbidity is handled.
- **Suspended solids (SS or TSS)**. Solids not dissolved; also called suspended matter. Little or no significance in water for domestic consumption where turbidity provides a proportional if not equivalent value but with easier determination.
- **Total solids**. All solids contained in the water sampled, determined by evaporation and drying.
- **Volatile solids**. Solids made up of organic chemicals. This test has little and unreliable significance in waters intended for domestic consumption. If high organic content is expected, a COD or BOD determination can be used for raw water evaluation (see DO, BOD, and COD in Chapter 3).

## Specific Conductance or Conductivity

Conductivity is a useful test in raw and finished water for quick determination of minerals. Specific conductance is a measure of the electric current in the water sampled carried by the ionized substances; therefore, the dissolved solids are basically related to this measure, that is also influenced by the good conductivity of inorganic acids, bases, and (salts*) the poor conductivity

---

*There are different laboratory methods for the determination of conductivity when samples with a high content of salt are examined, such as when brackish water (more than 1,000 mg/L) rather than raw and treated water for drinking is tested, since the upper limit of chlorides is 250 mg/L. In brackish and sea water, **salinity** is measured with method 2520 B for determination by conductivity, while density is measured with method 2520 C. But conductivity of raw or treated water is measured with method 2510 B. (These methods are described in the 1995 edition of *Standard Methods*—see Appendix A-VIII, "General References").

characteristic of organic compounds. Raw and potable waters normally register specific conductance from 50 to 500 micromhos/cm, with mineralized water registering values over 500 and even over 1,000 micromhos/cm. It is useful to repeat that specific conductance does not measure the dissolved or the total solids but indicates with a simple test the ability of the water examined to carry an electric current. Almost daily variation of solids concentration is expected. Tests can be conducted at reasonable intervals to correlate specific conductance with the TDS of the water supply of interest. In general, in dilute solution, the two parameters are more comparable than in the concentrated solutions (interferences due to the concentrations of ions).

Taking into consideration that the two parameters (TDS and specific conductance) are reported in different units, a ratio of one-half TDS absolute value in respect of specific conductance in dilute solutions is expected, with TDS increasing to 90% of the absolute value of specific conductance in a concentrated solution. Since electrical resistance is measured in OHMS, the electrical conductance is measured by its inverse, MHOS. The cm (centimeter) usually indicated in the specific conductance is the standard distance of 1 cm apart of two electrodes of 1 cm square. Temperature significantly influences the results. The standard test is conducted at 25°C. A chart can be easily prepared to relate sampled water at different temperatures while chilling the sample down to 0°C, and then reading the specific conductance while the sample reaches room temperature (25°C). *Standard Methods* (see General References) also provides information on precision and accuracy.

## Solids and Health Significance

High content of solids in water has been inversely correlated with increased morbidity and mortality rates (the possibility of potential danger of soft waters mainly relates to cardiovascular diseases). These mineral constituents have been related to hardness of water rather than total solids or dissolved solids in the public health studies recently conducted. For this reason, the influence of solids in public health is usually reported under *Hardness*. In the United States the TDS recommended upper limit has been 500 mg/L. These same recommendations have been used by the WHO, allowing a maximum of 1,500 mg/L.* The European Community has also reported a maximum admissible level of 1,500 mg/L with no guide number. The limitation of total dissolved solids in drinking water standards is dictated by the physiological effects that may appear with TDS values over 500 mg/L[†] but more likely over 1,000 mg/L.

---

*In the 1993 second edition of the *Guidelines*, WHO listed the dissolved solids at 1,000 mg/L, as a reason for consumer complaints (see Appendix A-II for the WHO Guidelines).
†The USEPA issued in 1989 a Secondary MCL = 500 mg/L for TDS.

*Standard Methods* lists the following methods for determination of **solids**:

- Total solids dried at 103°C–105°C
- Total dissolved solids dried at 180°C
- Total suspended solids dried at 103°C–105°C
- Fixed and volatile solids ignited at 550°C
- Settleable solids (Imhoff cone—volumetric—gravimetric)
- Total, fixed, and volatile solids in solid or semisolid samples

## TURBIDITY

Turbidity is the measure of the fine suspended matter in water, mostly caused by colloidal matter, officially reported in standard units or equivalent to milligrams per liter of silica of diatomaceous earth that could cause the same optical effect. The suspended matter causing turbidity is expected to be clay, silt, nonliving organic particulates, plankton, and other microscopic organisms, in addition to suspended organic or inorganic matter. Turbidity reporting units represent light-scattering and absorbing properties of suspended matter in water.

The interest of measuring turbidity is not merely due to the fact that small particles in water are objectionable; it is also the consequence of considering that these innocuous particles definitely tend to protect pathogens from the disinfection treatment of raw and drinking water. Organic matter and microorganisms are presumable to protect pathogenic organisms from the bactericidal action of disinfecting agents, while inorganic particles have little or no effect. Turbidity may also contain particles that are toxic or help to accumulate toxic substances in the water.

Turbidity, therefore, may be judged both a physical parameter—because it causes aesthetic and psychological objections by the consumer, and a microbiological parameter—because it may harbor pathogens and impede the effectiveness of disinfection.

It must be noted that turbidity can also be caused by precipitated calcium carbonate in hard waters, aluminum hydrate in treated waters, and precipitated iron oxide in corrosive water.

Since turbidity is removed by coagulation, sedimentation, and filtration, unexpectedly high readings of turbidity also can be used as an interpretation of failures in treatment plant operation.

Turbidity in the distribution system should be maintained under 5 units. While a very efficient water treatment plant should have turbidity of the effluent under 0.1 unit, an average water treatment plant should produce an effluent under 0.5 unit, with a goal of 0.1 Nephelometric Turbidity Unit (NTU). Old standards allowed a maximum of 10 units in the distribution system. Because of the continuous variation of particulate in raw water, the measure of turbidity is rather imprecise. It is clear by the summation of the statements indicated above that low turbidity may be correlated with low level of bacteriological contaminants, as a general assumption, prior to bacteriological examination.

The measurement of turbidity is simple and inexpensive, and can be used for daily surveillance manually or with continuous turbidity recorders. Given that the total coliform test is a very reliable routine test of drinking water quality, but is not an actual determination of pathogens in water, its use in combination with a turbidity reading and their joint evaluation provides an added safety factor to the judgment of water quality changes both at the source or during the distribution system sampling. In the preliminary evaluation of raw water, sanitary engineers were of the opinion that, when turbidity at the source of supply was under 10 units, disinfection only was required—with BOD less than 1.0, coliform under 50 MPN/100 mL monthly average and with acceptable chemical parameters. When turbidity at the source was over 40 units, conventional treatment was considered necessary.

In the United States turbidity has been regulated since 1942 by the U.S. Public Health Service. The Environmental Protection Agency National Interim Primary Drinking Water Regulations of 1975 (similar to the U.S. Public Health Service standards of 1962) selected a maximum contaminant level of 1 unit, allowing a maximum of 5 under certain circumstances (see more information on *Turbidity Regulations* in Chapters 8 and 9).*

### Standard Methods

The **Nephelometric method** (using *Nephelometric turbidity units*) is the preferred method based on precision, sensitivity, and applicability that takes into consideration a wide range of turbidity readings.

The **Visual method** (using *Jackson turbidity units*) becomes the preferable method when highly turbid waters are examined.

As stated above, there are a variety of materials that cause turbidity. A standard was selected for comparison using silica and setting **1 unit of turbidity equal to 1 mg SiO$_2$/L.**

The instruments used, commercial turbidimeters, give an indication of light scattered in one particular direction, normally at a right angle to the incident light. Comparison is made with a standard reference suspension; Formazin polymer is used as a primary standard under the same condition reference suspension. The higher the intensity of scattered light, the higher the turbidity.†

---

*For International Standards values of turbidity, see Appendixes A-II and A-III.
†See Method 2130 B Nephelometric Method for apparatus, reagents, dilutions, procedures, and reporting turbidity readings. See *Standard Methods* (reference Appendix A-VIII, "General References") and USEPA 1993, *Methods for Determination of Inorganic Substances in Environmental Samples*, EPA-600/R/93/100 Draft. Environmental Monitoring Systems Lab. Cincinnati, OH.

## Turbidity—Summary

| | |
|---|---|
| Caused by: | Clay, silt, organic particulates, plankton, and other microscopic organisms, ranging in size from colloidal to coarse dispersion |
| Concentration expected: | From almost zero (0.05 NTU) in distilled water to over 1,000 NTUs in highly turbid rivers |
| Indicator of: | Necessity of treatment: potential contamination by pathogens; poor treatment plant efficiency; coagulant dosage, filter run timing; contamination in the distribution system |
| Test: | A physical and microbiological parameter, simple, inexpensive, mandated by public health regulations; expressed in the nephelometric turbidity unit (NTU) |
| NTU: | Turbidity unit measured by the nephelometric standard method |
| Regulations: | 0.1 NTU as a goal; less than 1.0 as a standard; 5 NTU as an exception for potable water |

## SOLUBILITY

Solubility is *not* a physical, general, or chemical parameter of drinking water classification or evaluation. Ionization, concentration, normality, molarity, dissociation, buffer solutions are also *not* parameters used in standards. Nevertheless, they are terms frequently used in connection with water treatment, laboratory analyses, in chemical/physical studies of water, and in related technical publications.

Here these terms are defined, briefly explained, and reviewed as if "water" itself would be a physical or chemical parameter of drinking water.

# Water: Solutions—Ionization

## WATER: CHEMICAL AND PHYSICAL PARAMETERS

Water is an odorless, tasteless, colorless liquid, a chemical compound of hydrogen and oxygen. It can be represented by the formula $H_2O$ or as a structure of water molecules represented in two dimensions as:

$$H:\ddot{O}:H \quad \text{or} \quad \begin{matrix} H:\ddot{O}: & & \\ \ddot{H} & H & \\ H:\ddot{O}:H:\ddot{O}:H:\ddot{O}: & & \\ \ddot{H} & \ddot{H} & \\ :\ddot{O}:H & & \\ \ddot{H} & & \end{matrix}$$

or to indicate ionization of water:

$$H_2O \rightleftarrows H^+ + OH^-$$

Water has a freezing point of 0°C (32°F) and a boiling point of 100°C (212°F). Maximum density is at 4°C (36°F), when 1 milliliter = 1 gram.

One calorie increases the temperature of 1 gram of water 1°C. Water vapor pressure varies from 4.6 mm at 0°C to 760 mm of mercury at 100°C. Water expands upon freezing. Its chemical combinations can be summarized as follows:

Water + Salts = Hydrates ($5H_2O + CuSO_4 \rightarrow CuSO_4 \cdot 5H_2O$)
Water + Metal Oxides = Bases ($H_2O + Na_2O \rightarrow 2NaOH$)
Water + Oxides of Nonmetals = Acids ($H_2O + SO_3 \rightarrow H_2SO_4$)
Water + Certain Metals = Bases + Hydrogen
$\quad H_2O + HCl \rightarrow H_3O^+ + Cl^-$ ionization
$\quad H_2 + Cl_2$ moisture 2 HCl catalytic reaction

- *Distillation*: The process to obtain chemically pure water through heat and vaporization with subsequent condensation of water.
- *Occurrence*: The main interest of the engineer or environmentalist is "free" water; i.e., water not chemically combined, as in animal and plant tissue or with certain chemicals, but abundantly distributed in nature, as in oceans, lakes, streams, and underground aquifers. Water potentially usable for human consumption—with minimum or complete treatment—but not yet declared potable is classified throughout the book as **raw water**; after proper treatment and official certification by the Health Authorities, it is classified herein as **potable** or **finished drinking water**.

## Viscosity

Viscosity is a property of fluids that causes them to resist flowing as a result of internal friction from the fluids' molecules moving against each other. Viscosity is also defined as a resistance to change of form.

**Kinematic viscosity** is the ratio of viscosity to density. In the International System of Units (SI) (meter, kilogram, second or centimeter, gram, second) kinematic viscosity unit is the **stoke**:

$$\text{stoke} = \frac{(gm)}{\sec \times cm \times \text{density (at given temperature)}}$$

**Poise** is the unit of **absolute viscosity** in the SI.

$$\text{Poise} = \frac{\text{gm}}{\text{sec} \times \text{cm}}$$

**centipoise (cp)** is 0.01 poise
**centistoke** is 0.01 stoke

(For conversion factors see Appendix A-IX or the *CRC Handbook of Chemistry and Physics*, see Appendix A-VIII, "General References").

## Water Solutions—Definitions

A water solution is a homogeneous liquid made of the *solvent*, the substance that dissolves another substance; and the *solute*, the substance that dissolves in the solvent. In water chemistry, water is the solvent. The solute may dissolve up to a certain limit; such a limit is called *solubility* of the solute in the particular solvent (water) at a particular temperature and pressure. Stability of solutions is influenced by *temperature* and *pressure*, but not by filtration, because only suspended material can be eliminated by filtration or by sedimentation.

Solubility is also defined as the mass of substance contained in the solution that is in equilibrium with an excess of the substance. When this condition is reached, the solution is called *saturated*. Under particular circumstances a solution can be *supersaturated*; in this case more solute is in solution than is present in a saturated solution of the same substances at the same temperature and pressure.

*Concentration* of a solution can be defined according to use, as follows:

1. *Normality*—A normal solution contains one gram equivalent weight of the solute in a liter of a solution, or one gram molecular weight of the dissolved substance divided by the hydrogen equivalent of the substance (i.e., one gram equivalent) per liter of solution. A normal solution of an acid is one that contains 1.008 gram of replaceable hydrogen contained in the total volume of one liter of solution. A normal solution of a base is the one that contains 17.008 gram of $(OH)^-$ in a total volume of one liter of solution. Any fractional quantity or desired multiple of solutions of acids and bases can be prepared, such as 10N, 05N, 01N, etc.
2. *Weight of solute per volume of solution*—Usually expressed in grams per liter
3. *Percentage composition*—This is the weight of solute in 100 grams of solution.
4. *Molarity*—The number of moles of solute in a liter of solution. A solution containing one gram-molecular weight of the solute dissolved in enough solvent to make one liter of solution.
5. *Molality*—One mole of solute in 1,000 grams of solvent

*Standard solutions* are any solutions whose composition is known.

## Chemical Units and Conversions

$$\text{Weight in grams} = \text{molarity} \times \text{volume solution in liters} \times \text{gram–molecular weight}$$

$$\text{Parts per million} = \frac{\text{molarity} \times \text{gram. mol. wt.} \times 10^3}{\text{density of solution}}$$

$$\text{Milligrams per liter} = \text{molarity} \times \text{gram. mol. wt.} \times 10^3$$

To determine an unknown concentration, the following relation is used:

$$V_A N_A = V_B N_B$$
$$V_A, V_B = \text{volumes of solutions } A, B$$
$$N_A, N_B = \text{normalties of solutions } A, B$$

## Ionization (Dissociation)

In water chemistry, and particularly with reference to drinking water and its source, the original "Arrhenius theory" of the electrolytic dissociation is still acceptable, since we are dealing with weak electrolytes only.

This theory states that electrolytes (acids, bases, and salts) when dissolved in water break into smaller, electrically charged units called ions, as atoms or groups of atoms with positive or negative charge. Ionization proceeds as the substance dissolves in water; while the quality of the solute is the major important factor, the solvent also has a major role since ionization does not take place or in very limited amounts in other solvents.

The positively charged ions are called cations; these are metallic ions and hydrogen. The negatively charged ions are called anions; these are acid radicals or nonmetals.

The more dilute the solution (as is the case of water destined for human consumption), the more completely ionized is the solute. Weak electrolytes are partially ionized, while strong electrolytes are almost completely ionized.

Water ionization takes place as water solute dissolves in water solvent. It can be visualized as the transfer of proton (hydrogen ion) from one molecule to another resulting in a hydrated hydrogen ion and a hydroxyl ion.

$$H_2O + H_2O \rightleftarrows H_3O^+ + OH^-$$

What actually may happen is summarized or visualized in complex, modern theories, but for practical purposes ionization can be abbreviated as:

$$H_2O \rightleftarrows H^+ + OH^-$$

with the understanding that both ions are hydrated ions. For simplicity, assuming also that activity coefficients are unity, concentrations may be expressed as:

$$[H^+][OH^-] = K_w = 1.00 \times 10^{-14} \qquad (2\text{-}1)$$

In pure waters

$$[H^+] = [OH^-] \qquad (2\text{-}2)$$

because concentration must be equal. Therefore, substituting (2-2) into (2-1):

$$[H^+]^2 = 1.00 \times 10^{-14} \qquad [H^+] = 1.00 \times 10^{-7}$$

and

$$[OH^-] = 1.00 \times 10^{-7}$$

using for convenience of representation the logarithmic function:

$$pH = -\log [H^+] = 7.00$$
$$pOH = \log [OH^-] = 7.00$$
$$pH + pOH = 14.00.$$

These statements are true for pure water at 25°C.

To say pure water is to say neutral water.* Therefore;

pH = 7.00 is an indication of neutrality.
pH < 7.00 is an indication of acid solution.
pH > 7.00 is an indication of basic solution.

## Buffer Solutions

A buffer solution is a suitable mixture of salt and acid (or base) that regulates or stabilizes the pH of a solution. Another way of visualizing a buffer solution is to consider that when hydrogen ions are produced, they react with a base to increase the amount of conjugate acid; if hydrogen ions are consumed, the weak acid dissociates to give more hydrogen ions.

Also, it can be stated that in a buffered solution, the pH variation will take place slowly because of neutralization of added acid or alkali.

Ionization constants are listed in the *Handbook of Chemistry and Physics* (see Appendix A-VIII, "General References") and in many related textbooks. For example, acetic acid:

---

*See the section on pH above on page 27.

$$HAc \rightleftarrows H^+ A\bar{c}$$

$$\frac{[H^+][A\bar{c}]}{[HAc]} = K_A = 1.76 \times 10^{-5} \quad pK = 4.75 \text{ at } 25°C.$$

25°C hydrofluoric acid    $K = 3.53 \times 10^{-4}$    $pK = 3.45$
18°C hypochlorous acid    $K = 2.95 \times 10^{-5}$    $pK = 7.53$
25°C sulfuric acid          $K = 1.20 \times 10^{-2}$    $pK = 1.92$

## SECONDARY DRINKING WATER STANDARDS

Another way to classify the physical parameters is to list them under the "Secondary Drinking Water Standards," (SDWS) as used by USEPA. This is true for Color, Corrosivity, Odor, pH, and Total Dissolved Solids, but not for Turbidity, which can be classified as a microbiological parameter. USEPA issued final rules for SDWS parameters in 1979 and reclassified aluminum and silver in 1991, and fluoride in 1986. Specifically, the SDWS list includes aluminum 0.05—0.2; chloride 250; fluoride 2.0; foaming agents 0.5; iron 0.3; manganese 0.05; silver 0.10; sulfate 250; TDS 500; zinc 5 (all expressed in mg/L). In addition: Color 15 TCU; Corrosivity: noncorrosive; Odor 3 TON; pH 6.5–8.5.

## PARTICLE SIZE COUNT

Particle size counting has been used unofficially by some public water supply systems for many years. Once sufficient recorded data of sampling and counting have been established, the information collected may become useful in evaluating the natural waters as well as the process and finished water.

Particle counting and size distribution analysis can be utilized in designing treatment plants, in providing changes of operation, and in evaluating plant efficiency.

There is no related standard for potable water. As is true for all water parameters, standards may be promulgated when standardization is satisfactorily reached in laboratories.

Until 1993, the Standard Methods Committee did not have a proposed method. The 1995 (19th edition) of *Standard Methods* introduced the **Particle Counting and Size Distribution (Proposed)**.

One of the difficulties of particle counting is encountered in sampling; particularly in batch sampling versus continuous flow due to the necessity of avoiding mixing, settling, air contact, and variations in location of water sampling points. Dilution water must be accurately selected because of the difficulty in producing particle-free water.

*Standard Methods* includes three types of instruments:

- electrical sensing zone instruments
- light-blockage instruments
- light scattering instruments

All of the three methods mentioned above use the same basic reporting system. Data reporting can be listed as particle concentration (numerical values), or tabulated size distribution, or graphical size distribution, or example calculations for particular size distribution analysis.

## ADDITIONAL GENERAL OR PHYSICAL PARAMETERS

A few additional parameters have been added by *Standard Methods*, listed under Physical and Aggregate Properties. They are as follows:

**Flavor Profile Analysis (Proposed)** or FPA   A technique for the identification of sample(s) taste(s) and odor(s). A group of four or five trained panelists is used.

**Calcium Carbonate Saturation**   Calcium carbonate ($CaCO_3$) saturation indices are useful for evaluating the scale-forming and scale-dissolving tendencies of water. They are useful in corrosion control programs and in preventing $CaCO_3$ scaling in piping and equipment such as domestic water heaters (see *Corrosion Control* in Chapter 10, page 472).

**Oxidant Demand/Requirement (Proposed)**   Oxidants are added to water for disinfection, or oxidation of undesirable inorganic species (ferrous iron, reduced manganese, sulfide, and ammonia), as well as oxidation of organic constituents (taste and odor control and trihalomethane precursor; see THMs, Chapter 4, page 165).

## REFERENCES

APHA, AWWA, WPCF. 1995. **Standard Methods for the Examination of Water and Wastewater** (19th ed.). Washington, DC: American Public Health Association.

American Water Works Association. **Water Quality Analyses**. Vol. 4. Denver, CO.

ASTM. 1954. **Water. Annual Book of Standards**. Philadelphia.

Butler, J. N. 1964. **Solubility and pH Calculations**. Reading, MA: Wesley Publishing Co.

California Institute of Technology. 1963. **Water Quality Criteria**. Pasadena, CA: NTIS, Springfield, VA.

Culp, Wesner, and Culp. 1986. **Handbook of Public Water Systems**. New York: Van Nostrand Reinhold.

Fair, G. M. and J. C. Geyer. 1965. **Water Supply and Waste-Water Disposal**. New York: J. Wiley & Sons.

National Academy of Sciences. 1977. **Drinking Water and Health**. Washington, DC: Safe Drinking Water Committee.

New York State Department of Health. **Manual of Instructions for Water Treatment Plant Operators**. Albany, NY.

Salvato, J. A. 1992. **Environmental Engineering and Sanitation** (4th ed.). New York: J. Wiley & Sons.

Sawyer, C. N. 1960. **Chemistry for Sanitary Engineers**. New York: McGraw-Hill Book Co.

Skeat, W. O. 1969. **Manual of British Water Engineering Practice**. Vol. 3. **Water Quality and Treatment**. London: The Institution of Water Engineers.

USEPA. 1985. National Primary Drinking Water Regulations, 40 CFR Part 141, *Fed. Reg.* Vol. 50, No. 219.

WHO. 1984. **Guidelines for Drinking Water Quality**. Vols. 1 and 2. Geneva, Switzerland: WHO.

# Chemical Parameters—Inorganics

CRITERIA FOR EVALUATION AND STANDARDS SPECIFICATIONS

| Type A | Type B | Type C | Type D |
|---|---|---|---|
| Arsenic | Aluminum | Calcium | Antimony |
| Asbestos | Cyanide | Carbon Dioxide | Beryllium |
| Barium | Molybdenum | Chlorides | Cobalt |
| Cadmium | Nickel | Iron | Thallium |
| Chromium | Silver | Lithium | Tin |
| Copper | Sulfate | Magnesium | Thorium |
| Fluoride | Sodium | Manganese | Vanadium |
| Lead | Zinc | Oxygen Dissolved | |
| Mercury | | (DO, BOD, COD) | |
| Nitrate–Nitrite | | Phosphate | |
| (Nitrogen– | | (Phosphorus) | |
| Ammonia) | | Potassium | |
| Selenium | | Silica | |
| | | Bromine (Bromide) | |
| | | Chlorine | |
| | | Iodine (Iodide) | |
| | | Ozone | |

# Criteria for Evaluation and Standards Specifications

## INORGANIC CHEMICALS

Inorganic chemicals that may enter the water supply by natural causes or pollution are evaluated here to determine, comprehend, and justify the standards set by the Health Authority. Specifically, for each inorganic chemical element listed, the following pattern of evaluation is applied:

1. Chemistry
2. Environmental exposure
   (a) Raw water
   (b) Air and food
   (c) Water surveys
3. Health effects
4. Standards
5. Analysis
6. Removal from raw water, and general notes

## CHEMISTRY

Data on formulae, compounds, natural origin, solubility in water, and industrial and commercial use are summarized.

## ENVIRONMENTAL EXPOSURE

### Raw Water

Origin in the ground, potential pollutional sources, and maximum expected concentrations in surface or groundwaters, are briefly indicated.

### Air and Food

Pollutional sources into atmosphere; industrial contamination; expected respiratory intake; average contribution of food supply to daily dietary intake; and special food containing high concentration of the chemical in question: these data are examined here only as a potential contribution of the contaminant to the total dietary intake.

## HEALTH EFFECTS

Health effects information relates to occurrence of the specific inorganic chemical in drinking water; human organs that are the target of the chemical in question; animal studies and physiological results; carcinogenicity;

mutagenicity; factor of safety applicable and related to studies reported; and a summary.

The actual summary or final conclusion is normally reported in the following section, "Standards."

## STANDARDS

As a basis for the judgments derived in this section, the following general criteria were used:

I. In the author's opinion, the best summary of literature review and guidelines for interpretation is presently the USEPA issuance of proposed rules for "National Primary Drinking Water Regulations" based on the National Academy of Sciences (NAS) recommendations, written under the Federal Safe Drinking Water Act (SDWA), and reviewed by the National Drinking Water Advisory Council.

II. In consideration that it would be an impossible task to evaluate equally all known chemicals, or to analyze them with reliability, and impractible to enforce standards for all of them, inorganic chemicals are selected for health standards with the following limitations:

- Laboratory analysis can be performed for the contaminant in drinking water with precision and reliability.
- The detected presence of that contaminant in community drinking water or the potential to be found in drinking water (frequency of occurrence).
- The poisonous nature of that contaminant as related to the acute and chronic toxicity, carcinogenicity, toxicological concern (mutagen or teratogen).

III. The Acceptable Daily Intake (ADI) or the Adjusted Acceptable Daily Intake (AADI) must also be determined within limitations offered by selected assumptions: The following assumptions* can be considered reasonable in this field of application (enforceable standards or drinking water goals or recommended standards for community water supplies):

- The population considered is, in general, a large heterogeneous human population (the sensitive, the ill, and the elderly will be protected by the factor of safety).
- An adult of 70 kg (154 lb.) is computed in health-related formulas. When children are used, a 10 kg (22 lb.) child is calculated.
- The water intake used in formulas is of 2 liters per day (2.11 quarts).
- The target of the acceptable daily intake is the Highest No-Observed-Adverse-Effect-Level (NOAEL). This level is possibly determined by

---

*See additional comments, pages 54 and 155.

evaluation of epidemiologic studies on humans, compared with animal data, or based on studies on animals. In consideration that a factor of safety will always be used, sometimes Lowest-Observed-Adverse-Effect-Level (LOAEL) in mg/k of body weight per day can be used, raising the numerical value of the factor of safety.

- A *factor of safety* or an *uncertainty factor* will be used to arrive at the AADI. This is a judgmental factor; and as such, is subject to criticism for excessive conservatism or insufficient protection of the ill, the elderly, or infants. The National Academy of Sciences suggested the following guidelines (when no indication of carcinogenicity exists):

    A factor of 10 when good acute or chronic human exposure data are available and supported by acute and chronic data on other species.
    A factor of 100 when good acute or chronic data are available for one species, but human data are not.
    A factor of 1,000 when acute and chronic data in all species are limited and incomplete.

- Possible *synergistic actions* between chemicals are also evaluated in the factor of safety.
- An *additional factor of safety* under 10 can be used when AADI is derived from a LOAEL, rather than from a NOAEL.
- Some problems are unavoidable when a chemical is an *essential nutrient at low level and toxic at higher levels*. Mathematical precision is required for the dividing line when scientific judgment cannot be so precise.
- Human absorption through drinking water is assumed to be more efficient and, consequently, more harmful than absorption through food.
- In relating AADI to drinking water standards, it is necessary to take into consideration the air and food intake. Therefore, *the total estimated daily intake from air and food is deducted from AADI.*
- Whenever possible, evaluation or study of health effects should contain a short-term assessment standardized in one-day, ten-day, and longer periods of time for AADI determination.
- *Carcinogenicity* in its risk to HUMANS has been evaluated, reaching the following conclusions:

    1. *Sufficient evidence*—There is a causal relationship between the agent and human cancer.
    2. *Limited evidence*—A causal interpretation is credible; but chance, bias, or confounding alternative could not be adequately excluded.
    3. *Inadequate evidence*—Either there are insufficient pertinent data, or available studies show that the evidence is not valid or show no evidence of carcinogenicity.

- *Carcinogenicity* in ANIMALS is based on assessment of evidence from experimental animals; conclusions depend on whether the evidence is considered sufficient, limited, or inadequate:

1. *Sufficient evidence*—A causal relationship shows increased incidence of malignant tumors, possibly in multiple species or strains, or in multiple experiments using different dose levels or different route of administration or finally when unusual degree with regard of incidence, site, type of tumor, or age at onset has been recorded.
2. *Limited evidence*—Carcinogenic effect is noted but limited to a single species or inadequate dosage level or incompleteness of the study.
3. *Inadequate evidence*—The studies cannot be interpreted as showing either the presence or absence of a carcinogenic effect, or that the contaminant did not appear carcinogenic within the limits of the tests.
4. *No data*—No data are available for evaluation.

Carcinogenicity is also presented in Chapter 7, "Carcinogens," and in the introduction and evaluation of synthetic organic chemicals in Chapter 4, "Chemical Parameters—Organic Compounds" (page 155).

The following notes should be considered here to avoid repetition for each chemical examined.

In general, the No-Observed-Adverse-Effect-Level (NOAEL) or the Lowest-Observed-Adverse-Effect-Level (LOAEL) resulting from epidemiological/toxicological studies is expressed in mg/kg body weight/day. To arrive at a value in mg/L as used in Adjusted-Acceptable-Daily-Intake (AADI), the following formula is used:

$$AADI = \frac{(NOAEL \text{ in mg/kg/day})(70 \text{ kg adult})}{(factor \text{ of safety})(2 \text{ liters per day})}$$

$$\left[ \frac{mg}{kg/day} \times \frac{kg}{liter/day} \right] = [mg/liter]$$

The USEPA defined "provisional AADIs" when less than lifetime (presumably 70 years) studies were used. AADIs are normally used for noncarcinogenic chemicals. Proven or highly suspected carcinogenicity may lead to a maximum contaminant level of zero (MCL = 0) as a goal or recommended or final classification.

Additional information on this subject can be found in the *Introduction* of Chapter 4, "Chemical Parameters—Organic Compounds." Safety factors or uncertainty factors (UF) used by Health Authorities do not pretend to have mathematical precision.

Maximum Contaminant Levels (MCLs), recommended (RMCLs), or goals (MCLGs) or guidelines, are usually computed as follows:

RMCL = (AADI) − (contribution from food) − (contribution from air)

When insufficient data are available for air and food contamination, a 20% contribution from drinking water has been used in preparation of standards. This is a very conservative factor: The standard derived from water intake only is then divided by five.

In some cases, the World Health Organization used even more conservative factors, as low as 1%, when the contaminant had demonstrated to bioaccumulate to a high degree.

When the airborne contaminant is taken into account to evaluate the overall human intake, an assumption is made that a person breathes 20 cubic meters of air per day.

Some routine inorganic chemicals can be successfully analyzed directly in a water treatment plant, where results can immediately be interpreted by the plant senior operators and related operational adjustments implemented without delays (see Figure 3-1).

More information about the theory used in preparation of standard settings for all water quality parameters is provided in the introduction to Chapter 4, "Chemical Parameters—Organic Compounds," (see pages 158–162) and in Chapter 7, "Carcinogens."

## Inorganic Chemicals of Primary and Secondary Public Health Concern

Inorganic chemicals are cataloged here in four groups, indicated as Types A, B, C, and D. Such subdivision has no special recognition in chemistry or

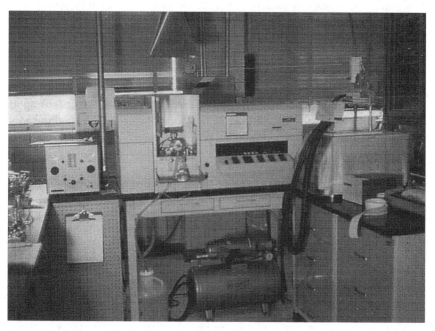

Figure 3–1.   Water purification plant. Poughkeepsie, NY—In the chemical laboratory (near the operators' control room and on the same floor) an Atomic Absorption Spectrophotometer is used to analyze inorganic chemicals. (Courtesy of the city of Poughkeepsie, NY.)

biochemistry nor has specific scientific merit. It should not be confused with official classification of carcinogens.

This subdivision has been selected by the author simply following public health criteria in setting priorities for standards based on higher toxicity and/or frequency of occurrence in drinking water.

A brief explanation for each group is listed as an introduction to each group, followed by a tabulation of parameters to include standards adopted or suggested by American/world/European public health officials. Following the general classification and summary, each inorganic chemical is examined in practical details, following the general *Criteria for Evaluation and Standards Specifications* outlined at the beginning of this chapter.

## Inorganic Chemicals—Type A

| ARSENIC | CHROMIUM | MERCURY |
|---------|----------|---------|
| ASBESTOS | COPPER | NITRATE-NITRITE |
| BARIUM | FLUORIDE | SELENIUM |
| CADMIUM | LEAD | |

**Inorganic Chemicals** considered in type A are water quality parameters closely scrutinized by the Health Authorities in consideration of their known or potential toxicity and their expected presence in drinking water (frequency of occurrence).

Water distribution system and house plumbing contribute to the presence at the consumer's tap of COPPER, LEAD, and CADMIUM. Cross-connections may introduce CHROMIUM, used for corrosion control in cooling towers. Water treatment may introduce COPPER and FLUORIDE. NITRATE (usually in well water) influences a sensitive population—babies.

ARSENIC, BARIUM, CADMIUM, and CHROMIUM, including ASBESTOS, have been studied for potential carcinogenicity. Proven toxicity has indicted MERCURY and SELENIUM. These eleven chemical parameters have been selected and listed in the proposed RMCLs issued by the USEPA in November 1985, publishing the National Primary Drinking Water Regulations (NPDWR), mandated by the SDWA.

Proposed RMCLs are guidelines for health goals leading to the issuance of enforceable standards (MCLs), which are expected to be very close to the RMCLs. USEPA also uses the term Maximum Contaminant Level Goals (MCLGs).

All type A parameters are regulated by the European Community water standards, with the exception of asbestos, and by the WHO with the exception of asbestos and barium.

# ARSENIC

## CHEMISTRY

A silvery-white, very brittle, crystalline, semi-metallic, solid chemical element, notorious for its toxicity to humans: symbol As; atomic weight 74.922; atomic no. 33; melting point 817°C (28 atm.); sublimes at 613°C; specific gravity 1.97, and 5.73 for the yellow and grey metallic elemental arsenic. Valence +3, +4, and +5. It is found in nature in sulfides (realgar and orpiment), in arsenide and sulfarsenides of heavy metals, oxides, $As_2O_3$, and arsenates. The most common mineral is mispickel or arsenopyrite ($FeSAs$). Arsenic is found in ores of copper, lead, zinc, iron, manganese, uranium, and gold. Arsenic is used in bronzing, pyrotechny, glassware, ceramics, dye manufacturing, agricultural insecticides, poisons, and medicines. Recently, it has been used in solid-state devices such as transistors and in laser material.

Arsenic (as As) is insoluble in water, but several arsenites are quite soluble.

## ENVIRONMENTAL EXPOSURE

### Raw Water

Groundwater is expected to contain higher arsenic concentrations than surface water. Because of its presence in geological materials, arsenic can be traced in water as originated by natural processes or by industrial activities—industrial waste, arsenical pesticides, and smelting operations.

### Air and Food

The above-mentioned industrial processes and coal burning cause the presence of arsenic in the atmosphere where it ranges from 0.0005–1.5 µg/mc with a median value of 0.01 µg/mc. The daily respiratory intake of arsenic is approximately 0.12 µg, of which 0.03 µg would be absorbed.

In food, arsenic is a common element in minute doses; found particularly in seafood, but also in meat, fish, poultry, and cereals. The FDA estimated for 1979 market basket survey that dietary arsenic adds up to 61.5 µg/day for adults.

### Water Supply Surveys

While water samples collected from the distribution system in the United States usually confirmed values of concentrations under 0.01 mg/L, well within the MCL of 0.05 mg/L, the detection of carcinogenic effects of arsenic in some remote areas of Taiwan, India, and, very rarely, in the United States has stimulated more interest, research, and reevaluation of MCL for arsenic.

Arsenic occurrence is highest in groundwater systems in the western United States.

Using the Federal Reporting Data Systems (FRDS), USEPA in 1985 conducted a survey collecting 982 samples of groundwater. With a detection level of 5 µg/L the survey found detectable concentration of arsenic in 64, equivalent to 7% of the 982 samples. Data collected from the 64 samples ranged from 5 to 48 µg/L with mean of 13 µg/L and a median of 9 µg/L.

## HEALTH EFFECTS

A poison in humans at 100 mg or more, arsenic has proved lethal at 130 mg (325 mg of sodium arsenite, equivalent to 187 mg as As). Accumulation in the body is expected to raise progressively at low intake level. It has been reported poisonous at concentrations of 0.21 mg/L and 0.4 to 10.0 mg/L in other cases; it has been reported safe at 0.05, 0.10, 0.15, 0.15–0.25 mg/L in drinking water. The trivalent compounds are the most toxic.

It is still questionable whether As is an essential element in human nutrition; it appears potentially essential in humans and even more in certain animals. It has been used in small doses to enhance growth. Research work is underway to clarify the potential essentiality of arsenic, already demonstrated in four mammalian species, according to the National Academy of Sciences (NAS). USEPA calculated a provisional AADI of 0.10 mg/L based on NOAEL of 3.74 mg/kg/day arsenate (2.8 mg As/kg/day), using an animal study and using an uncertainty factor of 1,000.

Characteristics of mutagenicity appeared in several tests (induced chromosomal aberrations) in vivo and in vitro systems.

Carcinogenicity of arsenic in animals has not been formally determined due to conflicting results. Target organs in humans are the skin, lungs, genital and visual organs as per observation on arsenic exposure. The International Agency for Research on Cancer (IARC) classified arsenic of questionable carcinogenicity in animals and sufficient evidence that inorganic arsenic compounds (arsenite and arsenate) are skin and lung carcinogens in humans.

## STANDARDS

| | |
|---|---|
| USPHS 1925 | = not issued |
| USPHS 1942 | = 0.05 mg/L |
| USPHS 1946 | = 0.05 mg/L |
| USPHS 1962 | = 0.05 mg/L |
| NIPDWR | = 0.050 mg/L |
| WHO guideline | = 0.05 mg/L |

European Community = 0.05 mg/L
RMCL (USEPA, 1985) = 0.050 mg/L
USEPA classification = "A" (carcinogenic in humans by inhalation and ingestion. However, this chemical has a potentially essential nutrient value.)
AADI (USEPA, 1985) = 0.10 mg/L (animal study—uncertainty factor of 1,000)
MCLG (USEPA, 1987) = 0.050 mg/L

A new arsenic rule has been under discussion in the scientific community, particularly after publications in 1968 and 1977 of studies on skin cancer in Taiwan. The first review by USEPA in 1984 confirmed the safety of arsenic at 0.05 mg/L as MCLG along with the MCL. Proposals to set MCLG at zero continued to be under discussion. Since MCL must be established as close as possible to MCLG, the final rule for arsenic must be reached after careful evaluation of the BAT (best available technology) but also taking cost into consideration. The practical quantitation level (PQL) is presently set at 2 µg/L for several existing methods. The new rule will be set between 2 and 20 µg/L. Time will be given to water utilities to monitor arsenic and evaluate the BAT and the cost to the consumer.

USEPA estimates that the national cost of compliance will vary from $22 million for annual cost and $140 million for capital cost for 20 µg/L, to $1 billion for annual cost and $6.225 billion for capital cost if the 2 µg/L would be adopted for MCL.

## ANALYTICAL METHODS

Approved methods for arsenic analysis are the following:

| | |
|---|---|
| Inductively coupled plasma | USEPA* (200.7); SM (3120B) |
| Inductively coupled plasma—mass spectrometry | USEPA (200.8) |
| Atomic absorption; platform | USEPA (200.9) |
| Atomic absorption; furnace | ASTM† (D-2972-93C) |
| | SM‡ (3113B) |
| Atomic absorption; hydride | ASTM (D-2972-93B) |
| | SM (3114B) |

---

*USEPA, "Methods for the Determination of Metals in Environmental Samples—Supplement 1," EPA-600/R-94-111, May 1994; NTIS publication #PB94-184942.
†*Annual Book of ASTM Standards*, 1994, Vols. 11.01 and 11.02. American Society for Testing and Materials, 1916 Race St., Philadelphia, PA 19103.
‡*Standard Methods for the Examination of Water and Wastewater*, 1992, American Public Health Association, 1015 15th St. NW, Washington, DC 20005. The 19th edition of Standard Methods (1995) also includes the "Silver Diethyldithiocarbamate Method."

## REMOVAL

USEPA has not as yet specified the best available technology for arsenic. In consideration of many pending studies on the possibility to reduce the concentration of arsenic below the level of 0.20 µg/L, several studies to improve existing technology will be produced in the future prior to issuing the *final* MCL. There are six treatment options under evaluation for treatment of arsenic: coagulation with filtration; lime softening; activated alumina; ion exchange; reverse osmosis; and electrodialysis. With the present technology, arsenic can be removed to below 20 µg/L. Using activated alumina, ion exchange, reverse osmosis, and electrodialysis, concentration below 2 µ/L can be reached.

---

## ASBESTOS

## CHEMISTRY

Any of several grayish amphiboles or similar mineral that separates into long, threadlike fibers or highly fibrous, hydrated silicate minerals can be classified as asbestos.* Because certain varieties do not burn, do not conduct heat or electricity, and are often resistant to chemicals, they have been and are used for making fireproof materials, electrical insulation, roofing, and fibers. Cement products, floor tiles, paper products, paint and caulking, textile and plastics are or were the major users of asbestos.

## ENVIRONMENTAL EXPOSURE

### Raw Water

It is difficult to assess the source of asbestos in raw water by routine or special surveys, mainly because of the complexity in the analytical method used to identify asbestos, the relatively short time that asbestos has been seriously considered and held responsible for asbestosis (a form of pneumoconiosis caused by inhaling asbestos fibers), and the unlikelihood of fibers in water being associated with asbestosis. The WHO reports expected values of asbestos in raw water to vary between less than 1 million and 10 million fibers per liter (avg. = 1 MFL).

---

*Six minerals have been classified as asbestos: chrysotile (Serpentine Group), crocidolite, anthophyllite, tremolite, actinolite, and amosite (Amphibole Group). General components are silica (40%–60%); oxides of iron, magnesium, and other metals [chrysotile = $Mg_3Si_{12}O_5(OH)_4$].

## Air and Nutrition

The National Academy of Sciences (NAS) reported in 1984 air concentrations of asbestos of 0.0004 µg/cm³ for a typical urban dweller—indoor and outdoor exposure combined—or 3,000 fibers/day at a ventilation rate of 20 mc for an adult male (USEPA, 1985) or less than 0.6 µg/day. No information is available about asbestos in food. Wine was reported to contain, in some cases, up to 64 MFL (USEPA [1985] 40 CFR Part 141). Beer and soft drinks recorded concentrations from 0.15–6.6 MFL. With less use of asbestos filters in the food industry, lower concentrations are presently expected.

## Water Surveys

The USEPA (1980) sampling of water distribution systems in 406 cities in 47 states reported that 29% were below detection limits, 53% were less than one, 8% with one to 10, and 10% had over 10 MFL. USEPA, in 1981, sampled 100 systems and found 0.08 MFL in 12 systems with values ranging from 0.385–1.071 MFL.

Asbestos evidently enters the distribution system from raw water, from air in open reservoirs, and asbestos-cement pipes in water mains and service lines.

A Canadian national survey concluded that 95% of the population is expected to intake daily less than 0.1 µg of asbestos.

## HEALTH EFFECTS

While ingestion of water is the primary intake of asbestos, some inhalation may take place through showers and humidifiers.

A high level of asbestos fed to animals produced minor effects in mucosal lining cells of the ileum and colon, rectum, and small intestine but no carcinogenic effects. Inhalation studies related lung tumors and mesothelioma developed in laboratory animals with various forms of asbestos.

Statistical evidence of occupational exposure to asbestos from inhalation has been reported showing increased risk of gastrointestinal cancer. Inhalation of *medium and long* fibers has resulted in mesotheliomas in rats, with no tumors caused by nonfibrous particulates. There is no evidence that asbestos-cement pipe may have caused cancer.

In September of 1974 the American Water Works Association Research Foundation issued a report with the following conclusions:

"No firm evidence shows that the proper use of asbestos-cement pipe poses a hazard to health by reason of ingestion of asbestos fibers. Calculations comparing the probable ingestion exposure in occupational groups to that likely to occur as a result of ingestion of potable water from asbestos-cement pipe systems suggest that the probability of risk to health from the use of such systems is small approaching zero."

## STANDARDS

The USEPA classification is "A."*

By inhalation, carcinogenic in humans and animals. By ingestion of intermediate (more than 10 micron length) range chrysotile asbestos, limited evidence in animals—benign polyps in male rats. However, the available epidemiologic/experimental data are inadequate to conclude that the chemical is carcinogenic via ingestion.

USEPA (1985) MCLG and MCL = 7 million fibers per liter—asbestos fibers exceeding 10 micron in length (proposed)
USEPA (1991) MCL = 7 million fibers per liter—asbestos fibers exceeding 10 micron in length (final rule; effective July 30, 1992)[†]

## ANALYSIS

The most reliable techniques are the separation of fibers and the quantitation by transmission electron microscopy and identification by X-ray diffraction. For economic reasons, these methods cannot be used routinely by water supply laboratories.

USEPA reports in its Nov. 13, 1985 (40 CFR Part 141) NPDWR that:

Three different approaches have been investigated to develop a simpler, faster, and cheaper measurement method for asbestos. These approaches depend upon light scattering properties of particulates (turbidimetric and magnetic alignment-light scattering methods), or surface properties of chrysotile asbestos, which is selectively extracted into isooctane in the presence of a surfactant (two phase liquid separation method). The most promising method is the one based on magnetic alignment-light scattering.

Optical microscopy may be used to screen water samples to measure fibers above a certain length. Fibers with a minimum length of 5 micrometers and about 0.5 micrometer in width can be measured using this method, with the analytical costs estimated at $50–$100 per sample and an initial outlay of approximately $1,500 for an optical microscope. This technique characterizes fiber shapes only and does not discriminate between asbestos and non-asbestos types. However, the technique could be useful for screening to determine which samples should be analyzed more intensively using transmission electron microscopy or for surveillance after the fibers in a water supply had been characterized by X-ray diffraction or transmission electron microscopy.

In 1993, the Standard Methods Committee approved Method 2570B, the "Transmission Electron Microscopy Method," for Asbestos in water (*Standard Methods*, 19th ed., 1995).

---

*Federal Register*, 11/13/85—40 CFR Part 141, National Primary Drinking Water Regulations.
†National Primary Drinking Water Regulations; Final Rule. *Fed. Reg.*, Jan. 30, 1991. USEPA regulated 26 synthetic organic chemicals and seven inorganic chemicals.

## REMOVAL/NOTES

It can be safely assumed that the portion of asbestos reaching the water treatment plant through the raw water can be substantially removed by the solid contact process, sedimentation, and filtration (conventional treatment). New technology for specific removal of asbestos fibers has not been developed pending final indictment of intermediate/long fibers of asbestos. Sand filtration is expected to eliminate over 90% of asbestos fibers.

The chrysotile fibers are usually 0.03–0.10 micron in diameter and less than 10 microns in length. The amphybolite forms are usually from 0.1–100 microns in length.

Conventional treatment should be particularly scrutinized when asbestos removal is of primary importance. Pilot plant performance should be evaluated for efficiency of removal. Important parameters are pH, coagulants, and filtering media.

More statistical and research work is necessary, particularly on metabolism. Less use of asbestos products and more environmental control of industrial production will reduce asbestos concentration in the environment.

In 1989, USEPA issued the "National Primary and Secondary Drinking Water Regulations; Proposed Rule" indicating for **asbestos** the Best Available Technology (BAT) for removal with "Coagulation/Filtration; Direct and Diatomite Filtration; and Corrosion Control."

The knowledge that asbestos had been used extensively in certain cases (for fire protection), be it in schools, public buildings, or multiple dwelling residential buildings, created fear and panic in American society in the 1980s. Consequently, Congress issued the Asbestos Hazard Emergency Response Act (1986) so that USEPA could order inspections of public and private schools leading to the elimination of asbestos products. The health problems in this case occur as a result of *inhalation* of asbestos fibers and have certainly no relation to problems that might result from ingestion of asbestos in drinking water.

---

```
BARIUM
```

## CHEMISTRY

A silvery-white, slightly malleable, metallic chemical element, found abundantly in nature as a carbonate or sulfate and used in alloys: symbol Ba; atomic weight 137.33; atomic no. 56; specific gravity 3.5; valence 2. It belongs to the alkaline earth group, chemically resembling calcium. It oxidizes easily and is decomposed by water and alcohol (poisonous when soluble). It is found in barite or heavy spar (sulfate) and in witherite (carbonate). Sulfates and carbonates are insoluble.

Barium is found in limestones, sandstones, and occasionally in soils. It may be brought to the surface as a by-product of the mining industry (oil, coal, muds) or in ambient air by combustion (aviation, diesel fuel) or manufacturing.

The industrial uses also include manufacture of paint, paper, ceramics, glass, special cement, X-ray diagnostic work, oil well drilling, fluids, rubber, linoleum, and rat poison.


## ENVIRONMENTAL EXPOSURE

### Raw Water and Water Surveys

By its chemical nature, barium is unlikely to be found in raw water as a barium ion. In sea water, it is recorded at 6.2 µg/L. Higher concentrations in groundwater are expected to be detected than in surface supplies. Maximum expected level to be found in community water supplies is 1 mg/L (7.3 mg/L was detected in a community water supply selected as a high barium area) in groundwater surveys—14% of 132 systems sampled contained levels greater than 250 µg/L and 1%–2% were over 1 mg/L. In surface water supplies surveys, 14% of 28 systems sampled contained barium greater than 250 µg/L but none with levels higher than 0.5 mg/L. In general, barium concentration is expected to be well below 0.1 mg/L.


### Air and Food

The mean level of barium in air ranges from 0.0015–0.94 µg/mc. An approximate respiratory intake level can be assumed at 0.03–22 µg/day.

The International Commission on Radiological Protection (ICRP) reports an average daily dietary intake of 750 µg per adult male from food and water, with the assumption that water contributes 80 µg/day. Presently, there is insufficient data to determine a level of barium in food supplies. Barium content is high in Brazil nuts (10,000 ppm) and pecans (1,000 ppm).


### HEALTH EFFECTS

Acute exposure to barium results in gastrointestinal, neuromuscular, and cardiac effects to animals and humans. A fatal oral dose for humans is reported at 550–600 mg, but in one case a man recovered after ingesting 7 g. Barium does not accumulate in bones, muscles, kidneys, or other tissues. It is excreted more rapidly than calcium. The USEPA calculated an AADI of

1.8 mg/L from the LOAEL of 100 mg/L of barium (5.1 mg/kg/day) with an uncertainty factor of 100.

High barium exposure in animals is associated with hypertension and cardiotoxicity. The WHO states: "There is no firm evidence of any health effects associated with the normally low levels of barium in water."

The International Agency for Research on Cancer (IARC) has not evaluated barium for potential carcinogenic effects.

## STANDARDS

| | |
|---|---|
| USPHS 1925 | = not listed |
| USPHS 1942 | = not listed |
| USPHS 1946 | = not listed |
| USPHS 1962 | = 1.0 mg/L |
| NIPDWR | = 1.0 mg/L |
| WHO guideline | = not established (did not consider necessary to establish a guideline value) |
| WHO (European) | = not established |
| European Community | = 100 µg/L |
| RMCL | = 1.5 mg/L |
| NAS-SNARL | = 4.7 mg/L |
| USEPA classification | = "D" (inadequate data to classify) |
| Russia | = 0.1 mg/L (MCL) |
| MCLG and MCL (USEPA, 1989) | = 5 mg/L (proposed) |
| MCLG and MCL (USEPA, 1991) | = 2 mg/L (final; effective 1/1/1993) |

## ANALYSIS

Reliable methods for laboratory analysis are the flame atomic absorption, furnace atomic absorption, and inductively coupled plasma emission spectrometry techniques, according to USEPA (Nov. 1985).

## REMOVAL/NOTES

Conventional water treatment as practiced in typical water plants (from prechlorination, conventional coagulation, sedimentation, aeration, rapid sand filtration, to intermediate and post-chlorination) has no effect on barium removal, with the exception of lime softening (pH = 10 to 11). Also, an ion exchange softening plant may reduce barium over 95%. Reverse osmosis is also listed among USEPA's proposed BAT.

Ambient water quality criteria to protect aquatic life and human health presently issued by USEPA limit discharges in water used, or potentially usable, for domestic water supplies to 1 mg/L of barium.

There is no evidence that barium is essential for human nutrition.

# CADMIUM

## CHEMISTRY

A soft, blue-white metallic chemical element occurring as a sulfide or carbonate in zinc ores; symbol Cd; atomic weight 112.41; atomic no. 48; specific gravity 8.65; valence 2. As an element it is insoluble in water; in nature cadmium appears in zinc, copper, and lead ores. Chlorides, nitrates, and sulfate of cadmium are soluble in water. (Cadmium will precipitate at high pH since carbonate and hydroxide are insoluble.)

Similar to zinc, it is used in electroplating, in many types of solders, batteries, television sets, and as a yellow pigment. It is also used in ceramics, photography, insecticides, and as an alloy with copper, lead, silver, aluminum, and nickel.

## ENVIRONMENTAL EXPOSURE

### Raw Water

Since it is found in low concentrations in rocks, coal, and petroleum, it is found in groundwater more than in surface water as a natural occurrence. Therefore, it may enter the water supply from mining, industrial operation, and leachates from landfill. Also, cadmium may enter the distribution system from corrosion of galvanized pipes. Raw water may contain normally less than 1 µg/L.

### Air and Nutrition

Cigarettes contain high levels of cadmium (1 ppm). Urban air has higher levels of cadmium: reported levels in air from 0.0005–0.01 µg/mc according to regulatory agencies in the United States. With 20 mc/day ventilation rate and a typical value of 0.01 µg/mc, a male adult respiratory intake is 0.2 µg/day. Exposure to cadmium dust should not exceed 0.15 mg/mc (15 min). Cadmium oxide fume exposure should not exceed 0.05 mg/mc.

Daily dietary intake of cadmium for an adult male averages between 10 and 50 µg, with grains, cereals, potatoes, meat, fish, and poultry as major cadmium suppliers. A tolerable weekly intake of 0.4–0.5 mg/individual was accepted by the United Nations Food and Agricultural Organization (FAO) and WHO in 1972.

### Water Surveys

The USEPA reported the results of federal surveys conducted between 1969 and 1980 of 707 groundwater supplies and 117 surface water supplies as:

Groundwater—27% above 2 µg/L (mean of positives: 3 µg/L).
Surface water—20% above 2 µg/L (mean of positives: 3.2 µg/L).

Soft water of low pH may register higher values when plumbing systems contain cadmium.

## HEALTH EFFECTS

Human beings reported nausea and vomiting at 15 mg/L, with no adverse effects at 0.05 mg/L. Severe toxic, but not fatal, symptoms are reported at concentrations of 10–326 mg (NAS, Vol. 4, 1982). The death of a boy within 1.5 hr was reported after ingestion of 9 g of cadmium chloride. The kidneys are the critical target organ by ingestion (renal dysfunction, hypertension, and anemia).

A. HUMAN STUDY—one-day assessment
   Uncertainty factor = 10
   0.043 mg/kg/day of NOAEL
   Consumption: 1 liter per 10-kg child = 43 µg/L
        2 liters per 70-kg adult = 150 µg/L
B. ANIMAL STUDY—ten-day assessments
   Uncertainty factor = 1,000 (+ additional factor of $10^{-1}$)
   0.08 mg/kg/day NOAEL (proteinuria in rats)
   Consumption: 1 liter per 10-kg child = 8 µg/L
        2 liters per 70-kg adult = 29 µg/L
C. AADI
   Endpoint: renal dysfunction
   Threshold: critical concentration of cadmium in the renal cortex
   Range: 50–300 µg/L
   LOAEL = 0.352 mg/day (in humans)
   Provisional AADI = 0.018 mg/L [0.352 ÷ (10 × 2)]
   Uncertainty factor = 10
   Consumption: 2 liters of water per day
   IARC = group 2B (limited evidence of carcinogenicity in humans, sufficient evidence for animals)
   Exposure: inhalation
   Ingestion: no evidence of carcinogenicity in animals or humans
   USEPA regulation: based on chronic toxicity

## STANDARDS

USPHS 1925–1942–1946 = not stated
USPHS 1962     = 0.01 mg/L
NIPDWR      = 0.01 mg/L

| | |
|---|---|
| WHO guideline | = 0.005 mg/L |
| European Community | = 0.005 mg/L |
| RMLC (1985) | = 0.005 mg/L |
| SNARL (NAS) | = 0.005 mg/L* |
| USEPA classification | = B1 = "Limited evidence in humans exposed to cadmium fumes, cancer in rats exposed to cadmium chloride aerosol, injection site tumors in animals given cadmium salts. However, regulating as "D" (see Barium) since there is inadequate evidence to conclude that the chemical is carcinogenic via ingestion." |
| MCLG and MCL (USEPA, 1989) | = 0.005 mg/L (proposed) |
| MCLG and MCL (USEPA, 1991) | = 0.005 mg/L (final; effective 7/30/1992) |

## ANALYSIS

Introducing cadmium, *Standard Methods* states that the cadmium concentrations of U.S. drinking water have been reported to vary between 0.4–60 µg/L, with a mean of 8.2 µg/L.

Reliable analytical methods have been determined and applied for cadmium in drinking water. Atomic Absorption Spectrometric Method is the preferred, with the Dithizone Method considered acceptable when an AA Spectrometer is unavailable. The USEPA (1985) approved the same methods reported for barium.

## REMOVAL/NOTES

There is no accepted, economically effective method for direct removal of cadmium at high concentration. Pilot plant study may be used attempting precipitation as carbonate and hydroxide at high pH. Every effort should be undertaken to locate the source of cadmium contamination and its partial or total elimination prior to consideration of a treatment solution.

The USEPA *Quality Criteria for Water* of 1976 recommends 10 µg/L for domestic water supply (based on health criteria). New York State limits to 0.3 mg/L cadmium content reaching all fresh waters with none in amounts that will be injurious to fish life or shellfish, or that would impair any designated uses of water.

---

*Based on 10 mg/L concentration of cadmium as the no-effect level in rats, resulting in 0.75 mg/kg, using a safety factor of 1,000, a 70-kg human consuming 2 L per day and assuming a 20% exposure from drinking water. SNARL is calculated as:

$$\frac{0.75 \text{ mg/L} \times 70 \text{ kg} \times 0.20}{1,000 \times 2 \text{ L}} = 0.005 \text{ mg/L (NAS, Vol. 4. 1982)}$$

The water treatment industry has confidence in cadmium removal by lime softening when concentrations are less than 0.5 mg/L. Less effective is the removal with ferric sulfate and alum clarification.

USEPA proposed in 1989 the following treatment processes as BAT for cadmium removal: ion exchange; reverse osmosis; coagulation/filtration; and lime softening.

> ### CHROMIUM

## CHEMISTRY

Defined as a grayish-white, crystalline, very hard, metallic chemical element with a high resistance to corrosion; used in chromium electroplating, in alloy steel (stainless steel), and in alloys containing nickel, copper, manganese, and other metals: symbol Cr; atomic weight 51.996; atomic no. 24; specific gravity 7.20; valence 2, 3, or 6. Principal ore is chromite—$FeCr_2O_4$. The most important compounds are the chromates of sodium and potassium ($K_2CrO_4$); the dichromates ($K_2Cr_2O_7$), and the potassium and ammonium chrome alums ($KCr(SO_4)_2 \times 12H_2O$); and lead chromate.

| | |
|---|---|
| chromous ion $Cr^{++}$ | metallic chromium |
| chromic ion $Cr^{+++}$ | } *trivalent*—more stable in general |
| chromite ion $CrO_3^{---}$ or $CrO_2^{--}$ | } but not in chlorinated water |
| chromite ion $CrO_4^{--}$ | } *hexavalent*—with tendency to be |
| dichromate ion $Cr_2O_7^{--}$ | } quickly reduced by organic species |

- chlorides, nitrate, and sulfate (trivalent chromic salts) are soluble in water;
- hydroxide and carbonates are quite insoluble; and
- sodium, potassium, and ammonium chromates (hexavalent chromates salts) are soluble; corresponding dichromates are quite insoluble.

Chromium is also used as a corrosion inhibitor in the textile, glass, and photographic industries.

## ENVIRONMENTAL EXPOSURE

### Raw Water

Found in the earth's crust from 10–200 ppm, chromium is a naturally occurring metal in drinking water. Chromite is a commercial chromium ore for the mining industry (runoff and leaching are potential sources of contamination). Chrome plating and chrome metallurgical and chemical operations may contaminate the atmosphere with chromium, in addition to fossil fuel combustion, solid waste incineration, and cement plant emission. Other chromium salt usage is found in the leather industry, paints, dyes, explosives, ceramics, and papers, leading to industrial pollution.

Rivers in the United States are reported to contain between 1 and 100 µg/L, averaging 10 µg/L. In a Canadian survey of surface and groundwaters in 1974, 99% of the samples showed concentrations below 50 µg/L.

## Air and Nutrition

Ambient air data are usually below 0.2 µg/mc. Maximum levels are between 0.01–13.5 µg/mc near emission sources. Chromium is contained in food from 20–600 µg/kg (meats, vegetables, mollusks, and crustaceans). Typical daily intake of 100 µg (range 5–500 µg/day) is expected in human diets.

The NAS (1974) reported autopsy results of humans showing the highest accumulation in lungs (more chromium from air than from water or food).

## Water Surveys

Data of 10% groundwater levels at above 5 µg/L (in positive samples 16 µg/L avg.; 49 max.) and 17% surface water with levels about 5 µg/L (in positive samples 10 avg.; 25 max.) were recorded by the USEPA between 1965 and 1980, sampling 795 groundwater supplies and 142 surface supplies. Daily exposure of chromium from water can be estimated to vary between 10–40 µg (in exceptional cases).

## HEALTH EFFECTS

Trivalent chromium may be nutritionally essential with a safe and relatively innocuous level of 0.20 mg/day. Hexavalent chromium has a deleterious effect on the liver, kidney, and respiratory organs with hemorrhagic effects, dermatitis, and ulceration of the skin for chronic and subchronic exposure. One-day exposure is safe at 1.4 mg/L for a child and 5.0 mg/L for an adult. In an animal study for hexavalent chromium, 60-day exposure was tested; a NOEAL = 14.4 mg/kg/day in rats. With an uncertainty factor of 100, the 10-day assessment is 1.4 mg/L (child) and 5.0 mg/L (adults).

With rats supplied with 25 mg/L of $Cr^{6+}$ for one year, NOAEL was 2.41 mg/kg/day. With an uncertainty factor of 500, the resulting AADI was 0.17 mg/L.

Chromium was classified by IARC in group 1 (carcinogenicity in humans and animals). A toxic dose for man was reported at about 0.5 g of potassium bichromate.

## STANDARDS

| | |
|---|---|
| USPHS 1925 | = not stated |
| USPHS 1942 | = 0 mg/L as hexavalent |
| USPHS 1946 | = 0.05 mg/L |
| USPHS 1962 | = 0.05 mg/L as hexavalent |
| SNARL (NAS) | = 0.05 to 0.20 mg/day via diet |
| NIPDWR | = 0.05 mg/L |
| WHO guidelines | = 0.05 mg/L (as $Cr^{6+}$ and total chromium) |
| European Community | = 0.05 mg/L (as $Cr^{6+}$ and total chromium) |
| RMCL (1985) | = 0.12 mg/L as total chromium |
| USEPA classification | = A = "based on data for $Cr^{+6}$—carcinogenic in humans by inhalation and rodents by intratracheal instillation. However, regulating as "D" since there is inadequate evidence to conclude that the chemical is carcinogenic via ingestion." |
| MCLG and MCL (USEPA, 1989) | = 0.1 mg/L (proposed) |
| MCLG and MCL (USEPA, 1991) | = 0.1 mg/L (final; effective 7/30/1992) |

## ANALYSIS

*Standard Methods* reports concentrations in drinking water to vary between 3 and 40 µg/L with a mean of 3.2 µg/L. Both the hexavalent and the trivalent state may exist in water, with the trivalent form being rather rare in potable water.

In potable water, the Colorimetric Method is preferable, but total chromium in water and wastewater also can be determined using *atomic absorption spectrometry by direct aspiration into an air-acetylene flame or for low concentration by chelation and extraction.* The USEPA (1985) recognized the available method recorded for barium and cadmium.

## REMOVAL/NOTES

Microstraining and clarification has an expected capacity to reduce chromium slightly (less than 33%). Chemical treatment with alum, lime, or ferrous or ferric sulfate may reach efficiency over 90% in removal with variations due to hexavalent vs. trivalent and operational range of pH. Insufficient experience is presently recorded for chromium removal through reverse osmosis and adsorption by activated carbon. Chromium is also contained in cigarettes (1.4 µg per cigarette).

The USEPA proposed in the 1989 Regulations the following treatment processes as BAT for chromium removal: Coagulation/Filtration; Ion Exchange; Lime Softening (Chromium III only); Reverse Osmosis.

In issuing the final rule in 1991 and 1992, USEPA listed as the best available technology for *chromium:* coagulation with filtration, lime softening (Cr III), and under specialized processes: ion exchange and reverse osmosis.

**COPPER**

## CHEMISTRY

A reddish-brown, malleable, ductile, metallic element, excellent conductor of electricity and heat. Symbol Cu; atomic weight 63.546; atomic no. 29; specific gravity 8.96; valence 1 or 2. The most important copper ores are sulfides, oxides, and carbonates. It is used in pipes, brass, domestic utensils, electrical industry, coins, and Monel. Oxide and sulfate are the most important compounds.

## ENVIRONMENTAL EXPOSURE

### Raw Water

Copper is very commonly found on the earth's crust as sulfides, oxides (cuprite, malachite, axurite, chalcopyrite, bornite), and rarely as metal. Consequently, it is found in surface water generally at concentrations below 20 µg/L. It may be detected in higher values from the consumer's faucet as a product of corrosion of brass and copper pipes. As an algicide in surface water, it may vary seasonably (spring/summer algae treatment) in surface supplies. Industrial sources can be listed for smelting and refining, copper wire mills, coal burning industries, electroplating, tanning, engraving, photography, insecticides, fungicides, and iron and steel producing industries.

In groundwater, it can be detected as an industrial pollutant.

### Air and Nutrition

Copper in the air can be the product of copper dust (industrial production), coal burning, and probably tobacco smoke.

Airborne concentration in rural areas = 0.01 µg/mc<br>
in urban areas = 0.257 µg/mc<br>
Maximum daily intake through inhalation = 1% of total.

In food it is found in organ meats, shellfish, nuts, dried legumes, cocoa (varying from 10–400 µg/g). A normal copper intake is 2 mg/day. Normal adult intake from food reported ranges between 2.0–4.0 mg/day. Consequently, copper from water can be significant in the total intake in certain areas (or with certain piping).

## Water Surveys

From surveys in 1967 reported by USEPA of 380 drinking water systems, levels of copper range from 1–1,060 µg/L; mean = 43 µg/L. Other surveys in the United States recorded a mean value of 134 µg/L with a maximum of 8 mg/L.

Soft water supplies with low pH (corrosion of copper pipes) produced tissue samples from long-time residents with higher levels of copper.

A special survey was conducted by USEPA in 1981 to measure copper from the consumer's tap (30-second flushed random daytime grab sample). Copper concentration exceeded 1 mg/L in 3% of the 772 samples and 0.2 mg/L in 19% of samples. The national average was 0.221 mg/L with the median of 0.04 mg/L.

Groundwater supplies were tested (NIRS—USEPA—1988) at random. Copper levels were below 0.06 mg/L in 85% of the 983 tap samples. Copper levels below 0.46 mg/L were registered in 98% of the samples (max. value = 2.37 mg/L; above 1 mg/L were detected in 1% of these samples). These results of analyses at the tap are of course much higher than the ones registered in the 1969 survey (USEPA—1969—CWSS) when samples were not specifically collected at the tap. In conclusion the USEPA (1988) reported that only 66 water suppliers in the U.S. would need to install treatment to lower copper levels to the MCLG of 1.3 mg/L.

## HEALTH EFFECTS

Copper is not a cumulative systemic poison. Doses up to 100 mg taken by mouth cause symptoms of gastroenteritis with nausea. Values of less than 30 mg, even for many days, have not caused poisoning. Poisoning from copper in water is normally avoidable because taste threshold concentration of copper is at 1.0–2.0 mg/L; levels at 5–8 mg/L make the water undrinkable, but poisoning occurs at higher concentrations. Also, green stains appear as a reaction of soap with copper at a concentration higher than 1.0 mg/L. Limited data are available on chronic toxicity of copper. Individuals with Wilson's disease (disorder of copper metabolism) are at additional risk from the toxic effects of copper.

Copper is considered an essential element for human nutrition since it is required in many enzymatic reactions. The daily requirement has been estimated at 2 mg/L.

The NAS (1977) stated: "The hazard to the general population from dietary copper up to 5 mg appears to be small. The interim copper limit of 1 mg/L of drinking water is based on consideration of taste, rather than toxicity." The USEPA (1985) considered a provisional AADI based on a

LOAEL of 5.3 mg/day, applying an uncertainty factor of 2, or AADI = 5.3 mg/2 × 2L = 1.3 mg/L.

No increase of tumor incidence appeared in studies based on orally administered copper compounds. The USEPA (1988)* confirmed the conclusions stated in its proposed regulations of November 1985. Those conclusions are part of the above-mentioned data.

## STANDARDS

| | |
|---|---|
| USPHS 1925 | = 0.2 mg/L |
| USPHS 1942 | = 3.0 mg/L |
| USPHS 1962 | = 1.0 mg/L |
| NIPDWR 1975 | = 1.0 mg/L |
| WHO guidelines | = 1.0 mg/L (1.5 mg/L "excessive")† |
| WHO (European) | = 0.05 mg/L (3.0 mg/L for 16 hours; new pipe contact) |
| European Community | = 0.1 mg/L guide (no max.) |
| RMCL (USEPA, 1985) | = 1.3 mg/L |
| NAS (1980) | = 2–3 mg/day = adequate |
| LOAEL (NAS—USEPA) | = 5.3 mg/day (factor of safety = 2) |
| AADI (provisional) (USEPA) | = 1.3 mg/L |
| USEPA classification | = Group D (inadequate data to classify) |
| MCLG and MCL (USEPA, 1988) | = 1.3 mg/L (proposed) |
| Secondary MCL (USEPA, 1989) | = 1.0 mg/L |
| MCLG (USEPA, 1991) | = 1.3 mg/L |
| USEPA (1991) Action Level | = 1.3 mg/L at the consumer's tap (final revised regulations for lead and copper according to the New Lead and Copper Rule as requested by the Safe Drinking Water Act—Revision of 1986)‡ |

## ANALYSIS

If the sample is not immediately analyzed following collection, the sample bottle should be acidified (0.5 mL + 1 HCl per 100 mL of sample) to avoid absorption by container. *Standard Methods* recommends the Atomic Absorption Spectrometric and Neocuproine methods because of their freedom from interferences. Also, for potable water, the Bathocuproine Method is an

---

*Federal Regulations, Vol. 53, No. 160—Aug. 18, 1988—40 CFR Parts 141 and 142 (drinking water regulations—MCLG and NPDWR for lead and copper; proposed rules).
†Not based on health effects but stains to laundry and plumbing fixtures.
‡See also Chapter 8 for the monitoring requirements and Chapter 10 in regard to corrosion problems of the distribution system as related to lead and copper.

acceptable method. USEPA (1985) recognized the availability of methods listed for barium.

## REMOVAL/NOTES

No single or economical method is available. Ion exchange is a potential method to be tested first in pilot plants. Elimination of pollutional sources is necessary; it must be first attempted, not only for economic reasons, but mainly to trace paths of pollution or excessive use of copper in pretreatment.

USEPA issued in 1988 the NPDWR for Lead and Copper and specified the following treatment techniques for achieving compliance for *copper:* Coagulation/Filtration; Ion Exchange; Lime Softening; Reverse Osmosis.

The responsibility of the USEPA standard setting is also the determination of the BAT for the removal of the contaminant up to the adopted maximum contaminant level goal or standard. The technologies accepted are:

- coagulation/filtration 60–95% removal
- ion exchange to 95% removal
- lime softening 90–96% removal
- reverse osmosis 90–99% removal

Lime softening is the most economical, with reverse osmosis the most expensive. In general consumer cost is significant in small plants. Sludge disposal of concentrated contaminant has also been considered by USEPA.

In consideration of the lack of possible control on private properties; that is, lack of responsibility of the public water supply system after connection with a private service line, the USEPA has recognized that in the case of lead and copper the monitoring point should be at the entry point of the distribution system. Corrosion control treatment may be necessary to reduce concentration of copper at the tap. The USEPA (1988) specified that, "A system that meets the no-action levels, i.e., 95% of samples at the tap with pH $\geq$ 8, lead levels at the tap of 0.010 mg/L or below as an average, and 95% of copper levels at the tap of 1.3 mg/L or below, would be considered to have minimally corrosive water and would be considered in compliance with the corrosion control portion of the treatment technique requirements."

## FLUORIDE

## CHEMISTRY

Fluorine is a halogen, defined as a corrosive, pale greenish-yellow gaseous chemical element, the most reactive nonmetallic element known, forming fluorides with almost all the known elements, organic and inorganic: symbol F; atomic weight = 18.998; atomic no. 9; density 1.696; valence 1.

Fluorine is found in fluorspar ($CaF_2$) and cryolite ($Na_2AlF_6$). Fluorine and its compounds are used in producing uranium and over 100 commercial fluorochemicals, including high-temperature plastics.

The free element has a characteristic pungent odor, detectable at concentrations below 20 ppb which is below the safe working level. The recommended maximum allowable concentration for an 8-hour time-weighted exposure is 0.1 ppm.

## ENVIRONMENTAL EXPOSURE

### Raw Water

In surface water from rivers, average concentrations have been recorded at 0.2 mg/L (0–6.5 range); in groundwaters an average of 0.3–0.4 for limestone and dolomites, shales and clays, as high as 8.7 average for alkalic rocks, as low as 0.1 average for basaltic rocks, and 9.2 (all in mg/L) in granitic rocks.

USEPA has reported the following estimated national occurrence of fluoride in public water supply systems.

| Population Served | Surface Water Systems with Fluoride (mg/L) | | | | Groundwater Systems with Fluoride (mg/L) | | | |
|---|---|---|---|---|---|---|---|---|
| | <1.0 | 1.0–2.0 | 2.0–4.0 | >4.0 | <1.0 | 1.0–2.0 | 2.0–4.0 | >4.0 |
| Less than 500 | 3,670 | 117 | 5 | 3 | 31,931 | 2,281 | 833 | 220 |
| 500 to 2,500 | 2,980 | 265 | 6 | 1 | 8,964 | 341 | 165 | 40 |
| 2,500 to 10,000 | 1,967 | 174 | 3 | 2 | 2,828 | 219 | 44 | 14 |
| More than 10,000 | 1,615 | 148 | 2 | 0 | 1,187 | 48 | 6 | 2 |

### Air and Nutrition

Fluoride in the atmosphere, in general, is at such a low level (below detection limits of 0.05 μ/mc) that it cannot contribute in any substantial amount to the total fluoride exposure. Industrial atmospheric emissions have caused adverse effects on animals and vegetation.

The typical diet also represents low intake of fluoride; it has been compared with a drinking water of 0.5 mg/L. In the case of water supplies with no significant natural fluoride and no fluoridation, food is the major source of fluoride. An average dietary intake is estimated at 0.0028–0.011 mg/kg for adults. Tea and fish are the exceptions (very high fluoride concentration).

### Water Surveys

The majority of the U.S. population receives artificially fluoridated water (see *Fluoridation* in Chapter 9, page 412). Fluoridated waters usually show concentrations within the range of 0.6–1.7 mg/L. Unfluoridated supplies usually report concentrations below 1 mg/L (see "Raw Water" above).

## HEALTH EFFECTS

A research of literature in regard to epidemiological studies of high concentration of fluoride in natural water can be summarized as follows:

- Dental fluorosis appears in a very small percentage when fluoride in drinking water is in the range of 1–2 mg/L.*
- Long-term intake of fluoride in concentrations higher than 4 mg/L may cause asymptomatic osteosclerosis in a small percentage of persons.
- Crippling fluorosis has been detected in individuals exposed to fluoride levels from 10–40 mg/L.
- Sharply reduced dental caries formation has been determined when the fluoride level is at least 0.8 mg/L, reaching maximum benefits around 3 mg/L.
- No carcinogenicity or other adverse effects have been detected.

As far as toxic reactions at high concentrations, fluoride with doses from 250–450 mg gives severe symptoms, causing death when reaching 4.0 g. A fatal dose also has been registered as low as 0.5 g/kg and as high as 2.5 g/kg.

More details of fluoride and public health are discussed in Chapter 9, "Public Health Regulations," in the section on fluoridation, page 412. A program to reduce the incidence of caries in teeth in a community and its implementation to supply fluoride with a treatment of water destined to the consumer's tap, and the monitoring of such treatment with sampling, analyzing, and reporting is defined as *fluoridation*.

## STANDARDS

| | |
|---|---|
| USPHS 1925 | = no standard |
| USPHS 1942 | = 1.0 mg/L |
| USPHS 1946 | = 1.5 mg/L |
| USPHS 1962 | = 1.4–2.4 mg/L max. related to temp. |
| USPHS 1962 | = 0.7–1.2 mg/L optimum related to temp. |
| NIPDWR | = 1.4–2.4 mg/L related to temp. |
| WHO guidelines | = 1.5 mg/L (depends on climatic condition and water consumption) |
| WHO (European) | = 0.7–1.7 mg/L related to temp. |
| European Community | = 0.7–1.5 mg/L related to temp. |
| Secondary MCL (1989) | = 2.0 mg/L† |
| RMCL (1985) | = 4.0 mg/L |

---

*When this limit is exceeded, the USEPA requires notification to consumers. (See Appendix A-VII, "Fluoridation Data.")

†Under the Safe Drinking Water Act statutory requirements, the USEPA issued in May 1985, "Fluoride; National Primary Drinking Water Regulations; Proposed Rule (40 CFR Part 141)" and in November 1985 the "National Primary Drinking Water Regulations; Fluoride; Final Rule and Proposed Rule (40 CFR Parts 141, 142, and 143)."

| NAS | = optimum levels for fluoridation should not be exceeded (See Appendix A-VII, "Fluoridation Data.") |
| USEPA classification | = no carcinogenicity |
| MCL (1986) | = 4.0 mg/L (final rule) |
| MCL (1993) | = 4.0 mg/L (confirmed)* |

In spite of the fact that the *final* rule for fluoride was issued in 1986, USEPA is requested by the amended SWDA to review drinking water regulations every three years. Accordingly, in 1990 USEPA requested a review of all recent studies, including adverse health effects, total exposure to fluoride, and BAT with related costs of removal. All informational material received by USEPA was reviewed by the National Research Council (NRC). The summary of this scientific investigation was published by the USPHS in 1991.[†] NRC's study was made public in 1993. NRC's conclusion was that the MCL issued in 1986 was still appropriate, pending additional research.[‡] See also the section on *Fluoridation* in Chapter 9, page 412, and Appendix A-VII, "Fluoridation Data."

## ANALYSIS

There are four reliable methods to determine the fluoride ion concentration:

- Ion Selective Electrode and Automated Ion Selective Electrode methods
- Colorimetric Method
- Complexone Method
- Alizarin Visual Method

*Standard Methods*, in the introduction to the determination of fluoride, reports: "Accurate determination of fluoride has increased in importance with the growth of the practice of fluoridation of water supplies as a public health measure. Maintenance of an optimal fluoride concentration is essential in maintaining effectiveness and safety of the fluoridation procedure. Among the methods suggested for determining fluoride ion ($F^-$) in water, the electrode and colorimetric methods are the most satisfactory."

---

*AWWA. Nov. 1991 *Journal*. Vol. 83. "Fluoride Regulation and Water Fluoridation," by F. W. Pontius, p. 20.
†Report of the Ad Hoc Subcommittee on Fluoride of the Committee to Coordinate Environmental Health and Related Programs (Feb. 1991). *Review of Fluoride Benefits and Risks.* Dept. of Health & Human Services, USPHS.
‡USEPA. "Notice of Intent Not to Revise Fluoride Drinking Water Standards." *Fed. Reg.*, 58:248:68826–68827 (Dec. 29, 1993).

## REMOVAL/NOTES

Activated alumina adsorption, reverse osmosis, and modified lime softening are proven methods of fluoride concentration reduction. Other potential methods are: adsorption using bone char and tricalcium phosphate, anion exchange resins, and electrodialysis.

The activated aluminum process as a reversible ion exchange medium has proved to be economically feasible even for small systems, but it requires a properly controlled operation; otherwise, quantities of aluminum can be found in the treated water. Typical of all ion exchange processes is the problem of disposal of the generated waste.

Reverse osmosis is a more expensive treatment but has the advantage of removing other inorganic ions, turbidity, and related bacteria and pathogens at the same time. Up to 95% removal of fluoride has been achieved.

Some amount of fluoride can be removed by alum coagulation, and, therefore, can be practiced when flocculation/sedimentation/filtration are available processes. However, large amounts of sludge could be a problem in sludge removal.

USEPA, April 1986 Rule (see *Fluoridation* in Chapter 9, page 412) mentions the evaluation of BTGA (Best Technology Generally Available) in the following water treatment technology:

- Activated Alumina Adsorption
- Reverse Osmosis (RO)
- Modified Lime Softening
- Adsorption Using Bone Char and Tricalcium Phosphate
- Anion Exchange Resins
- Electrodialysis

After evaluation, USEPA proposed activated alumina adsorption and RO as the BTGA. After consideration of cost for disposal of the generated wastes, the activated alumina adsorption is the most economical and reasonably affordable for medium and large populations served.

Assuming a 5.4 mg/L fluoride concentration (average system in need of treatment) for reduction to the standard 4 mg/L, USEPA (1986) estimates the following costs in dollars per 1,000 gallons:

| | Population Served | | |
|---|---|---|---|
| | *Small* 25–99 | *Medium* 2,500–5,000 | *Large* 10,000–100,000 |
| Activated Alum.—removal alone | 0.51 | 0.32 | 0.22 |
| Removal plus discharge to POTW | 0.53 | 0.33 | 0.23 |
| Removal plus controlled* discharge | 0.97 | 0.38 | 0.25 |
| Removal plus evaporation pond | 1.44 | 0.45 | 0.28 |
| Reverse Osmosis (RO) removal alone | 1.52 | 1.07 | 0.74 |
| Removal plus discharge to POTW* | 1.84 | 1.25 | 0.86 |

*Controlled long-term release using surge tank. POTW = Publicly Owned Treatment Work (for wastewater).

The USEPA in issuing secondary MCL (SMCL) in 1986, prescribed notification to the customers when fluoride concentration is higher than 2 mg/L. The language to be used in the public notification is prescribed by USEPA and contained in Appendix A-VII, "Fluoridation Data."

The SMCL of 2 mg/L was also supported by the American Medical Association and by the American Water Works Association (AWWA).

USEPA will consider variances* with an expected negative reply unless the water supply in question cannot utilize the BTGA due to unforeseen conditions. Public notification of customers in violation of SMCL should be required annually with separate notification of new billing units. The USEPA prescribes the required notice, contained in Appendix A-VII, "Fluoridation Data."

It has been estimated that, in the United States, there are approximately 1,300 public water systems with fluoride concentrations above 2 mg/L and approximately 300 systems with concentrations above 4 mg/L.

**LEAD**

## CHEMISTRY

A heavy, very soft, highly malleable, bluish-gray metallic element, poor conductor of electricity, and resistant to corrosion: symbol Pb; atomic weight 207.2; atomic no. 82; specific gravity 11.35; valence 2 or 4. Used primarily in storage batteries, lead is also used in cable covering, plumbing, ammunition, and in the manufacture of lead tetraethyl (used as an anti-knock compound in gasoline); as a radiation shield from nuclear reactors to X-ray equipment, in the glass industry, and in paints (recently the lead content has been curtailed due to potential lead poisoning). The nitrate and acetate are soluble salts.

The association of lead with lead poisoning (see "Health Effects" below) has eliminated the use of lead in products that may come in contact with children. Native lead occurs rarely in nature. The most known compounds are Galena ($PbS$), Anglesite ($PbSO_4$), Cerussite ($PbCO_3$), and Minim ($Pb_3O_4$). Average lead concentration in the earth's crust is 15 ppm in the United States.

## ENVIRONMENTAL EXPOSURE

### Raw Water

Concentration of lead in natural waters has been reported as much as 0.4–0.8 mg/L (mountain limestone and galena deposits encountered). Surface and ground raw waters range from traces to 0.04 mg/L, averaging about

---

*Guidelines for variances are contained in 40 CFR Parts 141, 142, and 143 issued by USEPA April 2, 1986 (NPDWR and NSDWR—Fluoride—Final Rule).

0.01 mg/L. Industrial and mining sources may contribute to some localized pollutional effect; but usually when high values of lead are detected in drinking water, the cause is always searched for in lead service lines and house plumbing.

## Air and Food

USEPA estimates human consumption in rural areas to be from 40 to 60 µg of lead per day due to the total environment (air, food, water, and dust). Atmospheric lead concentration varies from 0.000076 µg/mc in remote areas to 10 µg/mc near areas of lead use, with an average annual value below 1.0 µg/mc. USEPA estimates respiratory intake at 1.0 µg/day per adult and 0.5 µg/day for a 2-year-old child.* The WHO reports a typical range of 0.5–2 µg/mc for urban dwellers, equivalent to 6–9 µg/day (based on the assumption of a daily respired volume of air of 15–22.8 mc exposed to 1 µg of lead per cubic meter of air with a 40% retention).

In the atmosphere, 90% of lead intake was from leaded gasoline. The use of unleaded gasoline, initiated in 1974 in new cars, has greatly decreased the concentration of lead in urban air. The highest lead intake from foods is estimated at 19, 25, and 36 µg/day of Pb, respectively, for children, females, and males (adult) with an additional 7, 11, and 19, respectively, from water and beverages. In urban environments, the additional exposure is estimated by USEPA as 28, 91, and 36 µg/day, respectively, for adults, children, and a 6-month-old infant.* Children may consume 0.1 g of dust per day.

## Water Surveys

USEPA surveys found lead at levels above 5 µg/L in 75.5% of 1,200 groundwater supplies sampled, with a mean of 26 µg/L in "positive" supplies with a range of 5–380 µg/L. Approximately, the same percentage (76.2%) was observed as positive (above 5 µg/L) in 273 surface water supplies tested, with a mean of 24 µg/L and a range of 5–164 µg/L. Excluding Rural Water Survey (RWS), the above data are corrected as groundwater 76.3%–13 µg/L mean and 5–182 µg/L range, surface water 84%–14 µg/L mean and 5–32.5 µg/L range. (The percentage is for "positive" samples or more than 5 µg/L.) The NAS reported a 1962 survey of the 100 largest cities in the United States with 3.7 µg/L median, 62 µg/L maximum, and not detectable as a minimum.

It must be noted that the use of soft water of slightly acidic pH (Seattle, Boston, and New York) and the use of lead pipes in service and domestic water lines may provide higher concentrations of lead at the consumer's tap, particularly when the water use is minimal in the household (overnight still water in pipes). In most cases, higher lead content is expected at the con-

---

*USEPA (1985). *Occurrence of Lead in Drinking Water, Food, Air* and USEPA (1984). *Air Quality Criteria for Lead*, EPA-600/8-83-028B.

sumer's tap than at the treatment plant. Higher concentrations of lead are also expected in the early-morning samples, with minimum values in running water and intermediate value in composite samples.

The NAS (1977) summarized data from estimated food and water intake, stating that lead intake from drinking water constitutes one-tenth or less of that obtained from an ordinary diet. This conclusion is reached assuming an average concentration of 13 μg/L of lead in drinking water.

USEPA (1988) estimated that less than 1% of the community water systems have concentrations of lead greater than 0.005 mg/L.

## HEALTH EFFECTS

Health effects of a toxicologic nature are measured by blood lead levels. The effects are neurotoxic, which include irreversible brain damage. Such a toxic level is reached when the blood level exceeds 100–120 μg/dl. Severe gastrointestinal symptoms are associated with the encephalopathic symptoms. These symptoms start to be observed in adult lead workers at blood lead levels of 40–60 μg/dL.

The subpopulation to be carefully studied is represented by children, where encephalopathy and death are registered at a starting level of 80–100 μg/dL (blood). In nonfatal cases, permanent, severe mental retardation with other neurologic symptoms are observed at levels as low as 40–60 μg/dL. Adverse health effects are noted in children with blood lead levels of 40 μg/dL or higher with possible risk at levels as low as 15–30 μ/dL.

In rats a high level of lead may have produced renal tumors, but no higher incidence of tumors in lead-exposed populations has been determined. The International Agency for Research on Cancer (IARC) classified lead in group 3 (inadequate evidence for carcinogenicity in humans).

The NIPDWR set MCL at 0.05 mg/L, estimating the drinking water to contribute 25% to 30% of the lead normally ingested for a child and 33% of that in food for an adult.

USEPA has worked the following data to arrive at reasonable MCL:

- sensitive subpopulation: infant consuming formula
- 15 (to 20) μg/dL = level of concern
- conversion factor of 6.25 blood lead to water lead
- 15 μg/dl × 6.25 = 94 μg/L (1 L/day of water)
- factor of safety = 5 corresponds to 20 μg/L (as an alternate approach)

The health effects can be summarized as follows:

- Lead is not considered an essential nutritional element.
- Lead is a cumulative poison to humans. Acute lead poisoning is extremely rare.

- Typical symptoms of advanced lead poisoning are: constipation, anemia, gastrointestinal disturbance, tenderness and gradual paralysis in the muscles, specifically arms, with possible cases of lethargy and moroseness.
- A total intake of lead of 0.6 mg per day over an extended period may result in lifetime dangerous accumulation.
- There is no evidence to reach the conclusion that lead is carcinogenic or teratogenic in humans, and evidence of mutagenicity is scant (NAS, Vol. IV, 1982).
- As a piping material, lead corrodes easily and accumulates in concentrations higher than the maximum allowable limit, particularly in still water (lack of continuous water consumption).
- Its major effect in humans is impairment of hemoglobin and porphyrin synthesis (upset metabolism connected with utilization of chemicals).

## STANDARDS

| | |
|---|---|
| USPHS 1925–1942–1946 | = 0.1 mg/L |
| USPHS 1962 | = 0.05 mg/L |
| WHO guidelines | |
| (0.1 mg/L 1971) | = 0.05 mg/L (1984) |
| WHO (European) | = 0.10 mg/L |
| European Community | |
| (max.) | = 0.05 mg/L (for running water) |
| Netherlands & Germany | |
| (in pipes) | = 0.3 mg/L (water still for 24 h) |
| NIPDWR (1975) | = 0.05 mg/L |
| RMCL (1985 and | |
| 1988) USEPA | = 0.02 mg/L* |
| USEPA classification | = Group B2 (sufficient evidence in animals; insufficient basis to regulate as a human carcinogen via ingestion) |
| MCLG (USEPA, 1991) | = zero |
| MCL (USEPA, 1991) | = 0.015 mg/L as the action level for treatment technique (effective Dec. 7, 1992) |

The decision to lower the MCL from 0.05 to 0.02 mg/L and then to a final national primary drinking water regulation to "treatment technique," once

---

*NAS—1977 evaluation of lead in public health concluded that preliminary data suggested that the present limit of 50 µg/L may not provide a sufficient margin of safety, particularly for fetuses and young growing children in view of other sources of environmental exposure. A lower limit was therefore suggested to the USEPA.

the action level of 0.015 mg/L is reached, was adopted on June 7, 1991.* This was known as the Lead and Copper Rule that requires treatment.†

USEPA included in the "treatment technique": optimal corrosion-control treatment; source water treatment; public education, and lead service line replacement. For more information on the Lead and Copper Rule see Chapters 8, 9, and 10.

## ANALYSIS

Lead in drinking water can be reliably analyzed with the Flame Atomic Absorption, Inductively Coupled Plasma Atomic Emission Spectrometric Techniques (USEPA 1985).

*Standard Methods* lists the Atomic Absorption Spectrometric and the "Dithizone" methods with the remarks that the AA method "is subject to interference in the Flame Mode and requires an extraction procedure for the low concentrations common in potable water," while "the Electro-thermal AA method does not require extraction. The Dithizone method is sensitive and is preferred by some analysis for low concentrations."

## REMOVAL/NOTES

In general, reduction of lead concentration at the consumer's tap can be accomplished by the following:

• Raising the pH of treated water to reduce corrosivity.
• Progressively eliminating street service lead lines.
• Reducing chlorination levels by substituting the disinfectant or reducing the chlorine demand.
• Eliminating lead-soldered joints using tin–antimony solder instead of the traditional 50 : 50 lead–tin solder.

Because most lead poisoning (associated with children who eat paint chips from walls painted with lead paint) is originated by ambient air (traffic) and food, a general reduction of lead in the environment may lead to a decrease of lead content in the blood and more tolerance to a minimum quantity of lead in water.

The joint FAO/WHO Expert Committee on food additives established in

---

*Maximum Contaminant Level Goals and National Primary Drinking Water Regulations for Lead and Copper; Final Rule. *Fed. Reg.*, 56:110:26460–26564 (June 7, 1991).

Drinking Water Regulations; Maximum Contaminant Level Goals and National Primary Drinking Water Regulations for Lead and Copper; Final Rule; Corrections. *Fed. Reg.*, 56:135:32112–32113 (July 15, 1991). On June 30, 1994, USEPA issued a corrected version, namely *Fed. Reg.* 59:125:33860–33864.

†American Water Works Association. 1993. "Lead and Copper". 6666 West Quincy Ave. Denver, CO 80235.

1972 the provisional tolerable weekly intake of 3 mg of lead per person (not for children). This is equivalent to 0.05 mg/L with 25% water contribution.

Very sensitive to lead exposure are children, infants, fetuses in utero, and pregnant women. With the assumption that water of 0.1 mg/L causes an increase of 4–5 µg of lead for 100 mL of blood, and assuming that 30 µg/100 mL is the blood level maximum lead concentration recommended for children, a lower level of lead in water is advisable. The WHO reduced the 0.1 mg/L standard to 0.05 mg/L.

A worldwide average for adults is about 200 µg typical daily food intake (WHO—1984).

Old household paints contain lead; unsupervised young children may eat paint chips, soil, and dust. A determination of lead concentration from this source is practically impossible. USEPA estimated that 44% of the baseline consumption of lead by children is derived from a daily consumption of 0.1 g of dust.

Effects of lead in service lines and household plumbing are also examined in "Sampling and Monitoring" (Chapter 8); in "Lead Control Programs" (Chapter 9); and in *Distribution System* (Chapters 10 and 11).

---

## MERCURY

### CHEMISTRY

A heavy, silver-white, metallic chemical element, liquid at ordinary temperatures (the only common metal), mercury occurs rarely in a free state but usually in combination with sulfur. The chief ore is cinnabar ($HgS$). Symbol: Hg; atomic weight 200.59; atomic no. 80; specific gravity 13.546; valence 1 or 2.

Mercury is one of the least abundant elements in the earth's crust (listed 74th out of 90), but the natural degassing of the earth's crust is an important source of mercury in the environment. A poor conductor of heat that offers better conductivity in electricity, mercury is a rather inert chemical and insoluble in water. Therefore, in spite of the toxicity of mercuric salts, it received little attention in drinking water quality until 1950. It is used in laboratory work (thermometers, barometers, instruments) and in mercury-vapor lamps, mercury switches, and advertising signs.

Organic salts ($HgS$—sulfide) and organic mercury compounds (methyl mercury) are the basic forms in nature. But synthetic organic and inorganic salts of mercury are used industrially and commercially. Highly soluble in water are many of the mercuric and mercurous salts. The most important salts are mercuric chloride $HgCl_2$ (a corrosive sublimate), mercurous chloride—$Hg_2Cl_2$ (calomel), mercury fulminate $[Hg(ONC)_2]$—a detonator—and mercuric sulfide $HgS$ (paint pigment).

A major use of mercury is in electrical equipment. Mining, smelting, and fossil fuel combustion contribute to mercuric pollution.

## ENVIRONMENTAL EXPOSURE

### Raw Water

Mercury concentration in soil has been detected with a range from 30–300 ppb. In coal the range is 10–46,000 ppb. Sea water contains 0.03–2.0 µg/L. Groundwater has been recorded as below detection level (0.1 µg/L).

In surface waters, surveys show mercury concentrations of less than 2 µg/L (rivers) and only 4% with concentration in excess of 10 µg/L (small lakes and reservoirs). The NAS in 1977 reported a probable frequency of 10% with a minimum of 0.1 µg/L and a maximum of 10 µg/L.

### Air and Nutrition

Ambient levels of mercury range between 10–20 ng/mc. Near coal/power plants, the levels are as high as 1,000 ng/mc; and near mercury mines and fungicides fields, 10,500–15,000 ng/mc.

The USEPA estimated that 80% of the intake rate is absorbed, and an inhalation rate of 20 mc/day and an atmospheric concentration of 0.02 µg/mc results in 0.320 µg for an adult male.

It must be noted that practically no surveys were scheduled before 1970 in the United States, but investigation of environmental mercury took place in Sweden and Japan. After 1970, the result of concentration of 0.5 ppm in fish tissue caused monitoring of mercury in fish.

Food intake in the U.S. is estimated at 4.3 µg/day (3.8 expected from fish, meat, and poultry). The WHO (1984) reported 10–12 µg/day average. Methyl mercury is absorbed by fish and mammals more than inorganic mercury.

### Water Surveys

Mercury detected in potable water was predominantly in the form of inorganic mercury.

Maximum concentrations reported by the USEPA (1978–1980) are as follows:

- 30% groundwater had levels above 0.5 µg/L from 106 supplies sampled;
- 14% had above 2.0 µg/L;
- 32% surface water had levels of 0.5 µg/L from 31 supplies sampled; and
- 16% had above 2 µg/L.

## HEALTH EFFECTS

The principal target organ of inorganic mercury is the kidneys, with neurologic and renal disturbances. Methyl mercury compounds are very toxic to the central nervous system; they are also the major source of environmental contamination.

Fatal doses for man varies between 3–30 g. Four to 12 mg of mercury per day may be safe, and fatal doses of mercury would be 75–300 mg/day.

In animal studies, the NOAEL was determined to be 50 μg/kg/day; with an uncertainty factor of 1,000, and a ratio of 10 was used for subcutaneous/oral exposure. USEPA (1985) assumed a provisional AADI = 0.0055 mg/L (inorganic mercury).

The IARC did not evaluate the carcinogenic potential of mercury. No beneficial physiological function was determined in man. The WHO (1984) reported that long-term daily ingestion of approximately 0.25 mg of mercury, as methyl mercury, has caused the onset of neurological impairment.

## STANDARDS

| | |
|---|---|
| USPHS 1925 and 1942 | = no limits |
| USPHS 1962 | = no limits |
| NIPDWR (1975) | = 0.002 mg/L |
| WHO guidelines (1984) | = 0.001 mg/L |
| European Community | = 0.001 mg/L |
| RMCL (1985) | = 0.003 mg/L |
| NOAEL | = 50 μg/kg/day |
| AADI | = 0.0055 mg/L (diet 0.0043 mg/day, air 0.001 mg/day) |
| USEPA classification | = Group D ("inadequate data to classify") |
| MCLG and MCL (1989) | = 0.002 (proposed by USEPA) |
| (USEPA, 1991) MCLG and MCL | = 0.002 mg/L (final rule has been in effect since July 30, 1992) |

## ANALYSIS

*Standard Methods* advises to preserve samples by treatment with $HNO_3$ to reduce the pH to less than 2. For all samples the "Cold Vapor Atomic Absorption Method" is the method of choice, with the "Dithizone Method" as the method to be used in potable waters for high levels of mercury.

## REMOVAL/NOTES

It must be taken into consideration that the presence of mercury indicates pollution into the raw water from mining, industry, and commercial products discharge. Water treatment processes that may contribute to removal are:

coagulation–sedimentation, absorption, and ion exchange. Organic mercury is effectively reduced by granular activated carbon, inorganic mercury by coagulation with ferric sulfate (pH 7.0–9.5).

The aquatic food chain may absorb mercury from the toxic methyl and dimethyl mercury converted from inorganic and organic forms by microorganisms.

The USEPA proposed in 1989 the following treatment processes as Best Available Technology for **mercury** removal: Granular Activated Carbon; Coagulation/Filtration (with mercury influent concentration of less than 10 μg/L); Powdered Activated Carbon (same limits); Lime Softening (same limits); and Reverse Osmosis (same limits of mercury concentration of less than 10 μg/L).

## NITRATE—NITRITE

## CHEMISTRY

Nitrate is a salt or ester of nitric acid ($NO_3^-$) or an end product of the aerobic stabilization of organic nitrogen. Also, a product of conversion of nitrogenous material. Combined nitrogen should be considered a potential source of nitrate. Sources of nitrates are also mineral deposits (sodium and potassium nitrate), soils, seawater, and atmosphere.

Nitrite is a salt or ester of nitrous acid (nitrogen-dioxide—$NO_2^-$) formed by the action of bacteria upon ammonia and organic nitrogen. No significant concentration is found in surface water due to the prompt oxidation to nitrates. Combined nitrogen may be found concentrated in wastewater, landfills, and agricultural and urban runoff.

Nitrate is used as a fertilizer, as a food preservative, and as an oxidizing agent in the chemical industry. Nitrite is used in industry as a food preservative (sodium and potassium salts), particularly in meat and cheese.

## ENVIRONMENTAL EXPOSURE

Nitrates are particularly detectable in soil and consequently are widespread in the environment from food to atmosphere and water. Higher concentrations are expected where fertilizers are used, in decayed animal and vegetable matter, in leachates from sludge and refuse disposal, and in industrial discharges. Nitrates are always higher in concentration than nitrites.

### Raw Water

Lakes and reservoirs usually have less than 2 mg/L of nitrate measured as nitrogen. Higher levels of nitrates are found in groundwater ranging up to

20 mg/L, but much higher values are detected in shallow aquifers, polluted by housing (sewage) and/or excessive use of fertilizers.

In surface waters, the utilization by plants of nitrate as a fertilizer (photosynthetic action converts nitrates into organic nitrogen in plant cells) tends to reduce substantially the concentration of nitrates, leaving, therefore, the problem to groundwater.

## Air and Nutrition

Nitrogen oxide daily respiratory intake is reported by the USEPA as ranging from 25–70 μg with a maximum of 0.1 mg/day as N. An estimated dietary intake of nitrates is 100 mg/day (85%–90% from vegetables—beets, celery, lettuce, radishes, and spinach) and 9% from cured meats. Much lower levels are from potatoes, tomatoes, and peas. Food is the major source of human intake of both nitrates and nitrites.

## Water Surveys

In groundwater surveys, 56% of the 1,479 water systems sampled had nitrate/nitrogen levels above 0.3 mg/L with a mean value of 1.8 mg/L and 1.4% with levels above 10 mg/L. In 409 surface supplies sampled, 43% were with levels above 0.3 mg/L, the mean value was 1.6 mg/L and the maximum 21 mg/L. Levels above 10 mg/L with all values reported as nitrate/nitrogen were detected in five supplies (1.2%).

When nitrite is detected in potable water in considerable amounts, it is an indication of sewage/bacterial contamination and inadequate disinfection.

These surveys were conducted at random in the United States between 1969 and 1980 with data tabulated by USEPA (1985).

Nitrates, nitrites, and ammonia in water are indications of pollution.

## HEALTH EFFECTS

The NAS reports that the toxicity of nitrites is demonstrated by vasodilatory/cardiovascular effects at high doses and methemoglobinemia at lower doses. Partial reduction of nitrates to nitrites in humans takes place in saliva for all ages and in the gastrointestinal tract in infants during the first three months of life. Therefore, babies up to 3 months of age are more susceptible since they transform 100% of ingested nitrates into nitrites, while only 10% is expected in adults and children.

Nitrite acts in the blood to oxidize the hemoglobin to methemoglobin, which is not an oxygen carrier to the tissues, with consequent anoxia (methemoglobinemia). Large concentrations of nitrite in water may result in the potential formation of carcinogenic nitrosamines. Water consumption with concentrations of nitrate/nitrogen of 10 mg/L produced no methemoglobinemia cases. Only 2.3% of all cases appeared related to water with concentrations of nitrate between 10 and 20 mg/L (N). With no safety factor necessary, USEPA (1985) determined a provisional AADI of 10 mg/L (NOAEL = 10 mg/L).

The AADI for nitrite was determined assuming an uncertainty factor of 10, resulting in a provisional AADI = 1 mg/L nitrite/nitrogen. The USEPA concluded, based on the high risk subpopulation of infants, with NOAEL = 10 mg/L and no uncertainty factor, that a provisional AADI = 10 mg/L for nitrate/nitrogen; and the AADI = 1 mg/L for nitrite-nitrogen can be used as a guideline.

For a Carcinogenic Risk Assessment, USEPA classified nitrate/nitrite in "Group D" or "inadequate data to classify." It has been found to be carcinogenic in animals only when administered with nitrosatable compounds. More studies are necessary to better understand the metabolism of ingested nitrate in humans.

Several epidemiological studies have been examined by USEPA in 1990 to evaluate the possible relation of nitrate in drinking water and cancer. Searching for high concentrations of nitrates in drinking water and stomach or gastrointestinal cancer, studies were conducted in England, Colombia, Chile, Hungary, Italy, China, and Iran.* In setting the final determination for MCL in nitrate, USEPA reached the conclusion that the studies reviewed were unable to demonstrate an increase in cancer risks. Consequently, USEPA office of Science and Technology initiated actions for a study to be conducted by the National Academy of Sciences for reviewing epidemiological studies that have searched for a link between nitrate and cancer.[†]

## STANDARDS

| | |
|---|---|
| USPHS 1925–1942 | = no requirements for nitrates |
| USPHS 1962 | = 45 mg/L as nitrates |
| WHO (1963) | = 45 mg/L as nitrates for infants (no standard) |
| WHO (European) (1971) | = 23 mg/L as Nitrogen |
| European Community | = 25 mg/L as $NO_3$ as a guide |
| European Community | = 50 mg/L as $NO_3$ as a max. |
| European Community | = 0.1 mg/L as $NO_2$ as a max. |
| NIPDWR | = 10 mg/L as N |
| WHO guidelines (1984) | = 10 mg/L nitrate/nitrogen |
| WHO guidelines (1984) | = no value set for nitrite/nitrogen (no action required) |
| RMCL (1985) | = 10 mg/L nitrate/nitrogen = MCLG (USEPA, 1989) |
| RMCL (1985) | = 1 mg/L nitrite/nitrogen = MCLG (USEPA, 1989) |

---

*Studies in China and Iran were related only to esophageal cancer.
†AWWA. April 1993. *Journal*. Vol. 85. p. 12. "Nitrate and Cancer: Is There a Link?" by F.W. Pontius. Denver, CO.

EPA classification
   (1985)              = Group D
USEPA (1989) MCL   = 10 mg/L Nitrate/Nitrogen (proposed)
USEPA (1989) MCL   = 1 mg/L Nitrite/Nitrogen (proposed)
USEPA (1991) MCL   = 30 mg/L Nitrate/Nitrogen (final)
USEPA (1991) MCL   = 1 mg/L Nitrite/Nitrogen (final)
USEPA (1991) MCL   = 10.0 mg/L for total **Nitrate** + **Nitrite** (all final
                      rules became effective July 30, 1992)

## ANALYSIS

For nitrogen-nitrate in potable waters, *Standard Methods* lists the following methods:

- Ultraviolet Spectrophotometric Screening Method
- Nitrate Electrode Screening Method
- Cadmium Reduction Method
- Chromotrophic Acid Method
- Devarda's Alloy Reduction Method
- Automated Cadmium Reduction Method (tentative)

The screening techniques are used to determine approximate nitrate concentration.

USEPA (1994) published an advised methodology for nitrate to include ion chromatography; automated cadmium reduction; ion selective electrode; and manual cadmium reduction. For nitrite: ion chromatography; automated cadmium reduction; and spectrophotometric.

## REMOVAL/NOTES

Nitrates can be removed by anion exchange (over 90% removal tested). Chemical coagulation or lime softening proved to be noneffective techniques. Chemical reduction has given acceptable results. Biological denitrification and reverse osmosis (over 60% reduction tested) are also potential methods. All methods are practically uneconomical.

USEPA proposed in 1994 the following treatment processes as Best Available Technology for nitrate and nitrite: ion exchange and reverse osmosis.

---

**SELENIUM**

---

## CHEMISTRY

A gray (in crystalline hexagonal), deep red (in monoclinic selenium), non-metallic chemical element of the sulfur group, existing in many allotropic forms: symbol Se; atomic weight 78.96; atomic no. 34; specific gravity gray 4.79, vitreous 4.28.

Analogous to sulfur, as far as chemical combinations are concerned, selenium is used in its elemental form or in several salts for electronic and photocopy applications, glass manufacture, pigments, chemicals, pharmaceuticals, fungicides, electrical apparatus, and rubber industry. Selenium appears in soil as ferric selenite, calcium selenite, and elemental selenium. In soils at concentrations from 0.03–0.8 ppm (United States), selenium is more water soluble in more alkaline soil. Sedimentary rocks, such as shale, normally contain more selenium than limestone or sandstone. Selenium is generally produced recovering copper ores (limited use 600 tons in United States in 1981).

Selenium in water is expected in the selenite or selenate form (related to pH and salts).

## ENVIRONMENTAL EXPOSURE

### Raw Water

Selenium is expected to be found in raw water from soil contamination in areas rich in selenium. Otherwise, it appears in trace quantities only in public sewers due to industrial pollution. Normal concentration expected is below 10 µg/L. Maximum concentration, detected in watersheds, was 14 µg/L and 0.4 mg/L* in surface water with an average of 1 µg/L.

### Air and Nutrition

Daily respiratory intake of selenium has been reported at 0.02, 0.07, and less than 1 µg. It should not exceed 0.2 mg/mc.

Food intake of selenium is reported at very low levels. Dietary intake for an adult male was estimated at 152 µg/day: From grains and cereals 52%; 36% from meat, fish, and poultry; and 10% from dairy products. The WHO (1984) reported daily dietary intakes from 56 µg/day in New Zealand to over 320 µg/day in Venezuela.

### Water Surveys

USEPA reported 150 groundwater and 6 surface supplies containing selenium with concentration above 10 µg/L. A groundwater survey of 1969 (CWSS) indicated 97% of 671 systems with selenium levels ranging from 1–65 µg/L with mean values of detectable selenium of 2.7 µg/L. Surface water (1969–CWSS) surveys reported levels ranging from 1.0–10 µg/L (mean value 4.6 µg/L).

---

*Evident runoff from seleniferous soils.

## HEALTH EFFECTS

Toxic at high levels (accidental exposure), selenium is considered nutrition-ally essential at low levels; animal studies have confirmed its effectiveness in the prevention of certain endemic diseases. In animals it has caused acute and chronic toxicity at high levels, resulting in congenital muscle disease. Little knowledge is available for human poisoning doses, but it has been considered to be similar to arsenic in physiological reaction. In China, chronic toxicity was reported through a high selenium diet (daily average 4.8 mg; minimum 3.2 mg). As an essential element, selenium was considered at a minimum in-take level of 100–200 µg/day for an adult and 0.1 mg Se/kg in food for animals.

The NAS estimated a safe selenium intake between 0.5–0.20 mg/day for adults. Selenium appears to have a deleterious effect in dental caries rates. Elemental selenium is practically nontoxic, but hydrogen selenide is. A RMCL was based on provisional AADI of 0.106 mg/L with data on human exposure included.

A 10-day assessment of 144 µg/L for an adult was determined using 0.41 mg/kg/day as the NOAEL, with uncertainty factor of 100 (based on an animal study).

The AADI was calculated on the Chinese endemic selenium intoxica-tion in 0.106 mg/L, using 3.2 mg/day as a LOAEL with an uncertainty factor of 15. No carcinogenic effects have been determined in animals, but inhibi-tion of tumors (skin, liver, mammary glands, colon, lungs) was confirmed by many studies.

## STANDARDS

| | |
|---|---|
| USPHS 1925 | = no determination |
| USPHS 1942 | = 0.05 mg/L |
| USPHS 1962 | = 0.01 mg/L |
| NIPDWR | = 0.010 mg/L* |
| WHO guidelines (1984) | = 0.01 mg/L |
| WHO (European) | = 0.01 mg/L |
| European Community | = 0.01 mg/L max. |
| RMCL (1985) USEPA | = 0.045 mg/L |
| NAS (1977–1983) | = 0.05–0.20 mg/day safe for adults |
| EPA classification | = Group D (not classified, inadequate animal evidence) |
| Provisional AADI | = 0.106 mg/L (USEPA, 1985) |
| MCLG and MCL (USEPA, 1989) | = 0.05 mg/L (proposed) |
| MCLG and MCL (USEPA, 1991) | = 0.05 mg/L (final rule in effect since July 30, 1992) |

---

*Based on assumed intake of 0.7–7 mg/day of initial toxicity and a safe ingestion of 0.2 mg/day with a 10% max. intake from water.

## Summary of Type A—Inorganic Chemicals. All Values in Milligrams/Liter, but Asbestos Is in MFL.

| | AADI | MCLG | Carcinogenic | EEC | Nutritional Value | FINAL USEPA MCL |
|---|---|---|---|---|---|---|
| ARSENIC—As | 0.1 | 0.005 | A | 0.05 | YES | 0.05 |
| ASBESTOS | n.a. | 7 | A | n.i. | NO | 7 |
| BARIUM—Ba$^{++}$ | 1.8 | 2 | D | 0.1 | NO? | 2 |
| CADMIUM—Cd$^{++}$ | 0.018 | 0.005 | B1 | 0.005 | NO? | 0.005 |
| CHROMIUM—Cr | 0.17 | 0.1 | A | 0.005 | YES | 0.1 |
| COPPER—Cu$^{++}$ | 1.3 | 1.3 | D | 0.10 / 3.00(a) | YES | 1.3 |
| FLUORIDE—F$^-$ | | 4 | NO | 1.5* / 0.7* | YES | 4 |
| LEAD—Pb$^{++}$ | n.a. | zero | B2 | 0.05 | NO | TT† |
| MERCURY—Hg$^+$ | 0.005 | 0.002 | D | 0.001 | NO | 0.002 |
| NITRATE—NO$_3^-$ | 10.0 | 10 | D | (b) | NO | 10 |
| NITRITE—NO$_2^-$ | 1.0 | 1 | D | | NO | 1 |
| SELENIUM—Se | 0.106 | 0.005 | D | 0.01 | YES | 0.005 |

MFL = Million Fibers per Liter (MFL)
A = carcinogenic in humans
B1 = limited evidence in humans
B2 = sufficient evidence in animals
D = inadequate data to classify
(a) Lower value applicable to testing the running distribution system. Higher value at the consumer's tap after 12 hours of no use.
(b) The guide number is 25 and the maximum allowable is 50 for nitrate reported as NO$_3$.
*Related to air temperature (warm climate smaller value)
†TT = Treatment Technique
EEC = European Economic Community
AADI = adjusted acceptable daily intake
MCLG = maximum contaminant level goal
n.a. = not available (the low toxicity would yield a high value insignificant to concentrations expected in drinking water)
n.i. = not issued by EEC

## ANALYSIS

*Standard Methods* introduces the Atomic Absorption Spectrometric Method (the preferable method) and the Diaminobenzidine Method (colorimetric method; second choice because of the toxicity of the reagent) by stating that selenium concentration in most drinking waters is less than 10 µg/L, with concentrations as high as 0.5 mg/L expected from seepage from seleniferous soils.

USEPA (1994) selected the following methodology for selenium: Hydride-Atomic Absorption; ICP-Mass Spectrometry; Atomic Absorption, Platform; Atomic Absorption, Furnace.

## REMOVAL/NOTES

In summary of jar-test methods of treatment processes to remove trace metals from drinking water, selenium was rated*:

- Poor (0%–30% removal) with alum, lime, and activated carbon
- Fair (30%–60% removal) with excess lime
- Fair to Good (more than 50% removal) with ferric sulfate

Selenate removal was observed only in the Cation-Anion Exchange Column. Laboratory research has provided good results in removal of selenium (valence +4 and +6) by ion exchange and reverse osmosis. No available data have been reported on pilot or full-scale plants.

When setting the MCL in 1991, USEPA included the following treatment processes for the Best Available Technology for selenium removal: as conventional processes, coagulation–filtration for Se IV and lime softening; activated alumina and reverse osmosis for specialized processes.

## Inorganic Chemicals—Type B

| ALUMINUM | NICKEL | SODIUM |
|----------|--------|--------|
| CYANIDE | SILVER | ZINC |
| MOLYBDENUM | SULFATE | |

Inorganic Chemicals listed under type B are water quality parameters that have been examined extensively by the Health Authorities. When the concern is definitely their toxicity, their occurrence in potable water is rare, and when their occurrence at the consumer's tap is widely expected, their toxicity is of limited concern.

To issue standards instead of guidelines for these parameters would result in routine, expensive sampling of water supplies, while frequent sampling

---

*USEPA Symposium at Princeton University—AICE—Nov. 1973 (EPA-902/9-74-001).

should be limited to parameters of concern for toxicity or carcinogenicity or because the concentration is expected to vary vastly in a particular water supply.

Carcinogenicity is not a problem for these eight inorganic chemicals. Occurrence is not a problem for a trace element of low NOAEL and AADI as recorded for **aluminum, cyanide, molybdenum**, and **silver**.

In the case of **nickel, sulfate, sodium**, and **zinc**, frequency of occurrence in drinking water is a problem, but the NOAEL and AADI are much higher and safer for these parameters. **Zinc**, normally found in the distribution system, originates from corrosion of galvanized pipes (iron pipes coated with a thin layer of zinc).

## ALUMINUM

## CHEMISTRY

A silvery, light-weight, easily worked element that has excellent anticorrosion property, is nontoxic, and a good thermal conductor. Aluminum is alloyed with small amounts of copper, magnesium, silicon, manganese, and other elements to increase its strength and usefulness. Symbol Al; atomic weight 26.98; atomic no. 13; melting point 660°C; boiling point 2,467°C; specific gravity 2.70; valence 3.

Aluminum is found abundantly (minerals, rocks, and clay) only in combination. The most important compounds are aluminum oxide (alumina), the sulfate, and the soluble sulfate with potassium (alum). Bauxite, an impure hydrated oxide ore, is a major source of industrial production. Aluminum is the most abundant metal found on the earth's crust—8.1%—the third most abundant element. It is found in nature as aluminosilicates, such as clay, kaolin, mica, and feldspar.

It is used in the building industry, aircraft, utensils, electrical conductors, explosives, pigments, paints, and in the coagulation process (aluminum sulfate) in water treatment.

## OCCURRENCE IN WATER, AIR, AND FOOD

In finished drinking water tested for aluminum, the USEPA surveys resulted in the following tabulation in mg/L:

|  | Levels | Median |
|---|---|---|
| Groundwater—no coagulant | 0.014–0.290 | 0.031 |
| Surface Water—no coagulant | 0.016–1.167 | 0.043 |
| Surface Water—Alum coagulant | 0.014–2.670 | 0.112 |
| Surface Water—Iron coagulant | 0.015–0.081 | 0.038 |

Another survey produced aluminum concentrations from 0.073–0.104 µg/L for mean values and from 0.003–2.4 µg/L for individual samples. The NAS (1977) recorded from 380 finished water samples in the United States (1962–1967) a minimum of 3; mean of 85.9; maximum of 1,024 µg/L. In 1,577 raw surface waters in the U.S. (1962–1967) a minimum of 1; mean of 74; maximum of 2,760 µg/L for aluminum was reported.

Ambient air surveys over the continental United States reported values from 0.14–8.0 µg/mc. Aluminum is expected to be found in most food intake (food additives, antacids, and from aluminum cookware). Dietary intake varies from 1.53–100 mg/day with an average of 20 mg/day.

## HEALTH EFFECTS

Aluminum has been considered to be nontoxic, and in spite of clear knowledge of its metabolism has not yet been identified. It must be taken into consideration that the human body has a universal exposure to aluminum due to the abundance of this element in the environment. Large oral doses of aluminum, however, may develop into gastrointestinal tract irritation.

The report that high concentrations of aluminum have been found in some regions of the brains of patients who had died of Alzheimer's disease created circumstantial evidence that stimulated the need for further investigation of a role of aluminum in the aging process and in Alzheimer's disease.

In chronic dialysis patients, encephalopathic syndrome was also associated with aluminum as an etiologic agent. But the prolonged simultaneous intake of antacids containing aluminum for binding gastrointestinal phosphates must also be taken into consideration.

Studies in vitro or on animals to examine mutagenicity, carcinogenicity, and teratogenicity have produced negative results. The NAS has not calculated a SNARL for chronic exposure. Pending further study, the dialysis fluid should be used with softened water treated by reverse osmosis and deionization, in that order.

## STANDARDS

The WHO has recommended a guideline of 0.2 mg/L based upon discoloration; listed as of potential health significance but referred for consideration for aesthetic and organoleptic aspects. The AWWA has set a goal of 0.05 mg/L and a guide value of 0.05 mg/L; this level will assure an effective removal of coagulated material in the water treatment process with the purpose of reduction or settling of particles in the distribution system.

USEPA issued a secondary MCL of 0.05 mg/L in 1989 and a final Secondary Drinking Water Standard (SDWS) of 0.050–0.2 mg/L in 1991.*

*USEPA National Secondary Drinking Water Regulations. Final Rule. *Fed. Reg.* 56:20:3526, Jan. 30, 1991.

## ANALYSIS

Flame atomic absorption, furnace atomic absorption, and inductively coupled emission spectrometry techniques are reliable methods for analyzing aluminum, according to USEPA. *Standard Methods* (1985) recommends the Atomic Absorption Spectrometric Method and lists the "Eriochrome Cyanine R Method" for estimating aluminum with simpler instrumentation.

## REMOVAL/NOTES

Ion exchange and demineralization are potential methods for removal of aluminum from water for special cases. In treatment plants, aluminum is often added during the coagulation process; subsequently, it is found in higher concentration in treated water than in raw water.

When aluminum levels are above 0.1 mg/L in the distribution system, discoloration may result. This is also an indication of treatment problems (lack of pH control, poor coagulation, and filtration breakthrough).

---

## CYANIDE

---

## CHEMISTRY

Cyanide is a substance composed of a cyanogen group in combination with some element or radical. The most known is potassium cyanide, KCN, or sodium cyanide NaCN, a white, crystalline compound with an odor of bitter almonds. Cyanides are either organic (nitriles) or inorganic compounds.

In water, cyanide is identified as hydrocyanic acid—HCN—as cyanide ion—CN—or as simple cyanides, or metallocyanide or organic molecules.

Cyanides are used in plastics, steel, electroplating, and metallurgic industry as well as in synthetic fibers and chemicals. In finished water, cyanide is oxidized by chlorine to cyanate with a pH higher than neutral.

## EXPECTED OCCURRENCE (ENVIRONMENT)

Cyanide has not been measured in air and food. In drinking water surveys, the highest concentration detected was 8 µg/L with an average of 0.09 µg/L in 2,595 samples from 969 public water supply systems in the United States.*

Due to the presence of cyanides in wastewater and its potential pollution, concern must remain in spite of its chemical and biological degradability. A

---

*USEPA, 1970, reporting data from the community water supply survey (*Fed. Reg.* Vol. 50, No. 219, Nov. 13, 1985).

safe maximum total ingestion by humans has been estimated at less than 18 mg/day.

## HEALTH EFFECTS

The very poisonous nature of cyanides is well known. Lungs, gastrointestinal tract, and skin absorb cyanides. There is no evidence that CN ion can accumulate in the body, since small doses below intoxication level are disposed by the body by the natural detoxifying mechanism.

An AADI of 0.75 mg/L was calculated for a 10-day period by USEPA. A NOAEL of 10.8 mg/kg CN was derived from a study on rats; an uncertainty factor of 500, plus an additional 5 diet/water factor, produces a provisional AADI = 0.15 mg/L cyanide.

Experimental mutagenicity in salmonella typhimurium and bacillus subtilis produced negative results for potassium cyanide.

## STANDARDS

USEPA considered 3.77 mg/L a safe level for cyanide but with a factor of safety it is reduced to 0.2 mg/L.

The European Community set a maximum of 50 μg/L. The WHO concluded that 2.35 mg/L could be consumed in water but with a factor of safety, issued guidelines of 0.1 mg/L. Maximum limit was set at 0.01 mg/L as CN and also by the WHO—Europe. The USPHS (1962) set a recommended list of 0.01 mg/L and maximum of 0.2 mg/L. USEPA had decided in Nov. 1985 not to propose a RMCL for CN, mainly on grounds that cyanide is rarely detected in drinking water supplies.

USEPA reexamined CN as a contaminant in the Phase V Rule proposed in 1990. The search for the maximum contaminant level was finalized in 1992 (effective from Jan. 14, 1994).

MCLG AND MCL (USEPA) = 0.2 mg/L (final rule)*

## ANALYSIS

The colorimetric and automated electron techniques with distillation are reliable methods for analyzing cyanide. *Standard Methods* recommends preliminary treatment of samples related to the presence of interfacing substances. Distillation procedures are emphasized, and preservation of samples may require some preliminary field testing; consequently, sampling should be scheduled with laboratory concurrence in methodology.

---

*National Primary Drinking Water Regulations: Synthetic Organic Chemicals and Inorganic Chemicals. Final Rule. *Fed. Reg.* 57:138:31776 (Jul. 17, 1992).

USEPA (1994) listed the following methods: spectrophotometric; automated spectrophotometric (screening method for total cyanides); selective electrode (also for total cyanides); amenable spectrophotometric (for free cyanides). All are preceded by distillation.

## REMOVAL/NOTES

As stated above, cyanides are rarely found in raw water. Their presence may indicate a source of water pollution that must be traced and eliminated. It must be noted that toxicity is related to pH, and a deleterious effect of cyanide at level of pH = 6 may become innocuous at level of pH = 8 (due to formation of cyanates that will finally be decomposed to carbon dioxide and nitrogen gas).

Deterioration of cyanide is expected in natural streams, with decomposition further produced by bacterial action.

Time is also a factor in reduction of cyanides. Percolation through organic soil has produced better results than percolation through a sand column.

In industrial waste treatment of cyanide compounds, applications of alkaline chlorination are introduced to obtain first a cyanogen chloride (gaseous, toxic, and with limited solubility). The remaining cyanide chloride hydrolizes to a cyanate ion of limited toxicity at a high pH.

In processed food, heating destroys most of the originally small concentration of inorganic cyanide.

USEPA listed chlorination as a conventional process and ion exchange and reverse osmosis as specialized processes for the Best Available Technology for CN.

---

## MOLYBDENUM

---

## CHEMISTRY

A lustrous, silver-white, metallic chemical element, mostly used in alloys and particularly valuable to increase hardenability of tempered steel. Symbol Mo; atomic weight 95.94; atomic no. 42; specific gravity 10.2; melting point 2,617°C; sublimes at 4,507°C; valence 1, 3+ 4+ 5+ ?, or 6+.

Molybdenum does not occur as an element in nature; it is obtained by molybdenite ($MoS_2$) and minor commercial ores, such as wulfenite ($PbMoO_4$) or as a by-product of copper and tungsten mining operations. It is primarily used in metallurgy as metal and its salts.

Molybdenum is also used in nuclear energy, military use, in electrical products, petroleum industry, glass and ceramics industries, and production of pigments.

## ENVIRONMENT

The NAS (1977) reported very low concentrations in surface and ground-water with mean values of 60 µg/L.

USEPA surveys of drinking water supplies have produced a statistical median of 1.4 µg/L for molybdenum ranging from 0–68 µg/L. Other surveys produced mean values of 8.0 µg/L in 30% of positive samples, ranging in this group from 1.1–52.7 µg/L, with some exceptional readings of 1 mg/L.

According to NAS, the dietary intake of molybdenum varies from 0.1–0.46 mg/day.

An average daily intake of 0.18 mg was estimated as an expected normal consumption through food.

The NAS has concluded that molybdenum in drinking water is not a significant portion of the total human daily intake. Ambient air measurements ranged from 10–30 µg/mc in urban areas and from 0.1–3.2 µg/mc in nonurban areas (USPHS—1968) in the United States.

## HEALTH EFFECTS

Molybdenum is considered an essential trace element in humans. There is no apparent bioaccumulation of this element in animal or human tissues. Molybdenum is readily absorbed by the gastrointestinal tract with concentration in the liver, kidneys, and bones. The principal elimination is through the urinary tract. The target organs are the liver, kidneys, and sometimes adrenals, and spleen under acute toxic effects from exposure.

The minimum dietary requirements from molybdenum have not yet been established. The NAS has determined an adequate and safe daily intake of molybdenum to be from 0.15–0.5 mg. Chronic toxicity has been reported at levels from 10–15 mg.

In an animal study, a NOAEL of 27 mg/kg was determined, with a factor of safety of 100. Values of 0.27 and 0.95 mg/L for children and adults were determined by a 10-day assessment.

In a human study, a provisional AADI of 0.10 mg/L was determined using a NOAEL of 0.200 mg/L and uncertainty factor of 2.

The IARC did not classify molybdenum for potential carcinogenicity.

## STANDARDS

Both the NAS and WHO have not recommended a limit for molybdenum in drinking water. The USEPA had decided not to propose an RMCL in November 1985 based on insufficient data on the toxicity of the compound. The U.S.S.R. (now Russian Federation) set a limit of 0.5 mg/L for molybdenum in open waters, probably due to toxicological problems in livestock and consumption of molybdenum-rich forage.

*Molybdenum* was included in the USEPA Drinking Water Priority List (DWPL) of 83 potential contaminants as required by the SDWA as revised in 1986. USEPA's expected MCLG and MCL is 0.04 mg/L (1996). In 1993 WHO issued a guideline of 0.07 mg/L (see Appendix A-II, "World Health Organization Guidelines").

## ANALYSIS

The flame atomic absorption, furnace atomic absorption, and inductively coupled plasma atomic emission spectrometry techniques are reliable analytical methods for determining molybdenum in drinking water, according to the USEPA.

*Standard Methods* (1985–1995) lists determination of molybdenum by direct aspiration into a nitrous oxide-acetylene flame (atomic absorption spectrometry) with similar procedures used for aluminum, barium, beryllium, osmium, rhenium, silicon, thorium, titanium, and vanadium.

## REMOVAL

In consideration of the trace elements found in raw water, barring industrial pollutional source to be immediately localized, and based on the lack of enforceable standards of molybdenum pending further studies of interaction between molybdenum, iron, and fluorides, and molybdenum and copper, there is no need for or knowledge of an effective method for removal of this element from drinking water.

The expected Best Available Technologies for molybdenum are ion exchange, reverse osmosis, and nanofiltration.*

---

> **NICKEL**

## CHEMISTRY

A hard, silver-white, malleable, ductile metallic element used extensively in alloys and for plating because of its resistance to oxidation. Fair conductor of heat and electricity. Symbol Ni; atomic weight 58.69; atomic no. 28; specific gravity 8.90; melting point 1,453°C; boiling point 2,732°C; valence 0, 1, 2, 3.

Nickel is extensively used in making stainless steel, Invar, Monel, Inconel, and all corrosion-resistant alloys. It is also used in ceramics, special batteries, electronics, and space applications. Not commonly found in nature as a pure

---

*Phase VIb, SOCs and IOCs anticipated rule of USEPA, reported by F. W. Pontius, in "An update of the federal drinking water regulations," AWWA *Journal*, Vol. 87, Feb. 1995, Denver, CO.

metal, it occurs in sulfides, arsenides, antimonides, oxides, and silicates. Nickel salts are soluble in water, as are many of its compounds.

## ENVIRONMENT (WATER, AIR, AND FOOD)

The NAS (1977) recorded nickel concentration in tap water (Dallas, Texas) of 10 median, 5 minimum, and 23 µg/L maximum. Raw surface water showed nickel present in 16.2% of samples with a mean of 19; minimum of 1, and maximum of 130 µg/L. In the Hudson River, New York, nickel was detected at 71 µg/L. In summary, the NAS expected a frequency of detection between 25% and 66% with a minimum of 1 and a maximum of 530 µg/L. The NAS also concluded that most of the nickel in surface water and groundwater originates from human activities.

In finished water supplies, a mean concentration of nickel was calculated at 34.2 µg/L with a range from 1–490 µg/L. In another survey, an average value was 4.8 µg/L with maximum of 75 µg/L.

Ambient air levels of nickel were observed to range from 0.009–0.015 µg/mc for urban areas and 0.002–0.001 µg/mc* in nonurban localities. Only 0.04–0.16 µg of nickel could be inhaled per day at the above-mentioned concentration, and even using the maximum level, it would constitute only 0.1% of the dietary intake of nickel. Using the highest concentrations detected in urban areas, that maximum percentage could be 2% of the dietary intake. Dietary levels range from 165–900 µg/day, with average values between 400–500 µg/day. Nickel is common in many foods, and it is also introduced into food by stainless steel processing equipment.

New York City recorded 0.118 µg/mc in 1996, equivalent to 2.36 ng/day.

## HEALTH EFFECTS

It has been demonstrated that nickel is an essential element for animals, while nickel nutritional deficiency has not been recognized in humans. Interpolating animal studies to humans, a nutritional need for a human would be on the order of 50 µg/day. This need is adequately satisfied by the dietary intake, estimated between 165–500 µg/day.

Nickel has low toxicity comparable to zinc, manganese, and chromium: it does not accumulate in tissues. Extrapolating animal studies, it is assumed that toxic symptoms in humans could appear with a daily dose of 250 mg of soluble nickel. In normal circumstances (average concentration of nickel in drinking water) the nickel intake would be between 2%–6% of the usual dietary source. The NAS concluded that most drinking waters contribute a very small proportion of the daily nickel intake; and under present circumstances, no nickel deficiency is expected in the typical U.S. diet.

---

*Detection limit for nickel is 0.001 µg/mc.

## STANDARDS

The known fact that gastrointestinal absorption of nickel is very low, with 90% of nickel excreted in the feces in addition to the small percentage of nickel intake through drinking water, has guided the Health Authorities in the past and present to not set standards for nickel.

From animal studies, a NOAEL of 10 mg/kg/day with an uncertainty factor of 100, a 10-day value for a child and an adult of 1 mg/L and 3.5 mg/L, respectively, was calculated.

A provisional AADI of 0.350 mg/L was calculated from a NOAEL of 5 mg/kg/day, an uncertainty factor of 100, and an absorption efficiency of 0.20. Taking into consideration nickel's intake through food and air, USEPA has determined a guidance level of 0.150 mg/L.

The IARC has classified nickel in Group 2A: sufficient evidence of carcinogenicity in animals and limited evidence of carcinogenicity in humans (inhalation exposure). USEPA in its Nov. 1985 drinking water standards decided not to set a RMCL for nickel due to inadequate toxicological data and present knowledge of low oral toxicity. The WHO set no standards; the European Community set a 0.05 mg/L as a maximum.

USEPA included nickel in the "Phase V Rule," regulating five inorganic chemicals to determine MCLG and MCL.* The other four IOCs were antimony, beryllium, cyanide, and thallium.

MCLG and MCL (USEPA, 1992) = 0.1 mg/L (final)

The potential health effects connected to nickel are heart and liver damage.

## ANALYSIS

Reliable methods for nickel laboratory analysis are the flame atomic absorption, furnace atomic absorption, and inductively coupled plasma emission spectrometry technique, as per USEPA (1985).

*Standard Methods* (1985) lists the "Direct Aspiration into an Air-Acetylene Flame" determination (AAS); and for low concentration of nickel, the "Chelation with Ammonium Pyrrolidine Dithio Carbamate (APDC) and Extraction into Methyl Isobutyl Ketone (MIBK)"—AAS.

USEPA (1992 and 1994) selected, as approved analytical methods for nickel, methods 200.7, 200.8, and 200.9, listed in CFR 141.23 (k) (1).

*Standard Methods* (1995), in addition to previous methods, reports 3113 B, "Electrothermal Atomic Absorption Spectrometric Method" for nickel.

---

*USEPA National Primary Drinking Water Regulations. Synthetic Organic Chemicals and Inorganic Chemicals. Final Rule. *Fed. Reg.*, 57:138:31776 (Jul. 17, 1992), corrected Jul. 1, 1994 (59 *FR* 34320).

Also, suitable for the determination of micro quantities of "aluminum, antimony, arsenic, barium, beryllium, cadmium, chromium, cobalt, copper, iron, lead, manganese, molybdenum, selenium, silver, and tin. Also, gallium, germanium, gold, indium, mercury, tellurium, thallium, and vanadium, but precision and accuracy data are not yet available."

## REMOVAL/NOTES

In regulating wastewater discharges, many states in the United States have not specified limits for nickel. USEPA has recommended ambient water quality criteria to protect aquatic life and human health, specifying for nickel a 0.01 of the 96-h Lethal Concentration-50% kill ($LC_{50}$) for fresh water and marine aquatic life. Ohio and Colorado set a criteria of 0.2 mg/L, Illinois 1.0 mg/L, and Florida 0.1 mg/L.

Potential removal of nickel is through a cation exchange.

The Best Available Technology (BAT) for nickel advised by USEPA (1992 and 1994) was ion exchange, reverse osmosis, and lime softening. USEPA recognizes that the water purveyors may use a technology approved by the state. Moreover, lime softening is not BAT for small systems for variances unless treatment is currently in place.

---

### SILVER

---

## CHEMISTRY

A brilliant, white, hard, ductile, malleable, rare, metallic element, and an excellent conductor of heat and electricity. Symbol Ag; atomic weight 107.87; atomic no. 47; specific gravity 10.5; melting point 962°C; boiling point 2,212°C.

Silver is used in jewelry, alloys, photography, batteries, mirror production, electric/electronic components, electroplating, and as a disinfectant. In soil the average concentration of silver is 0.1 ppm; in the earth's crust, it ranges from 0.07–0.08 ppm. Chloride sulfide, phosphate, and arsenate are insoluble. Silver nitrate is highly soluble; silver sulfate is moderately soluble.

## ENVIRONMENTAL EXPOSURE (WATER, AIR, AND FOOD)

In 46% of the water supplies tested, silver was detected at levels ranging from 0.1–9 µg/L in groundwaters. In another survey, 10% of the groundwater supplies tested had values ranging from 30–40 µg/L. In surface water supplies, silver was measured in 54% of supplies sampled with values ranging from 0.1–4 µg/L. Granular activated carbon could be a main source of silver in treated water. No data are available for silver in the atmosphere; but at ground level, a concentration of 1 ng/mc is expected, equivalent to 0.02

μg/day intake level. Silver content in food has been estimated at 70 μg/day including fluids, with values of less than 10 ppb in meats to 2 ppm in seafood.

The NAS (1977) reported in potable waters of the 100 largest cities in the United States, a silver concentration of 0.23 μg/L, with a low of 0 and maximum of 7.0 μg/L. Another survey of drinking water produced a mean of 2.2, minimum of 0.3, and a maximum of 5 μg/L with a frequency of detection of 6.1%.

The maximum recorded exposure level for food and water is 0.18 mg/L.

## HEALTH EFFECTS

High concentration of colloidal silver can be fatal even with a single dose to humans and animals.

Chronic exposure of low-level silver may produce argyria, defined as a bluish-gray discoloration of the skin seen in people who have been treated with silver preparation or industrial poisoning over a long period of time, having a nonreversible effect.*

A provisional AADI based on clinical reports was derived by USEPA with an uncertainty factor of two (sensitive individual evaluated): AADI = 0.09 mg/L.

No mutagenic or carcinogenic effects were recorded in any study; consequently, the IARC did not classify silver for carcinogenicity.

There is no known adverse effect of silver in water beside argyria, a skin disease that does not impair the functioning of the body and, therefore, it can be considered a cosmetic defect. No nutritional or beneficial effect is recorded from ingestion of silver in trace quantity.

## STANDARDS

The USPHS restricted the concentration of silver in drinking water for the first time in the 1962 standards. Apparently, there was concern for potential use of disinfection treatment; the limit was set at 0.05 mg/L probably based on cosmetic reasons rather than public health. At this level it would take 27 years to produce the equivalent of 1 gram of silver that by intravenous injection may cause argyria, without taking into consideration the intestinal discharge during such prolonged use of water.

There is no limitation imposed for silver by WHO International or the European Community.

The USEPA adopted the USPHS standards in the "Primary" regulations but did not propose an RMCL in its 1985 regulations since health is not affected, pending a decision to select a "Secondary" MCL proposal.

The NAS (1977) stated that a primary standard would appear to be

---

*A continuous daily dose of 0.4 mg of silver may produce argyria (WHO, 1984).

unnecessary. A secondary MCL was issued by USEPA in 1989 to be 0.09 mg/L.

USEPA (1991) issued a Secondary Drinking Water Standard of 0.1 mg/L (final rule) for **silver.**

## ANALYSIS

Reliable laboratory techniques for analyzing silver are the flame atomic absorption technique and the inductively coupled plasma atomic emission spectrometry technique.

*Standard Methods* (1985) approved the atomic absorption spectrometry by "Direct Aspiration into an air-acetylene flame for silver advising for low concentration" by chelation (APDC) and extraction into MIBK (AAS).*

## REMOVAL/NOTES

Because silver is a precious metal, it is removed industrially and commercially in most silver processes, leaving a negligible problem in the environment.

While many states in the United States do not specify minimum requirements in their water quality standards, the USEPA recommends, in the ambient water quality criteria to protect aquatic life and human health, limiting silver to 50 $\mu$g/L for domestic water supply (health criteria) and 0.01, of the 96-h $LC_{50}$ as determined through bioassay using a sensitive resident species.

An ionic form of silver is removed by ion exchange in the treatment of industrial waste to recover this valuable material.

Other successful methods for silver removal are ferric sulfate coagulation (pH 7–9), alum coagulation (pH 6–8), lime softening, and excess lime softening.[†]

Silver salts can be used for water disinfection and as prophylactic agents due to the bactericidal properties.

---

*This method is applicable in determining the following metals: antimony, cadmium, calcium, chromium, cobalt, copper, iron, lead, magnesium, manganese, nickel, potassium, silver, sodium, thallium, tin, and zinc.

MIBK is the methyl isobutyl ketone reagent grade.
APDC is the ammonium pyrrolidine dithiocarbamate solution.

*Standard Methods*, 1995, lists this determination as Method 3111, Metals by Flame Atomic Absorption Spectrometry (AAS). (See *Standard Methods* in Appendix A-VIII, "General References".)

†*Manual of Treatment Techniques for Meeting the NIPDWR*, USEPA, Cincinnati, OH, May 1977.

## SULFATE

## CHEMISTRY

A salt or ester of sulfuric acid (the most important manufactured chemical) derived from elemental sulfur. Symbol S; atomic weight 32.06; atomic no. 16; valence 2, 4, and 6; a pale yellow, odorless, brittle, nonmetallic solid, insoluble in water. Indicated as $SO_4$—ion, sulfates are found in natural waters in the final oxidized stage of sulfides, sulfites, and thiosulfates or in the oxidized state of organic matter in the sulfur cycle; in all cases as a product of pollutional sources related to mining or industrial wastes.

Sodium, potassium, and ammonium sulfate are highly soluble in water. The $SO_2^-$ anion also occurs frequently in rainfall in or near metropolitan areas where sulfate is produced as a fossil fuel combustion by-product. Detergents add sulfate to sewage. Industrial pollution from tanneries, steel mills, sulfate pulp mills, and textile plants may contaminate raw water.

## ENVIRONMENTAL EXPOSURE (WATER, AIR, AND FOOD)

Both the NAS* and USEPA* reported drinking water surveys indicating the following concentration of sulfate:

| Supplies | Year | Min | Max | Median | Note in mg/L |
|---|---|---|---|---|---|
| 969 | 1970 | 1 | 770 | 46 | 3% more than 250 mg/L* |
| 625 | 1975 | — | 978 | — | 3.4% more than 250 mg/L* |

Ambient air concentrations also reported by the USEPA indicated a range of arithmetic mean from 4.5–13.3 µg/mc and a maximum range of 10.3–33.8 µg/mc. Metropolitan area rainfall may show concentrations sometimes greater than 10 mg/L.

No reliable data on occurrence of sulfate in foods are presently available.

## HEALTH EFFECTS

The NAS in their first review (1977) of sulfate concluded that the no-adverse-health-effect level was recorded at levels of less than 500 mg/L, whereas, the taste threshold may be as low at 200 mg/L (normal between 300 and 400 mg/L). No additional comment was recorded in the NAS review (1982, Vol. 4) of "Drinking Water and Health."

High levels of sulfate cause diarrhea and dehydration. Sulfate salts are absorbed by the intestine, excreted in the urine up to a cathartic dose of 1

---

*NAS, Vol. 1,—1977; EPA, 40 CFR, Part 141 (NPDWR, 1985). See Appendix A-VIII, "General References."

or 2 g (1,000 to 2,000 mg/L in a single liter). After a period of adjustment to unusually high doses with diarrhea and gastroenteritis occurring, particularly in infants, a tolerance even over 400 mg/L (an advisable maximum in certain situations) has been noted with the population apparently not suffering from routine ingestion of well waters with concentrations reported as high as 2,000–3,000 mg/L.

The USEPA did not calculate a provisional AADI, using a 400 mg/L as a guidance level to protect infants. Based on taste, a guidance level of 250 mg/L has been used by Health Authorities.

## STANDARDS

The USPHS *recommended* a sulfate limit of 250 mg/L from 1925–1962. USEPA issued secondary drinking water standards at the same level. European WHO standards also recommended the same level. The WHO International guidelines are at 400 mg/L concentration of sulfate based upon taste (measured as $SO_4^{--}$).

The USEPA did not issue an RMCL for sulfate in its Nov. 1985 standards in spite of the fact that high levels do present a health concern for the transient exposure. The European Community did not issue an MCL for sulfate (1980).

The WHO (1984) reported that fresh water concentration levels of 20–50 mg/L are expected in most of Europe, with an average of 23 public water supplies of major cities of 64 mg/L (range 9–125 mg/L). In England, the range was recorded in 4–303 mg/L (1969–1973).

The SDWA, as revised in 1986, required USEPA to propose and promulgate a National Primary Drinking Water Regulation (NPDWR) for **sulfate**. This decision was based on the fact that diarrhea and associated dehydration would constitute an adverse health effect from ingestion of high levels of sulfate. The target population in this case is constituted by infants and transient population. In 1990 USEPA issued a notice ragarding the proposal to use alternative levels of 400 mg/L and 500 mg/L for the MCLG for sulfate.

After a review of comments received, USEPA published an MCLG of 500 mg/L for sulfate and the same value for a proposed NPDWR (Dec. 20, 1994). A secondary maximum contaminant level (SMCL) for sulfate of 250 mg/L was also issued based on aesthetic effects, namely, taste and odor.

It was again noted that sulfate is not known to be mutagenic, carcinogenic, or teratogenic in mammals. Further studies performed on animals (pigs) and a very limited study on humans (volunteers) were released in January 1995.

In addition to the proposed MCL of 500 mg/L, USEPA proposed alternative options consisting of supplying bottled water for the target population or point-of-use devices (infants, new residents, and travelers must be protected). Other options limit the target population only to infants and the application of the Best Available Technology to be used for the central water supply system to reduce concentrations of **sulfate** below the 500 mg/L level.

## ANALYSIS

Reliable methods to analyze sulfate are the turbidimetric and automated colorimetric techniques, according to USEPA (1985). *Standard Methods* (1985–1995) advising Method "D" normally and Method "C" for concentrations between 1 and 40 mg/L of $SO_4$ and considering acceptable for concentrations above 10 mg/L for Method "A" and "B," lists the following standard methods:

- Gravimetric Method with Ignition of Residue—Method A
- Gravimetric Method with Drying of Residue—Method B
- Turbidimetric Method—Method C
- Automated Methylthymol Blue Method—Method D

## REMOVAL/NOTES

Ion exchange can be used for sulfate removal from municipal water supplies. Strong base resins are used and operated in the chloride cycle through regeneration with NaCl ($SO_4^{--} + 2(R^+Cl^-) \rightleftarrows (2R^+) SO_4^{--} + 2 Cl^-$. In the reverse osmosis process, calcium sulfate ($CaSO_4$) with calcium carbonate can cause scaling problems. In that case, the sulfate ion can be removed by softening, if practical.

Conventional treatment does not remove sulfate. Ion exchange and reverse osmosis are expensive treatments. Prior to the decision to install an unconventional treatment, the water supply system should examine the possibility of using dilution from other water supply sources or using groundwater obtainable from different aquifers. These alternatives should be evaluated and compared in cost with the ion exchange or reverse osmosis treatment.

If chemical coagulation/flocculation process is used with aluminum sulfate, concentration of the $SO_4$ ion in raw water is increased in the purification process with finished water containing an additional 20–50 mg/L of sulfate.

Corrosion of metals (distribution system or cooling water) may be aggravated by high sulfate concentration.

## SODIUM

## CHEMISTRY

A soft, silver-white, extremely active, alkaline element, found in nature only in combined form, the sixth most abundant element on earth. Symbol Na;

atomic weight 22.99; atomic no. 11; melting point 97.81°C; boiling point 883°C; specific gravity 0.97; valence 1.

Sodium compounds are used in paper, glass, soap, textile, petroleum, chemical and metal industries. The most important compounds are the following:

- common salt (NaCl)
- soda ash ($Na_2CO_3$)
- baking soda ($NaHCO_3$)
- caustic soda (NaOH)

Sodium phosphates, sodium thiosulfate, and borax ($NaB_4O_7 \cdot 10H_2O$) follow.

## ENVIRONMENTAL EXPOSURE (WATER, AIR, AND FOOD)

Sodium is a natural constituent of raw waters, but its concentration is increased by pollutional sources, such as rock salt treatment of road surfaces in below freezing temperature, precipitation run-off, and soapy solutions and detergents. The USEPA reported in 2,100 water supply systems tested levels of sodium ion ranging from 0.4–1,900 mg/L with 42% of supplies showing concentrations higher than 20 mg/L. In 5% of the water systems, the concentrations were higher than 250 mg/L. In another survey of 630 systems of finished waters, concentrations were found from less than one to 402 mg/L, with 42% sampled with levels over 20 mg/L; 3% with levels higher than 200 mg/L. Sodium may be added in water treatment mainly for pH control.* In food, USEPA reported daily levels from 1,600–9,600 mg for adults and 69–92 mg/kg/day for infants based on measurement of sodium excretion from urine (sodium is mostly excreted in urine). Food intake of sodium averages 90% or more of the total sodium intake, estimated at 2–8 g/day.

## HEALTH EFFECTS

Sodium is considered harmful in drinking water at high concentrations (even below threshold for taste) to persons suffering from cardiac, renal, and circulatory diseases. In an epidemiological study of students (tenth graders) from two communities in Massachusetts, one with low (8 mg/L) and one with high (107 mg/L) levels of sodium, E. J. Calabrese and R. W. Tuthill[†] reported higher blood pressure in the high-level-of-sodium community.

A similar study in the Netherlands confirmed that sodium intake influenced blood pressure.

---

*Potentially used in water treatment for fluoridation or pH control are sodium fluoride, sodium silicofluoride, sodium hydroxide, sodium carbonate, sodium bicarbonate, and sodium hypochlorite.
†"The Influence of Elevated Levels of Sodium in Drinking Water on Elementary and High School Students in Massachussets," (1980), *J. Environ. Pathol. Toxicol.* 4(2,3): pp. 151–165.

Other studies attempting to correlate sodium in drinking water to high blood pressure have failed to demonstrate this relationship.

The American Heart Association in 1968 recommended a water intake concentration not higher than 20 mg/L, evaluating a total intake of 500 mg by diet (nutritionally adequate), and considering 440 mg from food and 60 mg from water, drugs, and miscellaneous sources.

The NAS (1977) recognized the habitual intake of sodium in adults in the U.S. often exceeds body need by tenfold or more.

## STANDARDS

| | |
|---|---|
| USPHS 1925–1962 | = no max. contaminant level |
| USEPA 1976 | = no MCL |
| USEPA 1980 | = 20 mg/L recommended |
| WHO (European) 1971 | = no MCL—no recommendation |
| WHO (1984) | = no MCL but 200 mg/L based on taste |
| European Community (1980) | = 20 mg/L guide; 175 mg/L MCL |

USEPA (1985) did not issue a RMCL, or MCLG or MCL, but suggested a guide value of 20 mg/L of **sodium**, taking into consideration that segment of the population of possible high risk (genetic predisposition to hypertension, pregnant women, hypersensitive patients). At that time USEPA's intention was to issue in the future a secondary standard based on taste. The Secondary Drinking Water Standards of 1991 did not include sodium in their list. No regulatory values were issued in 1995, when a scheduled revision of the Safe Drinking Water Act of 1986 was submitted to the U.S. Congress.

WHO (1993) indicates sodium 200 mg/L as a potential complaint for taste but not an MCL. In the European Community (1980–1985) sodium guide level was advised at 20 mg/L.

## ANALYSIS

The current reliable analytical methods for analyzing sodium include the direct aspiration atomic absorption, furnace atomic absorption, and flame photometry, as approved by USEPA (1985).

*Standard Methods* (1995) lists as approved methods for **sodium**: the Direct Air-Acetylene Flame Method; Extraction Air-Acetylene Flame Method; Direct Nitrous Oxide-Acetylene Flame Method; and Extraction Nitrous Oxide-Acetylene Flame Method.

## REMOVAL/NOTES

Ion exchange and a demineralization process can be used for removal of sodium in the consumer's home; while in the municipal systems, limitation of pollutional sources is the first step for reduction of sodium content in potable water.

The NAS (1977) *Drinking Water and Health* issued the following "Summary and Recommendation" after extensive literature research:

The healthy population includes a large segment (15–20%) who are at risk of developing hypertension. There is evidence linking excessive sodium intake to hypertension, but for man the evidence is largely indirect. The risk of hypertension depends on genetic susceptibility and is influenced by other factors in addition to sodium intake. The development of hypertension is characterized by long latency and slowly rising blood pressure, entering the hypertensive range in middle life. People at greatest risk cannot be identified with certainty in advance. For most people, the contribution of drinking water to sodium intake is small in relation to total dietary intake.

Because drinking water is an obligatory dietary ingredient, concentrations should be maintained at the lowest practicable levels, and trends toward increasing concentrations of sodium in water supplies as a result of deicing and water-softening procedures should be discouraged. Optimal concentrations of sodium should be regarded as the lowest feasible.

Specification of a "no-observed-adverse-health-effect" level in water for a substance like sodium for which the effect is associated with total dietary intake and for which usual food intake is already greater than a desirable level is impossible.

The defining of upper allowable limits is inevitably arbitrary. Reduction in hypertension for a small segment of the U.S. population who are on severely restricted diets requires a total intake of sodium less than 500 mg/day. These persons need water containing less than 20 mg/liter sodium ion.

A larger proportion of the population, about 3%, is on sodium-restricted diets calling for sodium intake of less than 2,000 mg/day. The fraction of this that can be allocated to water varies, depending on medical judgment for individual instances. Knowledge of the sodium ion content of the water supply and maintenance of it at the lowest practicable concentration is clearly helpful in arranging diets with suitable sodium intake.

It appears that at least 40% of the total population would benefit if total sodium ion intake were maintained not greater than 2,000 mg/day. With sodium ion concentration in the water supply not more than 100 mg/liter, the contribution of water to the desired total intake of sodium would be 10% or less for a daily consumption of two liters.

Consumers in areas of hard water install water softener devices resulting in high concentration of sodium. Levels between 100 mg/L and 300 mg/L can be experienced.

Low sodium content has been recently prescribed in infant artificial feeding. Inhaled sodium in ambient air (and even in factory atmospheres) is insignificant when compared with sodium intake from the typical diet.

The taste thresholds for sodium vary from 20–420 mg/L according to the chemical compound tested. A guideline value of 200 mg/L can be used for taste but is not based on health effects.

$$\boxed{\textbf{ZINC}}$$

## CHEMISTRY

A bluish-white, metallic element, brittle at ordinary temperature but malleable at 100°C–150°C, fair conductor of electricity. Symbol Zn; atomic weight 65.38; atomic no. 30; specific gravity 7.13, melting point 419.6°C; boiling point 907°C, valence 2. Zinc is used as an alloy with brass, nickel, commercial bronze, soft solder, and aluminum solder. It is used in galvanization to prevent corrosion; in preparation of paints, rubber products, cosmetics, pharmaceuticals, floor coverings, plastics, printing inks, soap, storage batteries, textiles, electrical equipment, and other products. Low solubility in water is reported for carbonates, oxides, and sulfides of zinc, contributing to low zinc concentration in natural waters.

## ENVIRONMENTAL EXPOSURE (WATER, AIR, AND FOOD)

Many zinc salts are highly soluble in water; others are not (carbonate, oxide, sulfide). It is likely that the presence of zinc can be detected only in traces in natural waters; but industrial pollution, particularly in zinc-mining areas, may contribute to a concentration as high as 50 mg/L. In consideration that zinc at a low level (under 20 mg/L, for example) has no adverse physiological effects upon man, and is detected only in low concentrations in drinking water, no extensive surveys were conducted by Health Authorities.

The NAS (1977) reported a tap water survey in Dallas, Texas, showing the concentration of zinc with median, maximum, and minimum values of 0.011, 0.049, and 0.005 mg/L, respectively.

In three Swedish cities, raw water concentrations were recorded at 0.008–0.028 μg/L. In raw surface water, the NAS also reported a detection of 76.5% with a minimum of 0.002, mean of 0.064, and maximum of 1.18 mg/L. (1,577 supplies were tested in the United States.) Also, 380 finished water supplies, tested in the United States, indicated a frequency of detection of 77% with 0.003; 0.079; 2.01 mg/L concentrations as a minimum, mean, and maximum, respectively.

Ambient air concentration surveys were not conducted,* but freshly formed ZnO inhaled may cause a physiological disorder known as oxide shakes or zinc chills. An upper limit of 5 mg/mc (8 hr/40 hr work week) of zinc oxide is a max. concentration in zinc oxide working area.

No surveys were conducted of zinc daily intake from food because of lack of evidence of detrimental effects (as evaluated by the NAS). Normal human intake is estimated at 10–15 mg/day.

---

*WHO (1984) reported concentrations of zinc between 10–100 ng/mc in rural areas and 100–500 ng/mc in urban areas.

## HEALTH EFFECTS

USEPA has not identified any adverse effects that are caused by zinc and has not proposed primary or secondary standards. The NAS Safe Drinking Water Committee concluded that it is an extremely remote possibility that detrimental health effects may be generated by zinc in drinking water; on the other hand, there is a concern of borderline deficiencies in children's nutrition. Drinking water containing up to 20 mg/L of zinc has been consumed without reported ill effects.

A higher concentration of zinc (5–30 mg/L) is aesthetically objectionable in drinking water due to a milky appearance and a greasy film in boiling. Taste can be involved in concentrations higher than 20 to 30 mg/L with an astringent taste reaction.

The NAS (1977) considered zinc an essential trace element in human and animal nutrition. The recommended daily dietary allowances for zinc recommended by NAS in 1974 was of 15 mg/day for adults, 10 mg/day for children, and additional supplement for the pregnancy and lactation periods. The WHO reports 4–10 mg/day depending on age and sex.

## STANDARDS

The USPHS recommended a max. zinc concentration of 15.0 mg/L in the 1942 and 1946 standards, and 5.0 mg/L in the 1962 standards.

USEPA recommended 5 mg/L in 1980, and a SMCL = 5 mg/L in 1989. WHO (1971) recommended 5 mg/L with a maximum of 15 mg/L. WHO European (1970) used 5 mg/L. The European Community advised 0.1 mg/L, with a max. of 1.5 mg/L (16-hr storage in lead or copper lines). The WHO (1984) adopted a guideline of 5 mg/L based on taste consideration.

USEPA (1991) issued a "final status" for **zinc** as a Secondary Drinking Water Standard (SDWS) of 5 mg/L, confirming the final rule of 5 mg/L issued in 1980.

## ANALYSIS

*Standard Methods* (1985) points out that a concentration of zinc higher than 5 mg/L causes a bitter astringent taste and opalescence in alkaline waters. Concentrations of zinc in drinking waters in the United States are reported between 0.06–7.0 mg/L with a mean of 1.33 mg/L. Zinc, if associated with lead and cadmium, may indicate deterioration of galvanized iron and dezincification of brass pipes. The presence of zinc may also result from industrial pollution. The atomic absorption spectrometric method is recommended with the Dithizone Method I to be used for nonpolluted waters.

*Standard Methods* (1995) lists for **zinc**, as approved methods: the Direct Air-Acetylene Flame Method; Extraction Air-Acetylene Flame Method; and Inductively Coupled Plasma (ICP) Method.

## REMOVAL/NOTES

When detected at the consumer's tap, it is more likely that corrosion of piping has increased the concentration of zinc (similar to copper). Control of raw water to identify possible contamination and corrosion control measures may reduce the concentration to a reasonable level (see pH control, distribution system, corrosivity, copper, and lead).

Zinc can be removed by lime softening or cation exchange.

Air is a negligible source of zinc; drinking water contributes less than 0.4 mg/day (WHO, 1984). The first detection of zinc by the Taste Threshold Test is above 4 mg/L. It is a corrosion inhibitor; this is the reason it is found in metal pipes.

# Inorganic Chemicals—Type C

| | | |
|---|---|---|
| CALCIUM | MAGNESIUM | SILICA |
| CARBON DIOXIDE | MANGANESE | BROMINE |
| CHLORIDES | OXYGEN DISSOLVED | (BROMIDE) |
| IRON | (DO, BOD, COD) | CHLORINE |
| LITHIUM | PHOSPHATE | IODINE (IODIDE) |
| | (PHOSPHORUS) | OZONE |
| | POTASSIUM | |

Inorganic chemicals listed under type C are parameters that may have a high level of occurrence in drinking water, but are safe at the concentration expected, or their occurrence is rare and extremely limited when their toxicity is of concern, such as for the group of halogens examined here, **bromine**, **chlorine**, and **iodine**.

**Ozone** is listed here since it can be used for disinfection as are chlorine, bromine, and iodine.

**Oxygen Dissolved** is listed here for the association with ozone, iron, and manganese. **Iron, manganese, bromine, chlorine**, and **iodine** are primarily associated with piping problems or disinfection treatment.

---

CALCIUM

---

## CHEMISTRY

A silver-white, chemically active, metallic chemical element, found always in combination in limestone, marble, and chalk. One of the alkaline earth elements, fifth in abundance in the earth's crust (3%), reacts with water, essential constituent of bones and teeth. Symbol Ca; atomic weight 40.08;

## Summary of Type B—Inorganic Chemicals.
## All Values in Milligrams/Liter.

| | AADI | MCLG | Carcinogenic | WHO | Nutritional Value | Final MCL | Reasonable Range |
|---|---|---|---|---|---|---|---|
| Aluminum—Al | n.a | 0.05–0.2* | NO | 0.2 | ? | NO | 0.05–0.20 |
| Cyanide—CN | 0.15 | 0.2 | NO | 0.1 | NO | 0.2 | 0.05–0.2 |
| Molybdenum—Mo | 0.10 | 0.04 | NO | n.i. | YES | 0.04 | 0.04–0.2 |
| Nickel—Ni | n.i. | 0.1 | (a) | n.i. | YES | 0.1 | 0.1–0.2 |
| Silver—Ag | 0.09 | 0.10* | NO | n.i. | NO | NO | 0.05–0.20 |
| Sulfate—$SO_4$ | 400 | (b) | NO | 400 | NO | (b) | 250–500 |
| Sodium—Na | n.a. | 20 | NO | 200 | YES | NO | 20–200 |
| Zinc—Zn | n.a. | 5* | NO | 5 | YES | NO | 5–15 |

AADI = adjusted acceptable daily intake
MCLG = maximum contaminant level goal
n.a. = not available (the low toxicity would yield a high value insignificant to concentrations expected in drinking water)
n.i. = not issued by the Health Authorities
MCL = maximum contaminant level
(a) = questionable carcinogenicity (proved in animals)
(b) = USEPA issued a Secondary Drinking Water Regulation for sulfate of 250 and a MCLG = MCL = 500 mg/L (proposed)
WHO = World Health Organization (1984 guidelines)
Reasonable Range = minimum and maximum values used by Health Authorities
* = Secondary Drinking Water Regulations (USEPA will not issue a final rule)

atomic no. 20; melting point 839°C, boiling point 1,484°C; specific gravity 1.55; valence 2.

Most common compounds: limestone ($CaCO_3$), gypsum ($CaSO_4 \cdot 2H_2O$), fluorite ($CaF_2$); also calcium carbide ($CaC_2$), chloride ($CaCl_2$), cyanamide ($CaCN_2$), hypochlorite [$Ca(ClO)_2$], nitrate [$Ca(NO_3)_2$], and sulfide (CaS).

One of the major uses is as calcium oxide, quicklime (CaO), produced by heating limestone; then turned into slaked lime [$Ca(OH)_2$] by slow addition of water (or in moist air) for a more stabilized condition, and also usage in construction and chemical industries. Controlled heating of limestone produces the base material of Portland cement, widely used in construction. It is also used as a reducing agent in preparation of other metals and in treatment of ferrous and nonferrous alloys.

*Calcium Carbonate* (limestone, marble, calcite, chalk) is the most abundant compound of calcium. The solid is white or transparent (crystalline form), insoluble in water, but attacked by acids.

*Calcium Hydroxide*—As stated above, it is formed by adding water to calcium oxide. Slightly soluble in water, it is also used in water treatment. In water solution it is also called whitewash or milk of lime. It is a basic component of mortar where, by evaporation of water and contact with carbon dioxide of air, it returns into calcium carbonate form and strength.

*Calcium Sulfate*—Gypsum—$CaSO_4 \cdot 2H_2O$. Used in construction, industry, and as a fertilizer: chloride of lime—$CaOCl_2$—bleaching powder.

Bleaching powder is a widely used product, an unstable white powder that reacts with acids liberating chlorine. One commonly used product is the hypochlorite $Ca(OCl)_2$ marketed under the trade name of H.T.H.

See also Chlorination (calcium hypochlorite); Water Softening; Alkalinity; Hardness; pH; and Corrosion.

## EXPECTED OCCURRENCE (ENVIRONMENT)

Since there are no, or very limited, standards promulgated for calcium in drinking water because of no toxicity concern, there are no wide range surveys for determination of calcium in its elemental form. Values of hardness, alkalinity, and its advisable limits are quoted in Chapter 2, dealing with the physical parameters. The expected calcium concentration in raw or treated waters may vary from close to zero to several hundred mg/L.

## HEALTH EFFECTS

No special concern has been shown by the Health Authorities for the presence or absence (lack of nutritional values) of calcium. Consumption of water containing up to 1,800 mg/L of calcium has been reported harmless. When kidney or bladder stones may be suspected, high content of calcium and magnesium in drinking water should be avoided.

In consideration that calcium in drinking water has not been associated with any specific disease, an upper limit as a guide may be used, such as 75 mg/L in relation to hardness.

Calcium requirement in the diet should be at least equal to the calcium excretion, estimated at 10 mg/kg/day. It is normally expected that the nutritional dosage is provided by food, while hard waters may, nevertheless, provide a significant contribution.

## STANDARDS

The U.S. Public Health Service Drinking Water Standards did not contain upper limits for calcium in the 1925, 1942, 1946, and 1962 updating. USEPA did not include calcium in the parameters regulated by the National Primary Drinking Water Regulation of 1985.

The WHO (1963) has used 75 mg/L as a maximum acceptable limit for the International Standards, and 200 mg/L as an excessive limit. The European Community (1980) used 100 mg/L as a maximum guide number. The WHO (1984) did not issue standards for calcium.

USEPA (1991–1995) did not include **calcium** in the list of Secondary Drinking Water Standards or express the intention to regulate calcium.

## ANALYSIS

*Standard Methods* (1995) lists as approved methods for **calcium:** the Atomic Absorption Spectrometric Method; Inductively Coupled Plasma (ICP) Method; and EDTA Titrimetric Method.

## REMOVAL

Since calcium is a major component of hardness, calcium removal is associated to hard water treatment; therefore, chemical softening, reverse osmosis, electrodialysis, and ion exchange are applicable treatments.

---

## CARBON DIOXIDE

---

## CHEMISTRY

An odorless, colorless, tasteless, incombustible gas, heavier than air, moderately soluble in water. Formula: $CO_2$. It is normally contained in surface water in concentrations of less than 10 mg/L; much higher values may be encountered in groundwater supplies.

Since carbon dioxide is not harmful in the concentrations expected in surface or groundwaters, consequently the Health Authorities have not issued standards or guidelines. Nevertheless, it has interested the sanitary engineer for its effect on corrosivity, pH variation, and for its use—recarbonation of a supply—in the water softening treatment process. Because of its relationship to the type and degree of alkalinity, carbon dioxide in water has an effect also on coagulation and iron removal. Collection of samples must be done accurately due to the sizable loss of $CO_2$ by agitation of the sample. pH readings are usually obtained at the source.

*Standard Methods* lists the following as accepted methods:

- Nomographic Determination of Free Carbon Dioxide and the Three Forms of Alkalinity;
- Titrimetric Method for Free Carbon Dioxide; and
- Carbon Dioxide and Forms of Alkalinity by Calculation.

Total carbon dioxide is the sum of the free carbon dioxide and the carbon dioxide existing in the form of bicarbonate and carbonate ions as determined Nomographically (see *Standard Methods* in Appendix A-VIII, "General References)." See also pH; Alkalinity; Corrosion Control; and Water Treatment.

## CHLORIDES

Chlorides are compounds of chlorine. They remain soluble in water, unaffected by biological processes, therefore, reducible by dilution. Their concentration at higher levels than adjacent waters is an indication of pollution (usually chloride concentration under 10 mg/L is expected). Usually a higher concentration is due to ice and snow treatment on roads in the proximity of the surface water—reservoirs or streams.

Natural mineral origin can also be a cause of high chloride content. Sea water intrusion in well water used for communities along the sea may also be a major source of contamination. Industrial effluents (galvanizing plants, water softening plants, oil wells, refineries, and paper works) may also leach into groundwaters or streams. Ocean spray also can be carried inland in minute salt crystals or droplets. Sewage effluent contains a larger concentration of chlorides. In the past, sodium chloride has been used as a tracer in water flow since it is not toxic, not visible, not changed by the biological processes, not absorbed by soil, and easily measured. Dyes and radioactive material have substitute chlorides, but their limitations are obvious.

Chlorides by themselves have slight significance from a public health viewpoint. The limit of 250 mg/L established by health authorities long ago was not related to disease or perhaps not set by taste (mostly related to sodium ions) since concentrations above 500 mg/L or sometimes as high as 800 mg/L are detectable as a salty taste; taste also is related to the chemical composition of water. Sensitive persons have, nevertheless, detected a salty taste at 300 mg/L in water and 40 mg/L in coffee.

Plants are more sensitive than man to high chloride contents; therefore, 250 mg/L remains a good guide for chlorides, with a maximum of 100 mg/L when water is used for irrigation, and under 50 mg/L in industrial use. The USPHS Drinking Water Standards used 250 mg/L in the 1925, 1942, 1946, and 1962 regulations. USEPA did not include chlorides in the parameters regulated by the National Primary Drinking Water Regulations of 1985. The WHO (1963) listed 200 mg/L of chlorides, as Cl, as a maximum acceptable, and 600 mg/L as a maximum allowable. The European Community (1980) used a 25 mg/L as a guide (reported as Cl) and 200 mg/L as a level over which effects may be registered. It must be taken into consideration that individuals affected by heart and kidney disease should restrict water consumption with a high chloride concentration (see Sodium); but it also must be noted that chlorides from drinking water contribute to less than 2% of the total daily dietary intake.* WHO (1984) listed a guideline value of 250 mg/L.

USEPA (1979 and 1991) issued a Secondary Drinking Water Standard for chloride at 250 mg/L. Secondary Standards are selected for a contaminant when public health is not directly involved by this contaminant, but taste, odor, or appearance of the water may result in adverse effects on public welfare (consumers may discontinue the use of this water in necessary quantity, resulting in questionable health effects).

*Standard Methods* lists as acceptable methods for determination the Argentometric Method (Mohr Method) and the Mercury Nitrate Method— mg/L NaCl = mg/L Cl × 1.65. These two titrometric methods are relatively simple laboratory procedures. Recently, a third method (Potentiometric Method) as well as a fourth method (Automated Ferricyanide Method) has been added to the official analytical methods for the determination of chlorides.†

The chloride intake from ambient air is negligible. Chlorides are not removed by conventional treatment. Pollution control and dilution can be used to reduce chloride concentration. Actual removal can be accomplished through a demineralization process (reverse osmosis or electrodialysis).

**IRON**

## CHEMISTRY

A white, malleable, ductile, metallic chemical element, vital to animal and plant life; iron is the fourth most abundant element by weight in the earth's

---

*WHO (1984) reported an average daily intake of 6 g of chloride ion/day with a range as high as 12 g.
†*Standard Methods* (1995) listed also the Ion Chromatographic Method for **chlorides**.
  For WHO standards (1993), see Appendix A-II, "World Health Organization Guidelines."

crust. Symbol Fe: atomic weight 55.85; atomic no. 26; melting point 1,535°C; boiling point 2,750°C; specific gravity 7.87; valence 2, 3, 4, and 6.

Common ores are Hematite ($Fe_2O_3$); Magnetite ($Fe_3O_4$); Limonite ($Fe_2O_3 \cdot 3H_2O$); Diderate ($FeCO_3$); and Pyrite ($FeS_2$), or chemically as oxides, carbonates, silicates, chlorides, sulfates, and sulfides.

The pure metal is chemically very reactive and corrodes rapidly, particularly in moist air, but it is commonly found in combination with carbon (steel: 0.3% to 1.7% of carbon).

## ENVIRONMENTAL EXPOSURE

In surface water supplies, the presence of iron is almost exclusively due to corrosion of pipes (the most commonly used material for water piping) and storage tanks. In groundwater supplies, in addition to corrosion problems of the distribution system, high content of iron can be encountered due to the frequency of the elevated iron level in the earth strata related to the feeding aquifers.

Leaching of iron salts (acid mine drainage) and iron product industrial waste may be a pollutional source of iron.

Salts of ferric and ferrous compounds (such as chlorides) are highly soluble in water, but ferrous iron is readily oxidized to form insoluble ferric hydroxides that flocculate and settle; therefore, in raw water of surface supplies, iron is expected in low concentration.

In groundwaters when a significant amount of carbon dioxide and lack of oxygen is encountered, ferrous carbonate is dissolved as follows:

$$FeCO_3 + CO_2 + H_2O \rightarrow Fe^{++} + 2HCO_3^-$$

Ferric iron may become soluble under an anaerobic condition that reduces it to ferrous iron. Well aerated streams have a minimum of iron concentration because of the formation of hydroxides that precipitate.

A 1962 survey of finished water supplies of the 100 largest cities in the United States, provided a minimum concentration of 0, a median of 0.02, and a high of 1.3 mg/L; while in another survey of 380 finished waters between 1962 and 1967, the minimum was 0.002, the mean 0.069, and the maximum 1.92 mg/L. Raw water surveys for the same period of 1,577 supplies provided 0.001 as a minimum, 0.052 as a mean, and 4.6 mg/L as a maximum.

It is a known fact that iron in trace amounts is essential for nutrition. In cases of iron deficiency anemia, larger doses are taken for therapeutic reason. The daily requirement is 1–2 mg, with dietary ranges of 7–35 mg/day with an average of 16 mg/day. It is, therefore, evident that concentrations of 1.0 mg/L in violation of the drinking water standards would have no significant effect in the daily diet.

In evaluating iron in the distribution system, it must be taken into consideration that in conventional water treatment (coagulation) salts of iron are frequently used.

Iron from ambient air constitutes a negligible exposure. With a daily food contribution of 16 mg and a water contribution of 0.14 mg (or 2 mg with iron content of 1 mg/L), it is clear that water contributes to the total diet between less than 1%–15%.

## STANDARDS

Since standards for iron (or iron plus manganese) have been set for less than 0.3 mg/L, acceptability of water sources was a condition for meeting this concentration; while surface supplies usually have concentrations below the standard (0.05–0.20 mg/L), there are groundwaters (1.0–10 mg/L) that may need treatment to meet the standard at the distribution system. The objection to iron in the distribution system is not due to health reasons but to staining of laundry and plumbing fixtures and appearance. Taste and odor problems may be caused by filamentous organisms that prey on iron compounds (*frenothrix, gallionella*, and *leptothrix* are called "iron bacteria"), originating another consumer's objection (red water).

Well water containing soluble iron may remain clear while pumped out, but exposure to air will cause precipitation of iron due to oxidation, with a consequence of rusty color and turbidity. The presence of iron bacteria may clog well screens or develop in the distribution system, particularly when sulfate compounds in addition to iron may be subjected to chemical reduction. This problem is particularly observed in dead-end lines with consequent odor complaints and formation of black deposit made up by iron sulfide; nevertheless black deposits are also due to manganese* (see Manganese).

Solubility of iron is increased by a low pH ($<5$). High turbidity may help to keep acid-soluble iron in suspension.

In conclusion, iron in raw or potable water may be either ferrous or ferric or both and categorized as in solution, in colloidal state (subjected to organic matter), in inorganic or organic iron compounds, or in the form of coarse suspended or settled particles. To prevent formation of black deposit and iron bacterial growth, oxygen in the water should be maintained higher than 2 mg/L, and a free chlorine residual concentration should be higher than 0.2 mg/L. Copper sulfate treatment in open reservoirs before the water temperatures reach summer levels helps to prevent bacterial action. The maintenance of pH above 7.2 in the distribution system is also a positive factor in preventing the above-mentioned problems.

The 1925, 1942, 1946, and 1962 regulations of the U.S. Public Health Service always reported the maximum concentration for iron (or iron and manganese combined) as 0.3 mg/L. The USEPA did not include iron (or manganese) in the National Primary Drinking Water Regulations, but

---

*The thick slime is a combination of manganese oxides, bacteria, and hydrous iron.

maintained in the Secondary Drinking Water Regulations of 1989 the limit of 0.3 mg/L based on aesthetic and taste consideration.

WHO (1963) also adopted a 0.3 mg/L as a maximum acceptable level and 1.0 mg/L as a maximum allowable. The European Community adopted in 1980 a guide of 0.05 mg/L and a maximum of 0.20 mg/L. WHO (1984 and 1993) recorded a guideline of 0.3 mg/L.

USEPA (1979 and 1991) confirmed the original ruling for **iron** as a contaminant to be included in the Secondary Drinking Water Standards with a level of 0.3 mg/L as the final rule.

## ANALYSIS

Inductively Coupled Plasma Method was added after the 1985 edition and completed the three methods approved for **iron** included in the 1995 (19th) edition of *Standard Methods*.

## REMOVAL

Iron removal can be accomplished using aeration, sedimentation/filtration, pH adjustment, ozone or chlorine treatment, chemical precipitation/filtration, ion exchange (softening), potassium permanganate oxidation with precipitation/filtration, well water pretreatment with sequestering agent-like polyphosphates, and sodium silicates and lime softening.

## IRON—SUMMARY

| | |
|---|---|
| Definition: | A metallic chemical element vital to animal and plant life, abundant in the earth's crust, found chemically as oxides, carbonates, silicates, chlorides, sulfate, and sulfides |
| Caused by: | Corrosion of pipes; leaching of iron salts from soil and rocks and industrial pollution; by-product of iron bacteria, such as crenothrix, gallionella, and leptothrix |
| Concentration Expected: | 0.05–0.20 mg/L mean in raw or finished waters |
| Indicator of: | Presence of iron bacteria feeding in non-compounds; corrosion of galvanized pipes; lack of oxygen in water |
| Test: | Atomic absorption spectrometric and phenantroline methods |
| Regulation: | 0.30 mg/L for iron and manganese combined (secondary max. contaminent level); 0.05 mg/L as a guide; nontoxic at levels below rejection for taste and color; laundry stains major reason for regulation; essential element of nutrition. |

## LITHIUM

A soft, silver-white alkali metallic chemical element not abundantly found in nature; used in metallurgy, nuclear armament, storage batteries, special glasses, and ceramics; never found free in nature; known as the lightest of all metals. Symbol Li; atomic weight 6.94; atomic no. 3; melting point 181°C; boiling point 1,342°C; specific gravity 0.53; valence 1.

Most important minerals: lepidolite, spodumene, petalite, and amblygonite. If present in water, the maximum concentration expected is 10 mg/L; its mean concentration is 0.05 mg/L when present. Not considered in Primary Drinking Water Standards, suggested lithium maximum concentration is 5 mg/L. The biological function of lithium is not yet known. It has been used or proposed for usage as a tracer based on the fact that lithium is not expected in water and has been considered a safe chemical (lithium chloride).

In consideration of the industrial and commercial use of lithium or its salts (metallurgical processes, glass, and storage batteries), some potential concentration is expected to result in wastes. The Flame Emission Photometric Method has been accepted by *Standard Methods*, in addition to the Atomic Absorption Spectrometric Method (approved by the Standard Methods Committee in 1985).

In 1984, the WHO decided that no action was required to regulate lithium in drinking water.

## MAGNESIUM

A light, silver-white, malleable, ductile, metallic chemical element; it is the eighth most abundant element in the earth's crust; never found as a free element, it constitutes a large deposit as Magnesite (native magnesium carbonite) and the common rock forming Dolomite $[CaMg(CO_3)_2]$. Symbol Mg; atomic weight 24.3; atomic no. 12; melting point 648°C; boiling point 1,090°C; specific gravity 1.74; valence 2.

Used in flashlight photography, alloys, pyrotechnics, and incendiary bombs. In medicine magnesium is known as "milk of magnesia" as an hydroxide or as "Epsom Salts" as a sulfate. It is a nutritional element in animal and plant life. Most of magnesium salts are very soluble in water.

Magnesium has been considered as nontoxic to humans at the concentration expected in water (not rejected by taste). Magnesium salts have a laxative and diuretic effect particularly for individuals not accustomed to high dosage (a tolerance to magnesium may develop with time). The U.S. Public Health Service Drinking Water Standards of 1925 included an upper limit of 100 mg/L for magnesium. This limit became 125 mg/L in the 1942 and 1946 standards but was not included in the 1962 standards.

The WHO International Standards of Drinking Water (1963) list a maximum acceptable level of 50 mg/L and a maximum allowable level of 150 mg/L. The European Community Standards of 1980 adopted 30 mg/L as a guide and a maximum concentration admissible of 50 mg/L. Taste threshold for magnesium as sulfate has been reported as 100 mg/L and 500 mg/L for the average individual. As for the contribution of magnesium to hardness, see also *Hardness* in Chapter 2 and *Calcium*, page 31. High magnesium concentrations can be reduced by chemical softening, reverse osmosis, electrodialysis, and ion exchange. Magnesium is expected in raw water at a concentration of 1.8–62 mg/L, and in finished water between 0.8–49 mg/L, with a median concentration of 6 mg/L (NAS, Vol. 1, 1977, pp. 212 and 262). Sea water contains 1,350 mg/L of magnesium.

The NAS in the 1977 review of magnesium had concluded that, "In view of the fact that the concentrations of magnesium in drinking water less than that, that impart astringent taste pose no health problem and more likely to be beneficial, no limitation for reason of health is needed."

Magnesium in water can be determined by the Gravimetric Method, the Atomic Absorption Spectrophotometry, and by calculation with known total hardness and calcium hardness:

$$\text{mg Mg/L} = [\text{total hardness (as mg CaCO}_3\text{/L)} - \text{calcium hardness (as mg CaCO}_3\text{/L)}] \times 0.243)$$

## MANGANESE

A very brittle, grayish-white, metallic chemical element resembling iron but harder. Oxides, silicates, and carbonates are the most common manganese minerals (pyrolusite—$MnO_2^-$ and rhodochrosite—$MnC_3^-$). It is a very important contributor to alloys and widely used in the chemical industry. Symbol Mn; atomic weight 54.94; atomic no. 25; melting point 1,244°C; boiling point 1,962°C; specific gravity 7.3; valence 1, 2, 3, 4, 6, or 7.

The concentration of manganese in water is less than that of iron. A mean expected value is around 0.06 mg/L. When concentrations higher than 1 mg/L are found, then manganese-bearing minerals are contacted by water under "reduced" conditions or where bacteria (see *Iron* above) are active. In raw water concentrations were registered between 0.001–0.60 mg/L.

Manganese is an essential trace nutrient for plants and animals. Nutritional deficiency in man has not been evaluated as a health hazard. Also minimum requirements for nutrition have not yet been established. The WHO estimates an average daily requirement for normal physiological functions of 3–5 mg.

Manganese poisoning following work-related exposure is rare. High concentration of manganese dust or fumes and prolonged inhaling are necessary to cause chronic manganese poisoning (effects similar to Parkinson's disease).

It is very unlikely that a health hazard will be caused by manganese in the typical, expected concentration in raw and drinking water. Nevertheless, undesirable effects may be taste, staining of laundry, and discoloration of water, starting at 0.15 mg/L (see *Iron* above).

The U.S. Public Health Service Drinking Water Standards of 1925, 1942, and 1946 included manganese with iron for a combined maximum level of 0.3 mg/L, but in 1962 the regulations included, in addition, a maximum concentration of 0.05 mg/L for manganese. The USEPA adopted the 0.05 mg/L of the USPHS as recommendation, and issued a secondary standard in 1989. The WHO recommended 0.05 mg/L (maximum acceptable) and 0.50 mg/L as maximum allowable. The European Community (1980) used a guide value of 0.02 mg/L and a maximum of 0.05 mg/L. When manganese removal is necessary, aeration, superchlorination, pH adjustment, and chemical precipitation may be used. *Standard Methods* has specific recommendations for sampling of manganese; recognized analytical methods are:

- Atomic Absorption Spectrometric Method; and
- Persulfate Method—the periodate method, and the orthotolidine method for manganese in higher oxidation states were standard methods substituted by AA Spectrophotometry. Because of the similarity in natural condition and in the chemistry and bacterial action, manganese should be evaluated in conjunction with iron (see *Iron* and Iron and Manganese removal, pages 123 and 125).

*Standard Methods* (1995) also listed the Inductively Coupled Plasma Method for determination of **manganese**.

Ambient air tested for manganese recorded average values of 0.05 μg/mc and 0.3 μg/mc for industrialized and nonindustrialized areas, respectively. Food consumption is estimated between 2.0—8.8 mg/day by WHO (1984) with an average of 3.6 mg in North America. Manganese has been judged as one of the least toxic elements. It is not carcinogenic; there is a tendency to consider manganese as anticarcinogenic.

USEPA (1979 and 1991) listed **manganese** in the Secondary Drinking Water Standards with a level of 0.05 mg/L as the final rule.

## Dissolved Oxygen (DO)

### BIOCHEMICAL AND CHEMICAL OXYGEN DEMAND (BOD AND COD)

Oxygen is considered poorly soluble in water. Its solubility is related to pressure and temperature. In water supply technology, dissolved oxygen in raw water is considered as the necessary element to support life of fish and other aquatic organisms. It is also an indicator of water treatment process and is an important factor in corrosivity. The DO content is evaluated in comparison to its maximum level of solubility for a given pressure and temperature, defined as saturation. Because saturation is obtained from the atmosphere, oxygen is only a portion of the dissolved gases; but, fortunately, its content in water is approximately 38% of the dissolved gases (that is twice its percentage in air). In fresh water, dissolved oxygen reaches 14.6 mg/L at 0°C and approximately 9.1, 8.3, and 7 mg/L, respectively, at 20°, 25°, and 35°C, at 1 atm of pressure.

During summer months the rate of biological oxidation is highly increased, while, unfortunately, the oxygen dissolved concentration is at its minimum due to higher temperature. It must be noted that the popularity and frequency of the DO testing in sanitary engineering, and even more the BOD (Biochemical Oxygen Demand) and the COD (Chemical Oxygen Demand) are due to the necessity of evaluating particularly wastewater and industrial waste biological processes, stream aeration, and pollutional load in water pollution control surveys and programs. Less importance is given by the sanitary engineer to these tests when the water supply is involved with the exception of stream analysis, natural or artificial reservoirs limnological evaluation, aeration treatment process, and corrosion in the distribution system. Because oxygen in the water to be sampled is at a concentration below the saturation level, particular care is necessary in sampling to avoid atmospheric aeration of the sampled water. *Standard Methods* prescribes the pretreatment of the sample bottle to stop biological activity and maintain the DO of the water sampled, maintaining the sealed container as close as possible to the temperature of the sampled water.

In evaluating the health effects of chemical contamination, normally a high content of the chemical in question is always of serious concern; but in the case of DO, it is certainly not the maximum that is of concern (since it is limited by the saturation at a rather very low level of concentration) but the minimum for the effect on the aquatic life and the bad taste usually associated with dissolved oxygen of less than 2 or 3 mg/L. A pollutional load caused by algae bloom in open reservoirs may show sharp variations in DO and pH. A sudden loss of DO may cause organic decomposition with consequential odor problems (also associated with anaerobic decomposition).

No DO standards were ever issued by the Health Authorities, mainly due to the lack of toxicity. Moreover, the low concentration and the continuous variation in concentration (change in pressure and temperature) make the dissolved oxygen a parameter of limited importance from a health viewpoint, but its effect on corrosion has practical (piping life) and psychological results for consumers (discolored water and taste problems).*

Raw water evaluated for potential use as a drinking water supply is

---

*Anaerobic conditions cause reduction of nitrate to nitrite; sulfate is reduced to sulfide, normally associated with odor complaints.

normally sampled, analyzed, and tested for biochemical oxygen demand when turbid, polluted water is the only source available. *Standard Methods* recognizes, first, the importance of the DO test in water pollution and water treatment process control, connecting dissolved oxygen levels to the physical, chemical, and biochemical activities in natural and wastewaters. It then recognizes the Winkler or Iodometric Method and its modifications (Azide, Permanganate, Alum Flocculation, and Copper Sulfate—Sulfamic Acid Flocculation) and the Membrane Electrode Method as accepted methods to analyze dissolved oxygen.

## Biochemical Oxygen Demand (BOD)

BOD may be defined as the amount of dissolved oxygen demanded by bacteria during the stabilization action of the decomposable organic matter under aerobic conditions. This test, therefore, is a bioassay procedure to measure the oxygen consumed by living organisms utilizing the organic matter contained in the sample (more likely of wastewater) and the dissolved oxygen of the liquid.

Toxic substances must not be in the sample at the concentration to destroy bacterial action or depress the oxidation rate. Dilution of the sample is necessary to provide sufficient dissolved oxygen, when the sample of "waste" is too concentrated.

Standard temperature is maintained during the test, normally 20°C (average water condition). Normally BOD is reported as a 5-day BOD, since a large percentage of the total BOD (70%–80%) is exerted in the first five days of the test.

The BOD test is based upon determination of dissolved oxygen. Most river waters used as water supplies have a BOD of less than 7 mg/L; therefore, dilution is not necessary. *Standard Methods* prescribes all phases of procedures and calculations for BOD determination. A BOD test is not required for monitoring water supplies.

## Chemical Oxygen Demand (COD)

This is another parameter not requested for monitoring water supplies but sometimes used for evaluation of polluted raw water. Extremely useful in the determination of industrial wastes, and very practical in the determination of domestic waste and polluted waters, COD determination "provides a measure of the oxygen equivalent of that portion of the organic matter in a sample that is susceptible to oxidation by a strong chemical oxidant" as defined by *Standard Methods*. Reagents, procedures, and calculations are detailed in *Standard Methods*. The COD values are greater than the BOD values, but its handicap is the inability to show a difference between biologically oxidizable and biologically inert organic matter. Since COD can be determined in 3 hours versus the 5 days of BOD, sometimes the BOD test is substituted by COD for practicality.

In monitoring a specific source, tests can be run to determine empirically the relation between COD, BOD, organic carbon, or organic matter. Consequently, only COD tests can be run for a quick evaluation of the source in question.

**PHOSPHATE**

Mainly a salt of phosphoric acid, phosphate in nature is found in phosphate rock, in the mineral known as apatite, a tri-calcium phosphate. It is an important source of *phosphorus* (an element nonsoluble in water); also, it is the inorganic component of bones and teeth ($PO_4^{3-}$).

The large demand for chemical fertilizers worldwide has increased the production of phosphate. Phosphates are also used for production of special glasses, chinaware, baking powder, and detergents (see Chapter 4). Sodium, potassium, and ammonium phosphate are soluble, with the calcium compounds' solubility limited to monocalcium phosphate.

Water supplies may contain phosphate derived from natural contact with minerals or through pollution from application of fertilizers, sewage,* and industrial waste. Groundwaters, therefore, are more likely to have higher phosphate concentration.

Polyphosphates have been used to prevent scale formation and inhibit corrosion. Their discharges are considered as nutrients with consequent growth of plants and organisms. It has been estimated that the daily average human intake of phosphorus is approximately 1,500 mg, with 70% absorbed as free phosphate.

Phosphorus concentration in raw waters has been reported in 50% of the samples tested, with a mean concentration of 0.12 mg/L and a maximum of 5 mg/L. As part of finished water surveys, the USEPA reported that in 92% of the largest U.S. cities, phosphorus was not detected. In twelve large cities in Europe, the mean phosphate concentration reported was 0.32 mg/L and a maximum of 3.0 mg/L (0.10–1.0 mg/L of phosphorus—NAS, 1980, Vol. 3).

The human daily requirement of phosphorus is equal to that of calcium (NRC, NAS, 1974). The recommended dietary allowance (RDA) is 800 mg for adults. Dietary deficiency is not expected because of the abundance of phosphorus in the diet.

The NAS concluded in 1977 that the phosphorus level in drinking water is totally negligible as related to the daily diet because public drinking water contains a minimum amount of phosphorus.

*Phosphorus has not been regulated by the USPHS or by USEPA* (not even

---

*Municipal wastewaters contain levels of phosphates in the range of 15–30 mg/L as PO4.

in the Secondary Drinking Water Regulations). The WHO also did not issue regulations or guidelines for phosphorus, but the European Community (1980) issued a guide number of 0.400 mg/L and a maximum of 5 mg/L measured as $P_2O_5$.

*Standard Methods* lists phosphorus analysis as approved by the Standards Methods Committee in 1981 and introduced the examination stating: "Phosphorous occurs in natural waters and in wastewaters almost solely as phosphates. These are classified as orthophosphates, condensed phosphates, (pyro-, meta-, and other polyphosphates), and organically bound phosphates. They occur in solution, in particles or detritus, or in the bodies of aquatic organisms." *Standard Methods* lists three preliminary steps and the four methods for examination:

- Preliminary Filtration Step
- Preliminary Acid Hydrolysis Step for Acid-Hydrolyzable Phosphorus
- Preliminary Digestion Step for Total Phosphorus
- Vanadomolybdo Phosphoric Acid Colorimetric Method
- Stannous Chloride Method
- Ascorbic Acid Method
- Automated Ascorbic Acid Reduction Method (approved in 1985)

The WHO in 1980 concluded that there is no nutritional basis for the regulation of phosphorus levels in U.S. drinking water supplies.

Studies of eutrophication and problems related to shallow reservoirs subject to agricultural runoff and/or domestic liquid waste pollution have resulted in the indictment of phosphate as a cause of algal growth. Reduction of phosphate in municipal or industrial liquid wastes has been regulated. Feasible treatment can be reached with precipitation (lime at pH over 10) or with alum, sodium aluminate, or ferric chloride.

The USEPA 1976 Water Quality Standards—Criteria Summaries for Phosphorus—recommended a phosphorus criterion of 0.10 µg/L (elemental) phosphorus for marine and estuarine waters but no fresh-water criterion.

New York State (1979) criteria value for designated stream use required that "Concentration should be limited to the extent necessary to prevent nuisance growths of algae, weeds, and slimes that are or may become injurious to any beneficial water use." New Jersey adopted a max. of 50 µg/L as a total P for designated public water supplies.

## POTASSIUM

A soft, light, silver-white, waxlike, metallic, common chemical element of the alkali group, similar to sodium, that oxidizes rapidly when exposed to air. Potassium is the seventh most abundant element and constitutes 2.4% by weight of the earth's crust. It is never found free in nature. Most potassium minerals are insoluble. It decomposes in water, generating hydrogen and catching fire spontaneously. Chemically many potassium salts are very important: hydroxide, nitrate, carbonate, chloride, chlorate, bromide, iodide, cyanide, sulfate, and chromate.

Extensively used in fertilizers, in glass, and in a limited way in the chemical industry, potassium is definitely an essential nutritional element for humans, animals, and plants; it is nontoxic but acts as a cathartic in excessive concentration (1–2 g) with a threshold in taste for potassium of 340 mg/L in KCl. Four major biological functions are influenced positively by potassium.[*] Tap water surveys indicated values of potassium ranging from a minimum of 0.5 mg/L to a maximum of 8 mg/L with an expected mean concentration of 2 mg/L. Potassium is abundantly available in food. The daily intake estimated for adults in the U.S. is 4.6 g (range from 1.5–6 g). It has been estimated that potassium deficiency in adults may be associated with a daily intake under 2,000 mg (hypokalemia).

From food it is hardly possible to produce hyperkalemia[†] or potassium toxicity. Acute toxicity may be reached in adults with a dosage of 7–10 g/day (NAS, Vol. 3).

The effect of potassium in drinking water is negligible either to reach minimum requirements or to cause any health problem.

There are no USPHS Drinking Water Standards or USEPA primary or secondary maximum contaminant levels. Also, the WHO never issued standards for potassium. The European Community (1980) has issued a guide number of 10 mg/L as K and a maximum admissible level of 12 mg/L.

For **potassium**, *Standard Methods* (1995) lists the following methods: the Atomic Absorption Spectrometric Method, Inductively Coupled Plasma Method, Flame Photometric Method, and Potassium-Selective Electrode Method.

---

## SILICA (SILICON)

Silica is a hard, glassy mineral found in a variety of forms, such as quartz, sand, and opal.

Its formula is $SiO_2$, dioxide of silicon. Other silicate minerals are rock crystal, amethyst, agate, flint, jasper; also, granite, hornblende, asbestos, feldspar, clay, and mica.

Silicon makes up 25.7% of the earth's crust by weight, being the second most abundant element, exceeded only by oxygen. Silicon is one of the most useful elements in its application in concrete, brick, refractory; as silicates in enamel, pottery, and glass. Silica is used up in water by diatoms to produce

---

*NAS, *Drinking Water and Health*, Vol. 3, 1980.
†This deficiency is rather related to geriatric disability.

their cell walls. In natural waters, silica may be detected as the result of disintegration of rocks containing silica. In water silica appears as suspended particles, in a colloidal or polymeric state, and as silicates, iron, or silicic acids.

The NAS reported that the mean total silicon intake for adults in Great Britain was 1.2 g/day.

In drinking water surveys in large cities of the United States, silicon as silica was reported as zero for a minimum, 7.1 mg/L for a median, and 72 mg/L as a maximum. Raw waters may contain from 1–30 mg/L of silica. Recently it has been established that silicon is an essential nutritional trace element, confirmed only in animal studies, feeding dry the water-soluble sodium metasilicate ($Na_2SiO_3$). Essential dietary requirements have not been verified for humans. *No standards have been issued by the Health Authorities for silicon or silica.*

*Standard Methods* lists the following five acceptable methods:

- Atomic Absorption Spectrometric Method
- Gravimetric Method
- Molybdosilicate Method
- Heteropoly Blue Method
- Automated Method for Molybdate-Reactive Silica

With the AA method, standard metal solutions are prepared with sodium metasilicate ($Na_2SiO_3 \cdot 9H_2O$) dissolved in water.

At high concentrations, silica may be objectionable in a cooling tower and in boiler feed water makeup. Activated silica is used as a flocculant with alum to remove color and some organic matter in raw water. Silica removal can be accomplished in the coagulation process by adsorption on iron floc; water temperature is an important factor in efficiency.

## BROMINE (BROMIDE)

A chemical element of the halogen group, known as a reddish-brown, heavy, corrosive liquid that volatilizes in a vapor of disagreeable odor and is very irritating to mucous membranes (eyes and throat). Bromine is the only liquid nonmetallic element; with bleaching action, bromine is readily soluble in water. Inorganic bromides and bromates are highly soluble in water.

Bromine is used in fumigants, dyes, medicinals, sanitizers, water purification compounds (swimming pools), and gasoline (antiknock compounds). Halogens are all antiseptic and disinfectants. Taste threshold in humans ranges from 0.17–0.23 mg/L as bromide. The National Academy of Science (NAS) reports an acute fatal dose of inorganic bromate of approximately 57 mg/kg. A feeding of 500 mg/L to adults showed no significant adverse effect after four months.

The WHO in 1969 issued a 1.0 mg/kg for total inorganic bromide intake as a recommendation for an acceptable daily intake (ADI). Relating this value to water intake will result in a permissible bromide concentration of 7 mg/L. Disinfection with bromine requires a 2 mg/L concentration.

Drinking waters may contain bromide at a concentration of less than 1 mg/L. Some well water may contain higher values due to sea water intrusion.

The Standard Methods Committee approved in 1985 a reliable method to determine bromide by calculation from a calibration curve.

As a powerful oxidant and disinfectant, bromine could theoretically substitute chlorine in water disinfection, but insufficient research does not release the possible use as an equally effective bactericidal agent (see Chapter 10, "Disinfection"). No standards have been issued by Health Authorities.

## CHLORINE

### CHLORINE DIOXIDE, CHLORITE, CHLORATE, AND CHLORAMINES

Active member of the halogen group, chlorine is a greenish-yellow gas, combining directly with practically all elements; therefore, in nature it is found only in the combined state. Famous for its irritation to mucous membranes, its characteristic odor is detectable already at 2 to 3 mg/L and fatal at 1,000 ppm (chlorine in water). Chlorine is used extensively in the chemical industry and in the disinfection of drinking water.

Very soluble in water ionizing as follows:

$$Cl_2 + H_2O \rightarrow HOCL + H^+ + Cl^-$$
$$\text{Hypochlorous Acid HOCL} \rightleftharpoons H^+ + OCl^-$$

HOCl 96% pH = 6
HOCl 75% pH = 7          The maximum toxicity or disinfection effective-
HOCl 22% pH = 8          ness is reached with HOCl non-disassociated.
HOCl  3% pH = 9

Chlorine residual reacting with certain organic substances, particularly phenols, may produce very objectionable tastes and odors. Consumption of drinking water containing 50 mg/L of chlorine has produced no adverse effect.

Virucidal effectiveness of chlorination diminishes from HOCl—OCl⁻—Cl₂ to chloramines in the order listed. In contact with natural humic substances (known as the precursors), chlorine forms chloroform and other halomethanes (see THM).

*Standard Methods* first reduced and then dropped in the 14th and 15th editions the traditional orthotolidine procedures, first for lack of accuracy and then for the toxic nature of orthotolidine. Presently, the following methods are accepted:

- Amperometric Titration Method
- DPD Ferrous Titrimetric Method
- DPD Colorimetric Method
- Syringaldazine (FACTS) Method
- Iodometric Electrode Technique

as reported by the 19th edition of *Standard Methods* (1995) for **chlorine**.

Advice on the selection of methods, particularly related to expected concentrations and interferences, is given in *Standard Methods*. Potential contaminants of drinking water as a consequence of disinfection treatment are the following:

- Cholorine Dioxide—$ClO_2$
- Chlorite—$CLO_2^-$
- Chlorate—$ClO_3^-$
- Chloramines—$NH_2Cl$

An examination of the health aspects of the above-mentioned compounds of chlorine can be briefly summarized as follows.

## Chlorine Dioxide ($ClO_2$)

A greenish-yellow gas of irritating odor, potentially explosive in air, difficult to transport, but stable in solutions of pure water in closed containers. It has been considered a good substitute of chlorine for disinfection since it does not cause trihalomethanes, chloramines, or reaction with ammonia, but the chemistry of chlorine dioxide in water treatment is not yet fully developed.

From animal studies, a Suggested No-Adverse-Response Level (SNARL) of 0.38 mg/L can be derived with an assumed uncertainty factor of 100. The USEPA has recommended a disinfection treatment not to exceed 1 mg/L, while the former West Germany has suggested a maximum applied dosage of 0.3 mg/L.

No data are available for mutagenicity, teratogenicity, and carcinogenicity. *Standard Methods* (1995) lists the following approved methods for **chlorine dioxide**:

- Iodometric Method
- Amperometric Method I
- DPF Method
- Amperometric Method II

## Chlorite ($ClO_2^-$)

From animal studies, the NAS arrived at a calculated no-adverse-effect dose of 0.21 mg/L, with an uncertainty factor of 100. Concentration of chlorite in drinking water can result from an intended chlorine dioxide treatment.

## Chlorate (ClO$_3^-$)

From hypochlorite ion $3HClO \rightarrow 2HCl + HClO_3$. Sodium chlorate has been used as a herbicide as well as in leather, paper, and textile processing.

Chlorate is a minor end-product of disinfection treatment of drinking water through chlorine dioxide. Using concentrated solutions of chlorate in humans, a lethal dose is 71 mg/kg. The effect of dilution has not been determined, and there are no available data for mutagenicity, teratogenicity, and carcinogenicity.

## Chloramines (NH$_2$Cl)

Ammonia and hypochlorous acid combine to form chloramines in the following reaction:

$$HOCl + NH_3 \rightarrow NH_2Cl \text{ (monochloramine)} + H_2O$$

Also, dichloramine and nitrogen trichloride are formed. The major influencing factors in the formation and concentration of these combined forms are pH, temperature, initial chlorine-nitrogen ratio, chlorine demand, and reaction time.

Taking into consideration that raw water contains ammonia, usually both free and combined chlorine are expected to be found in potable water after chlorination.

Chloramines do maintain a chlorine residual but they are not very effective in disinfection as ozone, chlorine dioxide, and chlorine. The NAS has determined that not sufficient data are available to calculate a SNARL in humans for either acute or chronic exposure.

## Chloramino Acids

Hypochlorous acid not only reacts with ammonia in drinking water but also combines with amino acids and other amines to form organic haloamines.

The NAS reached the same conclusion for chloramino acids as for chloramines: sufficient data are not available to calculate a chronic Suggested No-Adverse-Response-Level (SNARL) in humans for either acute or chronic exposure.

## IODINE (IODIDE)

A less active member of the halogen family, iodine is a bluish-grayish-black, lustrous solid that sublimes into a blue-violet gas with an irritating odor (to eyes and mucous membranes). Iodine is a chemical element only slightly soluble in water. Symbol I, atomic weight 126.9, valence 1, 3, 5, or 7, solubility in water is 0.34 g/L.

The most common compounds are the iodides of sodium and potassium (KI) and the iodates ($KIO_3$). Iodine is used in medicine, dyes, photography; iodine compounds are frequently used in organic compounds, chemistry, and in medicine.

Iodine 127 is a stable isotope and Iodine 131 is an artificial radioisotope with a half-life of 8 days (see Chapter 6, "Radionuclide Parameters"). Only traces of iodine or iodides are found in raw waters. Sea water contains higher concentration.

Iodine has been used in the disinfection of swimming pool water, obtaining a free iodine residual between 0.2–0.6 mg/L. This water treatment is called *iodination* and iodine is recognized as capable of inactivating enteroviruses under controlled conditions.

## HEALTH EFFECTS OF IODINE/IODIDE/IODATE ($I_2$, $I^-$, $IO_3^-$)

Iodine is bactericidal, virucidal, and amebicidal at dilutions of 5 to 50 mg/L.

Iodine has been determined an essential trace element necessary for synthesis of thyroid hormone.

Daily requirements of iodine or iodide are 80–150 µg/adult. Goiter may result from iodine deficiency. In the United States, table salt has been supplemented with 100 µg of potassium iodide per gram of sodium chloride in order to prevent goiter.

Iodine is converted rapidly to iodide iron in an efficient absorption in the gastrointestinal tract in human metabolism. Acute poisoning from iodide is rare. Chronic iodide poisoning (Iodism) is more common than acute reaction. It has been estimated that in some areas of Japan, consumption up to 80 mg/day was detected without apparent ill effect; nevertheless, a limitation of 2 mg/day was suggested by some scientists. No ill effects appeared in military personnel who received doses of 12 mg/man/day for the first 12 weeks and 19.2 mg/man/day for the following 10 weeks.

The NAS in 1980 summarized the health effects of iodine and iodide as relatively low in toxicity with hypersensitivity registered. The SNARL for chronic exposure was calculated at 1.19 mg/L, considering a 20% intake from drinking water. The 7-day exposure was calculated at 16.5 mg/L. The *Standard Methods* first states for *iodide* (approved by the Standard Methods Committee in 1981): "only micrograms-per-liter quantities of iodide ($I^-$) are present in most natural waters. Higher concentrations may be found in brines, certain industrial wastes, and waters treated with iodine."

*Standards Methods* did not change determination of iodide and iodine procedures from the 1985 to the 1995 editions. Namely, for **iodide:**

- Leuco Crystal Violet Method, and
- Catalytic Reduction Method.

For **iodine**, the approved methods are:

- Leuco Crystal Violet Method, and
- Amperometric Titration Method.

## OZONE

Ozone is an allotropic form of oxygen consisting of an unstable pale-blue gas ($O_3$) with a penetrating odor produced by an electric discharge upon air or oxygen ($3O_2 \rightarrow 2O_3$). Ozone is very reactive chemically, but it decomposes readily ($2O_3 \rightarrow 3O_2$). It can be used as a bleaching agent (paper pulp, textiles, etc.), a germicide, air disinfectant, and in the tobacco, leather, and linoleum industries. The other extensive uses of ozone are in the disinfection of drinking water (*ozonation*) and in taste and odor removal.

Ozone occurs naturally in the atmosphere (estimated at 0.05 mg/L by volume at sea level) but in much larger concentration in the stratosphere caused by the ultraviolet radiation in oxygen of the air. Ozone is of bluish color when undiluted, bluish-black when liquid, and violet-black when solid.

The effectiveness of ozone as a disinfecting agent is such as to suggest an immediate substitution of chlorine by ozone, but the disinfecting power of ozone cannot be assured in the distribution system against post-treatment contamination (as provided by chlorine).

Ozone is 13 times more soluble than oxygen in water. Its half-life is 40 minutes at 14.6°C and pH = 7.6 but decreases rapidly at higher temperatures and pH. Ozone oxidizes ammonia in a negligible amount in typical water treatment disinfection. Ozone in air is a contaminant with the effect of impairment of lung function and possible acceleration of aging.

Because of the short life of ozone in water, potential toxicity and long-term effects have not concerned the Health Authorities. Recently many studies have been initiated—and few completed—to determine the health effects of by-products of ozonation.

For **ozone**, *Standard Methods* (1995) lists the Indigo Colorimetric Method as the approved method for determination of ozone residual.

The advantages and disadvantages of ozonation versus chlorination are described in water treatment disinfection principles and disinfection effectiveness in Chapter 10, "Water Treatment."

## Summary of Type C—Inorganic Chemicals. (All Values in Milligrams/Liter)

| | AADI | RMCL or MCLG | Carcinogenic | WHO | Nutritional Value | Reasonable Range |
|---|---|---|---|---|---|---|
| *CALCIUM—$Ca^{++}$ | n.a. | n.i. | NO | 75 | YES | 75–150 |
| *CARBON DIOXIDE—$CO_2$ | n.a. | n.i. | NO | n.i. | NO | n.a. |
| *CHLORIDES—$Cl^-$ | n.a. | 250 | NO | 250 | YES | 250–600 |
| *IRON—$Fe^{+++}$, $Fe^{++}$ | n.a. | 0.3 | NO | 0.3 | YES | 0.3–1.0 |
| *LITHIUM—$Li^+$ | n.a. | n.i. | NO | n.i | ? | 5 |
| MAGNESIUM—$Mg^{++}$ | n.a. | n.i. | NO | 0.1 | PROBABLY | n.a |
| *MANGANESE—$Mn^{++}$, $Mn^{++++}$ | n.a | 0.05 | NO | 0.1 | YES | 0.05–0.10 |
| *DISSOLVED OXYGEN—D.O. | n.a. | n.i. | NO | 10–12 | ? | n.a |
| *PHOSPHATE—$PO^{---}$ | n.a. | n.i | NO | n.i. | YES | 0.4–5.0 |
| *POTASSIUM—$K^+$ | n.a. | n.i. | NO | n.i. | YES | 10–12 |
| *SILICA—$SiO_2$ | n.a. | n.i. | NO | n.i. | PROBABLY | n.a. |
| BROMINE—$Br^-$ | 50 Bromate | n.i. | ? | n.i. | NO | n.a. |
| | 7 Bromide | n.i. | ? | n.i. | NO | n.a. |
| CHLORINE—$Cl^-$ | 50 | n.i. | ? | n.i. | NO | n.a. |
| CHLORINE DIOXIDE—$ClO_2^-$ | n.a. | 1.0 | ? | n.i. | NO | n.a. |
| IODINE—$I^-$ | 1.19 | n.i. | NO? | n.i. | YES | 0.3–1.0 |
| OZONE—$O_3$ | n.a. | n.i. | NO? | n.i. | ? | 0.5–1.0 |

AADI      = adjusted acceptable daily intake
RMCL     = recommended maximum contaminant level
MCLG     = maximum contaminant level goal
n.a.       = not available (the low toxicity would yield a high value insignificant to concentrations expected in drinking water)
n.i.        = not issued by the Health Authorities
MCL      = maximum contaminant level
WHO      = World Health Organization (1984 guidelines)
Reasonable Range = minimum and maximum values used by Health Authorities
*No health effects at the concentrations expected in drinking water.
**N.B. Final MCls are not expected to be issued by USEPA.**
**N.B. Chlorides, iron, and manganese are now listed by USEPA as Secondary Drinking Water Standards.**

# Inorganic Chemicals—Type D

ANTIMONY  TIN
BERYLLIUM  THORIUM
COBALT  VANADIUM
THALLIUM

**Inorganic chemicals** listed under type D are parameters very likely to be found in very small concentrations in raw or drinking water. There is no toxicity at the low concentrations that may normally appear in water destined to reach the consumer's tap.

No carcinogenicity has been determined in any of these chemicals but some is suspected in antimony, beryllium, cobalt, thorium, and vanadium.

## ANTIMONY

A metallic chemical element, very brittle, bluish-white with metallic luster, not abundant in nature, rarely found native, more frequently combined as sulfide, **stibnite** ($Sb_2S_3$); very useful as an alloy, in semiconductor technology, and in rubber, textile, paint, and glass industries. Symbol Sb; atomic weight 121.75; atomic no. 51; specific gravity 6.69; valence 0, $-3$, $+3$, or $+5$. It is considered toxic in air when concentrations are higher than 0.5 mg/mc (40-hour work week).

While trichloride, sulfate, potassium tartrate, and pentachloride are soluble in water, antimony tends to precipitate and be removed by sedimentation. Required antimony in human nutrition has not been determined, but its toxicity is known. With similarity in symptoms to arsenic poisoning, doses of 100 mg have been demonstrated to be lethal. In consideration of the practical impossibility of finding concentration of antimony in raw or drinking waters, the Health Authorities did not issue limiting standards for antimony, in spite of its toxicity as a metal and presence in toxic compounds, except for the European Community (1980) which issued a max. admissible level of 10 µg/L.

In 1992 (effective from 1994) USEPA issued final contaminant levels for **antimony:** MCLG = 0.006 mg/L and MCL = 0.006 mg/L (final rule), indicating also as conventional processes coagulation and filtration and reverse osmosis. The health effects of concern were decreased growth and longevity.*

For the analysis of **antimony**, *Standard Methods* (1995) prefers the Electrothermal Atomic Absorption Spectrometric Method because of its sensitivity; but the Atomic Absorption Spectrometric Method and the Inductively

---

*National Primary Drinking Water Regulations: Synthetic Organic Chemicals and Inorganic Chemicals, Final Rule. *Fed. Reg.* (Jul. 7, 1992).

Coupled Plasma Method are also accepted. Carcinogenicity, mutagenicity, and teratogenicity of questionable results were recorded and the NAS reached no conclusion.

## BERYLLIUM

A chemical element, member of the alkaline earth metals, steel gray in color, very light with excellent thermal conductivity, nonmagnetic, good alloy component to increase strength in several metals including copper and nickel. Symbol Be; atomic weight 9.01; atomic no. 4; specific gravity 1.85; boiling point 2,970°C, valence 2.

It is found in many minerals; mostly produced by reducing beryllium fluoride with magnesium metal. Beryllium has been used in industry only during the last 30 years. Besides its use as an alloy it is used in nuclear reactors, aircraft, and space technology.

Beryllium is an air contaminant regulated by industrial hygiene officials (beryllium dust causes skin and pulmonary ailments to exposed workers) to prevent berylliosis (occupational standard in the United States is 2 μg/mc). Only the chloride and nitrate are soluble in water; therefore, beryllium is not expected to occur in raw waters. Preliminary studies have indicated no harmful effects when taken internally.

In drinking water surveys, beryllium was confirmed in approximately 1% of water supplies tested; and, in those cases, the minimum concentration was reported to be as low as 0.02 μg/L with a mean of 0.1 and a maximum of 0.17 μg/L. Testing groundwater and surface raw waters, the highest concentration observed was 1.22 μg/L. The lowest concentration observed was 0.01 μg/L with an 80% probable frequency of detection, as reported by NAS (1977). Actually, groundwaters are not particularly suspected of potential concentration of beryllium because of its absorption by the clay of the soil strata. *Standard Methods* estimates a beryllium concentration range of 0.01–0.7 with a mean of 0.013 μg/L.

Food is not a significant source of beryllium. In human metabolism, it appears that beryllium is only slightly absorbed by the digestive tract and is excreted rapidly.

In addition to the above, health effects can be summarized as follows:

- Benefits of beryllium in human nutrition have not been determined.
- Beryllium is relatively harmless when ingested in food and water, except at a very large, continuing dosage.
- Since concentration of beryllium in water is naturally minimal, there is no present need to issue standards.
- Beryllium has caused cancer in animals, but so far has proved not to be carcinogenic to humans, and, if potentially so, it could be of low potency.

USEPA issued final rules for **beryllium**, effective in 1994, based on possible cancer and potential damage to bones and lungs. Both MCLG and MCL were set at 0.004 mg/L. As far as the best available technology for removal is concerned, USEPA cited coagulation and filtration and lime softening (not applicable to small systems for variances unless treatment is already installed) and activated alumina, ion exchange, and reverse osmosis as specialized processes.*

*Standard Methods* describes the AAS Method (using beryllium nitrate, with caution in laboratory handling for its extreme toxicity) by "direct aspiration into a Nitrous-Oxide-Acetylene Flame" and by "Chelation with 8-Hydroxyquinoline and Extraction into Methyl Isobutyl Ketone."

Listed also as a standard method for micro quantities is the "Electrothermal Atomic Absorption Spectrometry" as the method of choice, but a colorimetric method is also listed as "Aluminon Method."

## COBALT

A hard, brittle, metallic chemical element resembling iron and nickel in appearance, found in various ores. Symbol Co; atomic weight 58.93; atomic no. 27; specific gravity 8.9; valence 2 or 3. it has been used as alloys, in metal electroplating, in glass, porcelain, and enamels.

Cobalt 60 is an artificial isotope used as a gamma ray source. Raw water concentration of cobalt was found to be 0.1 µg/L in 3 cities in Sweden; while in 1,577 raw surface waters in the United States the minimum was one, the mean 17, and the maximum 48 µg/L with a frequency of 2.8% detection. In 380 drinking water supplies† (distribution system), cobalt was reported as 22, 26, and 29 µg/L, respectively, and a 0.5% frequency. In spite of the fact that cobalt is limited to green leafy vegetables in human diets, its concentration is such as to state that of total intakes: the water intake is negligible. In metabolism, cobalt is considered an essential nutrient (part of vitamin B12 molecule).

Health effects can be summarized as follows:

- Cobalt is generally of low toxicity, but the mechanism of the toxic action should be better understood.
- Nutritional benefits to humans should be reevaluated after further studies, but it is judged an essential element since cobalt is a component of vitamin B12.
- Concentrations higher than 1 mg/kg of body weight may be considered a health hazard to humans.
- Since the concentration of cobalt as related to the potential toxicity in

---

*National Primary Drinking Water Regulations: Synthetic Organic Chemicals and Inorganic Chemicals. Final Rule. *Fed. Reg.* (Jul. 7, 1992).
†Reported by NAS (1977).

water is negligible, *Health Authorities have not issued maximum contaminant levels*, with the exception of Russia—1 mg/L.

*Standard Methods* (1995) advises for determination of **cobalt** the use of the Atomic Absorption Spectrometric Method or the Inductively Coupled Plasma Method.

**TIN**

A silvery-white, malleable, metallic chemical element of high crystalline structure, mostly found in cassiterite ($SnO_2$). Symbol Sn; atomic weight 118.69; atomic no. 50; specific gravity 5.75 (gray) or 7.31 (white); valence 2, 4. Tin is used as an alloy (tinfoils, solders, plates), particularly for tin cans. Also, tin is used in glass, porcelain, fungicides, and insecticides. The mineral is rather scarce in the world of demand and supply, so a substitution for tin is used whenever possible in the metallurgic industry.

Tin is not found naturally in raw waters; therefore, its presence is due to industrial pollution. Fortunately, many tin salts are not water soluble. Because of scarcity of tin ores, lack of solubility of tin salts, and the low potential toxicity, very limited surveys were performed to determine the concentration of tin in raw or drinking waters. Moreover, in the past, there were no reliable methods to determine the presence of tin in water. Limited surveys indicated concentrations between 1.1–2.2 µg/L for distribution system waters and 0.8–30 µg/L for selected natural sources. Sea water concentration measures 0.2–0.3 µg/L as tin.

From a metabolic viewpoint, the USFDA in 1975 classified tin as "poorly absorbed from the alimentary tract and most ingested tin is excreted via the feces. The tin that is absorbed is found mainly in the liver and lungs with small traces in other tissues." Other scientists found all ingested tin was excreted almost equally between urine and feces.

Human dietary intake of tin is difficult to assess with accuracy. It has been reported as low as 3–4 mg/day or between 1–30 mg/day:* very negligible amount when compared with a potential safe level of over 500 mg/day.

Health effects may be summarized as follows:

- There is no conclusion reached on the essentiality of this trace element in human nutrition.† Limited animal studies tend to confirm that tin enhances growth in rats.

---

*Schroeder, et al. (1964) "Abnormal Trace Elements in Man: Tin," *J. Chron. Dis.* 17:483–502.
†NAS: USEPA, McKee and Wolf, "Water Quality Criteria," California Institute of Technology. (See Appendix A-VIII, "General References.")

- Inorganic tin is relatively nontoxic, but organic tin compounds have been used as fungicides, bactericides, and insecticides.
- Symptoms of toxicity (typical food chemical poisoning) were reported from liquids in cans where concentrations of tin higher than 1,000 ppm were analyzed.
- Based on the very low expected concentration of tin in water and the relatively high tolerance in human diet, the Health Authorities have not been concerned or found it necessary to issue maximum contaminant levels.

*Standard Methods* recognizes the Atomic Absorption Spectrometry Method (AAS) either with the direct aspiration into an air-acetylene flame or by the electrothermal AA spectrometry.

## THORIUM

When pure, thorium is a silvery-white, radioactive, metallic chemical element, air stable; it changes into a grey and then black material when contaminated by oxides. Symbol Th; atomic weight 232.04; atomic no. 90; specific gravity 11.72; valence +2(?), +3(?), +4. It is used in the electronics industry as a nuclear fuel (nuclear breeder cores) and in the manufacturing of incandescent lamps. In human metabolism, since the diet or drinking water contains minute doses of thorium (less than 1% mg/mL), it tends to be distributed to the skeleton.

Health aspects may be summarized as follows:

- The salts of thorium have been cataloged as chemical and radiological hazards. Long-term effects produce malignant neoplasms (radioactive rather than toxic effects expected).
- The International Commission on Radiological Protection has set the following standards for thorium:

0.5 mg/L soluble thorium 232, and
0.5 to 9 mg/L for insoluble compounds.

The NAS in 1980 (Volume III) stated "since thorium produces cancer in humans, estimates for exposure limits will not be calculated."

The USEPA has not yet listed thorium in the category of zero maximum contaminant level in "Primary Drinking Water Standards" of November 1985.

*Standard Methods* lists the Atomic Absorption Spectrometry by Direct Aspiration into a Nitrous Oxide-Acetylene Flame to calculate concentration of thorium ion to be reported in micrograms per liter. Water soluble thorium nitrate is used in standard solutions.

# VANADIUM

A bright, white, soft, ductile, abundant, chemical element found in 65 different minerals: symbol V; atomic weight 50.94; atomic no. 23; specific gravity 6.11; valence 2, 3, 4, or 5.

Particularly useful in nuclear production, it is also used as alloy (vanadium steel or ferrovanadium), in glass manufacturing and photography. Vanadium and its compounds have been known in the chemical industry for their toxicity (respiratory disturbances). A finished water survey of the 100 largest cities in the United States recorded a median value of less than 4.3 and a high concentration of 70 µg/L.* In another survey of 1,577 raw surface waters in the United States,† vanadium was recorded at 2 as a minimum, 40 as a mean, and 300 µg/L as a maximum, with a frequency of detection of 3.4%. In another finished water survey from 1962–1967, vanadium was measured while investigating 380 water supplies, at concentrations of 14 as a minimum, 46.1 as a mean, and 222 as a maximum, all in micrograms per liter (µg/L), with a frequency of 3.4%.

Solubility in water is recorded for vanadium pentoxide and sodium metavanadate at 0.07 and 21.1 g/100 mL, respectively. It forms a vanadyl cation (VO) from soluble salts; it also forms an anion vanadate ($VO_4$) leading to the conclusion that vanadium salts in wastewater solution will become a potential source of contamination.

Vanadium as a chemically pure element is not found naturally. Health effects may be summarized as follows (NAS, 1977):

- Absorption in the human body is extremely low.
- Requirements in human nutrition have not been proven, but it is suspected (suggested as protective against atherosclerosis).
- Acute vanadium toxicity in man is primarily respiratory.
- There is no evidence of chronic oral toxicity.

Vanadium content in food is very low, probable daily average 20 µg or less.† Vanadium in drinking water may be significant when considering the total diet due to the small contribution of vanadium in food. If vanadium is proven as a beneficial nutritional trace element, the daily contribution in drinking water will be considered beneficial.

Based on the above health effect considerations, the Health Authorities have not issued maximum contaminant levels for drinking water.

*Standard Methods* lists the Atomic Absorption Spectrometry by Direct Aspiration into a Nitrous Oxide-Acetylene Flame and the Gallic Acid Method as reliable for analysis of vanadium.

---

*NAS-1977.
†NAS, *Drinking Water and Health*, Vol. 3, Sept. 1980.

## Summary of Type D—Inorganic Chemicals.

| | AADI | MCLG | Carcinogenic | WHO | Nutritional Value | FinaL MCL |
|---|---|---|---|---|---|---|
| ANTIMONY—Sb | n.a. | 0.006 | ? | n.i. | ? | 0.006 |
| BERYLLIUM—Be | n.a. | 0.004 | (a) | n.i. | ? | 0.004 |
| COBALT—Co | n.a. | n.i. | ? | n.i. | YES | NO |
| THALLIUM—Tl | n.a. | 0.00005 | ? | n.i. | ? | 0.002 |
| THORIUM—Th | n.a. | n.i. | YES? | n.i. | NO? | n.i. |
| TIN—Sn | n.a. | n.i. | NO | n.i. | ? | NO |
| VANADIUM—V | n.a. | n.i. | ? | n.i. | YES? | n.i. |

AADI  = adjusted acceptable daily intake
MCLG = maximum contaminant level goal
n.a.   = not available (the low toxicity would yield a high value insignificant to concentrations expected in drinking water)
n.i.   = not issued by the Health Authorities
MCL   = maximum contaminant level
(a)    = questionable carcinogenicity (proved in animals)

# THALLIUM

Thallium is a rare, soft, and malleable metallic chemical element. Exposed to air, thallium exhibits first a metallic luster, but turns into a bluish-gray tinge resembling lead in appearance. Symbol Tl; atomic weight 204.38; atomic no. 81; specific gravity 11.85 (20°C.); valence 1 or 3. It is found in the rare minerals crooksite, lorandite, and hutchinsonite. It is also found in minute quantities in pyrites from where it is recovered while producing sulfuric acid. Thallium is also obtained from the smelting of lead and zinc ores. Thallium oxide, in presence of water, forms an hydroxide.

Thallium is a suspected carcinogen to humans.

As a thallium sulfate it was used as a rodenticide and an ant killer. It is prohibited in this country since 1975 as a household insecticide and rodenticide.

Thallium has been used in photocells as infrared detector (in thallium bromide-iodide crystals), in glass manufacturing, in the electronics industry, as an alloy, and in pharmaceutical products to treat skin infections.

Thallium was not finally evaluated by USEPA until the final rule was issued in 1992, based on potential drinking water health effects on kidneys, liver, and brain. Effective January 17, 1994, MCLG = 0.00005 mg/L and MCL = 0.002 mg/L.

With this final rule, USEPA also indicated the use of ion exchange and activated alumina as the "best available technology" for the removal of thallium.

For the determination of **thallium**, *Standard Methods* advises the use of the Atomic Absorption Spectrometric Method or the Inductively Coupled Plasma Method.

## REFERENCES

APHA, AWWA, WEF. 1995. **Standard Methods**. 19th edition, Washington, D.C.: American Public Health Association.

American Water Works Association. 1995. **SDWA Advisor**. Regulatory Update Service. Denver, CO: A.W.W.A.

American Water Works Association. 1995. **Journal**. Vol. 87. No. 2. An Update of the Federal Drinking Water Regulations. Denver, CO: A.W.W.A.

APHA, AWWA, WPCF. 1985. **Standard Methods**. 16th edition. Washington, D.C.: American Public Health Association.

American Water Works Association. **Seminar Proceedings**. Ozonation. Denver, CO: A.W.W.A.

Boggess, W. R. and B. G. Wixson. 1979. **Lead in the Environment**. Austin, TX: Castle House Publications Ltd.

California Institute of Technology. 1963. **Water Quality Criteria**. Pasadena, CA: NTIS, Springfield, VA.

Cox, C. R. 1964. **Operation and Control of Water Treatment Processes**. Geneva, Switzerland: WHO.

Drinking Water Research Foundation. **Safe Drinking Water**. Chelsea, MI: Lewis Publishers, Inc.

Culp, Wesner, and Culp. 1986. **Handbook of Public Water Supplies**. New York, NY: Van Nostrand Reinhold.

Nalco Chemical Company. 1988. **The Nalco Water Handbook**. 2d edition. New York, NY: McGraw-Hill Book Co.

National Academy of Sciences. 1977 and 1980. **Drinking Water and Health**. (Vols. 1 and 3). Washington, D.C.: National Academy Press.

New York State Dept. of Health. 1989. New York State Sanitary Code, Part 5, Drinking Water Supplies, Albany, NY.

Sawyer, C. N. 1960. **Chemistry for Sanitary Engineers**. New York, NY: McGraw-Hill Book Co.

USEPA. 1982. Water Programs. National Interim Primary Drinking Water Regulations, 47 *Fed. Reg.* 10999. Washington, D.C.

USEPA. 1983. **Methods for Chemical Analysis of Water and Waste**. Cincinnati, OH: Office of Research and Development.

USEPA. 1983. National Revised Primary Drinking Water Regulations, Federal Regulations, (Vol. 46, No. 194). Washington, D.C.

USEPA. 1984. Proposed Guidelines For the Health Assessment of Suspect Developmental Toxicants, *Fed. Reg.*, (Vol. 49, No. 227). Washington, D.C.

USEPA. 1984. Proposed Guidelines For Exposure Assessment, *Fed. Reg.*, (Vol. 49, No. 227). Washington, D.C.

USEPA. 1985. Proposed Guidelines For the Health Risk Assessment of Chemical Mixtures, *Fed. Reg.*, (Vol. 50. No. 6). Washington, D.C.

USEPA, N.P.D.W.R. Synthetic Organic Chemicals, Inorganic Chemicals and Microorganisms, *Fed. Reg.*, (Vol. 50, No. 219). Washington, D.C.

USEPA. 1986. Safe Drinking Water Act. 1986 Amendments, EPA 570/9-86-002. Washington, D.C.

USEPA, N.P.D.W.R. Drinking Water Regulations, Maximum Contaminant Level Goals. *Fed. Reg.*, (Vol. 53, No. 180). Washington, D.C.

USEPA. 1989. National Primary and Secondary Drinking Water Regulations, Proposed Rule, *Fed. Reg.*, (Vol. 54, No. 97).

USEPA. 1992. N.P.D.W.R.; Synthetic Organic Chemicals and Inorganic Chemicals, Final Rule, *Fed. Reg.* (Jul. 19, 1992).

USEPA. 1994. N.P.D.W.R. and MCLG: Lead and Copper. Final Rule. *Fed. Reg.* (Jun. 30, 1994).

# Chemical Parameters— Organic Compounds

## INTRODUCTION

ORGANIC CHEMICALS are subdivided differently from a general chemical viewpoint, as they are viewed in the concern and interest of the environmental health engineer or worker. In chemistry **organic compounds** may be listed as aliphatic hydrocarbons, aliphatic halogen derivatives, aliphatic sulfur compounds, aromatic hydrocarbons, aromatic halogen compounds, aromatic oxygen derivatives, aromatic sulfur derivatives, and so forth. A different approach is applied to identify, evaluate, classify, and finally issue standards for **contaminants** already known or in the process of being identified in drinking water.

In 1975 the National Academy of Sciences (NAS) finalized* the preliminary investigation of chemicals in the environment stimulated by the Environmental Protection Agency (USEPA), searching for scientific guidance.

Chemicals were reviewed from a practical social benefit and risk-benefit analysis and then evaluated from the general principles of toxicology and the utilization of data regarding acute, subchronic, and chronic effects.

Chemical carcinogenesis was then evaluated, looking also into potential hazards of reproduction.

This preliminary work was immediately followed by the specific evaluation of chemicals in drinking water in NAS (1977), *Drinking Water and Health.*

It must be noted that NAS (1977) recognized that the Safe Drinking Water Committee work was only the beginning of a very large task since the organic contaminants identified in drinking water so far constitute a small percentage of the total organic matter present in water. Even the volatile organic compounds (VOCs), where 90% of the VOCs are actually identified and quantified, represent only 10% of the total organic material in water.

**Pesticides** (herbicides, insecticides, fumigants) received top priority in the evaluation. **Detergents** had been studied from the very beginning in relation to the specific environmental problems clearly visible in the euthrophication and nutrients contaminants of wastewater discharges. **Phenols** have an important role in taste and odor control and disinfection by chlorination. **Trihalomethanes** (THMs) received a thorough investigation in the early 1970s because of the suspected carcinogenicity of compounds formed by disinfection through chlorination.

## SYNTHETIC ORGANIC CHEMICALS (SOCs)

Of high toxicity and suspected carcinogenicity and/or potential for occurrence in drinking water, Synthetic Organic Chemicals are of immediate interest to Public Health Authorities. To investigate them for toxicity evaluation and standards promulgation, it was first necessary to reduce their

---

*NAS—1975, *Principles for Evaluating Chemicals in the Environment.*

number to a reasonable list of priorities (see the section on *Evaluation of Synthetic Organic Chemicals* on page 159).

## VOLATILE COMPOUNDS (VOCs)

Volatile organic compounds are lightweight compounds that vaporize and evaporate easily, as the name indicates. VOCs belong to the synthetic (man-made) chemicals. They have been placed in a separate category by the Safe Drinking Water Act (SDWA), by legislative action, and by USEPA's issuance of standards and monitoring regulations. Some of them are the most frequently detected contaminants in drinking water connected with hazardous waste sites.

The SDWA requires publication of maximum contaminant levels for chlorinated organic compounds and aromatic organic compounds (see *THMs* and *VOCs*, in this Chapter). The first list of regulations for VOCs appeared in 1985, after advance notice of proposed rule-making in 1982 and 1984. The final rule for VOCs was issued in 1987. In 1990 the USEPA published a list of 36 unregulated VOCs and 17 additional potential contaminants that may become of state interest for monitoring and future investigation. Some VOCs will be regulated in the 21st century, after evaluating many research studies upom their completion. In general, VOCs are used in solvent and degreasing compounds as raw materials.

## DEFINITIONS AND NOMENCLATURE

The following are brief introductory notes for the reader not familiar with the symbolic representation used to identify VOCs and SOCs. These chemical contaminants are relatively new in the drinking water standards.

### Organic Chemistry

The section of chemistry dealing with compounds in which carbon is present as a principal element and in combination usually with hydrogen, oxygen, halogen, and, in natural compounds, with minor elements, such as nitrogen, phosphorus, and sulfur.

### Organic Compounds Properties

Organic compounds are usually combustible with lower melting and boiling point. They have high molecular weight. They may be a source of food for bacteria.

Their chemical reaction is normally slow (molecular rather than ionic activity). For a given formula there may be several compounds (isomerism). They have low solubility in water, poor conductivity in water solutions, and high solubility in organic solvents. Not only the chemical properties are different from the inorganic compounds, but the vast number of organic chemicals is frightening from an environmental control viewpoint.

## Synthetic Compounds

Organic compounds can be defined as synthetic compounds when they are produced by manufacturing process or are not natural compounds nor product of fermentation.

## Aliphatic Compounds

These are acyclic or "open-chain" compounds; they contain homologous series such as:

| | |
|---|---|
| $CH_3 \cdot CH_3$ | Ethane |
| $CH_3 \cdot CH_2 \cdot OH$ | Ethanol (ethyl alcohol) |
| $CH_3 \cdot CHO$ | Ethanal (acetaldehyde) |
| $CH_3 \cdot CO \cdot OH$ | Ethanoic (acetic acid) |

In this group are contained hydrocarbons, alcohols, aldehydes and ketones, acids, esters, ethers, alkyl halides, amines, amides.

## Aromatic Compounds

These are "closed-chain" or ring compounds. In this group are contained hydrocarbons, phenols, alcohols, aldehydes, ketones, acids, amines, and nitro compounds.

**Hydrocarbons**   These are organic compounds containing only carbon and hydrogen. The simplest is methane $CH_4$.

**Alcohols**   These are hydrocarbon derivatives in which one or more hydrogen atoms have been replaced by the OH group.

**Aldehydes**   These are hydrocarbon derivatives in which one or more hydrogen atoms have been replaced by the oxygen atom $R \cdot CHO$.

**Ketones**   Same as the aldehydes, but in the ketones the carbonyl group (C:O) is linked to 2 carbon atoms—$R \cdot CO \cdot R$.

**Acids**   These are hydrocarbon derivatives in which one or more of the hydrogen atoms have been replaced by a carboxyl group (CO·OH).

**Esters**   These are derivatives of inorganic or organic acids.

**Ethers** These are hydrocarbon derivatives in which two alkyl groups are attached to an oxygen atom, $R \cdot O \cdot R$.

**Alkyl Halides** These are compounds in which one hydrogen atom in an alkane has been replaced by a fluorine, chlorine, bromine, or iodine atom.

**Amines** These are hydrocarbon derivatives in which a hydrogen atom has been replaced by ammonia derivatives—$RNH_2$; $R_2NH$; $R_3N$.

**Amides** These are hydrocarbon derivatives in which a hydrogen atom has been replaced by the $CO \cdot NH_2$ (amide group).

## NOMENCLATURE

With a million compounds to be named and properly classified, the task would be very complicated and chaotic, particularly if each nation used its own system; but since 1892 standardization was accomplished by the International Union of Chemistry in Geneva, Switzerland, an organization now called IUPAC (International Union of Pure and Applied Chemistry).

The following general rules are applied:

- The largest carbon chain determines the root of the name, i.e.,

$$\overset{5}{CH_3}-\overset{4}{CH_2}-\overset{3}{CH_2}-\overset{2}{CH_2}-\overset{1}{CH_2}- = \textbf{Penthyl}.$$

- Each homologous series shall have a characteristic ending, i.e., paraffins: "**-ane**."
- The carbon atoms in the longest chain are numbered for the purpose of locating a substituent group so as to give the smallest number to a locating number.
- The $CH_3$ group branching off is a methyl group. One H atom removed from an alkane is called **alkyl** group; for example, dropping the ending -ane and adding **-yl**, so from methane, ethane, propane, etc., are named:

  | | |
  |---|---|
  | Methyl | $CH_3-$ |
  | Ethyl | $C_2H_5$ or $CH_3 \cdot CH_2-$ |
  | Propyl | $CH_3 \cdot CH_2 \cdot CH_2-$ |
  | Phenyl | $C_6H_5-$ |
  | Tolyl | $CH_3C_6H_4-$ |
  | Benzyl | $C_6H_5 \cdot CH_2-$ |
  | Xylil | $(CH_3)_2C_6H_3-$ |
  | Styryl | $C_6H_5CH{=}CH-$ |

- With the parent name given by the carbon atoms, the branches or side chains are attached as prefixes directly to the parent name. The number of the carbon atom is used (where the branches are attached). The number is repeated, preceded by a comma, when there are two branches attached to the same carbon. A hyphen is used after the numbers to introduce the parent name; for example, 2,3,4-trimethylpentane.

Hydrocarbons having carbon to carbon double bonds are called; **Alkenes, Alkadienes**, and **Alkatrienes:**

$CH_3CH = CHCH_3$       2-Butene

$CH_2 = CHCH = CH_2$       1, 3-Butadiene

$CH_2 = CHCHCH = CH_2$       3-Propyl-1, 4-Pentadiene
     |
     $C_3H_7$

# Principles of Evaluation of Synthetic Organic Chemicals

## SELECTION OF SIGNIFICANT PARAMETERS

In consideration of the gigantic number of synthetic organic chemicals (SOCs) that could be examined for significance, adverse effects, and potential occurrence in raw or potable water, some intelligent limitation to the number of chemicals to be regulated or analyzed is necessary.

In the United States drinking water supply over 700 SOCs have been identified. In 1986, American scientific and regulatory experience suggested the use of the following criteria for proposal of standards for SOCs:

• Actual and potential pollutional occurrence
• Significance of human exposure
• Associated health effects
• Availability of reliable laboratory analysis

A USEPA recommended maximum contaminant levels (RMCLs) list of 26 synthetic organic chemicals appeared in its 1985 regulations.*

The evaluation is subdivided in this handbook into the following components:

• Chemical composition
• Use (industrial/commercial/home)
• Environmental data (occurrence in air/food/water)
• Health effects (comments)

---

*USEPA—40 CFR—Part 141—Nov. 13, 1985—NPDWR—Synthetic Organic Chemicals: Final and Proposed Rule. USEPA—40 CFR—Part 141 and 142—NPDWR—Volatile Synthetic Organic Chemicals; Final and Proposed Rule.

- Standards
- Analytical methods (laboratory's approved analyses)
- Notes: comments on removal and/or contaminant peculiarities

It must be noted here that SOCs enter surface and underground supplies through several sources of pollution, such as:

- Wastewater discharges (domestic and industrial);
- Rural and urban runoff;
- Lecheate from landfills; and
- Runoff from vegetation and agricultural products chemically treated.

Under normal circumstances only a minimum concentration of SOCs is expected in raw water; this has been confirmed by water supply surveys (see Fig. 4-1).

The major problem of the water purveyors is the expensive and elaborate task of analyzing SOCs. The choice is between treatment that could show certain reliability or excessive laboratory work. To contain these chemicals within acceptable limits Granular Activated Carbon (GAC) treatment or aeration (PTA) or elimination of notoriously suspected sources of water supply may be used.

A forecast of probable contamination can be accomplished through an industrial toxic chemical inventory from production, transportation, storage, manufacturing, utilization, and disposal. For example, in this direction the New York State Department of Environmental Conservation (NYSDEC) conducted the first survey in 1977 using the following major components:

- air pollution (dust, acid rain, manufacturing plants, etc.)
- accidental spills (improper storage)
- heavy chemical storage or use
- halogenated hydrocarbons
- halogenated organics
- pesticides
- aromatic hydrocarbons
- substituted aromatics
- ketones and aldehydes
- plastics

The New York Department of Environmental Conservation (NYDEC) in 1978 conducted also a gasoline spill survey cataloging 202 spills for a total of 2,027,178 gallons of which only 50,370 were cleaned up.* A similar survey was conducted for reported spills involving sodium, cyanide, acetone, chromic acid, heavy metal sludge, hydrocarbon waste, paint, pesticides, dry cleaning solvents, and PCBs. A total of 99 spills were recorded with a volume of 190,108 gallons with only 8,402 gallons cleaned up.

---

*"Organic Chemicals and Drinking Water" by Kim, N. K. and Stone, D. W., N.Y.S. Dept. of Health, 1980 (Rev. 1981).

| Parameter | Quantification limit μg/L | Positives No. | Positives Percent | Median of +μg/L | Max. μg/L |
|---|---|---|---|---|---|
| Tetrachloroethylene | 0.2 | 34 | 7.3 | 0.5 | 23 |
| Trichloroethylene | .2 | 30 | 6.4 | 1 | 78 |
| 1,1,1-Trichloroethane | .2 | 27 | 5.8 | .8 | 18 |
| 1,1-Dichloroethane | .2 | 18 | 3.9 | .5 | 3.2 |
| 1,2-Dichloroethylenes (cis and/or trans) | .2 | 16 | 3.4 | 1.1 | 2 |
| Carbon tetrachloride | .2 | 15 | 3.2 | .4 | 16 |
| 1,1-Dichloroethylene | .2 | 9 | 1.9 | .3 | 6.3 |
| m-Xylene | .2 | 8 | 1.7 | .3 | 1.5 |
| o-+p-Xylene | .2 | 8 | 1.7 | .3 | .9 |
| Toluene | .5 | 6 | 1.3 | .8 | 2.9 |
| 1,2-Dichloropropane | .2 | 6 | 1.3 | .9 | 21 |
| p-Dichlorobenzene | .5 | 5 | 1.1 | .7 | 1.3 |
| Bromobenzene | .5 | 4 | .9 | 1.8 | 5.8 |
| Ethylbenzene | .5 | 3 | .6 | .8 | 1.1 |
| Benzene | .5 | 3 | .6 | 3 | 15 |
| 1,2-Dichloroethane | .5 | 3 | .6 | .6 | 1 |
| Vinyl chloride | 1 | 1 | .2 | 1.1 | 1.1 |
| 1,2-Dibromo-3-chloropropane | 5 | 1 | .2 | 5.5 | 5.5 |
| 1,1,2-Trichloroethane | .2 | 0 | | | |
| 1,1,1,2-Tetrachloroethane | .2 | 0 | | | |
| 1,1,2,2-Tetrachloroethane | .5 | 0 | | | |
| Chlorobenzene | .5 | 0 | | | |
| n-Propylbenzene | .5 | 0 | | | |
| o-Chlorotoluene | .5 | 0 | | | |
| p-Chlorotoluene | .5 | 0 | | | |
| m-Dichlorobenzene | .5 | 0 | | | |
| o-Dichlorobenzene | .5 | 0 | | | |
| Styrene | .5 | 0 | | | |
| Isopropylbenzene | .5 | 0 | | | |

Figure 4–1. Occurrence Data (from 466 random samples) Groundwater Supply Survey (GWSS)—USEPA. 1975 (concentrations expressed in micrograms per liter). *Fed. Reg.*, Vol. 49, No. 114. June 12, 1984.

In a survey conducted in 1978 in New York State, the ten most commonly found organic chemicals in 39 wells tested were tabulated as follows:

| Contaminant | Percent Positive | Max. Level Detected (in μg/L) |
|---|---|---|
| bis (2-ethylhexyl) phthalate | 92 | 170.0 |
| toluene | 85 | 10.0 |
| di-n-butyl phthalate | 54 | 470.0 |
| trichloroethylene | 46 | 19.0 |
| ethylbenzene | 44 | 40.0 |
| diethyl phthalate | 33 | 4.6 |
| trichlorofluoro methane | 28 | 13.0 |
| antracene/phenantracine | 18 | 21.0 |
| benzene | 15 | 9.6 |
| butyl benzyl phthalate | 13 | 38.0 |

In another New York State (NYS) survey in 1977 for pesticides and herbidices (endrin, lindane, methoxychlor, toxaphene, 2, 4-D, and 2, 4, 5-TP silvex) it was concluded that surface supplies did not contain these contaminants.

## RELATIVE RISKS—RISK APPROACH

The probability of a specific adverse effect occurring to an individual over a lifetime is defined as *risk* or *risk ratio*. The Health Authority's available epidemiological reports or causes of death reports with a locally known factor of life expectancy and living population can produce the "risk" or "risk ratio."

$$\frac{d \times L}{P} = risk \quad \text{with } d = \text{number of deaths per year or}$$
$$\text{average for several years}$$
$$L = \text{average lifetime}$$
$$P = \text{population included in survey}$$

For example, the NYS Health Department reported the following leading causes of death in the state (population, 18 million; life expectancy, 70 years) for the period 1974–1977:

| Cause of Death | d | Estimated Lifetime Risk | Risk Ratio |
|---|---|---|---|
| Cardiovascular Disease | 70,413 | 27 in 100 | 0.27 |
| Cancer | 36,297 | 14 in 100 | 0.14 |
| Stroke | 13,700 | 5 in 100 | 0.05 |
| Pneumonia | 5,716 | 2 in 100 | 0.02 |
| Accidents | 5,592 | 2 in 100 | 0.02 |

Voluntary risk is related to voluntary activities such as smoking or driving/riding an automobile. Drinking contaminated water can be listed as an involuntary risk. The explanation of tolerable level of risk to the public is a difficult task in public relations. Zero levels of MCL are also difficult to sell to the scientific community.

## TOXICOLOGICAL APPROACH

In determining risk and toxicity of organic chemicals in drinking water, known facts are evaluated first; such as dose-response extrapolations from performed animal studies for carcinogen compound and the no-observed-adverse-effect levels for noncarcinogens.

In lieu of reliable data a certain correlation with other chemicals may be temporarily considered pending ongoing or scheduled research work. *Acute toxicity* is defined as poisonous effects occurring shortly after the administration of a single dose or multiple doses of a substance. Death is a clear outcome of acute toxicity, but permanent damage of organs is also a result of acute toxicity.

*Chronic toxicity* is defined as an injury that persists because it is irreversible or progressive or because the rate of injury is greater than the rate of repair during a prolonged exposure period.

## EPIDEMIOLOGICAL STUDIES

All scientific epidemiological reports are considered a solid foundation in the evaluation process since they may report reliable data on human exposure with a control group of similar formation. Nevertheless, there are limitations on the reliability of data due to the variation of concentration of the substance under study, the temporary exposure to more than one chemical, the working population (industrial hygiene) made up predominately of healthy adult males, etc.

## ANIMAL STUDIES

An intelligent correlation between small animals and humans can be attempted, but it will be always questioned. A weight ratio can be used: mg/kg body weight basis. A weight for body surface area can be used: mg/square meter. In spite of uneasiness of such comparison, animal studies are always usable to provide guidelines.

## CARCINOGENESIS

The Health Authorities in the United States use the prudent approach of assuming that no threshold value exists in cancer induction. A threshold level is defined as an exposure level below which no toxic effect is observed because body mechanisms are capable of protecting a person from injury. A threshold level may theoretically always exist, but it may be limited by age, illness, and other known and unknown factors (see Fig. 4-2).

Cancer risk values may be calculated, but limitations may be set as hypotheses by the Health Agency (see Fig. 4-3). For example, the National Academy of Sciences (NAS) did not set an acceptable risk in 1977, but used a confidence level of 95% in stating a risk for one μg/L of a substance. At the same time the NYS Department of Health started to use as an insignificant risk the value of 1 in 1 million for a lifetime statistical assurance level of 95%.

The NAS in 1980 reevaluated epidemiological studies, the problems of risk estimation, the toxicity of selected drinking water contaminants, and the contribution of drinking water to mineral nutrition in humans in Volume 3 of *Drinking Water and Health.*

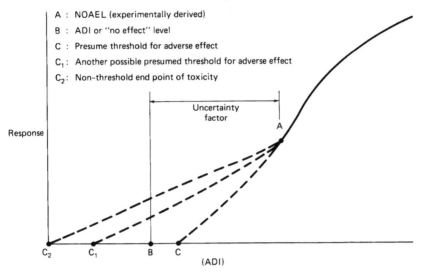

Figure 4–2. Noncarcinogenic Effects. From the Environmental Protection Agency—40 CFR Part 141. National Primary Drinking Water Regulations. Volatile Synthetic Organic Chemicals. Proposed Rulemaking.

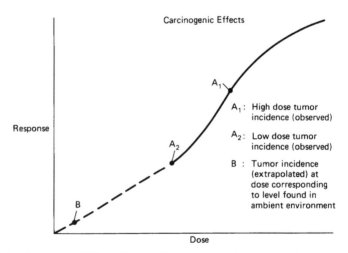

Figure 4–3. Carcinogenic Effects. *Fed. Reg.* Vol. 49, No. 114, June 12, 1984. (NAS) and USEPA. Representation of typical dose—response curve for animal experiments. $A_1$ and $A_2$ are determined by actual research work. B is mathematically extrapolated, estimated to occur at an exposure level below those experimentally applied. This point and points on the dotted line allow for projection of an associated cancer risk.

The NAS conclusion was that for noncarcinogenic toxicity "the preferred procedure is to make a risk estimate based on extrapolation of low dose levels from experimental curves obtained from much larger doses where effects can be readily measured. When such curves are not available, the ADI (acceptable daily intake) approach should be used until better data are available. In this approach safety factors are applied to the highest no-observable-effect dose found in animal studies . . . the ADI approach is not applicable to carcinogenic toxicity and that high dose to low dose extrapolation methods should be used for carcinogens. . . . Extrapolation from animals to humans should incorporate information on comparative pharmacokinetic data between the two species."

In the absence of such data the subcommittee suggests that the extrapolation be based on the surface area rule.

## NOAEL AND SAFETY FACTORS

The No-Observed-Adverse-Effect-Level (NOAEL) is the highest level that does not produce a toxic effect. From this level a guideline can be set for noncarcinogens. The WHO advised in 1975 to elaborate and evaluate the following data:

- "Biochemical aspects, including the kinetics of absorption, tissue distribution and excretion, biological half-life, effects on enzymes, metabolism, etc."
- Carcinogenicity, mutagenicity, neurotoxicity, potentiation, reproduction, teratogenecity, etc.
- Acute toxicity—LD 50s.
- Short-term studies, which generally include the classic Subacute 90-Day Toxicity Test. These studies generally extend from weaning to sexual maturity, usually 3 months in rodent species and 1–2 years in dogs and monkeys. The safety factors in the test compiled by WHO in 1975 ranged from 6–2,500 with a mean of 254 and a median of 100.

The NAS, evaluating 39 substances, used safety factors from 10–1,000 with a mean of 713 and a median of 1,000.

When there are no studies available, chemical similarities can be carefully examined, and tentative conclusions may be reached, as for example that vinyl fluoride has a high probability to be carcinogenic.

## LIMITATION OF CONTAMINATION PROBLEMS

A watershed investigation must be conducted every four years very thoroughly and twice a year as a general inspection in order to evaluate all the potential sources of contamination by synthetic organic chemicals. A more complicated task is investigation regarding groundwaters because of slow movements of water in aquifers and complexity of water transfer from stratum to stratum. Elimination of particular underground supply or surface stream may be the most economic solution of a synthetic, organic chemical contamination. This option must be analyzed against an estimate of treatment facility cost.

### Aeration

This can be particularly effective for volatile compounds, and its effectiveness is proportional to air-to-water ratio, contact time, and the temperature of the water and air. (See Chapter 10, "Water Treatment.")

### Granular Activated Carbon (GAC) Treatment

It appears by utilization and pilot studies at specific sites that GAC is the most reliable absorber of synthetic organic chemicals. While some existing sand filter beds can replace the media with GAC, it is more likely that insufficient filter beds are the problem, caused by increased water demand and aging of plants, leaving preferable solution of post conventional filtration "Contactors" for maximum utilization of GAC removal potential. Aeration is in general the least expensive treatment solution, particularly for plants larger than 4,000 cubic meters/day (1 million gallons/day).

As previously stated, synthetic organic compounds may contaminate a water supply at an extremely low concentration; moreover, the present knowledge of toxicity and possible threshold limits to being "safely usable" is limited at best and sometimes extremely limited, leaving only the ability to hypothesize the contamination potential. Furthermore, the additional cost of treatment is in all cases rather expensive in relation to the total cost of water billed to the consumer. Sometimes abandoning a specific groundwater source or a surface stream is a locally unrealistic solution. Therefore, risk and cost benefit relationship may be another factor to be evaluated in decision making. (See Chapter 10, "Water Treatment," for more information.)

## DETERGENTS (SURFACTANTS)

Detergents, also called *surfactants* or *surface active agents* or *syndets*, are synthetic chemical compounds or cleansing substances produced to emulsify dirt, such as soap, but made from **aromatic sulfonates**, the **alkyl sulfates**, etc.

Many household detergents are aromatic sulfonates. Benzene is alkylated with a mixture of unsaturated hydrocarbons producing an alkylbenzene that is sulfonated and then converted into salt.

The WHO issued in 1963 a maximum acceptable concentration of alkyl benzyl sulfonates (ABS; surfactants) of 0.5 mg/L and a maximum allowable of 1.0 mg/L. In evaluating raw water the same standards reported a maximum allowable concentration of 0.5 mg/L based on the maximum sensitivity of the accepted analytical procedures when the standards were issued.

In drinking water quality control the detection of surfactants is an indication of pollution. In water pollution control efforts have been made during the last 30 years to use biodegradable surfactants and enzymes and limit the content of phosphorus to curtail eutrophication (the result of growth of plants and animals promoted by nutrients in slow-moving bodies of water such as lakes and ponds). In present industrial production, two-thirds of the total surfactants used are **anionic** (negatively charged, containing 90% of the detersive properties) and **cationic** (positively charged, for disinfection and fabric softening). The remaining third is made up of nonionics.

Raw sewage may contain from 1–20 mg/L. Raw water as source of water supply prior to treatment is expected to contain less than 0.1 mg/L.

*Standard Methods* lists the Surfactant Separation By Sublimation; Anionic Surfactants as MBAS (ethylene blue active substances); and Nonionic Surfactants as CTAS (cobalt thiocyanate active substances) for analytical methods.

Foaming Agents (MBAS), nontoxic or carcinogenic at levels expected are usually regulated by "recommended" standards:

| | |
|---|---|
| USEPA (1980) | 0.5 mg/L |
| USEPA (1989) | 0.5 mg/L as Secondary MCL |
| WHO (1971)* | 0.2 mg/L recommended |
| | 1.0 mg/L max. permissible level |
| European Community† | 0.2 mg/L max. admissible |

USEPA (1995) confirmed the classification of *foaming agents* with a Secondary MCL of 0.5 mg/L as a final rule; consequently, no final MCL will be issued.

USPHS drinking water standards recommended no limits in 1925, 1942, and 1946, but in 1962 the alkyl benzene sulfonates recommended concentrations were set at 0.5 mg/L. This limit appears to have been selected only for foaming at the tap, not based on toxicological data.

It must be noted that conventional water treatment is not expected to remove surface-active material. Consequently, problems with foaming, influence on turbidity readings, interference in the coagulation process, and potential production of taste and odor may be expected with ABS concentrations higher than 0.5 mg/L.

---

*WHO (1984) set no guideline value with the remark that there should not be any foaming or taste and odor problems.

†Listed as surfactants (reacting with methylene blue) and unit of measurement expressed in 200 μg/L (lauryl sulfate).

## PHENOLS

A phenol is defined as a colorless, crystalline substance contained in tars of coal and wood or synthetically manufactured, known for its characteristic odor.

Phenols can be defined as a group of compounds in which one or more hydrogen atoms in the aromatic nucleus have been replaced by the hydroxyl group (OH).

The formula $C_6H_5 \cdot OH$ or a monohydroxy derivative of benzene, is also commonly called carbolic acid.

**Phenols** are used extensively in medicinals, dyes, resins, and other commercial products; once they were mainly used as disinfectants.

A **phenol** is highly soluble in water; it may enter water supplies by industrial waste pollution. The acute toxicity of phenol is known (1.5 grams may be fatal), but consumption of phenol in high doses in drinking water is not expected since the threshold level is well below the toxic level. Consequently, the Health Authorities showed limited concern in standard setting.

The major concern in drinking water is the avoidance of bad taste that may easily develop, particularly in connection with the disinfection process of chlorination (**chlorophenol** compounds are formed, known for their taste and odor objection). Limitations of phenolic compounds were issued by USPHS in 1942, 1946, and 1962 drinking water standards at 0.001 mg/L as phenol concentration based on taste resulting from chlorination and not because of toxicologic concern.

It is important to note that phenol solutions are decomposed by bacterial and biological actions in streams. Some studies have recorded that 1 mg/L of phenol was biologically dissimilated in 1–7 days under aerobic conditions. FDA limits phenols in bottle water with a standard of 0.001 mg/L with no standards issued by USEPA in their "primary" or "secondary" drinking water standards.*

*Standard Methods* lists the Gas-Liquid Chromatographic Method and two analytical procedures using the 4-Aminoantipyrine Colorimetric Method.

The WHO (1984) set no standard but suggested to maintain the total phenol content of water below 1 µg/L. Taste thresholds for mono-, di-, tri-, and tetrachlorophenols have been recorded in the range 0.1–1 µg/L.

The WHO (1984) toxicological review of phenol and chlorophenols maintains these contaminants still under observation until mutagenicity and carcinogenicity is determined. In the United States, the NCI has found that 2,4,6-trichlorophenol increased tumor incidence.

The European Union listed among the "substances undesirable in excessive amounts" *phenols* (phenol index), the maximum admissible concentra-

---

*Pentachlorophenol (1985) Secondary MCL = 0.03 mg/L.

tion of 0.5 µg/L of $C_6H_5 \cdot OH$, excluding phenols that do not react with chlorine.

## HALOGENATED CHLORO-ORGANIC COMPOUNDS—TRIHALOMETHANES (THMs)

Since the early 1970s the Health Authorities have been particularly under pressure to issue standards for control, reduction of concentration, and tentative identification of acceptable levels of THMs. The cause of concern was the presence of halogenated chloro-organic compounds in drinking water and the carcinogenicity suspected. The present and the even more threatening future use of more polluted water, requiring higher level of chlorination for disinfection and the consequent formation of THMs, demanded extensive studies in toxicity and carcinogenicity. Pending final results in research, water consumers and water suppliers called for the identification of THMs reduction technology and the selection of recommended limits (MCLs).

For THMs or total THMs (TTHMs) the following organic compounds were identified:

1. Trichloromethane (Chloroform)   $CHCl_3$
2. Bromodichloromethane   $CHBrCl_2$
3. Dibromochloromethane   $CHBr_2Cl$
4. Tribromomethane (Bromoform)   $CHBr_3$
5. Dichloroiodomethane   $CHCl_2I$
6. Bromochloroiodomethane   $CHClBrI$
7. Chlorodiiodomethane   $CHClI_2$
8. Dibromoiodomethane   $CHBr_2I$
9. Bromodiiodomethane   $CHBrI_2$
10. Triiodomethane (Iodoform)   $CHI_3$

Major concern was **chloroform** concentration since preliminary studies supported the opinion that chloroform was a suspected carcinogen in animals.

The above-mentioned organic compounds are clearly derivatives of methane ($CH_4$), but methane gas is not in question but rather the reaction of chlorine in raw or treated water with certain compounds mainly defined as **humic acids,** part of organic material associated with decaying vegetation—they are also called the *precursors.*

$$\text{Chlorine} + \text{Precursors} \rightarrow \text{Chloroform} \times \text{Other THMs}$$

It is therefore clear from the above that public health concern and water surveys should first scrutinize surface water (rivers, lakes, ponds). Particular attention and priority should be given to water supply systems where prechlorination is practiced or multiple chlorination where no conventional treatment is provided.

Since chlorination has been extensively used in surface water treatment during the last 50 years, this problem did evidently exist from the beginning of disinfection with chlorine. Lack of identification was particularly attributable to the minimum emphasis in organic compounds, determination in laboratory work in general and lack of specific analytical procedures. Of course more pollution may also signify more chlorine demand and consequently higher concentration of TTHMs.

From the list of the 10 TTHMs, *chloroform* and *bromodichloromethane* are more frequently expected in drinking water, with the THMs listed as 3 and 4 frequently found, while 5 is rarely encountered.

Since here halogens are dealt with (chlorine, bromine, iodine, fluorine) one thinks next to fluoridation or fluoridation/chlorination, but fortunately there are no fluorine-containing THMs.

In 1975, USEPA conducted the National Organic Reconnaissance Survey (NORS) for THMs frequency and concentration determination in drinking water in the United States. NORS results can be summarized as follows:

*CHLOROFORM*   Range: 0.2–311 μg/L*
Median: 20 μg/L

**Chloroform** concentration of more than 105 μg/L was registered in 10% of the water systems surveyed.

In 1976 a similar survey was organized by USEPA, the National Organic Monitoring Survey introducing the seasonal factor in THMs concentration.

It must be mentioned at this point that, unfortunately, the identification of THMs by gas chromatographic procedure required trained laboratory operators and the "precursors" could not be identified as such, so the general organic content was considered instead, leaving undefined the variable relationship of precursors to the general organic content. In 1976 USEPA issued an official position encouraging water utilities to initiate action to reduce chloroform concentration (see Fig. 4-4). Such reduction is possible moving the point of chlorine application to a stage of the treatment process where the organic content is already partially reduced; for example reducing prechlorination and increasing an intermediate chlorination following the water treatment process of coagulation, flocculation, and preliminary sedimentation (or sedimentation prior to filtration). Also reduction of prechlorination and intermediate chlorination in favor of final chlorination are steps favorable to the reduction of THMs.

---

*"Interim Treatment Guide for the Control of Chloroform and Other Trihalomethanes," Water Supply Research Division, Municipal Environmental Research Laboratory, Cincinnati, OH, 45268 (June 1976).

The use of **ozone** and **chlorine dioxide** in disinfection of raw water may lead to total or partial elimination of THMs.

The adoption of Granular Activated Carbon (GAC) treatment tends to eliminate organic matter, but the cost of installation and reactivation of GAC is a significant economic problem.

It is evident that the removal of chloroform, once formed, requires a difficult treatment process, leaving consequently first choice to the prevention of chloroform formation. NORS identified the following statistics from nine utilities having high concentration of chloroform:

|                          | Range            | Average    |
|--------------------------|------------------|------------|
| Chloroform concentration | 103–311 µg/L     | 177 µg/L   |
| Total chlorine dose      | 4.3–18 mg/L      | 9.0 mg/L   |
| Combined residual        | 0–1.7 mg/L       | 0.5 mg/L   |
| Free residual            | 0–2.7 mg/L       | 1.2 mg/L   |
| Chlorine demand          | 2.8–15.7 mg/L    | 7.3 mg/L   |

Besides the adsorption on powdered and granular activated carbon or the use of chlorine dioxide or ozonation, another process for the removal of chloroform is **aeration**.

Chlorine dioxide produces a reliable residual for disinfection but the potential toxicity of the organic by-products (for example chlorite) advises caution until related studies are satisfactorily conducted.

It is repeated here that it is more practical to increase chlorination after settling and before filtration process than to increase concentration at the effluent or final chlorination, since normally a lack of contact time necessary for an efficient disinfection is expected. This is due to insufficient storage of the treated water or too-close proximity of the first water user to the point of final chlorination.

Moreover active chlorine will help disinfect the filters. The disadvantage of intermediate chlorination is the reduction of prechlorination with the consequent formation of algae, slime, and related raw water organisms influencing treatment. Before this problem becomes critical, shock pretreatment (higher dosages of chlorine for limited time) can be used but at that time formation of THMs is expected.

As previously stated, the direct removal of chloroform can be achieved through the use of adsorption on powdered activated carbon, aeration, and ozonation. Powdered activated carbon and ozonation require high installation cost for treatment with consequent high removal costs. Granular activated carbon is an effective process since other organic contaminants may also be removed (dieldrin, lindane, 2,4,5-T, DDT, and parathion) but regeneration is in general required monthly to be effective in THMs removal. Aeration has been tested for the removal of chloroform. When treatment plants are already equipped with aeration for removal of iron, manganese, or hydrogen sulfide, such treatment takes place prior to chlorination, there-

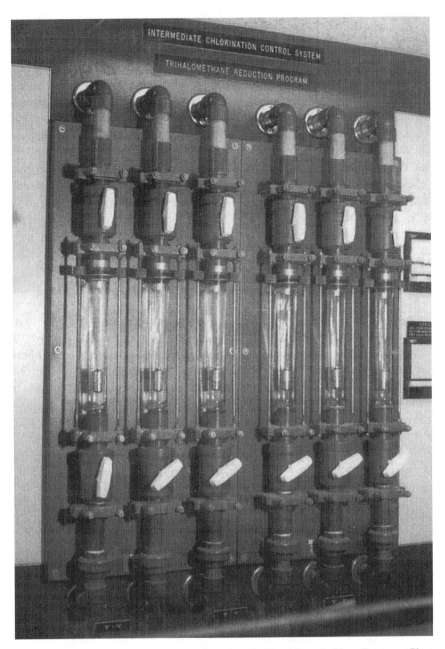

Figure 4-4. To control Trihalomethanes' formation, the Poughkeepsie Water Treatment Plant reduces prechlorination in favor of intermediate chlorination. Two injection points per settling basin were added, following the solids—contact basin and before aeration and filtration. Shown are the flow control devices located in the operator control rooms. Quick feeding adjustments are possible to relate dosage to variation in turbidity readings. (Courtesy of the City of Poughkeepsie, NY.)

fore it could not be effective in THMs removal (THMs not yet formed). But if a plant is equipped for taste and odor control following chlorination, some chloroform would be removed to the atmosphere through aeration.

In conclusion, THMs removal is difficult and expensive once these chemicals are formed.

Ozonation or treatment with chlorine dioxide is effective in THMs removal but cost, residual disinfectant and end-products or organic by-products are potentially inherent problems. Existing plant operation can be improved also to achieve a goal of THMs reduction: This can be achieved with improved coagulation and settling effectiveness through the use of coagulant and flocculant aids and more frequent jar testing. Precursors will therefore be reduced, at least temporarily until plant upgrading is built, possibly combining with reduced prechlorination and increased final chlorination.

A careful microbiological investigation should follow any change of chlorination practice to assure continuous or improved bacteriologic safety of the drinking water.

The USEPA in 1979 issued the final rule for the control of trihalomethanes in drinking water as part of revised National Interim Primary Drinking Water Regulations establishing a maximum contaminant level of 0.10 mg/L for TTHMs and regulations for monitoring with comments on possible reduction or removal of TTHMs. Also included was a dissertation on comments received and analytical methods.*

## CURRENT RULES AND REGULATIONS IN THE UNITED STATES FOR TRIHALOMETHANES

As stated, USEPA issued on November 29, 1979 (Vol. 44, No. 231—*Fed. Reg.*) the amendment to the National Drinking Water Regulations (Interim—Primary) to establish a national policy and MCLs.

Such regulations reaffirmed that THMs were frequently found in drinking water with chloroform concentrations ranging from 0.001–0.0540 mg/L.

For the determination of the MCL, the USEPA reviewed in its regulations all aspects of the related health effects. They may be summarized here as follows:

- Short-term toxic responses to THMs in drinking water are not documented, but prolonged administration of chloroform at relative high doses to rats and mice produced oncogenic effects.
- It is safe to assume that there is a potential health risk to humans subjected to drinking water containing chloroform consumed daily at a

---

*Federal Register*—Vol. 44, No. 231 (Nov. 29, 1979) Rules and Regulations—USEPA—Part III—NPIDWR: Control of Trihalomethanes in Drinking Water; Final Rule. Based on the July, 1976 USEPA, ANPRM entitled "Control Options for Organic Chemicals in Drinking Water" (41 *Fed. Reg.* 28991 *et seq.*).

concentration higher than 0.10 mg/L, assuming an intake of two liters of water per day for a period of 70 years.

- Since chloroform has been reported as a carcinogen in rodents at high doses, pending more reliable epidemiological studies and considering a potential human risk, the level of chloroform in drinking water should be reduced as much as possible technologically and economically.
- The maximum contaminant level presently remains at 0.1 mg/L, but the USEPA is continuously reviewing MCLs for THMs. More restrictive MCLs are expected.
- The present USEPA regulations and the latest Feb. 28, 1983 (Vol. 48, No. 40) rule—see references at the end of this Chapter) establishing a MCL at 100 ppb are applicable to communities serving more than 75,000 persons. This limitation has been questioned by the National Drinking Water Advisory Council, which recommended to apply such limitations to suppliers also serving between 10,000–75,000 persons. The Council also recommended to leave jurisdiction to States as to apply an MCL to utilities serving less than 10,000 persons.

The USEPA 1979 regulations also contain descriptions of analytical methods for the determinations of **Trihalomethanes**. The first method described is the **Purge and Trap Method** applicable for the determination of chloroform, dichlorobromomethane, dibromochloromethane, and bromoform in finished drinking water. The TTHMs are determined by adding up these four components.

The second method described is the **Liquid/Liquid Extraction Method** applicable only to the determination of chloroform, bromodichloromethane, chlorodibromomethane, and bromoform.

The USEPA (1979) also prescribes the determination of MTP or the **Maximum Total Trihalomethanes Potential**. This determination is based on a sample taken from a point of the distribution system that provides the maximum development of TTHM, or maximum residence time. It must be pointed out that production of maximum values of TTHM are influenced by pH, temperature, reaction time, and the presence of the disinfectant residual.

Current laboratory technology for efficient and reliable analysis of THMs and VOCs requires very automated, computerized, expensive equipment at a cost of approximately $100,000—when the best available analyzers are used (see Figure 4-5).

The WHO in 1984 has set no guideline level for trihalomethanes, but used a guideline value of 30 µg/L for chloroform, with the remarks that disinfection efficiency must not be compromised when controlling chloroform content. This remark was dictated by the fear of reduction of effective chlorination in the developing countries where waterborne diseases cause thousands of deaths every day.

The chloroform guideline value also came with the note " . . . computed from a conservative hypothetical mathematical model which cannot be

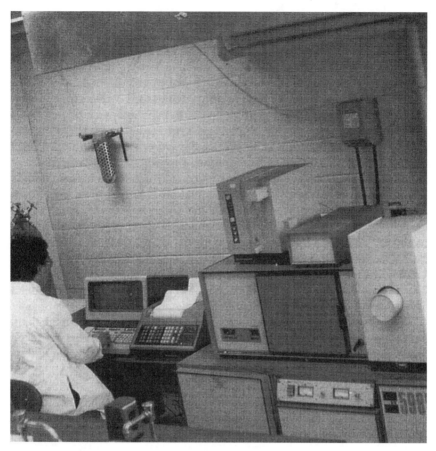

Figure 4–5.  In the New York City Department of Health's First Avenue Laboratories, a gas chromatograph/mass spectrometer unit is used to analyze water samples from the N.Y.C. distribution system, both from surface supplies and groundwater supplies. (Water wells supply only the southern section of the borough of Queens.) Analysis for trihalomethanes and volatile organic chemicals are completed herewith and results evaluated by the Bureau of Public Health Engineering. New York City surface supplies receive only multiple chlorination for disinfection and treatment (beside pH control and fluoridation) leaving the finished water subject to THMs formation. (Courtesy of N.Y.C. Dept. of Health.)

experimentally verified, and values should therefore be interpreted differently. Uncertainties involved may amount to two orders of magnitude (i.e., from 0.1–10 times the number)." The WHO also reported that THMs limits have been set in several countries from 25 to 250 μg/L.

USEPA (1995) confirmed the final rule of 0.10 mg/L as MCL for the **Trihalomethanes**. This rule has been in effect from Nov. 29, 1981, for systems serving more than 75,000 people, but became effective Nov. 29, 1983, for systems serving more than 10,000 people but less than 100,000.

USEPA (1993) approved for TTHM compliance monitoring using analytical methods 502.2 and 524.2.*

For monitoring, compliance, and public notification of TTHM, see Chapter 8, "Sampling and Monitoring."

## ANALYTICAL METHODS FOR TRIHALOMETHANES AND CHLORINATED ORGANIC SOLVENTS

*Standard Methods* (1995) listed the following methods recommended for the measurements of THMs and chlorinated organic solvents:

- Liquid-Liquid Extraction Gas Chromatographic Method
- Micro Liquid-Liquid Extraction Gas Chromatographic Method

This second method was developed to analyze at the same time also the following *disinfection by-products* in raw or treated water:

- Monochloroacetic acid (MCAA)
- Monobromoacetic acid (MBAA)
- Dichloroacetic acid (DCAA)
- Trichloroacetic acid (TCAA)
- Bromochloroacetic acid (BCAA)
- Dibromoacetic acid (DBAA)
- 2,4,6-Trichlorophenol (TCPh)

Other disinfection by-products commonly reported are:

- Bromodichloroacetic acid (BDCAA)
- Dibromochloroacetic acid (DBCAA)
- Tribromoacetic acid (TBAA)
- Dihaloacetic acids (DHAAs)
- Chloral hydrate (CH)
- Dihaloacetonitriles (DHANs)
- Dichloroacetonitrile (DCAN)
- Bromochloroacetonitrile (BCAN)
- Dibromoacetonitrile (DBAN)
- 1,1,1-Trichloropropanone (111-TCP)
- Chloropicrin (CP)
- Cyanogen chloride
- Dissolved Organic Halide (DOX)

---

*National Primary Drinking Water Regulations: Analytical Techniques; Final Rule. *Fed. Reg.* (Aug. 3, 1993).

When USEPA issued the *proposed rule* for Disinfectants and Disinfection By-Products* in 1994, the following limits were proposed for D/DBP:

MCLG = not applicable; MCL = 0.080 mg/L for Stage 1, and
MCLG = not applicable; MCL = 0.040 mg/L for Stage 2.

Proposed "Stage 1" is expected to establish with the final rule maximum residual disinfectant levels (MRDLs) for chlorine (4.0 mg/L), mono-chloramine (4.0 mg/L), and chlorine dioxide (0.8 mg/L). See *Disinfectants— Disinfection By-Products* in Chapters 8, 10, and 11.

The "Stage 2" final rule will be issued after the elaboration of data obtained from the Information Collection Rule (ICR) (see Chapter 11). A setting of TTHMs = 40 µg/L and THAAs = 30 µg/L is expected with the best available technology as precursor removal with chlorination. THAAs is the total of monochloracetic acid, dichloracetic acid, trichloracetic acid, and dibromoacetic acid.


## VOLATILE SYNTHETIC ORGANIC CHEMICALS

In issuing standards or guidelines for volatile synthetic organic chemicals (VOCs) in drinking water, the Health Authorities should be guided by the scientific research work on the toxicity and carcinogenicity of the contami-nants, but other factors are equally important.

The contaminant in water must be quantified through a reliable labora-tory analysis capable to determine the very low level concentration ex-pected in finished waters; moreover, a well organized survey must report the occurrence of the contaminant at random and in selected water supplies where potential occurrence is expected due to the environmental specific conditions and, consequently, at levels higher than the minimum detectable level (quantification limit).

Other important factors influencing decision on the contaminant are ambient air concentration, food intake, inhalation versus ingestion factors, the persistence of the specific contaminant in the environment, the effect of an existing or future, partial, or total ban on the use of the contaminant in the environment, the reasonableness of excessively high factor of safety when related to high cost of removal or practical impossibility of removal from a water system.

The Health Authorities should be free to issue standards based on profes-sional judgement of facts and reliable and sufficiently complete research work taking into consideration the above factors. The Health Authorities should not be pressured or guided by legislative action in selecting or dictating investigative criteria, the list of contaminants, deadlines for stand-

---

*National Primary Drinking Water Regulations: Disinfectants and Disinfection By-Products; Proposed Rule. *Fed. Reg.*, Vol. 59, No. 145 (July 29, 1994).

ard settings, or similar political actions that seek to prejudge and influence the professional task of public health experts who are obliged to interpret at times incomplete or insufficient data, but who are guided by scientific experience.

Examining chemical contaminants, particularly the organic compounds that show carcinogenicity confirmed in animals in a limited way, the Health Authorities are inclined to issue standards of zero as the maximum contaminant level (MCLs). This value cannot be determined by laboratory technique no matter how precise, accurate, and refined the expert water laboratory technician can be. Health Authorities could then adopt the lowest value achievable in a typical certified laboratory. These standards would be consequently changed every time a more refined method is approved after experimentation with a new technique.

Another impossibility is the monitoring of the contaminants on a daily basis and throughout the distribution system. So, reality must be faced and, whenever possible, reality should be measured, evaluated, and interpreted.

Another way to evaluate MCLs at a "noncomfortable" level, that is, above zero in spite of suspected carcinogenicity, is to assume that there is a small concentration of the contaminant unable to develop cancer, because of human metabolism or body defenses. This is certainly a reasonable approach but scientific evidence is not available at this time to assume a 100% safety below that threshold value.

A mathematical approach can be used to measure this area of questionable safety between a zero MCL and a threshold limit. The risk assessment or carcinogenic risk estimates as concept of risk evaluation was presented to USEPA by NAS (1977). The risk estimates are expressed as probability of cancer after a lifetime consumption of one liter per day containing Q parts per billion of the compound in question. For example, a 5 μg/L concentration constantly determined or assumed to remain in a particular water supply for over 70 years, assuming a daily consumption of 2 L, results in $1 \times 10^{-6}$ Q = $2 \times 5 \times 10^{-6}$ = $1 \times 10^{-5}$ or one excess case of cancer for every 100,000 persons exposed. For a population of 7 million persons, this is equivalent to 70 excess lifetime deaths for cancer or one per year (see Table 4-1).

## U.S. DRINKING WATER REGULATIONS

An intention to issue regulations for **volatile synthetic organic chemicals** was published on March 4, 1982, by the USEPA (47 *Fed. Reg.* 9350 *et seq.*). On June 12, 1984, the USEPA issued a "Proposed Rulemaking" (40 CFR 141) establishing recommended maximum contaminant levels for the following parameters:

Trichloroethylene
Tetrachloroethylene
Carbon tetrachloride
1,1,1-Trichloroethylene
Vinyl chloride
1,2-Dichloroethane
Benzene
1,1-Dichloroethane
p-Dichlorobenzene

On Nov. 13, 1985, the USEPA issued "Final Rule and Proposed Rule" for VOCs (NPDWR-40 CFR Parts 141 and 142, *Fed. Reg.* Vol. 50, No. 219). The maximum contaminant levels in Table 4-2 summarize the final RMCL (recommended) and the proposed maximum contaminant levels (MCLs). The final MCLs were promulgated by USEPA in 1987 (40 CFR Parts 141 and 142-NPDWR-Synthetic Organic Chemicals—Monitoring of Unregulated Contaminants; Final Rule—*Fed. Reg.* Vol. 52, No. 130). The final maximum contaminant levels are also contained in Table 4-2 and expressed in micrograms per liter. USEPA (1985 and 1987) also reported acceptable reliability

### Table 4-1.  Volatile Organic Compounds—Cancer Risk Estimates.

| Contaminant | Concentrations in Drinking Water Corresponding to a $10^{-5}$ Risk[3] (Values Expressed in $\mu g/L$[1]) | | |
| --- | --- | --- | --- |
| | NAS[4] | CAG[5] | USEPA |
| Benzene | N.C. | 13 | 10 |
| Vinyl chloride | 10 | 0.15 | 0.2 |
| Carbon tetrachloride | 45 | 2.7 | 3 |
| 1,2-Dichloroethane | 7.0 | 3.8 | 4 |
| Trichloroethylene | 45 | 26 | 30 |
| 1,1-Dichloroethylene | N.C.[2] | 0.61 | N.C. |
| Tetrachloroethylene | 35 | 6.7 | N.C. |

[1]$\mu g/L$ = micrograms per liter
[2]N.C. = not calculated
[3]$10^{-5}$ risk = one theoretical probability of cancer death in 100,000 population
[4]NAS = National Academy of Sciences
[5]CAG = Carcinogen Assessment Group (an advisory board to USEPA)
NAS and CAG data reported by *Fed. Reg.* Vol. 50, No. 219, 11/13/85
USEPA data reported in *Fed. Reg.* Vol. 52, No. 130, 7/8/1987
RISK ESTIMATES were calculated at the 95% confidence limit using multistage model for the known or probable human carcinogens, with the assumptions of 2 L of water ingested daily over a lifetime of 70 years for a 70-kg adult.

Table 4-2.  USEPA Standards for VOCs Final
MCLG; Proposed and Final MCLs.

| Contaminant | Final MCLG | Proposed MCL | Final MCL |
|---|---|---|---|
| Benzene | zero | 5 | 5 |
| Vinyl chloride | zero | 1 | 2 |
| Carbon tetrachloride | zero | 5 | 5 |
| 1,2-Dichloroethane | zero | 5 | 5 |
| Trichloroethylene | zero | 5 | 5 |
| p-Dichlorobenzene | 75 | 5[a] | 75 |
| 1,1-Dichloroethylene | 7 | 7 | 7 |
| 1,1,1-Trichloroethane | 200 | 200 | 200 |
| Tetrachloroethylene | zero | — | 5 |

Values: expressed in micrograms/liter (μg/L)
MCLG: maximum contaminant level goal
MCL: maximum contaminant level (enforceable standards)
VOCs: Volatile Organic Chemicals
Zero MCLG: no safe level assumed—contaminant is suspected for carcino-
genicity
Final MCL first issued in 1987 with exception of tetrachloroethylene (MCL,
1989)
[a]Secondary MCL = 5 μg/L (USEPA, 1989).

in existing analytical methods to determine VOCs concentrations. Cost of
sample analysis was estimated (1987) as approximately $150–$200.

The MDL, the method detection limit or the minimum concentration of a
substance that can be measured, has been determined to range from 0.2–0.5
μg/L for the VOCs contained in Table 4-2 as presently prioritized by USEPA.

The methods recommended by USEPA (1987) for laboratory determination of VOCs are:

- USEPA Method 502.1—Volatile Halogenated Organic Compounds in
  Water by Purge and Trap Gas Chromatography.
- USEPA Method 502.2—Volatile Organic Compounds in Water by
  Purge and Trap Gas Chromatography with Photoionization and Elec-
  trolytic Conductors in Series.
- USEPA Method 503.1—Volatile Aromatic and Unsaturated Organic
  Compounds in Water by Purge and Trap Gas Chromatography.
- USEPA Method 504—1,2-Dibromomethane and 1,2-Dibromo-Trichlo-
  ropropane in Water by Microextraction and Gas Chromatography.
- USEPA Method 524.1—Volatile Organic Compounds in Water by
  Purge and Trap Gas Chromatography/Mass Spectrometry.
- USEPA Method 524.2—Volatile Organic Compounds in Water by
  Purge and Trap Capillary-Column Gas Chromatography/Mass Spec-
  trometry.

## ANALYTICAL METHODS FOR VOLATILE ORGANIC COMPOUNDS

For the VOCs, *Standard Methods* (1995, 19th edition) listed the following approved methods:

"A": Purge and Trap Packed-Column Gas Chromatographic/Mass Spectrometric Method I for: Benzene; Vinyl Chloride; Carbon Tetrachloride; 1,2-Dichloroethane; Trichloroethylene p-Dichlorobenzene; 1,1-Dichloroethylene; 1,1,1-Trichloroethane; Tetrachloroethylene. For these VOCs, also listed is the Purge and Trap Capillary-Column Gas Chromatographic/Mass Spectrometric Method, listed here as "B."

## *STANDARDS METHODS* (1995) OTHER ORGANIC COMPOUNDS

The remaining organic compounds are listed by *Standard Methods* in the following 15 groups. The approved method recommended for each group is listed with a letter for simplicity to identify the approved methods for each group of organic parameters. Each method is identified with a letter from "C" to "V" in the second list.

1. Volatile Aromatic Organic Compounds: "C"; "D"; "E"; "F"
2. Volatile Halocarbons: "G"; "H"; "K"; "E"
3. EDB and DBCP: "J"; "E"; "I"
4. Trihalomethanes and Chlorinated Organic Solvent: "L"; "E"; "I"
5. Disinfection By-Products: "M"
6. Disinfection By-Products Aldehydes: "N" (Proposed)
7. Extractable Base/Neutrals and Acids: "F"
8. Phenols: "L"; "F"
9. Polychlorinatedbiphenyls (PCB): "L"; "F"
10. Polynuclear Aromatic Hydrocarbons: "P"; "F"
11. Carbamate Pesticides: "Q"
12. Organochlorine Pesticides: "R"; "S"; "F"
13. Acidic Herbicide Compounds: "M"
14. Glyphosate Herbicides: "T"
15. Methane: "U"; "V"

Denomination of Advisable Methods—*Standard Methods* (1995)

C. Purge and Trap Gas Chromatographic Method I
D. Purge and Trap Gas Chromatographic Method II
E. Purge and Trap G.C./M.S. Method
F. Liquid-Liquid Extraction G.C./M.S. Method

G.  Purge and Trap Packed-Column G.C. Method I
H.  Purge and Trap Packed-Column G.C. Method II
I.  Purge and Trap Gas Chromatographic Method
J.  Liquid-Liquid Extraction G.C. Method
K.  Purge and Trap Capillary-Column G.C. Method
L.  Liquid-Liquid Extraction G.C. Method
M.  Micro Liquid-Liquid Extraction G.C. Method
N.  PFBHA Liquid-Liquid Extraction G.C. Method
P.  Liquid-Liquid Extraction Chromatographic Method
Q.  High Performance Liquid Chromatographic Method
R.  Liquid-Liquid Extraction G.C. Method I
S.  Liquid-Liquid Extraction G.C. Method II
T.  Liquid Chromatographic Post-Column Fluorescence Method
U.  Combustible-Gas Indicator Method
V.  Volumetric Method

G.C. = Gas Chromatographic
M.S. = Mass Spectrometric

## MONITORING REQUIREMENTS

USEPA (1987) adopted the policy of making the effective dates of require-ments phased in a four-year period. Priority is given to the initial sampling that requires analyses of one sample every three months per source for a year for both surface and groundwater systems. The states have the preroga-tive to reduce or increase frequency of sampling according to the VOCs results as they appear in the first samples.

Repeat monitoring frequency is proposed by USEPA every five years as a minimum for systems not considered "vulnerable" (potential contamina-tion or actual detection of contaminants) leaving to the states the duty to confirm vulnerability every year.

A VOCs detection starts at 0.5 µg/L concentration.

While the final rule became effective as of January 9, 1989 in general, the monitoring requirements became effective on Jan. 1, 1988 for water systems serving populations of more than 10,000 persons, January 1, 1989 for popu-lations between 3,300–10,000, and January 1, 1991 for populations of less than 3,300 persons.

USEPA (1987) also listed the "Unregulated Contaminants" (see Table 4-3). The frequency of sampling for first determination and repeat monitor-ing is the same as for VOCs, leaving again to the states jurisdiction on possible additions or substitutions of contaminants in the lists.

In regulating drinking water quality parameters and compliance the ma-jor Health Authority in the United States is the Environmental Protection

## Table 4-3. Unregulated Contaminants.

*List 1: Monitoring Required for All Systems.*

| | |
|---|---|
| Bromobenzene | 1,1-Dichloroethane |
| Bromodichloromethane | 1,1-Dichloropropene |
| Bromoform | 1,2-Dichloropropane |
| Bromomethane | 1,3-Dichloropropane |
| Chlorobenzene | 1,3-Dichloropropene |
| Chlorodibromomethane | 2,2-Dichloropropane |
| Chloroethane | Ethylbenzene |
| Chloroform | Styrene |
| Chloromethane | 1,1,2-Trichloroethane |
| *o*-Chlorotoluene | 1,1,1,2-Tetrachloroethane |
| *p*-Chlorotoluene | 1,1,2,2-Tetrachloroethane |
| Dibromomethane | Tetrachloroethylene |
| *m*-Dichlorobenzene | 1,2,3-Trichloropropane |
| *o*-Dichlorobenzene | Toluene |
| *trans*-1,2-Dichloroethylene | *p*-Xylene |
| *cis*-1,2-Dichloroethylene | *o*-Xylene |
| Dichloromethane | *m*-Xylene |

*List 2: Required for Vulnerable Systems*

Ethylene dibromide (EDB)
1,2-Dibromo-3-Chloropropane (DBCP)

*List 3: Monitoring Required at the State's Discretion*

| | |
|---|---|
| Bromochloromethane | *n*-Propylbenzene |
| *n*-Butylbenzene | *sec*-Butylbenzene |
| Dichlorodifluoromethane | *tert*-Butylbenzene |
| Fluorotrichloromethane | 1,2,3-Trichlorobenzene |
| Hexachlorobutadiene | 1,2,4-Trichlorobenzene |
| Isopropylbenzene | 1,2,4-Trimethylbenzene |
| *p*-Isopropyltoluene | 1,3,5-Trimethylbenzene |
| Naphthalene | |

Agency. Based on studies and reports submitted by the Safe Drinking Water Committee on behalf of the National Academy of Sciences and other reviewing committees and hearings, USEPA selected the nine compounds in issuing standards for volatile organic chemicals. Beside toxicity and carcinogenicity, the frequency of occurrence of VOCs in drinking water was a major factor in the selection (see Table 4-4).

Table 4-4. Groundwater Supply Survey. The 1975 Preliminary
Survey to Demonstrate the Presence of These Contaminants
in Drinking Water Supplies.*

| | Summary of GWSS Occurrence Data (random sample: n = 466) | | Summary of State Occurrence Data[1] | |
| | Positives | | | |
| Compound | Number | Percent | Number of Samples | Number of Positives |
|---|---|---|---|---|
| Tetrachloroethylene | 34 | 7.3 | 3,636 | 628 |
| Trichloroethylene | 30 | 6.4 | 4,228 | 624 |
| 1,1,1-Trichloroethane | 27 | 5.8 | 3,330 | 715 |
| 1,1-Dichloroethane | 18 | 3.9 | 2,628 | 177 |
| 1,2-Dichloroethylene | 16 | 3.4 | 1,249 | 197 |
| Carbon Tetrachloride | 15 | 3.2 | 2,646 | 368 |
| 1,1-Dichloroethylene | 9 | 1.9 | . . . | . . . |
| p-Dichlorobenzene | 5 | 1.1 | . . . | . . . |
| Vinyl chloride | 1 | 0.2 | 1,793 | 126 |

*Results reported by USEPA, Nov. 13, 1985; 40 CFR Parts 141 and 142, *Fed. Reg.* Vol. 50, No. 219.
[1]The State data, representing a collection of available data from various State agencies, are normally in response to contamination incidents, and are not considered to be statistically representative of national occurrence.

# Benzene
# $C_6H_6$

## Contaminant

A volatile, flammable, colorless liquid with an ethereal odor. Basically a commercial solvent and a product by chemical processes related to petroleum refining, coal tar distillation and coal processing.

## Use

**Benzene** is normally used as a chemical intermediate in the manufacturing of styrene, cyclohexane, detergents, pesticides, synthetic rubber, aviation fuel, dye, paints, pharmaceuticals, gasoline, and photographic chemicals.

## Occurrence in the Environment

In gasoline **Benzene** accounts for 0.8% by volume normally, with some higher concentration occasionally. In ambient air of gasoline stations, benzene varies between 0.001–0.008 mg/L. The highest concentration reported in drinking water was 10 μg/L. Summarizing reported values in drinking water, NAS (1980, Vol. 3) recorded from surveys conducted by the USEPA 0/111, 7/113 (0.4 μg/L), and 4/16 (0.55 μg/L) as positive samples for benzene per number of cities sampled.

## Health Effects

According to USEPA (*Fed. Reg.* Vol. 49, No. 114, June 12, 1984, and Vol. 50, No. 219, Nov. 13, 1985), carcinogenic effects in humans have been documented: myelocytic anemia, thrombocytopenia, and leukemia. Toxic effects include the central nervous system, hematologic and immunologic effects.

A NOAEL of 1 mg/kg was determined in animals, so using an uncertainty factor of 1,000, the ADI resulted in 0.025 mg/L (100% exposure from drinking water, 70-kg adult and 2 liters of water per day and 5/7 correction factor for weekly feeding days). The IARC, NAS, and USEPA classify **Benzene** as a contaminant with sufficient evidence of carcinogenicity in humans.

## Standards

| | |
|---|---|
| USEPA Final MCL | = 0.005 mg/L (July 1987) |
| USEPA RMCL | = zero (Nov. 1985) |
| USEPA proposed MCL | = 0.005 mg/L (Nov. 1985) |
| WHO recommended 1984 | = 0.010 mg/L as a guide (see Notes below) |
| EEC 1980 | = No guideline |

USEPA (1995) did not revise the MCLG = zero and MCL = 0.005 mg/L for **Benzene**, a final rule in effect from Jan. 9, 1989.

## Analytical Methods

**Benzene** can be analyzed as other volatile aromatic and unsaturated organic compounds in water by purge-and-trap gas chromatography (photoionization detector). Also, all regulated VOCs in the above-mentioned regulations can be determined by purge-and-trap gas chromatography/mass spectrometry (GC/MS).

## Notes

Minimum detectable concentration of **Benzene** in water is registered at 0.5 mg/L, with taste detectability at 4.5 mg/L. NAS (1977, Vol. 1) recognized only the possibility of carcinogenicity of benzene in water. Because all data were related to industrial exposure to benzene, more toxicologic studies were recommended prior to the issuance of drinking water standards by USEPA.

The USEPA final rules for public notification include the following language to be used by public water supply system responsible managers:

"*Benzene.* The United States Environmental Protection Agency (EPA) sets drinking water standards and has determined that benzene is a health concern at certain levels of exposure. This chemical is used as a solvent and degreaser of metals. It is also a major component of gasoline. Drinking water contamination generally results from leaking underground gasoline and petroleum tanks or improper waste disposal. This chemical has been associated with significantly increased risks of leukemia among certain industrial workers who were exposed to relatively large amounts of this chemical during their working careers. This chemical has also been shown to cause cancer in laboratory animals when the animals are exposed at high levels over their lifetimes. Chemicals that cause increased risk of cancer among exposed industrial workers and in laboratory animals also may increase the risk of cancer in humans who are exposed at lower levels over long periods of time. EPA has set the enforceable drinking water standard for benzene at 0.005 parts per million (ppm) to reduce the risk of cancer or other adverse health effects which have been observed in humans and laboratory animals. Drinking water which meets this standard is associated with little to none of this risk and should be considered safe."

The WHO (1984) issued a guideline value for benzene of 10 $\mu$g/L under the organic constituents of health significance. This value was computed "from a conservative hypothetical mathematical model." The WHO recognized that drinking water concentrations of benzene did not exceed 1 $\mu$g/L and all the studies involved inhalation exposure. Because benzene tends to affect the entire organism (systemic), WHO guidelines were set on data for leukemia applied to a linear multistage extrapolation model.

Total uptake per person of benzene in urban environments has been estimated at about 125 mg/year; food is considered to have contributed with 90 mg/year. Normal water concentration of less than 1 $\mu$g/L would contribute to less than 1 mg/year (1 $\mu$g/L $\times$ 2 L/day $\times$ 365 days = 0.730 mg/year).

**Vinyl Chloride
or Chloro Ethylene
or Monocholoroethylene
$CH_2 = CHCl$**

*Contaminant*

Commercially synthesized by the halogenation of ethylene; slightly soluble in water (less than 0.11% by weight). CAS #75.01-4.

*Use*

The major use in the U.S. is in the production of polyvinyl chloride (PVC) resins for the building and construction industry (pipes, conduit, floor covering).

USEPA banned the sale of propellents and all the aerosols containing vinyl chloride monomer due to human and animal carcinogenicity. A distribution system constructed with PVC pipes showed concentrations of 1.4 µg/L.

*Occurrence in the Environment*

USEPA in 1975 tested the finished water of 10 cities registering a concentration of 5.6 µg/L in Miami and 0.27 µg/L in Philadelphia. States recorded 2 positives out of 646 samples collected with a max. = 17 ppb.

*Health Effects*

Industrial exposure to this contaminant in workers confirmed hepatic angiosarcoma when liquified vinyl chloride under pressure was handled, probably in concentrations from 1,000 to several thousand ppm (NAS, 1977). Lesions of skin, bones, liver, spleen, and lungs were reported after chronic exposure. NAS (1983, Vol. 5) recorded the no-observed-effect-level at less than 1.7 mg/L. Mutagenicity and teratogenicity studies did not produce conclusive results (mutagen in vitro), but carcinogenicity in animals was proved by inhalation (NAS, 1977 and 1983).

Studies on rats through ingestion of **Vinyl Chloride** were used by NAS (1977) to determine risk estimates (see *Methods* *of Evaluation* in this chapter and Chapter 7, pages 173 and 357). Assuming a concentration of 1 µg/L the estimated risk for man is $3.0 \times 10^{-7}$ Q with an upper 95% confidence estimate at the same concentration of $4.7 \times 10^{-7}$ Q; Q = compound concentration in ppb for a lifetime consumption of 1 L/day.

The USEPA (1984) summarized the toxicologic effects from high dosage exposure in congestion and edema of the lungs and hyperemia of the kidney and liver with minor effects on the central nervous system, pulmonary insufficiency, cardiovascular manifestation, and gastrointestinal symptoms.

In spite of the established carcinogenicity, an AADI was calculated in 0.06 mg/L from animal studies (rats); an uncertainty factor of 1,000 was applied with 100% exposure from drinking water (70-kg adult and 2 L/day consumption).

For Cancer Risk Estimated for VOCs, see Table 4-1 showing the USEPA proposed rule. The IARC, NAS (1983), and USEPA confirmed the carcinogenicity of vinyl chloride to humans.

### Standards

| | |
|---|---|
| WHO (1984) | = No guideline (but potential health significance) |
| USEPA (1985) RMCL | = zero |
| USEPA (1987) Final MCL | = 0.002 mg/L |

USEPA (1995) did not revise the MCLG = zero and MCL = 0.002 mg/L for **Vinyl Chloride** (Chloro Ethylene or Monochloroethylene), the final rule in effect since Jan. 9, 1989.

### Analytical Methods

The purge-and-trap gas chromatography method (electrolytic conductivity detector) is the recommended analytical method.

### Notes

Due to registered evidence on toxicity and carcinogenicity (confirmed however by inhalation only in humans), and the existing ban by USEPA since 1974, **Vinyl Chloride** has been properly defined as far as drinking water standards are concerned. The continued interest remains in monitoring the finished water as prescribed by current regulations.

USEPA's final rules for public notification include the following terminology to be used by public water supply system managers:

"*Vinyl chloride.* The United States Environmental Protection Agency (EPA) sets drinking water standards and has determined that vinyl chloride is a health concern at certain levels of exposure. This chemical is used in industry and is found in drinking water as a result of the breakdown of related solvents. The solvents are used as cleaners and degreasers of metals and generally get into drinking water by improper waste disposal. This chemical has been associated with significantly increased risks of cancer among certain industrial workers who were exposed to relatively large amounts of this chemical during their working careers. This chemical has also been shown to cause cancer in laboratory animals when the animals are exposed at high levels over their lifetimes. Chemicals that cause increased risk of cancer among exposed industrial workers and in laboratory animals also may increase the risk of cancer in humans who are exposed at lower levels over long periods of time. EPA has set the enforceable drinking water standard for vinyl chloride at 0.002 part per million (ppm) to reduce the risk of cancer or other adverse health effects which have been observed in humans and laboratory animals. Drinking water which meets this standard is associated with little to none of this risk and should be considered safe."

# Carbon Tetrachloride
# or Tetrachloromethane
# $CCl_4$

## Contaminant

Classified as persistent in the environment, but insoluble in water. CAS #56-23-5.

## Use

Its major use is the manufacture of chlorofluoromethanes with secondary use in fire extinguishers, solvents, cleaning agents, and in grain fumigants.

## Occurrence in the Environment

Results of national and state surveys can be summarized as follows:

|  | Max. | Median (µg/L) |
| --- | --- | --- |
| GWSS (random) 3.2% | 16 | 0.4 |
| GWSS (nonrandom) 3.1% | 15 | 0.5 |

STATES' DATA *368 positives* out of 2,646 sampled.
COMBINED NATIONAL DATA: concentration range between 0.5 and 30 µg/L.

## Health Effects

In humans, acute exposure caused renal insufficiency or severe renal damage. Chronic exposure registered gastrointestinal upset and nervous system symptoms (drowsiness, fatigue). NAS (1977) computed the risk estimate between 4.5–5.4 $\times$ $10^{-8}$ Q using Q = 5 µg/L as a compound concentration in ppb for a lifetime consumption of 1 L/day; for 2 L per day the value is 5 $\times$ $10^{-7}$ to produce one excess case of cancer for every 2 million persons lifetime or 0.014 per year.

The NAS (1977) calculated that mutagenicity and teratogenicity were not established* but carcinogenicity in animals

---

*One microbial mutagenicity assay was reported in NAS—1983. (Vol. 5).

produced hepatomas (malignant tumor of the liver) in mice. The NAS (1980, Vol. 3) calculated a SNARL of **one-day** in 14 mg/L with 100% exposure from 2 L of water intake and 1,000 safety factor and animal data of 0.25 mL/kg (0.4 g). **Seven-day** was reduced to 2 mg/L mathematically. **Chronic exposure** was not calculated because of the carcinogenicity of **Carbon Tetrachloride.**

USEPA (1984) calculated an AADI of 0.025 mg/L. IARC, NAS, and NCI calculated that there is sufficient evidence of carcinogenicity identified in **Carbon Tetrachloride.** The USEPA Carcinogen Assessment Group (CAG) calculated risk at $10^{-6}$; 4 μg/L (see also VOCs, Table 4-1).

### Standards

USEPA (1985) RMCL = zero
USEPA proposed MCL = 5 μg/L
USEPA Final (1987) MCL = 0.005 mg/L
WHO guideline = 3 μg/L (tentative)
EEC = no standard or guideline
USEPA classification: Group B2 (sufficient evidence of carcinogenicity in animals and inadequate evidence in humans)

USEPA (1995) did not revise the MCLG = zero and the MCL = 0.005 mg/L for **Carbon Tetrachloride** (Tetrachloromethane); these final rules have been in effect since Jan. 9, 1989.

### Analytical Methods

Purge-and-trap gas chromatography (electrolytic conductivity detector) is the acceptable method for determination of **Carbon Tetrachloride.**\*

### Notes

See also the summary tables introducing VOCs and *Methods of Evaluation* section, this Chapter.

USEPA issued final rules for public notification (40 Parts 141, 142, and 143—Oct. 28, 1987—*Fed. Reg.* Vol. 52, No. 208). The owner or operator of a public water system is expected

---

\*The 1988 supplement to the 16th edition of *Standard Methods* details the **purgeable organics in water** under the Purge-and-Trap Packed-Column Gas Chromatography/Mass Spectrometry Method and the Purge-and-Trap Capillary-Column GS/MS Method.

to include the following language specified for **Carbon Tetrachloride:**

"The United States Environmental Protection Agency (EPA) sets drinking water standards and has determined that carbon tetrachloride is a health concern at certain levels of exposure. This chemical was once a popular household cleaning fluid. It generally gets into drinking water by improper waste disposal. This chemical has been shown to cause cancer in laboratory animals such as rats and mice when the animals are exposed at high levels over their lifetimes. Chemicals that cause cancer in laboratory animals also may increase the risk of cancer in humans who are exposed at lower levels over long periods of time. EPA has set the enforceable drinking water standard for carbon tetrachloride at 0.005 parts per million (ppm) to reduce the risk of cancer or other adverse health effects which have been observed in laboratory animals. Drinking water which meets this standard is associated with little to none of this risk and should be considered safe."

PVC packaging material may yield concentrations up to 14.8 mg/kg in food, consequently, rules have been issued in many countries to limit the vinyl chloride monomer (WHO, 1984).

Air concentrations up to 8.8 mg/mc have been detected near manufacturing plants and 1–3 mg/mc in the air in the interior of new automobiles.

## 1,2-Dichloroethane
## or
## Ethylene Dichloride
## $C_2H_4Cl_2$ ($ClCH_2$-$CH_2Cl$)

### Contaminant

An industrial solvent with a pleasant odor and sweet taste that degrades biologically with difficulty; soluble in water in ratio of 1:20. CAS #107-106-2.

### Use

Its application is in manufacturing of vinyl chloride and tetra-ethyl lead; also used as a component of insecticide, fumigant, paint, varnish, tobacco, and in chemical synthesis.

### Occurrence in the Environment

USEPA water supply surveys of 80 systems in 1974 resulted in 26 positive systems with concentrations varying from 0.2–6 µg/L. A concentration was recorded at 8 µg/L in New Orleans finished water.

### Health Effects

Toxic effects in man were observed and summarized by the NAS (1977) as exposure to high vapor concentration producing eye, nose, throat irritation, followed by central nervous system depression and injuries to the liver, kidneys, and cardiovascular system. Weakly mutagenic with no data on carcinogenicity or teratogenicity. The NAS (1977) concluded that data for standard setting were not available. The NAS (1980) recognized that toxicological effects had been reported due to inhalation or injestion, but practically all studies on **Dichloroethane** were based on inhalation. No 1-Day or 7-Day SNARL could have been established. Also the NAS stated that chronic exposure to **Dichloroethane** is potentially hazardous. From animal studies by inhalation could be derived a 100 ppm of dichloroethane as a NOAEL, but extrapolation to a chronic ingestion is not scientifically a safe criterion.

NAS (1983) suggested mutagenicity in microbial mutagenicity assays; no teratogenicity was recorded, but carcinogenic properties were demonstrated in limited studies in animals. USEPA (1984) calculated an AADI = 0.260 mg/L based on a NOAEL of 405 mg/mc (100 ppm) under the assumption of 100% exposure from drinking water, 2 L/day intake, a 70-kg adult, and 1,000 used as a factor of safety.

IARC concluded that there is sufficient evidence of carcinogenicity in animals. USEPA's CAG and NCI bioassay data provided the regulatory agency to conclude a risk estimate of 5 µg/L and $10^{-5}$ projected upper limit (see VOCs Cancer Risk Estimate, Table 4-1, page 175).

## Standards

| | |
|---|---|
| USEPA (1985) RMCL | = zero |
| USEPA (1985) MCL | = 0.005 mg/L |
| USEPA (1987) MCL | = 0.005 mg/L Final |
| WHO (1984) recommended | = 10 µg/L |
| IARC | = Group B2 = sufficient evidence of carcinogenicity in animals. |
| USEPA | = sufficient evidence of carcinogenicity in animals and inadequate evidence in humans. |

USEPA (1995) did not revise the MCLG = zero and the MCL = 0.005 mg/L for **1,2-Dichloroethane** (Ethylene Dichloride); these final rules have been in effect since Jan. 9, 1989.

## Analytical Methods

The acceptable method is the purge-and-trap gas chromatography (electrolytic conductivity detector) used for most volatile halogenated organic compounds in water. (See the *Standard Methods* 1988 Supplement—"Purgeable Organics in Water" and/or the 1995 edition.)

## Notes

For better evaluation of carcinogenicity and toxicity, refer to the introduction and summary tables in the VOC section, page 173.

USEPA issued final rules for public notification (40 Parts 141, 142, and 143—Oct. 28, 1987—*Fed. Reg.* Vol. 52, No. 208).

The owner or operator of a public water system is expected to include the following language specified for **1,2-Dichloroethane**:

"The United States Environmental Protection Agency (EPA) sets drinking water standards and has determined that 1,2-dichloroethane is a health concern at certain levels of exposure. This chemical is used as a cleaning fluid for fats, oils, waxes, and resins. It generally gets into drinking water from improper waste disposal. This chemical has been shown to cause cancer in laboratory animals such as rats and mice when the animals are exposed at high levels over their lifetimes. Chemicals that cause cancer in laboratory animals also may increase the risk of cancer in humans who are exposed at lower levels over long periods of time. EPA has set the enforceable drinking water standard for 1,2-dichloroethane at 0.005 parts per million (ppm) to reduce the risk of cancer or other adverse health effects which have been observed in laboratory animals. Drinking water which meets this standard is associated with little to none of this risk and should be considered safe."

The threshold odor limit is 2 mg/L (WHO, 1984). Food intake is negligible (less than 1% of total intake water intake more than 99%; WHO, 1984).

# Trichloroethylene
## $ClHC = CCl_2$

## Contaminant

A nonflammable liquid used as a solvent and in solvent extraction, practically insoluble in water. Other names include: *Trichloroethene; Ethinyl Trichloride; Tri-clene; and Trichloroethene*. CAS #79-01-6.

## Use

Used in dry-cleaning operations, in organic synthesis, and in refrigerants and fumigants; and as a degreasing solvent in metal industries.

## Occurrence in the Environment

Besides an industrial, pollutional entry, **Trichloroethylene** can be formed by chlorination of water. USEPA (1975) in a survey of 10 cities recorded positive results in 5 supplies with a range of 0.1–0.5 µg/L. Some higher concentrations were reported in contaminated groundwater (100 µg/L).

## Health Effects

Since **Trichloroethylene** can be used as an anesthetic, exposure to this contaminant results in central nervous system depression and unconsciousness. NAS (1977) concluded that **Trichloroethylene** has low acute toxicity. No teratogenicity was reported, and limited carcinogenicity was demonstrated in animals (heptocellular carcinoma in one strain of mice). The NAS in 1983 reported evidence of mutagenicity in only one microbial mutagenicity assay.

Carcinogenic Risk Estimates were reported by the NAS (1977) as $0.36$–$1.1 \times 10^{-7}$ Q. The upper 95% confidence estimate of risk is respectively $0.55$–$1.6 \times 10^{-7}$ Q. The NAS (1983) estimated $3.3 \times 10^{-7}$ Q (upper 95% confidence estimate).

Because of limited studies and the difficulty of relating scientifically animal studies to humans, such estimates are really indicative rather than mathematical factors of extrapolation. NAS (1983) confirmed these calculations restating that they are based on "limited evidence." In 1980 the NAS reported several additional studies on **Trichloroethylene** but insufficient to evaluate toxicity, mutagenesis, and carcinogenesis for chronic exposure. A 1-DAY SNARL of 105 mg/L, and a 7-DAY SNARL of 15 mg/L were calculated.

USEPA (1984) calculated an AADI of 0.257 mg/L, based upon a minimal effect level of 300 mg/mc or 55 ppm with an uncertainty factor of 1,000 (with 100% exposure from water, 70-kg adult, 2 L/day consumption).

IARC concluded that **Trichloroethylene** shows limited evidence of carcinogenicity. The USEPA Carcinogen Assessment Group (CAG) used the linearized nonthreshold multistage model to calculate projected excess cancer risk estimates based on NCI bioassay data (18 $\mu$g/L $\times$ 10$^{-5}$).

### Standards

USEPA (1985) RMCL = zero
USEPA (1985) MCL  = 5 $\mu$g/L
USEPA (1987) MCL  = 0.005 mg/L Final
WHO guideline      = 30 $\mu$g/L (tentative)

Carcinogenic classification: USEPA (Risk Assessment Forum) = B2—sufficient animal evidence of carcinogenicity and inadequate human evidence.

IARC's classification: 3—chemical cannot be classified as to carcinogenicity to humans.

USEPA (1995) did not revise the MCLG = zero and the MCL = 0.005 mg/L for **Trichloroethylene** that have been in effect since Jan. 9, 1989.

### Analytical Method

The purge-and-trap gas chromatography method (electrolytic conductivity detector) is the recommended method.

### Notes

USEPA (1985) ranked **Trichloroethylene** in category I ("known or probable human carcinogens: strong evidence of carcinogenicity"), admitting, however, that in this case the evidence is weaker than for other chemicals included in that category.

USEPA issued final rules for public notification (40 Parts 141, 142, and 143—Oct. 28, 1987—*Fed. Reg.* Vol. 52, No. 208). The owner or operator of a public water system is expected to include the following language specified for **Trichloroethylene**:

"The United States Environmental Protection Agency (EPA) sets drinking water standards and has determined that trichloroethylene is a health concern at certain levels of exposure. This chemical is a common metal cleaning and dry cleaning fluid. It generally gets into drinking water by improper waste disposal. This chemical has been shown to cause cancer in laboratory animals such as rats and mice when the animals are exposed at high levels over their lifetimes. Chemicals that cause cancer in laboratory animals also may increase the risk of cancer in humans who are exposed at lower levels over long periods of time. EPA has set forth the enforceable drinking water standard for trichloroethylene at 0.005 parts per million (ppm) to reduce the risk of cancer or other adverse health effects which have been observed in laboratory animals. Drinking water which meets this standard is associated with little to none of this risk and should be considered safe."

In the southeastern section of the New York City borough of Queens, where groundwater is distributed to half a million consumers, concentrations of trichloroethylene from 1–15 µg/L were recorded between 1979–1986 from samples collected from the distribution system. Well water showed positive concentrations in 18% of the wells. The water supply system was then operated by a private company (franchised), the Jamaica Water Supply Co., which used groundwater only in that section of Queens and in the western part of Nassau County, NY. During the following years this company eliminated some wells and provided treatment for the removal of contaminants. As of May 29, 1996, the Jamaica Water Supply Co. franchise was removed and now the operation is provided by the New York City Department of Environmental Protection (DEP), which provides the surface water supply to the rest of the city. DEP is presently operating the wells but is planning to supply this section of Queens with surface water in the future.

The WHO (1984) listed trichloroethylene under **Trichloroethene** (1,1,2-trichloroethene or TCE). The WHO recognized the negligible importance of TCE in food, while inhalation exposure was considered as confined to a small industrial population.

A NOMS 1976 survey reported positive results in 14% of cities tested with an approximate average of 13 µg/L among positives.

## Tetrachloroethylene
$$Cl_2C{=}CCl_2$$

### Contaminant

An organic solvent insoluble in water. Other names include: PCE and Tetrachloroethene. CAS #127-18-4.

### Use

It is used as a solvent, heat-transfer medium, and in the manu-facture of fluorocarbons. As a solvent, PCE is particularly use-ful in the dry-cleaning industries.

### Occurrence in the Environment

USEPA surveyed 10 water utilities, recording positive results in 8 varying from 0.07–0.46 µg/L. Finished water in New Or-leans contained 5 µg/L. **Tetrachloroethylene** is also generated in small quantities during the water treatment process of chlorination. Atmospheric concentrations exceeding 200 ppm can lead to narcosis and death (NAS, 1983). USEPA drinking water surveys were conducted in 1984. **Tetrachloroethylene** is widespread in the environment (WHO, 1984).

### Health Effects

NAS (1977) concluded that no data were available to classify **tetrachloroethylene** for standard settings. The NAS (1980) reviewed additional studies: a SNARL for 1-Day exposure was derived at 172 mg/L, using a factor of safety of 100, a 70-kg man, a 2 L/day intake, and a "minimum toxic dose" of 0.3 mL/kg (0.49 g/kg) based on the tissues of rats; similarly a 7-Day exposure of 24.5 mg/L was calculated. The NAS (1983) also reported studies on animals with no toxic effects at rela-tively high oral toxic doses.

USEPA (1984), interpreting high-dose noncarcinogenic ef-fects from acute or long-term exposure to **Tetrachlo-roethylene**, centered the target organs as the central nervous system, depression and fatty infiltration of the liver and kid-ney, with concomitant changes in serum enzyme activity lev-els indicative of tissue damage.

An AADI = 0.085 mg/L was derived from an animal study where NOAEL = 10 mg/mc (15 ppm) was recorded. The factor of safety used was 100.

CAG calculated the cancer risk estimate at 10 $\mu$g/L and $10^{-6}$ upper limit. NAS (1980) estimated the lifetime risk at $6.2 \times 10^{-8}$ with upper 95% confidence resulting in $1.4 \times 10^{-7}$. The NAS (1983) reported insufficient data were available for risk estimates.

## Standards

| | |
|---|---|
| USEPA (1985) RMCL | = not promulgated |
| USEPA (1987) Final MCL | = not promulgated (but on the list of 83 contaminants to be regulated by the SDWA, 1986). |
| WHO guideline (1984) | = 10 $\mu$g/L (tentative) |
| IARC | = Group 3 (chemical cannot be classified as to carcinogenicity to humans) |
| USEPA | = B2 guideline (carcinogenicity—similar to group 3) |
| USEPA (1989) MCLG | = zero |
| MCL | = 5 $\mu$g/L |

USEPA (1995) did not revise the MCLG = zero and the MCL = 0.005 mg/L for **tetrachloroethylene**; these final rules have been in effect from July 30, 1992.

## Analytical Methods

The purge-and-trap gas chromatography method (USEPA Method 502.1) is the recommended method.

## Notes

In consideration that MCLs will not be issued by USEPA in the immediate future, it is advisable to use the WHO guideline of 10 $\mu$g/L.

The WHO (1984) listed tetrachloroethylene as **Tetrachloroethene** (1,1,2,2-tetrachloroethene) also known as *perchloroethylene* or PCE; the WHO reported significant concentration of PCE in food in England (fish and butter); air concentrations were reported generally low.

The NOMS (1976 survey) reported positive values in less than 10% of drinking water samples with a range of 0.2–3.1 $\mu$g/L. In Switzerland contaminated groundwater registered high values of 1 mg/L.

# 1,1-Dichloroethylene
## Vinylidene Chloride
## 1,1-DCE
$$CH_2{=}CCl_2$$

## Contaminant

A volatile liquid (boiling point of 31.7°C), practically insoluble in water. Other name: 1,1-Dichloroethene (WHO list). CAS #75-35-4.

## Use

Used as an intermediate in the synthesis of copolymers (food packaging films and coatings). One of the three isomers of DCE and the most used in the chemical industry.

## Occurrence in the Environment

Positive samples in drinking water were reported in 197 cases out of 1,245 State occurrence sampling data and a 3.4% occurrence in the GWSS federal survey. Concentrations in public drinking water systems range from 0.2–6.3 µg/L (USEPA,1985) but usually less than 1 µg/L.

## Health Effects

Exposure to a high dose of **Dichloroethylene** has caused liver and kidney injury to animals. Hepatoxicity is the most commonly reported toxic effect.

Mutagenicity, teratogenicity, and carcinogenicity have shown some indication of positive results but insufficiently to categorize **Dichloroethylene** as carcinogenic to humans.

CAG used the linearized, no-threshold, multistage model to calculate projected excess cancer estimate extrapolated from high dose animal studies ($Q = 2.4$ µg/L; $10^{-5}$). Using a NOAEL of 10 mg/kg and 100% exposure from drinking water, an AADI = 350 µg/L was derived by USEPA (1984) for **1,1-Dichloroethylene**.

NAS (1983) derived a SNARL of 0.1 mg/L (NTP bioassay in rats and mice, 1982) resulting from a NOAEL of 2 mg/kg, uncertainty factor of 100, 20% exposure to drinking water for a 70-kg man, 2 L/day intake, and 5-day/week related to 7 days exposure:

$$\frac{2 \text{ mg/kg} \times 70 \text{ kg} \times 0.2}{100 \times 2 \text{ L}} \times \frac{5}{7} = 0.10 \text{ mg/L}$$

USEPA (1984) held in abeyance the classification of **Dichloroethylene** as far as carcinogenicity is concerned. USEPA (1985) confirmed liver and kidney damage and reported central nervous system depression and sensitization of the heart.

### Standards

| | |
|---|---|
| USEPA (1985) (as a carcinogen) | = Group C* |
| IARC | = Group 3* |
| WHO (1984) guideline | = 0.3 μg/L† |
| USEPA (1985) RMCL | = 7 μg/L |
| USEPA (1987) Final MCL | = 0.007 mg/L |

USEPA (1995) did not revise the MCLG = 0.007 mg/L and MCL = 0.007 mg/L for **1,1-Dichloroethylene** (vinylidene chloride); these final rules have been in effect since Jan. 9, 1989.

### Analytical Methods

The acceptable method is the purge-and-trap gas chromatography method (electrolytic conductivity detector).

### Notes

The threshold limit value is 10 ppm (40 mg/mc [NAS-1983]). No definite studies in humans have been reported. SNARL or AADI and consequent standards are based on carcinogenic effects not considered.

USEPA issued final rules for public notification (40 Parts 141, 142, and 143—Oct. 28, 1987—*Fed. Reg.* Vol. 52, No. 208). The owner or operator of a public water system is expected to include the following language specified for **1,1,-Dichloroethylene**:

"The United States Environmental Protection Agency (EPA) sets drinking water standards and has determined

---

*Equivocal evidence of carcinogenicity.
†WHO used a low-level guideline based on the assumption of 1,1-DCE potential carcinogenicity.

that 1,1-dichloroethylene is a health concern at certain levels of exposure. This chemical is used in industry and is found in drinking water as a result of the breakdown of related solvents. The solvents are used as cleaners and degreasers of metals and generally get into drinking water by improper waste disposal. This chemical has been shown to cause liver and kidney damage in laboratory animals such as rats and mice when the animals are exposed at high levels over their lifetimes. Chemicals which cause adverse effects in laboratory animals also may cause adverse health effects in humans who are exposed at lower levels over long periods of time. EPA has set the enforceable drinking water standard for 1,1-dichloroethylene at 0.007 parts per million (ppm) to reduce the risk of these adverse health effects which have been observed in laboratory animals. Drinking water which meets this standard is associated with little to none of this risk and should be considered safe."

# 1,1,1-Trichloroethane
## or Methyl Chloroform
## or Chlorotene
## Cl₃C—CH₃

*Contaminant*

Nonflammable liquid, insoluble in water. CAS #71-55-6.

*Use*

Used as a solvent for fats, waxes, and resins.

*Occurrence in the Environment*

Inhalation through occupational exposure is the major environmental concern. The USEPA Groundwater Supply Survey of 1975 resulted in 5.8% of positives from 466 random samples with a median of 0.8 µg/L and a max. of 18 µg/L.

*Health Effects*

Most of the research studies conducted on animals were related to inhalation and on that basis the NAS (1980 and 1984) concluded that it was difficult to make judgments concerning the health effects following oral exposure. Carcinogenicity studies were negative, but limited evidence of **1,1,1-Trichloroethane** was recorded for carcinogenicity in mice but not in rats, leaving at least temporarily some suspicion of carcinogenicity in humans (NAS, 1984). Questionable data were also obtained for mutagenicity (NAS, 1984 and USEPA, 1985).

USEPA (1984) summarized dose exposure in humans and animals as depression of the central nervous system, increase in liver weight, and cardiovascular changes.

An AADI of 1.0 mg/L was calculated in an inhalation study exposing mice to **1,1,1-Trichloroethane**. USEPA (1985) classified this contaminant in category III (inadequate evidence of carcinogenicity in animals).

### Standards

```
USEPA (1985) RMCL     = 200 µg/L
USEPA (1987) Final MCL = 200 µg/L
WHO and EEC           = no standards
IARC                  = Group 3 (inadequate evidence
                        of carcinogenicity in animals)
USEPA (1985)          = Group D (inadequate animal
                        evidence)
```

USEPA (1995) did not revise the MCLG = 0.2 mg/L and MCL = 0.2 mg/L of **1,1,1-Trichloroethane** (methyl chloroform or chlorotene); these final rules have been in effect since Jan. 9, 1989.

### Analytical Methods

The purge-and-trap gas chromatography method (electrolytic conductivity detector) is applicable for 1,1,1-**Trichloroethane** detection in water, as well as for other volatile halogenated organic compounds.

### Notes

USEPA (1987) reported a taste and odor threshold for **1,1,1-Trichloroethane** at 1 mg/L.

USEPA issued final rules for public notification (40 Parts 141, 142, and 143—Oct. 28, 1987—*Fed. Reg.* Vol. 52, No. 208). The owner or operator of a public water system is expected to include the following language specified for **1,1,1-Trichloroethane:**

"The United States Environmental Protection Agency (EPA) sets drinking water standards and has determined that 1,1,1-trichloroethane is a health concern at certain levels of exposure. This chemical is used as a cleaner and degreaser of metals. It generally gets into drinking water by improper waste disposal. This chemical has been shown to damage the liver, nervous system, and circulatory system of laboratory animals such as rats and mice when the animals are exposed at high levels over their lifetimes. Some industrial workers who were exposed to relatively large amounts of this chemical during their working careers also suffered damage to the liver, nervous system, and circulatory system. Chemicals which cause adverse effects among exposed industrial workers and in laboratory animals also may cause adverse health effects in humans who are exposed at lower levels over long periods of time. EPA has set the enforceable drinking water standard for 1,1,1-trichloroethane at 0.2 parts per million (ppm) to protect against the risk of these adverse health effects which have been observed in humans and laboratory animals. Drinking water which meets this standard is associated with little to none of this risk and should be considered safe."

# p-Dichlorobenzene

## Contaminant

Also called PDB, it is produced commercially by chlorination of benzene or chlorobenzene at high temperature and in the presence of catalysts. Soluble in water at 70 mg/L at 25°C (max.). CAS #106-46-7.

## Use

Used as insecticidal fumigant but mostly used as a space deodorant and sanitizer, and for moth larvae and insect control; also a solvent in the chemical industry.

## Occurrence in the Environment

The 1975 Groundwater Supply Survey (GWSS) recorded for PDB a percentage of positives of 1.1 with a median of 0.7 µg/L and a max. of 1.3 µg/L from random sampling. From the nonrandom sampling, GWSS recorded 0.8% of positives, with a median of 0.7 µg/L and a max. of 0.9 µg/L.

## Health Effects

NAS in 1977 (*Drinking Water and Health*, Vol. 1) calculated an ADI of 0.0134 mg/kg/day with a NOAEL = 13.4 mg/kg/day in rats but reported a serious lack of data in toxicity, teratogenicity, mutagenicity, and carcinogenicity in spite of extensive use of PDB since 1915.

The NAS (1983, Vol. 5) summarized the most recent evidence on **p-Dichlorobenzene** confirming the previous calculation of a suggested no-adverse-response-level (SNARL) for chronic exposure at 0.094 mg/L. More studies were recommended for acute and subchronic toxicity from oral exposure observing preliminary responses of potential epatic toxicity.

USEPA (1984) reported human response to high concentration of **p-Dichlorobenzene** resulting in anorexia, nausea, yellow atrophy of the liver, and blood dyscrasias.

An AADI of 3.75 mg/L for PDB was calculated following an animal study on rats from a NOAEL of 150 mg/kg/day with a total factor of safety of 1,000.

## Standards

USEPA (1987) Final MCL = 0.075 mg/L

WHO and EEC = no standards

IARC = inadequate evidence of carcinogenicity. Same criteria were adopted by USEPA pending finalization of ongoing studies on carcinogenicity.

USEPA (1989) SMCL = 0.005 µg/L (PDB)

USEPA (1989) SMCL = 0.01 mg/L (*p*-Dichlorobenzene)

USEPA (1995) MCLG = 0.075 mg/L and MCL = 0.075 mg/L for **p-Dichlorobenzene**, indicating possible cancer as health effects.

## Analytical Methods

The purge-and-trap gas chromatography method (electrolytic conductivity detector) is applicable for *p*-dichlorobenzene detection in water, as used for other volatile halogenated organic compounds.

## Notes

The threshold for DCB is in the range of 0.01 mg/L. USEPA final rules for public notification include the following language for **para-Dichlorobenzene**:

"The United States Environmental Protection Agency (EPA) sets drinking water standards and has determined that para-dichlorobenzene is a health concern at certain levels of exposure. This chemical is a component of deodorizers, moth balls, and pesticides. It generally gets into the drinking water by improper waste disposal. This chemical has been shown to cause liver and kidney damage in laboratory animals such as rats and mice when the animals are exposed to high levels over their lifetimes.

Chemicals which cause adverse effects in laboratory animals also may cause adverse health effects in humans who are exposed at lower levels over long period of time. The EPA has set the enforceable drinking water standard for para-dichlorobenzene at 0.075 parts per million (ppm) to reduce the risk of these adverse health effects which have been observed in laboratory animals. Drinking water which meets this standard is associated with little to none of this risk and should be considered safe."

The WHO (1984) reviewed the drinking water quality of dichlorobenzenes in general and decided not to set health-related guideline values, but in consideration of an odor threshold concentration between 0.1–3.0 µg/L, a guideline value could be adopted based on "aesthetic quality." Such value would vary for each isomer of the 10 contaminants considered.

## VOLATILE ORGANIC CHEMICALS SUMMARY

USEPA regulated nine contaminants, finalizing the enforceable standards on July 8, 1987. Maximum contaminant levels are summarized in Table 4-2.
USEPA (1992) issued final rules for the following contaminants listed as Volatile Organics (solvents):

| Contaminant | MCLG (mg/L) | MCL (mg/L) |
|---|---|---|
| cis-1,2-Dichloroethylene* | 0.07 | 0.07* |
| 1,2-Dichloroproprane | zero | 0.005 |
| Ethylbenzene* | 0.7 | 0.7 |
| Monochlorobenzene* | 0.1 | 0.1 |
| o-Dichlorobenzene* | 0.6 | 0.6 |
| Styrene* | 0.1 | 0.1 |
| Tetrachloroethylene* | zero | 0.005 |
| Toluene* | 1 | 1 |
| trans-1,2-Dichloroethylene | 0.1 | 0.1 |
| Xylenes (total)* | 10 | 10 |

### Removal of VOCs

The Best Available Technology (BAT) for the removal of the ten regulated contaminants (VOCs) is specified by USEPA, listing granular activated carbon and packed-tower aeration treatments with the exception of vinyl chloride for which only packed-tower aeration is listed.

# Synthetic Organic Chemicals (SOCs)

This distinction between synthetic and volatile chemicals is maintained because the American standards, as set by the USEPA and the SDWA, use this distinction in standard promulgation, sampling programs, and monitoring requirements.

Chemically speaking, both volatile organic compounds and the synthetic organic compounds listed are synthetic organic chemicals.

The following Table 4-5 lists the USEPA's final rules for SOCs. Final rules must be reviewed every three years by USEPA, according to the Safe Drinking Water Act as modified in 1986.

*These contaminants are described in this chapter under *Synthetic Organic Chemicals (SOCs)*.

### Table 4-5. USEPA Maximum Contaminant Level (MCL) and Maximum Contaminant Level Goal (MCLG). Final Rule for Synthetic Organic Chemicals.

| | MCLG mg/L | MCL mg/L |
|---|---|---|
| Acrylamide | zero | TT* |
| Alachlor | zero | 0.002 |
| Aldicarb | 0.001 | 0.003 Delayed |
| Aldicarb sulfone | 0.001 | 0.002 Delayed |
| Aldicarb sulfoxide | 0.001 | 0.004 Delayed |
| Atrazine | 0.003 | 0.003 |
| Carbofuran | 0.04 | 0.004 |
| Chlordane | zero | 0.002 |
| 2,4-D | 0.07 | 0.07 |
| Dibromochloropropane (DBCP) | zero | 0.0002 |
| cis-1,2-Dichloroethylene | 0.07 | 0.07 |
| trans-1,2-Dichloroethylene | 0.1 | 0.1 |
| 1,2-Dichloropropane | zero | 0.005 |
| Dinoseb | 0.007 | 0.007 |
| Endrin | 0.002 | 0.002 |
| Epichlorohydrin | zero | TT |
| Ethylbenzene | 0.7 | 0.7 |
| Ethylene dibromide (EDB) | zero | 0.00005 |
| Heptachlor | zero | 0.0004 |
| Heptachlorepoxide | zero | 0.0002 |
| Lindane | 0.0002 | 0.0002 |
| Methoxychlor | 0.04 | 0.04 |
| Monochlorobenzene | 0.1 | 0.1 |
| Pentachlorophenol | zero | 0.001 |
| Polychlorinate biphenyls (PCBs) | zero | 0.0005 |
| Simazine | 0.004 | 0.004 |
| Styrene | 0.1 | 0.1 |
| 2,3,7,8-TCDD (Dioxin) | zero | $3 \times 10^{-8}$ |
| Toluene | 1 | 1 |
| Toxaphene | zero | 0.005 |
| 2,4,5-TP (Silvex) | 0.05 | 0.05 |
| 1,2,4-Trichlorobenzene | 0.07 | 0.07 |
| 1,1-2-Trichloroethane | 0.003 | 0.005 |
| Trichloroethylene | zero | 0.005 |
| Xylenes (total) | 10 | 10 |

Note: Final rules are contained in the following *Fed. Reg.*:
USEPA: National Primary Drinking Water Regulations; Synthetic Organic Chemicals. Final Rule. *Fed. Reg.* (1/30/91).
USEPA: N.P.D.W.R.; Synthetic Organic Chemicals. Final Rule. *Fed. Reg.* (7/17/92).
USEPA: N.P.D.W.R.; Synthetic Organic Chemicals. Final Rule. *Fed. Reg.* (7/1/94).
*TT = Treatment Techniques (see Acrylamide)

# Acrylamide
## $H_2C$═$CHCONH_2$

## Contaminant

Cat. Propenamide, CAS #79-06-1.*
Extremely soluble in water, acrylamide monomer is a common contaminant of polyacrylamide (0.05%).

## Use

In water, soluble polymers improve fluid extraction from wells or as a flocculant in water treatment and food processing. Acrylamide is also applied in the manufacturing of paper, dye, adhesives, soil conditioners, and permanent press fabrics.

## Occurrence in the Environment

No routine monitoring data, reliable and applicable to the standard setting, are available. It may appear in drinking water from contamination and/or water treatment.

## Health Effects

In animal anatomy it may lead to peripheral neuropathy, atrophy of skeletal muscles, testicular atrophy followed by neurotoxic effects; probably carcinogenic in animals.

| 1-Day Assessment | 10-Day Assessment |
|---|---|
| Factor of safety = 100 | Factor of safety = 100 |
| NOAEL = 15 mg/kg | NOAEL = 3 mg/kg/day |
| 10-kg child = 1.5 mg/L | 10-kg child = 0.3 mg/L |
| 70-kg adult = 5.25 mg/L | 70-kg adult = 1.05 mg/L |
| AADI = 0.007 mg/L (based on NOAEL = 0.2 mg/kg with uncertainty factor of 1,000 for subchronic to chronic exposure) | |

*Chemical classification number by "Chemical Abstracts" publications used here for more reliable identification of organic compounds.

### Standards

```
CEE/WHO              = no standards or guidelines
USEPA classification = Group B2 (carcinogenic in two spe-
                       cies)
EPA (1985) RMCL      = zero (proposed)
USEPA (1989) MCLG    = zero
MCL (Proposed)       = left to treatment technique
```

USEPA (1991) issued final rules (phase II)* with MCLG = zero and MCL = Treatment Techniques (TT). The Best Available Technology is obtained with proper practice of polymer addition. **Acrylamide** in drinking water is expected to be an impurity found in coagulant aids. TT are used because of the lack of simple and reliable analytical methods for measuring this contaminant. The water systems must certify on an annual basis that these chemicals do not exceed the specified levels (0.005% dosed at 1 mg/L) based on dosage and percentage of the compound in a coagulant aid or other water treatment chemicals.

### Analytical Methods

No analytical methods have been officially approved.

### Notes

Removal of acrylamide can be partially reached through careful management of water coagulant usage and avoidance of mining contamination.

The WHO (1984) reported acrylamide under the groups of organic compounds of potential health significance (introduced during treatment) but with the conclusion that no further action is required. **Acrylamide** is listed in the SDWA of 1986.

Polymer Additions Practices are indicated by USEPA in their 1989 NPDWR as BTA for treatment/removal of **Acrylamide**.

---

*National Primary Drinking Water Regulations. Synthetic Organic Chemicals and Inorganic Chemicals; Monitoring for Unregulated Contaminants; National Primary Drinking Water Regulations Implementation; National Secondary Drinking Water Regulations. Final Rule. *Fed. Reg.* (Issued 1/30/91, but effective from 7/1/91).

# Alachlor
## 2-Chloro-2, 6-Diethyl-N-(methoxymethyl) Acetanilide

*Contaminant*

Slightly soluble in water. CAS #15972-60-8.

*Use*

Used as a herbicide primarily on corn and soybeans.

*Occurrence in the Environment*

It may enter the water supply from runoff in areas where the herbicide is applied. In general and for orientation purposes, 5% of positive samples may be expected varying from 0.1–1 µg/L (max. 16 µg/L) during spring and summer months.

*Health Effects*

Skin and eye irritation potential after acute exposure. Questionable teratogenicity: high doses caused maternal and feto-toxicity in rats.

```
10-Day Health Advisory
NOAEL     = 150 mg/kg/day
10-kg child = 15 mg/L
70-kg adult = 52.5 mg/L
```

Carcinogenicity is still questionable in animals with some positive results (target organs: lungs, stomach, and thyroid).

*Standards*

IARC Classification: Group 3 (inadequate evidence of carcinogenicity in humans and animals).
USEPA Classification: Group B2 (strong evidence of carcinogenicity in rats/mice).

```
USEPA (1985) RMCL     = zero (proposed)
USEPA (1989) MCLG     = zero
```

CCE/WHO                        = No standards
USEPA (1989) MCL (Proposed)   = 2 µg/L

USEPA (effective from 7/30/92) issued final rules for **Alachlor** with MCLG = zero and MCL = 0.002 mg/L.

## Analytical Methods

The Solvent Extraction–Gas Chromatography Technique is a reliable method.

## Notes

Environmental management and control of this herbicide, pending a probable ban will suffice for removal. USEPA listed in the 1989 NPDWR the BAT for **Alachlor** removal is Granular Activated Carbon.

# Aldicarb
# Aldicarb Sulfoxide and
# Aldicarb Sulfone (Temik)

## Contaminant

2-Methyl-2-(methylthio)    Propanal-O-[(Methylamino)    Carbonyl] Oxime. CAS #116-06-3; 1646-87-3; 1646-88-4.
   Highly soluble in water.

## Use

Registered pesticide for insects, mites, and nematodes; also used in nursery plantings and greenhouse crops.

## Occurrence in the Environment

Estimated food consumption: 100 μg/day/adult. In the Long Island survey of well waters, 29% had a total residue higher than the detection limit of 1 μg/L. In other selected well water surveys, 4% had concentrations higher than 10 μg/L (max. concentration, 50 μg/L).

## Health Effects

Aldicarb is readily absorbed and rapidly excreted by animals in laboratory animal studies.

| | |
|---|---|
| NOAEL | = 0.125/mg/kg/day (reliable animal study) with an uncertainty factor = 100. Results in: |
| AADI | = 0.042 mg/L (adult) |
| AADI | = 0.012 mg/L (child) |
| ADI | = (NAS, 1977–1980) = 0.001 mg/kg/day (food) |
| ADI | = (NAS, 1983) = 0.007 mg/L |
| SNARL | = 7 μg/L (with uncertainty factor of 1,000). No mutagenic potential noted; negative results for carcinogenicity. |
| USEPA ADI | = 0.003 mg/kg/day |

Inhibition of Cholinesterase (enzyme important in the functioning of the nervous system) is the "target organ." Sulfoxide is slightly more potent than the parent compound.

### Standards

| USEPA (1985) RMCL | = 0.009 mg/L for total aldicarb residues (mostly sulfoxide and sulfone). As a potential human carcinogen, it is presently classified in Category III (inadequate or no evidence of carcinogenicity in animals). |
|---|---|
| EEC/WHO | = no standards at present (total pesticides = 0.005 mg/L used as max. admissible concentration by EEC). |
| USEPA (1989) Proposed MCLG | = 0.01 mg/L for aldicarb and aldicarb sulfoxide with MCLG = 0.04 mg/L for aldicarb sulfone (same values for MCL). |

USEPA (1991) issued a final rule for **Aldicarb** (Temik) with MCLG = 0.001 mg/L and MCL = 0.003 mg/L. Also for **Aldicarb Sulfone** a MCLG = 0.001 mg/L and mg/L and MCL = 0.002 mg/L and for **Aldicarb Sulfoxide** a MCLG = 0.001 mg/L and MCL = 0.004 mg/L. In May 1992, USEPA postponed the effective date of these three MCLs.

### Analytical Methods

High Pressure Liquid Chromatographic Technique is the advisable, reliable method.

### Notes

Environmental management control and monitoring of Temik to reduce or eliminate contaminant concentration. In the 1989 NPDW Regulations, USEPA listed Granular Activated Carbon as the BAT for **Aldicarb, Aldicarb Sulfoxide**, and **Aldicarb Sulfone**.

# Carbofuran

## Contaminant

2,3-Dihydro-2,2-Dimethyl-7-Benzofuranol-Methylcarbamate.
CAS #1563-66-2.
Solubility in water is 700 mg/L.

## Use

Carbofuran is used as an insecticide and nematocide—mostly used on corn.

## Occurrence in the Environment

Detected in groundwater in New York and Wisconsin at the levels of 1–50 µg/L. No air and food survey data are presently available (peanuts from Arkansas contained up to 25 µg/kg).

## Health Effects

Rapidly absorbed and excreted, **Carbofuran** is a potent inhibitor of cholinesterase (enzyme related to functioning of the nervous system) typical of the carbamate pesticides.
From animal studies, a NOAEL = 0.5 mg/kg/day was determined providing an AADI = 0.18 mg/L for adult, applying an uncertainty factor of 100. No carcinogenicity has been determined.

## Standards

| | |
|---|---|
| USEPA (1985) RMCL | = 0.036 mg/L. |
| USEPA classification | = Group E (no evidence of carcinogenicity to humans) |
| WHO (world) | = No standards |
| EEC (European) | = 0.005 mg/L (total pesticides MCL) |
| USEPA (1989) | = 0.04 mg/L (proposed) for both MCL and MCLG |

USEPA (effective from 7/30/92) issued a final rule for **Carbofuran** (Furadan 4F) with MCLG = MCL = 0.04 mg/L.

### Analytical Methods

The solvent extraction-gas chromatography/mass spectrometry and the solvent extraction-high pressure liquid chromatography techniques are reliable methods for analysis.

### Notes

Environmental management of this pesticide is the probable "best solution" for "removal" of **Carbofuran** from potable water.

USEPA (1985) RMCL was based upon the AADI of 0.18 mg/L with a 20% assumed drinking water contribution.

No data on **Carbofuran** in ambient air are available. In the 1989 NPDW Regulations, the USEPA listed the Granular Activated Carbon as the BAT for **Carbofuran** removal.

# Chlordane
## $C_{10}H_6Cl_8$

### Contaminant

1,2,4,5,6,7,8,8-Octachloro-2,3,3*a*,4,7,7*a*-Hexahydro-4, 7-Methanoindene.

Also listed as 1,2,4,5,6,7,8,8-Octachloro-4,7-Methano-3*a*,4,7, 7*a*-Tetrahydroindane. CAS #57-74-9

Considered a mixture of stereoisomers and other chlorinated analogs, including Heptachlor. Basically an insecticide from the polycyclic chlorinated hydrocarbons called cyclodiene insecticides.

### Use

Perhaps the most used of all insecticides for termite control; previously used for soil insects and ants.

### Occurrence in the Environment

Maximum ambient air levels have been reported at 204 ng/mc. Maximum allowable level in raw agricultural products is 300 µg/kg. In water, it has been found occasionally at levels of 0.01 µg/L but in overall surveys, very rarely.

Since it is possible to find **Chlordane** in hazardous waste sites, it is a potential contaminant of groundwaters.

### Health Effects

Chronic exposure involves neurotoxicity and liver effects. Based on an animal study, with an uncertainty factor of 100, LOAEL was calculated at 0.063 mg/L (child) and 0.22 mg/L (adult) as a 10-day assessment. In another animal study with NOAEL = 0.075 mg/L/day, an uncertainty factor of 100 and an AADI of 0.03 mg/L was calculated.

Carcinogenicity mainly related to liver was observed in limited animal studies. Mutagenicity was also determined in human cells. Negative teratogenicity was proven in rats.

Most uses of **Chlordane** were banned in 1980 for this pesticide and herbicide.

### Standards

| | |
|---|---|
| USEPA (1985) RMCL | = zero (proposed) |
| WHO guidelines | = 0.3 µg/L (ADI = 0.001 mg/kg/day —total isomers, 1984 |
| Canadian standards | = 0.007 mg/L |
| European | = 0.005 mg/L MCL—total pesticides |
| EPA classification | = Group B2 (carcinogenic in mice) |
| IARC classification | = Group 3 (equivocal evidence of carcinogenicity) |
| USEPA (1989) MCL | = 2 µg/L (proposed) MCLG = zero |

USEPA (effective from 7/30/92) issued the final rule for **Chlordane** with MCLG = zero and MCL = 0.002 mg/L.

### Analytical Methods

Analytical Methods approved by USEPA are listed for **Chlordane** as Methods 505, 508, 525.2. *Standard Methods* (1995) lists the approved methods for chlordane in detail. See Chapter 8 regarding monitoring of VOCs and SOCs in regard to sampling, analytical methods, and approved laboratory requirements.

### Notes

Strict environmental regulations in use of **Chlordane** and its disposal control will practically eliminate the need for potential water treatment. Production and use have been decreased notably in recent years. The GAC is BAT (effective removal— USEPA, 1989).

WHO (1984) recorded that drinking water and ambient air are relatively insignificant sources of **Chlordane**. The guideline is based on an ADI = 0.001 mg/kg, with a 70-kg adult, 2 L/day and 1% allocation to drinking water (equivalent to a factor of safety of 100); the guideline = 0.35 µg/L rounded off to 0.3 µg/L.

# Dibromochloropropane
## $CH_2BrCHBrCH_2Cl$

## Contaminant

1,2-Dibromo-3-Chloropropane (DBCP). CAS #96-12-8.
DBCP is moderately soluble in water (1 g/L).

## Use

Primarily used as a soil fumigant for nematode control on crops. Presently, DBCP has been banned.

## Occurrence in the Environment

In groundwaters there have been detected levels ranging from 0.02–20 μg/L; higher levels were found in hazardous waste sites locally influencing drinking water to 0.08 mg/L, groundwater to 95 mg/L, and soil to 2,600 mg/L.

## Health Effects

Acute oral exposure to DBCP in rats affected renal function, heptacellular necrosis, and testicular and epididymal atrophy. Human male antifertility has been attributed to exposure to DBCP.

| 1-Day Assessment | | 10-Day Assessment | |
|---|---|---|---|
| LOAEL | = 20 mg/kg/day | NOAEL | = 0.5 mg/kg/day |
| Uncertainty factor | = 1,000 (animal) | Uncertainty factor | = 100 |
| Child | = 0.02 mg/L | Child | = 0.05 mg/L |
| Adult | = 0.7 mg/L | Adult | = 0.175 mg/L |
| | | AADI | = not determined (insufficient data) |
| | | NAS (SNARL) | = not calculated |

Mutagenicity and carcinogenicity was observed in rats. The target organs include: liver, kidney, and stomach.

### Standards

```
USEPA (1984) RMCL  = zero (proposed)
WHO (world)        = no standards
EEC (European)     = 0.005 mg/L (MCL—total pesticides)
USEPA Classification = Group B2* (carcinogenic in
                       rats/mice)
IARC Classification  = Group 2B (strong evidence of car-
                       cinogenicity)
USEPA (1989) MCL   = 0.2 µg/L (proposed), MCLG = zero
```

USEPA (effective from 7/30/92) issued a final rule for **Dibro-mochloropropane** (DBCP) with MCLG = zero and MCL = 0.0002 mg/L. This pesticide was canceled in 1977.

### Analytical Methods

The purge-and-trap gas chromatography/mass spectrometry techniques are reliable laboratory methods for DBCP. *Standard Methods* (1988–1995) lists for DBCP the Liquid-Liquid Extraction Gas Chromatographic Method.

### Notes

The authorized use was cancelled in the United States. In general, a reduction of DBCP concentration is expected in consequence of the ban.

The threshold for taste and odor is 0.01 mg/L. The WHO (1984) still has this pesticide under "detailed examination," but with existing information a decision was reached not to issue a specific guideline.

In the 1989 NPDW Regulations USEPA proposed GAC and Packed Tower Aeration (PTA) as the BAT for the removal of DBCP.

---

*"Usually a combination of sufficient evidence in animals and inadequate data in humans."

## Ortho-, Meta-Dichlorobenzene
## $C_6H_4Cl_2$

### Contaminant

1,2-Dichlorobenzene and 1,3-Dichlorobenzene. CAS #95-50-1, 541-73-1.

*o-* and *m-*Dichlorobenzene are solvents with low vapor pressure.

### Use

Used as a chemical component in the production of organic compounds, including pesticides and dyes.

### Occurrence in the Environment

No reliable survey is available or reported for food and water contamination. Milk from nursing mothers was found with a mean level of contamination of 9 µg/mL. Ambient air contamination was reported by the EPA as 0.0, 6.6, and 350 ng/mc for rural, urban, and near source areas dominated by **Dichlorobenzene**. The estimated respiratory intake for an adult male is between 0–6.2 µg/kg/day for *Ortho-* and between 0–5.3 for *Meta-*Dichlorobenzene.

In water it was detected in surface supplies only. USEPA estimated that 99.3% of the population served by public drinking water supplies are receiving water with zero or less than 0.5 µg/L, tested for the ortho-compound.

### Health Effects

Target organs are the central nervous system, lung, kidney, and liver damage, and blood dyscrasias (a blood abnormality) for the ortho-isomer. The meta-isomer appears to be similar in toxicity.

For a 10-day assessment, a NOAEL of 125 mg/kg/day for a mouse and rat, with an uncertainty factor of 100 (since the no-effect level was identified) with a time of study correction factor of 5/7 (feeding time per week). For a chid it was derived in 8.9 mg/L and 31.2 mg/L for an adult.

The AADI = 3.12 mg/L was determined similarly but with an uncertainty factor of 1,000.

There are minimum indications of mutagenicity and carcinogenicity.

### Standards

| | |
|---|---|
| USEPA (1985) RMCL | = 0.62 mg/L for the ortho-isomer |
| USEPA (1985) RMCL | = not established for the meta-isomer |
| Classification | = Group D (negative or inadequate data) |
| IARC classification | = Group 3 (inadequate or no evidence of carcinogenity in animals) |
| WHO guidelines | = 0.3 μg/L (organoleptic considerations) |
| NAS SNARL (tentative) | = 0.3 mg/L (for the ortho-isomer) |
| USEPA (1989) MCL and MCLG | = 0.6 mg/L (proposed) for the ortho-isomer |
| SMCL (1989) | = 0.01 mg/L (proposed) for the ortho-isomer |
| SMCL (1989) | = 0.005 mg/L (proposed) for the p-Dichlorobenzene |

USEPA (effective from 7/30/92) issued a final rule for the **Ortho-Dichlorobenzene** with a MCLG = MCL = 0.6 mg/L, reminding of the potential health effects on liver, kidney, and blood cell damage.

### Analytical Methods

The purge-and-trap gas chromatography and the purge-and-trap gas chromatography/mass spectrometry techniques are acceptable methods for analyzing o- and m-Dichlorobenzene.

### Notes

Odor threshold has been recorded between 0.01–0.03 mg/L.

USEPA 1980 Ambient Water Quality Criteria specified a 2.6 mg/L for an ortho-isomer maximum level. General criteria for removal of synthetic organic compounds are applicable (see *Introduction*—same chapter).

The WHO (1984) concedes that toxicologic data on the chlorinated benzenes are scarce. USEPA (1987–88) included *o*-**Dichlorobenzene** but not *m*-**Dichlorobenzene** in the list of VOCs compulsory monitoring.

In the 1989 NPDW Regulations USEPA proposed Granular Activated Carbon and Packed Tower Aeration (PTA) as the Best Available Technology for the removal of the ortho-isomer.

## *CIS*-and-*TRANS*-1,2-Dichloroethylenes
## ClCH=CHCl

### Contaminant

Also known as an **Acetylene Dichloride**, basically a solvent made up by a mixture of *cis-* and *trans*-isomers (halogenated hydrocarbons). Solubility; 0.77% and 0.63%, respectively. CAS #156-59-2, 156-60-5.

### Use

Used as a low-temperature extraction solvent for organic materials, and as a chemical intermediate in the synthesis of the compounds.

### Occurrence in the Environment

No reliable data are available on food surveys. USEPA estimated respiratory intake for an adult male will vary between zero (rural) and 9 µg/kg/day (max. exposure). In limited water surveys 2.2% of the U.S., the population is estimated to receive more than 0.5 µg/L and 0.2% of the population could receive levels greater than 20 µg/L.

### Health Effects

Target organs are the liver and kidney, and effects on humans are similar to a general anesthetic and narcotic.

For a *1-day assessment* from animal studies with an LOAEL = 400 mg/kg and an uncertainty factory of 1,000, results for a child in 4.0 mg/L and for the adult in 14 mg/L for *cis*-1,2-Dichlorethylene and 1,1-Dichloroethylene.

For a *10-day assessment*, a NOAEL = 100 ppm (10 mg/kg) and an uncertainty factor of 100, a child limit is equal to 1 mg/L, and an adult limit is 3.5 mg/L (limits reached actually from 1,1-Dichloroethylene toxicological tests).

For a *trans*-1,2-Dichloroethylene: uncertainty factor = 100 NOAEL which is 200* ppm *inhaled* by mature female rats.

*Equivalent absorbed dose is 27.2 mg/kg (USEPA).

*1-Day assessment*:       child = 2.7 mg/L

                          adult = 9.45 mg/L

*10-Day assessment*: by assumption of similarity with 1,1-Dichlorethylene in lieu of actual study—

                          child = 1 mg/L

                          adult = 3.5 mg/L

An AADI equal to .035 mg/L with an uncertainty factor of 1,000, was derived by USEPA (1989) from studies on 1,1-Dichlorobenzene.

Some mutagenicity was observed in *cis*—but not *trans*—1,2-Dichloroethylene. No long-term studies have been undertaken for carcinogenicity evaluation.

### Standards

| | |
|---|---|
| RMCL | = 0.07 mg/L (USEPA 1985) |
| EPA classification | = Group D (not classifiable as to human carcinogenicity) |
| WHO | = no standards for 1,2-Dichloroethylene (but 0.3 µg/L for **1,1-Dichloroethene**) |
| EEC | = 5 µg/L MCL for total pesticides |
| MCL and MCLG (1989) | = 0.07 mg/L (proposed) for *cis*-1,2-Dichlorethylenes (USEPA) |
| MCL and MCLG (1989) | = 0.1 mg/L (proposed) for *trans*-1,2-Dichloroethylenes |

USEPA (effective from 7/30/92) issued the final rule for ***cis*-1,2-Dichloroethylenes** with MCLG = MCL = 0.07 mg/L and for ***trans*-1,2-Dichloroethylenes** with MCLG = MCL = 0.1 mg/L.

### Analytical Methods

The purge-and-trap gas chromatography and the purge-and-trap gas chromatography/mass spectrometry techniques are reliable methods for laboratory determination of *cis*- and *trans*-1,2-Dichloroethylene.

### Notes

The origin of **Dichloroethylenes** in drinking water is assumed to be the transformation of other chlorinated hydrocarbons. Granular activated carbon filtration is questionable in effectiveness.

In the 1989 NPDW Regulations USEPA proposed GAC and PTA as the BAT for the removal of the dichloroethylenes.

# 1,2-Dichloropropane
## CH₃CHClCH₂Cl

*Contaminant*

Also known as **Propylene Dichloride**. Basically a solvent for fats, oils, waxes, gums, and resins. Solubility is 0.26%. CAS #78-87-5.

*Use*

Soil fumigation for nematodes and insects; as a solvent and in dry cleaning fluids.

*Occurrence in the Environment*

For ambient air concentration and for food survey there are no official data available.

In drinking water surveys **1,2-Dichloropropane** appeared 1.3% in a random survey and 1.5% in a special survey with a mean concentration of 3.7 µg/L in both cases. In a groundwater survey in California 16% of the wells tested had detectable levels of **1,2-Dichloropropane**.

*Health Effects*

Target organ is the liver; also in animal studies the kidneys and lungs.

From the NAS a determination of LOAEL is equal to 8.8 mg/kg/day with an uncertainty factor of 1,000 provided a *10-day assessment* of 0.3 mg/L for an adult and 0.09 mg/L for a child, but AADI was not provided due to insufficiency of data.

Suspected mutagenicity was observed as well as suspected carcinogenicity in animals (mice) with a consequent USEPA classification of **1,2-Dichloropropane** in Group C. (Equivocal evidence of carcinogenicity.)

*Standards*

USEPA (1985) RMCL = 6 µg/L (proposed) and based upon a $10^{-5}$ cancer risk level (see *Risk*, page 158).

EEC                  = 5 µg/L (MCL for total pesticides)
WHO                  = no guidelines
USEPA (1989) MCL     = 5 µg/L (proposed) and MCLG = Zero

USEPA (effective from 7/30/92) issued the final rule for **1,2-Dichloropropane** with MCLG = zero and MCL = 0.005 mg/L.

### Analytical Method

The purge-and-trap gas chromatography and the purge-and-trap gas chromatography/mass spectrometry techniques are reliable methods for **1,2-Dichloropropane** analysis.

### Notes

Curtailment in use may follow confirmation of carcinogenicity by pending studies.
   For the removal of **1,2-Dichloropropane** the BAT proposed by USEPA in their 1989 NPDW Regulations are the Granular Activated Carbon (GAC) and Packed Tower Aeration (PTA).

# 2,4-D
## (2,4-Dichlorphenoxyacetic Acid)
### $ClC_6H_3OCH_2COOH$

## Contaminant

Also listed under **Chlorophenoxys** or **Chlorinated phenoxy acid herbicides**. CAS #94-75-7.

## Use

Mainly as a **herbicide**, very soluble in its variety of salts, esters, and derivatives. Registered for aquatic weed control in open waters, rivers, and streams.

## Occurrence in the Environment

**In Food**: Minimum or negligible amounts are detected. For example, in a FDA 1970–73 survey, a concentration range of 10–130 $\mu$g/kg was found in leafy vegetables and 14 $\mu$g/kg in potatoes.

**In Air**: Maximum values reported (USEPA, 1985): 4 $\mu$g/mc in New York and Utah (maximum values; less than 0.1 $\mu$g/mc as a max. in specific area).

**In Water**: In general surface water surveys indicated no concentrations in excess of the minimum quantification limit of 0.01 $\mu$g/L. States occasionally reported concentrations of 0.025 $\mu$g/L. (USEPA, NORS, NSP, NAS, 1977).

**Total Human Intake**: In adult males the USEPA estimates a theoretical concentration range of 0.00033–0.45 $\mu$g/kg assuming an air concentration range of 0.001–0.05 and a water concentration range of 0–10 $\mu$g/L.

## Health Effects

Acute oral toxicity is in general considered moderate. For rats and mice, 580 and 1,625 mg/kg were reported, respectively, in acute oral toxicity tests (NAS, 1977). Target organs are the liver and kidney.

Progressive Symptoms: muscular incoordination, paralysis, stupor, and coma and death in animals.

*One-day assessment*: NOAEL is 3.85 mg/L from animal studies with a factor of safety of 1,000 (1.1 mg/L for a child).

*Ten-day assessment*: 1.1 mg/L adult; 0.30 mg/L for a child.

NOAEL preliminary data (interim standards)—1 mg/kg with an uncertainty factor of 100.

The AADI is 0.35 mg/L preliminary (based on the assumption that 20% of total exposure comes from water).

*Mutagenicity*: lack of reliable studies in mammalian mutagenicity.

*Carcinogenicity and Teratogenicity*: Insufficient demonstration of carcinogenicity in animals; disagreements in the results of subchronic and chronic toxicity and minimum evidence in animal teratogenicity (NAS 1977; USEPA 1985). USEPA (1985) RMCL = 0.07 mg/L.

### Standards

| | |
|---|---|
| USEPA (1975) | = 0.1 mg/L interim standards |
| FAO/WHO | = ADI—0.3 mg/kg |
| USEPA (1985) | = recommended MCL = 0.07 mg/L. For carcinogen risk assessment Group D (not classified—inadequate animal evidence of carcinogenicity). |
| EEC | = no standards |
| WHO | = 0.1 mg/L as a guideline* (FAO/WHO AADI = 0.3 mg/kg) |
| Canadian MCL guide | = 0.1 mg/L |
| USEPA (1989) MCL | = 0.07 mg/L (proposed) = MCLG |

USEPA (effective from 7/30/92) issued a final rule for **2,4-D** with MCLG = MCL = 0.07 mg/L, reminding of the potential damage to liver and kidney.

### Analytical Methods

*Standard Methods* lists 2,4-D under the organic constituents (pesticides-organic) in the section entitled **Chlorinated Phenoxy Acid Herbicides**. After preparation of sample, the methyl esters are determined by **gas chromatography**.

Results are reported in micrograms per liter as methyl ester without correction for recovery efficiency.

---

*WHO (1984) also stated that 2,4-D may be detectable by taste and odor at lower concentrations (≥0.05 mg/L).

**Notes**

Detection of extremely low levels of 2,4-D in drinking water has been registered and, therefore, a potential for higher concentrations exists. Moreover, a high dose concentration may have health effects in humans. This contaminant has been selected by NAS and the USEPA for primary standards promulgation.

GAC and PTA are proposed by USEPA (1989) to remove 2,4-D.

## Epichlorohydrin (ECH)
## 1-Chloro-2,3-Epoxypropane
## $CH_2OCHCH_2Cl$

### Contaminant

Also known as a **Halogenated Alkyl Epoxide** or **Chloropropylene Oxide** or EPI. CAS #106-89-8.

### Use

Soluble in water and organic solvent, ECH is mainly used to produce glycerin; also used as raw material for epoxy and phenoxy resins and flocculants. It is considered a contaminant of polymers in water treatment and food processing. The drinking water contamination concern is due to the fact that ECH was or may be used for interior coating of water tanks and pipes. ECH has been used as an insecticide and as a component of agricultural chemicals.

### Occurrence in the Environment

Presently in the United States, no monitoring data are available on human intake from ambient air, food, or water supplies. Industrial hygienic standards in the U.S. have been lowered recently from 19 to 2 mg/mc for a 40-hr workweek. European equivalent data vary from 1 (Russia) to 18 mg/mc (West Germany) (NAS, 1980).

### Health Effects

ECH is a strong mutagen, showing evidence of carcinogenicity in animals; no data are presently available on its carcinogenicity. There are limited studies on ECH metabolism.

Target organs are: the nasal turbinates, lungs, kidneys, liver, and central nervous system. In industrial workers also eye and throat irritation, nausea, and vomiting were observed.

Acute effects (oral $LD_{50}$ values for ECH) range from 90 to 240 mg/kg in animals.

### Standards

Using a NOAEL of 2 mg/kg with an uncertainty factor of 100, a resulting 10-day assessment of 0.14 mg/L and 0.5 mg/L for a child and an adult, respectively, was calculated by USEPA (also protective for 1-day exposure). Considering that the threshold for odor is 0.5–1.0 mg/L, and the threshold for irritant action is 0.1 mg/L, animal studies have been conducted with 10 ppm exposure levels resulting in a LOAEL of 2.16 mg/kg. Using an uncertainty factor of 1,000 a provisional AADI of 0.076 mg/L has been determined by USEPA.

Due to the carcinogenicity in animals and IARC classification of Group 2B (inadequate evidence of carcinogenicity in humans, sufficient evidence of carcinogenicity in animals), and the similar EPA classification in Group B2, the EPA proposed standards of 1985 recommended: *RMCL of 0* for ECH.

USEPA (1989) (proposed control of this contaminant of polymers in water treatment): MCLG = zero (proposed)

USEPA (1991) issued final rules for **Epichlorohydrin** with MCLG = zero and MCL = Treatment Techniques (TT). This MCL is selected based on the assumption that the Best Available Technology is obtained with the proper practice of polymer addition. The water systems must certify on an annual basis that this chemical does not exceed the specified levels (0.01% dosed at 20 mg/L) based on dosage and percentage of the compound in a coagulant aid or other water treatment chemicals.

### Analytical Method

The direct injection gas chromatographic technique is recognized by USEPA as a reliable method.

### Notes

Health effects are indeed registered on inhalation studies, not on ingestion of water or food. The metabolism of ECH is not well known. Because of the carcinogenicity in animals through inhalation studies, USEPA derived their 1985 proposed standard of zero considering the likelihood that **Epichlorohydrin** be detected in drinking water following its use in coating of tanks and water pipes.

In the 1989 NPDW Regulations, the USEPA proposed the Polymer Addition Practice (PAP) as the BAT for the removal of **Epichlorohydrin**.

# Ethylbenzene
## $C_6H_5C_2H_5$

## Contaminant

A flammable liquid, soluble in water, alcohol, benzene, ether, acetone, and other organic solvents. Solubility is 0.015%. CAS #100-41-4.

## Use

Mainly used for the production of styrene (aromatic liquid used in the production of synthetic rubber and plastics), and a solvent.

## Occurrence in the Environment

Based on limited data, USEPA reports ambient air concentration between 2–5 µg/mc in various areas. No data are presently available for food.

Water surveys limited to selected and nonselected groundwater sources were reported by USEPA with mean concentration of 0.87 µg/L in positive samples and 0.78 µg/L for nonselected cases. Positive samples were detected in less than 1% in both cases. Potential contamination from hazardous waste sites is expected.

## Health Effects

Normal metabolism has been observed with accumulation in adipose tissue. Mild toxicity followed acute exposure. Target organs: liver, kidney, and nervous system.

*Mutagenicity*: ethylbenzene does not appear to be a mutagen.

*Carcinogenicity*: reliable data are not available since it has not been adequately tested.

## Standards

*One-day assessment* has been established for adults with an uncertainty factor of 10 at 72.5 mg/L.

*Ten-day assessment* is simply calculated at 7.25 mg/L. Starting from a NOAEL of 136 mg/kg/day and using an uncertainty factor of 1,000, AADI was derived at 3.4 mg/L.* Consequently, *USEPA (1985) recommended an MCL of 0.68 mg/L* assuming a 20% of exposure via the drinking water.

EEC/WHO = No standards or guidelines.

Group D was used for classification of carcinogenicity for **Ethylbenzene** (inadequate or no evidence of carcinogenicity in animals—USEPA, 1985).

USEPA (1989) (proposed) MCL = 0.7 mg/L, SMCL = 0.03 mg/L, MCLG = 0.7 mg/L.

USEPA (1991) issued final rules for **Ethylbenzene** with MCLG = MCL = 0.7 mg/L effective from 7/30/92.

### Analytical Method

The purge-and-trap gas chromatography technique is considered adequate by USEPA.

### Notes

Taste and odor threshold are, respectively, 0.1 and 0.2 mg/L.

---

*Adjusted for 7 days versus 5 days of animal feeding.

## Ethylene Dibromide
## BrCH$_2$CH$_2$Br

*Contaminant*

**1,2-Dibromomethane** or **Ethylene Bromide** or EDB. CAS #106-93-4.

A highly volatile pesticide, clear, colorless, heavy liquid with distinctive odor.

*Use*

Used as an additive in unleaded gasoline, EDB is also used as a fumigant (soil, stored grain, field logs, tobacco fields, and termites). Also used as solvent for resins, gums, and waxes. Most uses of EDB have been banned in the U.S. from 1984. This soluble compound has a half-life of 8 years.

*Occurrence in the Environment*

**In Food**: Concentrations as high as 5.4 ppm have been found in grain products.
**In Air**: Concentrations of concern were detected as related to consumption of unleaded gasoline. Urban areas show concentrations of 200 ng/mc with extreme values of 1,500 ng/mc.
**In Water**: There are limited surveys of groundwater with results indicating concentrations of 0.1–2.0 μg/L. Some selected wells show concentrations from 0.02 to 560 μg/L.

*Health Effects*

Lethal dose for animals ranges from 55–420 mg/kg. Target organs include: lungs, liver, spleen, kidney, and the central nervous system.

One animal study produced a LOAEL of 7.8 mg/kg/day. From this value a 10-day assessment of 0.027 mg/L was derived for adults with an uncertainty factor of 10,000.

*Carcinogenicity*: demonstrated in animals, including mutagenicity.

### Standards

EEC (1980) MCL = 5 μg/L for total pesticides
WHO            = no guidelines

In consideration that EDB has been found in drinking water and that EDB has developed cancer in animals, *USEPA (1985) proposed a standard of zero for the MCL.* 1989 MCL = 0.05 μg/L (proposed).
1989 MCLG = zero.
IARC classification is Group 2B and USEPA classification for carcinogenicity is Group B2 confirming the evidence of carcinogenicity in animals only. Risk estimate has been determined by NAS (1980) and CAG (1984).
USEPA (effective from 7/30/92) issued final rules for **Ethylene Dibromide** with MCLG = zero and MCL = 0.00005 mg/L.

### Analytical Methods

Analytical Methods approved by USEPA are listed for **Ethylene Dibromide** as Methods 504 and 551. *Standard Methods* lists the approved methods for EDB in details and comments. See also Chapter 8 for monitoring.

### Notes

EDB is known for acute toxicity. Nevertheless a NOAEL of 27 μg/L has been determined for adults using an extremely high value of the uncertainty factor 10,000.
In the 1989 NPDW Regulations, the USEPA proposed GAC and Packed Tower Aeration (PTA) as the BAT for the removal of **Ethylbenzene**.

## Heptachlor
## $C_{10}H_7Cl_7$

### Contaminant

**1,4,5,6,7,8,8-Heptachloro-3a,4,7,7a-Tetrahydro-4,7-Methano-indene** that rapidly oxidizes to the **Epoxide 1,4,5,6,7,8,8-Hep-tachloro-2,3-Epoxy 3a,4,7,7a-Tetrahydro-4,7-Methanoindene**. Insecticide of the group of polycyclic chlorinated hydrocarbons called **Cyclodiene** insecticides. CAS #76-44-8.

### Use

Since April 1, 1976, most of **Heptachlor** registrations were suspended in the U.S. Until 1974 **Heptachlor** was used for corn crops protection but mainly as an insecticide for termites, ants, and agricultural and garden insects. Presently, the registration is limited to subterranean termites.

### Occurrence in the Environment

**In Food: Heptachlor** has been detected very widely in concentrations varying from 0.2–2 µg/L with over 50% occurrence in milk, meat, fish, and poultry.
**In Air**: USEPA surveys of 1970–72 produced ambient air values of **Heptachlor** in the range of mean value for positive samples of 1.0 ng/mc with max. equal to 27.8 ng/mc. (occurrence in 42% of samples).
**In Water**: Positive samples were detected in groundwater in concentrations of 0.01–1.0 µg/L; in selected areas approximately 50% of samples showed concentration above 0.01 µg/L.

### Health Effects

Acute intoxication of **Heptachlor** targets the central nervous system (tremor, convulsion, paralysis, and hypothermia).
From a NOAEL in an animal study of 1.0 mg/kg/day and a factor of 1,000, a 0.035 mg/L was calculated for adults.
From another animal study a LOAEL of 0.075 mg/kg/day

was derived and using an uncertainty factor of 1,000 an AADI of 0.001 mg/L was derived for adults.

Carcinogenicity studies produced inadequate evidence in animals; mutagenicity also was questionable in limited studies. **Heptachlor Epoxide** studies indicated some evidence of carcinogenicity in animals.

### Standards

USEPA (1985) proposed a MCL of zero for both **Heptachlor** and **Heptachlor Epoxide** based on animal studies for carcinogenicity and the presence of these chemicals in drinking water.

WHO (1984) proposed a guideline value of 0.1 μg/L as 1% of ADI (FAO/WHO).

IARC classified **Heptachlor** in **Group 3** (inadequate evidence in humans; limited in animals) but did not classify the **Heptachlor Epoxide**.

USEPA (1989) MCL = 0.4 μg/L (proposed) for **Heptachlor** (MCLG = 0).

MCL = 0.2 μg/L (proposed) for *Heptachlor Epoxide* (MCLG = 0).

USEPA (effective from 7/30/92) issued final rules for **Heptachlor** with MCLG = zero and MCL = 0.0004 mg/L and for **Heptachlor Epoxide** with MCLG = zero and MCL = 0.0002 mg/L.

### Analytical Method

USEPA (1985) considers the solvent-extraction gas chromatography and solvent-extraction gas chromatography/mass spectrometry techniques as adequate.

### Notes

A detection and odor threshold of 0.02 mg/L has been reported for **Heptachlor**.

**Heptachlor Epoxide** ($C_{10}H_9Cl_7O$) is a degradation product of **Heptachlor**, also an effective insecticide and more toxic than **Heptachlor**.

*Standard Methods* (1988) lists both **Heptachlor** and **Heptachlor Epoxide** applicable to the Extractable Base/Neutrals and

Acids—Liquid-Liquid Extraction GC/MS Method and also to the Organochlorine Pesticides and PCSs—Liquid-Liquid Extraction Gas Chromatographic Method.

Granular Activated Carbon (GAC) is the only treatment method proposed by USEPA in the 1989 NPDW Regulations to remove both **Heptachlor** and **Heptachlor Epoxide**.

# Lindane (BHC)
## $C_6H_5Cl_6$

## Contaminant

**Benzene Hexachloride** or **1,2,3,4,5,6-Hexachloro-Cyclohexane** or **Gamma-Benzene Hexachloride**. The commercial insecticide is defined as a product containing at least 99% gamma isomer (the most reactive of the isomers). Other names are: **Gamma-HCH (Gamma-Hexachlorocyclohexane)** ($\gamma$-HCH or $\gamma$-BHC). CAS #58-89-9.

## Use

Home use is in shampoos for elimination of head lice and fleas and lice on pets. The commercial use is registered for several agricultural and livestock applications. Also used in buildings, clothes, and water (mosquito control).

## Occurrence in the Environment

**In Food**: FDA reported concentrations in positive samples varying from traces to 8.0 µg/kg (12.7% of food tested).
**In Air**: In the 1970–72 study, mean value in ambient air was reported as 0.9 ng/mc with a max. of 11.7 ng/mc. In the 1980 survey, the respective levels were 0.01 ng/mc and 1.5 ng/mc (less than 1% positive). Tolerance is equal to 0.5 mg/mc.
**In Water**: Very limited data exists. In the Rural Water Survey one out of 71 groundwater tested exceeded the minimum quantification limit of 0.002 µg/L. In one State well water concentration above 0.01 µg/L (detection limit) was recorded in 58.3% of samples. In the United States, the highest reported level was 0.1 µg/L.

## Health Effects

Acute exposure results in neurologic and behavioral effects in animals. Subchronic and chronic studies show liver and kidney degeneration.

From animal studies a NOAEL of 12.3 mg/kg was used with an uncertainty factor of 100 to derive a 4.3 mg/L for adults.

From another study a provisional AADI was calculated in 0.01 mg/L using 1,000 for uncertainty factor and a NOAEL of 0.6 mg/kg/day.

**Lindane** does not appear, so far, to have mutagenic characteristics. Carcinogenicity in animals has not been proven to satisfactory evidence. NAS (1980) is indicating some evidence of carcinogenicity in mice. Pending further studies EPA is using **Group C** for carcinogenicity classification (limited and insufficient evidence).

### Standards

USEPA (1975)                  = 0.004 mg/L
USEPA (1985) (proposed) MCL = 0.2 µg/L
WHO guideline                 = 3 µg/L

IARC classified **Lindane** in Group 3* and USEPA in Group C—equivocal evidence of carcinogenicity.

USEPA (1989) MCL = 0.2 µg/L (proposed) = MCLG

USEPA (effective from 7/30/92) issued final rules for **Lindane** with MCLG = MCL = 0.0002 mg/L, reminding of potential health effects on liver, kidney, nervous, immune, and circulatory organs.

### Analytical Method

The solvent-extraction gas chromatography technique is considered adequate by USEPA (1985). Analytical methods approved by USEPA are listed under Methods 505, 508, 525.2.

### Notes

The NAS (1980) calculated a 7-day exposure safe at 3.5 mg/L. Published reports indicate no significant mutagenic potential for **Lindane**. Different isomers have been found to have different metabolism.

WHO (1984) issued a guideline value of 0.3 µg/L based on ADI[†] of 0.01 mg/kg body weight. **Gamma HCH** or **Lindane** is

---

*Inadequate evidence.
†"Pesticide Residues in Food—Report 1977," *FAO*, Plant Production and Protection Paper, Rome, 1978.

defined as a broad-spectrum insecticide of the group of cyclic chlorinated hydrocarbons.

WHO (1984) also reported that in Germany **Lindane** was present in all surface-water samples taken in the range of 0.005–7.1 µg/L.

GAC is the proposed BAT (USEPA, 1989).

# Methoxychlor
$Cl_3CCH(C_6H_4OCH_3)_2$

## Contaminant

Generic name for **2,2-bis-(p-Methoxyphenyl)-1,1,1-Trichloro-ethane**. CAS #72-43-5.

Used as an insecticide that mainly substituted DDT during the last 40 years. Beside the lack of toxicity, **Methoxychlor** has a half-life in water of 46 days. Other names: **Methoxy DDT; DMDT; 1,1(2,2,2-Trichloroethylidene) BIS (4-Methoxy-benzene)**.

## Use

Used as an insecticide of very low mammalian toxicity for home and garden, for fly control, and tree disease; also regis-tered for 87 crops.

## Occurrence in the Environment

**In Food: Methoxychlor** was detected in 1% of domestic food samples (2–6 µg/kg in a whole milk sample).
**In Air**: No data available for ambient air concentrations (toler-ance = 10 mg/mc).
**In Water**: In general surveys in the U.S. no detection of **Methoxychlor** was encountered; but in selected areas of heavy use mean value of 0.033 µg/L was recorded in 46% of samples of rural water supplies. In another similar survey the mean concentration was 0.023 µg/L in 64% of samples.

## Health Effects

Metabolism of **Methoxychlor** shows rapid excretion by ani-mals. NAS (1977) reports no conclusive evidence of chemical intoxication in humans; negative results in mutagenicity; no information on teratogenicity and inadequate data on carcino-genicity in animals.

USEPA (1985) reported toxicological effects in animals when high dose levels were administered, influencing the central nervous system and kidney and liver changes.

*One-Day assessment* was determined at 22.4 mg/L for adults based on a LOAEL of 640 mg/kg/day and an uncertainty factor of 1,000.

*Ten-Day assessment* for **Methoxychlor** in man and animal studies a 7.0 mg/L was calculated from a NOAEL of 2.0 mg/kg/day with an uncertainty factor of 10 (2.0 mg/L for a 10-kg child).

AADI was calculated from another animal study using a NOAEL of 5 mg/kg/day, an uncertainty factor of 100, deriving for an adult a value of 1.7 mg/L.

USEPA (1985) concluded that **Methoxychlor** is not carcinogenic, mutagenic, or teratogenic (evidence from animal studies).

### Standards

USEPA interim standards 0.1 mg/L

USEPA (1985) recommended standard 0.34 mg/L (based on AADI of 1.7 mg/L and a 20% intake via drinking water)

WHO recommended a guideline value of 0.03 mg/L (1984) based on ADI = 0.1 mg/kg

Carcinogenicity: Group D—inadequate animal evidence. Inconclusive results. (USEPA, 1985)

EEC = no specific standards or guidelines (5 μg/L = MCL = total pesticides)

USEPA (1989) = 0.4 mg/L (proposed MCL) = MCLG

USEPA (effective from 7/30/92) issued the final rules for **Methoxychlor** with MCLG = MCL = 0.04 mg/L, reminding of toxicity with potential health effects on liver, kidney, nervous system, and reproductive organs.

### Analytical Methods

The solvent-extraction gas chromatography technique has been considered adequate by USEPA (1985).

*Standard Methods* describes the principle, apparatus, reagents, procedure, calculation, precision, and accuracy under Organochloride Pesticides (1985 and 1988 editions).

### Notes

GAC is the proposed BAT in the 1989 NPDWR.

The odor threshold detection value has been recorded at 4.7 mg/L.

**Methoxychlor** has been detected in drinking water in areas of high applications of this chemical. In spite of the limited survival in water, **Methoxychlor** is a potential contaminant and, therefore, it should be monitored and regulated by the Health Authorities.

Sampling finished drinking water from the Mississippi and Missouri Rivers, no concentration of **Methoxychlor** was detected in 500 samples analyzed.

# Monochlorobenzene
## $C_6H_5Cl$

## Contaminant

Also **Chlorobenzene** is a solvent with a characteristic of nearly no solubility in water (.0049%). Other names include: **Phenyl Chloride; MCB**. CAS #108-90-7.

## Use

Used in the manufacture of aniline, insecticides, dyes, and as a solvent in cleaning operations.

## Occurrence in the Environment

**In Food**: Insufficient data available
**In Air**: Survey of ambient air registered in the U.S. for rural/remote 0.0; for urban/suburban 1,500; for source dominated areas 140 ng/mc. (or 0–32 µg/kg/day for respiratory intake).
**In Water**: Water surveys indicated zero or less than 0.5 µg/L in 99.9% of the public drinking water systems; the remaining 0.1% registered results between 0.5–5 µg/L (all from groundwater). WHO (1984) reported a maximum of 10 µg/L in drinking water.

## Health Effects

In humans **Chlorobenzene** is irritating to the respiratory system and is a central nervous system depressant. The acute oral $LD_{50}$ in rats is of 2,910 mg/kg. Also liver damage and kidney changes have been reported. The mutagenicity, carcinogenicity, and teratogenicity have given very limited data to confirm or deny negative results.

*Ten-Day assessment* was derived in animals by inhalation, using a NOAEL of 18 mg/kg/day, was calculated in 6.3 mg/L for adults (UF = 100). AADI was derived from another animal study with a NOAEL of 125 mg/kg/day, using an uncertainty factor of 1,000 a value of 3.0 mg/L was determined (using a corrective factor of 5/7).

### Standards

USEPA (1985) recommended an MCL as primary regulation: 0.06 mg/L (this was based on the above-mentioned AADI with an additional uncertainty factor of 10, assuming a 20% contribution from drinking water).

USEPA in the proposed Guidelines for Carcinogen Risk Assessment classified **Monochlorobenzene** in **Group C**—increased occurrence of neoplastic nodules of the liver in high-dose male rats.

WHO guideline is 0.3 µg/L based on 10% of threshold odor concentration ("Aesthetic Quality," no health significance).

Russian guideline is 0.02 mg/L based on odor and taste.

USEPA (1989)  MCL  = 0.1 mg/L (proposed)
SMCL          = 0.1 mg/L (proposed)

USEPA (effective from 7/30/92) issued the final rules for **Monochlorobenzene** with MCLG = MCL = 0.1 mg/L for this solvent and pesticide.

### Analytical Method

USEPA (1985) recognized the purge-and-trap gas chromatography and the purge-and-trap gas chromatography/mass spectrometry technique as reliable methods.

### Notes

The neoplastic nodules of the liver in male rats, detected in one study, will leave open for further determination of carcinogenicity any primary classification.

WHO (1989) recorded no data available for mutagenicity and teratogenicity and no information indicating carcinogenicity for **Monochlorobenzene**. NOAEL range was recorded at 0.001–14.5 mg/kg; assuming the maximum value ADI = 0.0015–0.015 mg/kg of body weight using a minimum of 1,000 and a maximum of 10,000 of safety factor or ADI = 0.05–0.5 mg/L and with a 10% allocated to water consumption AADI = 5–50 µg/L, while the threshold concentration in water is 30 mg/L.

In the 1989 National Primary and Secondary Drinking Water Regulations, the USEPA proposed GAC and PTA as the BAT for the removal of **Monochlorobenzene**.

## Polychlorinated Biphenyls (PCBs)

### Contaminant

A colorless, stable organic chemical with a biphenyl nucleus and two or more substituent chlorine atoms. They can also be classified as mixed isomers from 10 classes of chloro-biphenyls containing 209 possible isomers, insoluble in water but soluble in many organic solvents. CAS #1336-36-3.

### Use

With a limited initial use as hydraulic fluids, they became major chemicals used in capacitors and transformers and, consequently, commonly found in the environment. Their heat-resistant properties created a large demand of chlorin-ated biphenyls. Use was discontinued in the U.S. in 1976, based on ecological damage from water pollution.

### Occurrence in the Environment

See *Notes*—this section.

### Health Effects

The IARC determined that there is inadequate evidence of carcinogenicity in humans with insufficient evidence of car-cinogenicity in animals (mice and rats). The NAS (1980) did consider that their toxicity is a function of water solubility, volatility, bioaccumulation, photostability, and biodegradabil-ity. Moreover, some PCB mixtures are contaminated by chlo-rinated naphthalenes and chlorinated dibenzofurans which have separate toxicological input. In spite of no solubility in water, PCBs have been detected in surface water and in drink-ing water. The NAS reviewed, in 1980, all research works available evaluating acute, subchronic, and chronic effects as well as mutagenicity, carcinogenicity, and teratogenicity of PCBs. A suggested no-adverse-response level (SNARL) was determined of a 24-hour exposure in 0.35 mg/L with an uncer-tainty factor of 100. A 7-day exposure, consequently, was determined at 0.05 mg/L. Chronic exposure was not calculated. A conclusion was reached that there is no evidence in the U.S. that dietary PCBs have any deleterious effects on health, but "there is a growing concern about the long-range effects of the contamination of our ecosystem with these chemicals" (NAS, 1980). PCBs accumulate in bottom sediments, consequently, they contaminate fish and other animals.

### Standards

A maximum concentration of 1 ppb (1 µg/L) was considered a reasonable standard. The USEPA proposed, in November 1985, a RMCL of zero, based on carcinogenicity of PCBs (Arochlor 1260) in animals. USEPA (1989) MCL = 0.5 µg/L (proposed); MCLG = zero.

WHO (1984) issued a "no further action required" based on potential health significance evaluation.

USEPA (effective from 7/30/92) issued the final rules for **PCBs** with MCLG = zero and MCL = 0.0005 mg/L.

### Analytical Method

The solvent-extraction gas chromatography technique is a reliable method for water analysis in laboratory for the **Polychlorinated Biphenyls**.

### Notes

The FDA estimated a total adult intake of PCBs to be 0.93 µg/day (0.37 milk; 0.52 meat + fish; 0.03 fats and oils) and evaluated a 1977, 1978, 1979 daily intake, respectively, of 0.016; 0.027; 0.014 µg/kg of body weight.

No information is available on PCBs from ambient air.

In drinking water (NOMS 1976–1977) PCBs were found in 6% of groundwater supplies at level of 0.1 µg/L. In surface water, 2% of supplies were less than or equal to 1.4 µg/L.

**Hazardous Waste Sites** were cited by complaints for PCBs disposal ground with concentrations in runoff, sediments, leachate (up to 200 µg/L in liquid, up to 444,000 µg/L in soil).

**Toxicity** of PCBs in humans affects the liver and produces chloracne; in animals enlargement of the liver and liver cells with decreased reproductive function was observed.

*Standard Methods* (1988) described the Liquid-Liquid Extraction GC Method for "Organochlorine Pesticides and PCBs."

GAC is the proposed BAT for the removal of PCBs according to 1989 USEPA, NPDWR.

# Pentachlorophenol (PENTA) (PCP)
## $C_6Cl_5OH$

*Contaminant*

Slightly soluble in water (0.008%). CAS #87-86-5.

*Use*

Used as a herbicide, defoliant, insecticide, fungicide but primarily as wood preservative. Agricultural uses include seed treatment.

USEPA in 1984 initiated action to cancel all registration of its use for nonwood preservative use.

*Occurrence in the Environment*

**Total Intakes** determined by FDA for adults, infants, and toddlers were 0.010, 0.005, and 0.009 µg/kg/day (1974–1979).

In Water it may occur in surface supplies, but rarely with streams and rivers surveys showing **Pentachlorophenol** in concentrations of 0.01–16 µg/L.

*Health Effects*

**Pentachlorophenol** is rapidly absorbed and excreted via the urine. The target organs are liver, kidney (brain, spleen, and fat) and the central nervous system, showing fetotoxicity and adverse effects on reproduction.

*One-Day assessment* was calculated in 1.0 mg/L for child and 3.5 mg/L for adult with an uncertainty factor of 100 and an animal study NOAEL of 10 mg/kg/day.

*Ten-Day assessment* from another animal study, with same corrective factor and a NOAEL of 3 mg/kg/day, was calculated at 0.3 mg/L for child and 1.1 mg/L for adult.

AADI was calculated from a NOAEL of 3 mg/kg/day and an uncertainty factor of 100 to record 1.1 mg/L for adult.

Mutagenicity was noted in limited studies. Carcinogenicity was excluded in preliminary studies.

### Standards

USEPA (1985) recommended a MCL of 0.2 mg/L (assuming a 20% contribution through drinking water).

WHO (1984) used a guideline of 10 μg/L.

IARC classified **Pentachlorophenol** in **Group 3**—inadequate evidence of carcinogenicity in humans and animals.

USEPA issued similar classifications recorded in **Group D**.

USEPA (1989) proposed MCL = 0.2 mg/L; SMCL = 0.02 mg/L.

USEPA (effective 1/01/93) issued the final rules for **Pentachlorophenol** with MCLG = zero and MCL = 0.001 mg/L. This contaminant is still usable as a wood preservative but nonwood uses were banned in 1987.

### Analytical Methods

*Standard Methods* (1995) advised use of the Liquid-Liquid Extraction Gas Chromatography Method and Liquid-Liquid Extraction Gas Chromatography/Mass Spectrometry Method for the indicated phenol compounds.

For **Pentachlorophenol** is also advised the Micro-Liquid-Liquid Extraction Gas Chromatography Method in the group of acidic herbicide compounds.

### Notes

Odor threshold has been recorded at 0.9 mg/L at 30°C and 1.6 mg/L at 20°C. Taste threshold at 0.3 mg/L.

Commercial grade of **Penta** contains also other contaminants listed by the USEPA as hexachloro-*p*-dioxin, heptachloro-*p*-dioxin, octachloro-*p*-dioxin, hexachlorodibenzofuran, heptachlorodibenzofuran, and octachlorodibenzofuran.

WHO (1984) reported levels of **Pentachlorophenol** in drinking water well below 1 μg/L. In 1978 in the River Rhine in the Netherlands, levels were reported from 0.15–1.5 μg/L.

In the 1989 National Primary and Secondary Drinking Water Regulations, USEPA proposed the use of GAC as the only treatment for the reduction of **Pentachlorophenol**.

# Styrene
## $C_6H_5CH{=\!=}CH_2$ #13495 CRC

## Contaminant

Also known as **Vinyl Benzene, Phenylethylene, Cinnamene, Styrene** monomer. Minimum solubility in water (0.031%); soluble in many organic solvents. CAS #100-42-5.

## Use

Used extensively in the manufacture of polystyrene plastics, resins, insulators, and synthetic rubber.

## Occurrence in the Environment

**In Food**: No data available but suspicion of presence in food is due to packaging, when polymers and resins of **Styrene** are used.

**In Air**: In spite of the many tests on inhalation, the ambient air data are lacking. In "source dominated areas" concentrations of 2.3 µg/mc were detected.

**In Water**: **Styrene** was not detected in 1,000 samples of groundwater and 100 samples of surface water (USEPA, 1985).

## Health Effects

The acute toxicity of **Styrene** is relatively low. Symptoms reported in nonlethal doses were reduced gain in weight, increased liver and kidney weight, and lung congestion. Most of the tests conducted in humans and animals were with inhalation rather than ingestion.

*One-Day assessment* determined in animals with an uncertainty factor of 100 from a NOAEL of 270 mg/kg/day resulted in 94.5 mg/L in adults.

*Ten-Day assessment* in another animal study provided a NOAEL of 200 mg/kg/day; combined with an uncertainty factor of 100 the adult safe level results in 70 mg/L. AADI was determined from another animal study yielding a NOAEL of

200 mg/kg/day that combined with an uncertainty factor of 1,000 produced a safe level of 7 mg/L.

A carcinogenicity study has indicated a questionable evidence in one out of several animal studies, but positive tests were recorded for mutagenicity. Inadequate evidence is still catalogued by IARC and USEPA leaving, therefore, the possibility to establish an MCL above zero. NAS (1977) registered an ADI of 133 mg/kg/day that can be translated with an uncertainty factor of 1,000 and a 20% via water into a safe value of 0.9 mg/L.

### Standards

USEPA (1985) proposed an MCL of 0.14 mg/L (based on AADI of 7 mg/L) with an additional factor of 10 and a 20% drinking water contribution.

IARC classified **Styrene** in **Group 3**—inadequate evidence for carcinogenicity in humans, limited evidence in animals, and sufficient evidence for activity in short-term tests.

USEPA (1985) classified **Styrene** in **Group C** as guideline for carcinogenic risk assessment (nonconclusive studies in animals).

| | |
|---|---|
| WHO/EEC | = no standards or guidelines |
| USEPA (1989) (proposed) MCL | = 0.005–0.1 mg/L |
| SMCL | = 0.01 mg/L |
| MCLG | = zero–0.1 mg/L |

USEPA (effective from 7/30/92) issued final rules for **Styrene** with MCLG = MCL = 0.1 mg/L, reminding of potential health effects on the liver and nervous system.

### Analytical Method

The purge-and-trap gas chromatography technique is listed as an acceptable method by USEPA (1985).

### Notes

The recorded values for odor and taste threshold are, respectively, between 10–60 ppm and 0.005–0.773 mg/L.

It is important to note that actual detection of **Styrene** in drinking water surveys is not registered. The insolubility of **Styrene** reported by NAS (1977) or the slight solubility reported by USEPA (1985) in 0.32 mg/L is certainly a major factor for the difficulty of detecting this chemical. Potential contamination from hazardous waste is, therefore, the only present potential source.

In the 1989 NPDW Regulations USEPA proposed GAC and PTA as the BAT for the removal of **Styrene**.

In the 1989 setting of maximum contaminant levels, both MCL and MCLG, USEPA proposed a parameter of 0.1 mg/L based on a Group C carcinogen classification. Based on a B2 classification USEPA used a parameter of zero for MCLG and a parameter of 5 µg/L (0.005 mg/L) for MCL.

It is interesting to note that selection of two different values in setting standards is an unusual decision on the part of the Health Authority. Indeed mathematical precision in an area of many variables in a field of incomplete testing on animals is also questionable. It is hoped that the Health Authority consider issuing standards limited by a lower and higher value. Particularly in cases of legal interpretation, a standard within two reasonable and safe levels would provide a scientific explanation, so necessary to interpret a standard usually based on a questionable factor of safety.

# Toluene
# Methyl Benzene and
# Phenylmethane
# $C_6H_5CH_3$

## Contaminant

An aromatic solvent, soluble in organic solvents, very slightly soluble in water (0.05%). Produced in petroleum refining and coal tar distillation. CAS #108-88-3.

## Use

Used in the manufacture of benzene derivatives, perfumes, dyes, medicines, and solvents. As a solvent for paints, coatings, gums, oils, and resins.

## Occurrence in the Environment

**In Food**: No information available.
**In Air**: USEPA summarized in 1985 a mean respiratory intake of **Toluene** for adults in 83 µg/day using a value of ambient air concentration of 3.6 µg/mc.
**In Water**: Not high toxicity but presence of **Toluene** in drinking water originated investigation of the NAS in 1977 and 1980, leading the USEPA in standards recommendation in 1985. In reality several surveys of drinking water recorded only occasional detection of **Toluene**. The CWSS recorded 2 groundwater systems with 0.505 and 0.56 µg/L. Three surface water systems had concentrations of 0.52, 0.72, and 1.62 µg/L. Meanwhile, the GWSS recorded levels ranging from 0.5–2.9 µg/L in six selected systems.

## Health Effects

The NAS (1977, 1980) recorded that acute human exposure to **Toluene** suggests a narcotic effect (central nervous system depression). Kidney and liver damage, cardiac sensitization and blood dyscrasias are also listed in **Toluene** toxicological effects. Practically all tests were performed via inhalation or evaluating **Toluene**-exposed industrial workers.

*Mutagenicity*: The only data available are studies on workers exposed to **Toluene** where a significant increase in chromosome aberrations was not recorded. *Teratogenicity* was not recorded positive in animals, but some evidence was recorded in female industrial workers. *Carcinogenicity* was not demonstrated in the available studies, very limited indeed so far.

*One-Day assessment* was derived from human study, using therefore an uncertainty factor of 10, and arriving through respiratory inhalation to a concentration of 63 mg/L for adults.

*Ten-Day assessment* was calculated from the same studies at 21 mg/L for adults and 6 mg/L for children. AADI animal studies produced a NOAEL of 1,130 mg/mc by inhalation; assuming an uncertainty factor of 100 and a 50% pulmonary absorption, 10.1 mg/L was determined.

### Standards

USEPA (1985) proposed MCL at 2.0 mg/L based on AADI of 10.1 mg/L, assuming 20% contribution from drinking water.

NAS (1977) reported that other than central nervous system depression, the inhalation of **Toluene** at less than 2,000 ppm has produced no adverse effects.

NAS (1980) calculated a 0.34 mg/L for chronic exposure.

USEPA (1985) classified **Toluene** as D (negative in inhalation and microbial bioassays) for carcinogenicity evaluation.

WHO/EEC = no standards or guidelines
USEPA (1989) (proposed) MCL = 2 mg/L = MCLG
SMCL (proposed)          = 0.04 mg/L

USEPA (effective from 7/30/92) issued the final rule for **Toluene** with MCLG = MCL = 1 mg/L for this solvent and gasoline additive that may affect liver, kidney, nervous system, and circulatory organs.

### Analytical Method

The purge-and-trap gas chromatography technique is included by USEPA in its 1985 standards.

### Notes

The threshold for **Toluene** in drinking water is recorded as 1 mg/L for odor.

NAS (1984) recorded **Toluene** under the "Organic Compounds for which no guideline value is recommended."

*Standard Methods* (1988) listed the Purge and Trap GC

Method and the Purge and Trap Capillary-Column GC/ MS Method.

NAS (1980) calculated the SNARL value as follows: assumptions: 70-kg human consuming 2 L/day and inhaling 10 mc/day with 30% absorption from the lung into the blood and a 20% of the toluene coming from drinking water and a factor of safety of 1,000:

$$\frac{1{,}130 \text{ mg/mc} \times 10 \text{ mc/day} \times 0.3 \times 0.2}{1{,}000 \times 2 \text{ L}} = 0.34 \text{ mg/L}$$

Removal of **Toluene** with the GAC and the PTA was proposed by USEPA in the 1989 NPDW regulations.

# Toxaphene

## Contaminant

Defined as a complex mixture of largely uncharacterized **Chlorinated Camphene** derivatives with an approximate overall empirical formula of $C_{10}H_{10}Cl_{18}$. CAS #8001-35-2.

Solubility in water is approximately rated at 0.4 mg/L (less than 0.001%).

## Use

Used as an insecticide (the most used organochlorine insecticide that has shown tendency of accumulation and persistence in the environment). Currently, USEPA limits the use of **Toxaphene** in agriculture but originally it was used on food and fiber crops.

## Occurrence in the Environment

**In Air**: Was reported at maximum level of 8.7 µg/mc in early studies (1970–1978). Presently, lower levels are expected in the U.S. due to reduction in use.
**In Food**: Concentration from 10–173 µg/kg was detected in agricultural product from 10%–30% of product tests.
**In Water**: No **Toxaphene** concentrations higher than the MCL established in 1975 by USEPA in 5 µg/L were reported in compliance data of surveys to evaluate NIPDWR. Low concentrations were recorded in streams and rivers but not in raw surface or groundwater supplies.

## Health Effects

With the understanding of **Toxaphene** metabolism not clearly known in humans or animals, it can be reported that the target organ is definitely the liver in chronic or subchronic exposure. NAS (1977) reported no evidence of carcinogenic action in animals with negative results also in reproduction, teratogenicity, and mutagenicity. NAS also reported that considerable information was available in laboratory animals and various aquatic organisms, so that an ADI of 1.25 µg/kg/day was calculated.

NAS (1977) also concluded that the possibility that large quantities of **Toxaphene** residues could be found in drinking water was not great.

USEPA (1985) used a LOAEL of 4 mg/kg/day, an uncertainty factor of 100, to derive a *One-Day assessment* of 1.75 mg/L for the adult and 0.5 mg/L for the child. The Ten-Day health advisory reported in the proposed 1985 standards by USEPA from the same LOAEL was 0.28 mg/L for the adult and 0.08 mg/L for the child using an uncertainty factor of 500. Based on studies conducted in 1979 by the National Cancer Institute, some evidence of carcinogenicity of **Toxaphene** in animals was documented.

### Standards

USEPA issued in the interim regulations (NIPDWR) a standard of 5 µg/L based on available data (still acceptable if based on toxicity only), revised in 1985 the recommended MCL to zero based on potential carcinogenicity in animals. The IARC has classified **Toxaphene** in Group 2—inadequate evidence of carcinogenicity in humans and adequate evidence of carcinogenicity in animals.

USEPA guidelines similarly classified **Toxaphene** in B2—carcinogenic in rats/mice.

WHO or EEC has not determined specific guidelines for **Toxaphene** in drinking water.

USEPA (1989) (proposed) MCL = 0.005 mg/L and MCLG = zero.

USEPA (effective 7/30/92) issued final rules for **Toxaphene** with MCLG = zero and MCL = 0.0003 mg/L with potential public health effects connected to cancer.

### Analytical Method

The solvent-extraction gas chromatography technique was recommended by USEPA (1985). The approved menthods for the determination of **Toxaphene** are listed by USEPA as Methods 508 and 525.2. *Standard Methods* lists **Toxaphene** under "Organochlorine Pesticides and PCBs."

### Notes

Organoleptic threshold of **Toxaphene** is rated at 0.14 mg/L.

NAS (1977) concluded that **Toxaphene** is or was a widely

used organochlorine insecticide that apparently has not caused a great deal of environmental harm. Only the potential carcinogenicity confirmed in animals (rats/mice) caused USEPA to restrict the applications and recommend a zero standard.

GAC is the only treatment for the removal of **Toxaphene** that has been recommended by USEPA in the 1989 National Primary and Secondary Drinking Water Regulations (*Fed. Reg.* Vol. 54, No. 97, 5/22/89).

# 2,4,5-TP Silvex

## Contaminant

**2-(2,4,5-Trichlorophenoxy) Propionic Acid.** CAS #93-72-1.
A herbicide for weed and brush control with short persistence in the environment, particularly due to photolysis and bacterial degradation.

## Use

Use as an herbicide has been curtailed in the U.S. by USEPA's decision since 1979 to terminate (prior approvals for existing stocks) the expected use for weed and brush control in forests, rights of way, pasture, irrigation canals, and other waterways.

## Occurrence in the Environment

**In Air**: No data are available.
**In Food**: The very limited data consist of 97 µg/kg concentration in unwashed fruit, reduced to 36 µg/kg after 4 months of storage.
**In Water**: It can be summarized that concentrations above the minimum detectable level of 0.1 µg/L were practically never encountered with exceptions such as that from data collected from 15 surface water systems in Florida where concentrations from 0.03–0.08 µg/L were registered. A USGS survey of groundwater supplies in Florida registered values from 0.04–0.06 µg/L.

## Health Effects

NAS (1977) reported no available data of toxicological observation in man; the oral $LD_{50}$ of **2,4,5-TP** was reported at 650 mg/kg and 500 mg/kg in rats. The NAS also summarized no-adverse-effect doses up to 5 mg/kg/day and 6.8 mg/kg/day in dogs and rats, respectively. Also, ADI was calculated at 0.75 mg/kg/day.
USEPA (1985) stressed the concern that **2,4,5-TP** is contaminated in various degrees by 2,3,7,8-TCDD, a highly toxic poly-

chlorinated dibenzo-*p*-dioxin. In subchronic exposure to **2,4,5-TP** in animals, histopathological changes in the liver and kidney were recorded in the animals.

A NOAEL of 2 mg/kg/day was used to calculate a 10-day assessment with an uncertainty factor of 100 (animal study) resulting in 0.2 mg/L for the child and 0.7 mg/L for the adult.

USEPA (1985) calculated and AADI of 0.26 mg/L. Present studies tend to indicate a lack of mutagenicity and carcinogenicity in animals, pending final conclusions.

## Standards

USEPA—National Interim Drinking Water Regulations were set at 0.01 mg/L.

USEPA (1985) recommended a maximum contaminant level of 0.052 mg/L based upon the AADI of 0.26 mg/L with 20% drinking water intake contribution. This decision was based on results of chronic adverse health effects reports and that **2,4,5-TP** was detected in several drinking water systems.

WHO or EEC has set no guidelines for 2,4,5-TP.

USEPA (1989) (proposed) MCL = 0.05 mg/L = MCLG

USEPA (effective 7/30/92) issued final rules for **2,4,5-TP (Silvex)** with MCLG = MCL = 0.05 mg/L. Target organs for potential health effects are liver and kidney.

## Analytical Methods

*Standard Methods* describes procedures for analysis of **2,4,5-TP** under section "Chlorinated Phenoxy Acid Herbicides" indicating the gas chromatography procedure. USEPA (1985) lists the "derivatization—gas chromatography technique" as an acceptable method for analysis.

## Notes

Phenoxy Acid Herbicides are very effective herbicides, toxic at relatively low concentrations. The most used are 2,4-D (see details under 2,4-Dichlorophenoxyacetic Acid), Silvex, or 2,4,5-TP, and 2,4,5-*T* (2,4,5-Trichlorophenoxyacetic Acid—$C_6H_2Cl_3OCH_2CO_2H$), whose use has been curtailed drastically in the U.S. whenever there is a potential for water pollution.

The laboratory analyses are standardized with the same methods.

In the 1989 National Primary and Secondary Drinking Water Regulations USEPA proposed GAC as the only removal treatment for 2,4,5-TP **(Silvex)** as BAT.

# Xylene
## Dimethylbenzene
### $C_6H_4(CH_3)_2$

*Contaminant*

CAS #1330-20-7 is a mixture of:

*Ortho*—CAS #95-47-6—1,2-dimethylbenzene
*Meta*—CAS #108-38-3—1,3-$C_6H_4(CH_3)_2$
*Para*—CAS #2106-42-3—1,4-$C_6H_4(CH_3)_2$

In the evaluation of this contaminant in drinking water the mixture of the above mentioned isomers is considered. Formed in petroleum, coal tar, and coal gas distillation, it is considered between insoluble and slightly soluble in water (0.017%; 0.02%; 0.02%, respectively).

*Use*

Used in gasoline; as a solvent in many organic liquids, including pesticides; for protective coatings, and in pharmaceutical products.

*Occurrence in the Environment*

The NAS (1980) recorded that over 4 million workers are believed to have been exposed to xylenes.
**In Air**: Reported by USEPA in ambient air from 0.4 µg/mc in rural areas to 3 µg/mc in selected vulnerable areas for *o*-Xylene; but the combined level of *m*- and *p*-xylene varies between 0.4 µg/mc (rural) to 73 µg/mc (vulnerable).
**In Food**: No data available.
**In Water**: 3% of groundwater systems show detectable levels (GWSS, 1975) with a maximum of 0.75 mg/L; 6% of surface water systems tested (CWSS—USEPA, 1985) with a maximum of 5.2 µg/L.

*Health Effects*

The NAS (1977) concluded that insufficient data are available to determine the long-term oral toxicity and also mutagenicity, carcinogenicity, and teratogenicity of **Xylene**.

NAS (1980) working on recent data on inhalation calculated a 1-day suggested no-adverse-response level (SNARL) at 21 mg/L based on human data, factor of uncertainty of 10.2 L/day intake, 0.64% retention, 3.3 mc done by a 70-kg adult air intake with 200 mg/mc of **Xylene** concentration, and 100% exposure from drinking water. Seven-day exposure was calculated from inhalation studies of rats and dogs (6 hr/day; 5 days/week; 13-week test) with no adverse effects at 3,500 mg/mc, assuming u.f.* 1,000, resulting in 11.2 mg/L. A chronic SNARL could not be calculated.

USEPA (1985) summarized the health effects of **Xylene** with targets the central nervous system and the liver. One-day value of 42 mg/L reported by USEPA following a human study, u.f. = 10. A 27 mg/L was calculated for a 10-day value from an animal study (inhalation).

From an animal study on **Xylene** a provisional AADI of 2.2 mg/L was reported by USEPA (1985) with a u.f. = 1,000 and NOAEL = 337 mg/mc.

WHO did not report data on contamination of **Xylene** either for lack of information on food/drinking water intake (inhalation studies only are available) or for lack of detection of **Xylene** in drinking water or the combination of these two factors.

*Standards*

| | |
|---|---|
| USEPA (1985) RMCL | = 0.44 mg/L based on AADI = 2.2 mg/L + 20% drinking water contribution |
| USEPA (1985) | = Group D—insufficient information on carcinogenicity |
| WHO/EEC | = No standards or guidelines |
| USEPA (1989) MCL | = 10 mg/L (proposed) = MCLG |
| SMCL | = 0.02 mg/L (proposed) |

USEPA (effective from 7/30/92) issued final rules for **Xylenes** (total) with MCLG = MCL = 10 mg/L. The potential health effects are on liver, kidneys, and nervous system.

*u.f. = uncertainty factor = safety factor. See NOAEL and safety factors in this chapter.

## Analytical Methods

The purge-and-trap gas chromatography technique is considered a reliable method. (USEPA, 1985).

*Standard Methods* (1988) described **Xylene** under the Purgeable Organics in Water-Purge and Trap Packed-Column—GC/MS Method and under the Purge and Trap Capillary Column-GC/MS Method and under the Purgeable Aromatic Unsaturated Organics in Water—Purge and Trap GC Method, also under the Purgeable Halocarbons in Water-Purge and Trap Capillary-Column GC Method.

## Notes

For taste and odor **Xylenes** can be detected in surface water between 0.3–1.0 mg/L.

Removal of total **Xylenes** with the GAC and the PTA was proposed by USEPA in the 1989 NPDWR as the BAT.

## Atrazine

### Contaminant

2-Chloro-4-Ethylamino-6-Isopropylamino-S-Triazine    (NAS, 1977) 6-Chloro-N-Ethyl-(1-Methylethyl)-1,3,5-Triazine-2,4-Diamine (USEPA, 1985). CAS #1912-24-9.
  Slightly soluble at 0.0033%.

### Use

Used as a herbicide; 96% of domestically supplied **Atrazine** is used in corn and soybeans (USEPA, 1985).

### Occurrence in the Environment

**In Air**: No data available.
**In Food**: Limited and insignificant data available.
**In Water**: NORS (USEPA) reported from surface supply finished drinking water a single reading of 0.1 µg/L of **Atrazine** concentration. From another survey 29% of surface supply systems were in excess of the quantification limit of 0.1 µg/L with range 0.1–2.9 µg/L.
  From river water (Northwestern Ohio), **Atrazine** concentration ranged from 0.087–15.9 µg/L with an average of 6.76 µg/L. In the Mississippi River it was detected between 4.7–5.1 µg/L.
  In groundwater supplies in three midwestern States, **Atrazine** tested positive with concentration typically in the range of 0.8 µg/L. (All data reported by USEPA, 1985.)

### Health Effects

NAS (1977) reported low chronic toxicity with no significant increase in cancer evidence, but in a limited study; an ADI of 0.0215 mg/kg/day was calculated based on a LOAEL = 21.5 mg/kg/day and an uncertainty factor of 1,000.
  USEPA (1985) from an ADI of 0.0215 mg/kg/day calculated an AADI of 0.75 mg/L.

### Standards

| | |
|---|---|
| USEPA (1985) RMCL | = not issued (inadequate data) |
| USEPA (1989) MCL | = 3 µg/L = MCLG |
| IARC | = no classification issued on carcinogenicity and no evaluation of **Atrazine** |
| USEPA (1985) | = Group D (inadequate evidence in animal studies) |
| New York State (1988) MCL | = 5 µg/L for POC |
| MCL | = 50 µg/L for UOC |
| MCL | = 100 µg/L for POC + UOC |
| POC | = Principal Organic Contaminant |
| UOC | = Unspecified Organic Contaminant (like **Atrazine** + other unspecified contaminant) |

USEPA (effective from 7/30/92) issued the final rules for **Atrazine** at MCLG = MCL = 0.003 mg/L. Target organs are heart, mammary glands, and reproductive organs for this herbicide.

### Analytical Methods

Analytical methods for analyzing **Atrazine** in drinking water include the solvent-extraction gas chromatography technique (USEPA, 1985). Approved methods by USEPA (1991) are 505, 507, and 525.2.*

### Notes

Same MCLs are applicable for Atranex and Crisazina. USEPA (1991) prescribed GAC as the Best Available Technology for **Atrazine**.

---

*USEPA issued Analytical Methods for Regulated Drinking Water Contaminants with a proposed rule in the *Fed. Reg.*, 12/15/93; a Notice of Data Availability in *Fed. Reg.*, 7/14/1994; and a Final Rule in *Fed. Reg.*, 12/5/94.

## Dioxin
## 2,3,7,8-TCDD

### Contaminant

**2,3,7,8,-Tetrachlorodibenzo-*p*-Dioxin**. CAS #828-00-2.

This contaminant is not commercially produced but is formed during chemical production or chemical pyrolisis.

A chemical formed from **1,2,4,5-Tetrachlorobenzene** and known as **2,4,5-Trichlorophenol (2,4,5-TCP)** is contaminated with **Dioxin** or **2,3,7,8-TCDD**.

During production of several herbicides **2,4,5-TCP** is used, and the contamination continues in the herbicides, prominently **2,4,5-Trichlorophenoxyacetic Acid (2,4,5-T)** and **2,4,5-TP** or **Silvex**.

**2,3,7,8-TCDD** is highly toxic, fortunately for the water works this contaminant is practically insoluble (0.2 ppb).

### Use

Dioxin is not used as a commercial product but enters the environment through the use of herbicides (see above).

### Occurrence in the Environment

**In Air**: After extensive application of Silvex, ambient air may reach values as high as 0.1 ng/mc.

**In Food**: Maximum concentrations in fish and shellfish were recorded between 1 and 700 parts per trillion ($10^{-12}$). In localized cattle 4–70 parts per trillion were registered.

**In Water**: Not detected and not expected to show in drinking water because of the physical/chemical characteristics of TCDD (insoluble).

**In Hazardous Waste**: TCDD was found in soil in concentrations as high as 0.5 mg/kg and at 20 mg/L in material defined as "nonaqueous phase liquids" in the dump.

### Health Effects

TCDD is not found in raw or drinking water. Consequently, TCDD should not be considered as a water quality parameter, but its very high toxicity requires careful scrutiny and health concern. *One-Day* ADI was calculated in $1 \times 10^{-3}$ µg/L per child and $3.5 \times 10^{-3}$ per adult, from a LOAEL = 0.1 µg/kg and u.f. = 1,000 (animal study).

*Ten-Day*: The same value of above divided by 10.

Chronic exposure to TCDD has also been calculated by USEPA (1985) from an animal study producing a LOAEL = 0.001 µg/kg feeding 50 parts per trillion of TCDD; with an u.f. of 1,000 and an AADI = $3.5 \times 10^{-8}$ mg/L registered.

TCDD has been implicated in causing chloracne, hyperpigmentation, alteration of liver function, and porphyria cutanea tarda in humans (USEPA, 1985).

Conflicting results were reported in mutagenicity, but USEPA (1985) stated that animal studies had demonstrated that TCDD is a potent animal carcinogen.

### Standards

IARC: Carcinogenicity = Group 2B (inadequate evidence in humans, sufficient evidence in animals, and inadequate evidence for activity in short-term tests)

USEPA (1985)  = Group B2 (carcinogenic in animals tested)

Maximum contaminant levels are as follows: USEPA (1985) RMCL = not regulated; reason: "2,3,7,8-TCDD has not been detected in drinking water supplies. The compound is not mobile in runoff or soils and has not been found in groundwater or surface water that is a potential source of drinking water" (USEPA, 1985).

USEPA (effective from 7/30/92) issued the final rules for **2,3,7,8-TCDD (Dioxin)** at MCLG = zero and MCL = 0.00000003 mg/L. Potential health effects: cancer.

### Analytical Methods

USEPA (1985) listed among other available methods for analyzing 2,3,7,8-TCDD in drinking water, the Solvent Extraction GC/MS Technique.

## Endrin
## $C_{12}H_{10}OCL_6$

### Contaminant

**1,2,3,4,10,10-Hexachloro-6,7-Epoxy-1,4,4a,5,6,7,8,8a-Octa-hydro-endo-1,4-endo-5,8-Dimethano-Naphthalene.**
CAS #72-20-8.

**Endrin** is listed among Pesticides, Chlorinated Hydrocarbons, and Cyclodienes (Aldrin, Dieldrin, Endrin, Chlordane, Heptachlor, and Heptachlor Epoxide).

**Endrin** is slightly soluble in alcohol and practically insoluble in water.

### Use

Used as insecticide and rodenticide; since 1976 its use has been reduced by USEPA's questioning and limitations. Since 1985, **Endrin** can be considered no longer in use in the U.S. for lack of production.

### Occurrence in the Environment

In surveys of rivers in the U.S., **Endrin** concentrations between 0.008–0.214 ppb were recorded by USEPA (NAS, 1977). **Endrin** was found positive in 46% of grab samples in 1964 (preusage of USEPA limitations). **Endrin** in food was detected at 0.001 mg or less in the daily dietary intake (summarized by NAS, 1975).

### Health Effects

Literature on **Endrin** quotes many studies showing toxicity in animals, usually at levels of 3–10 mg/kg of body weight. In humans **Endrin**, as the other cyclodienes, was recorded as producing typical symptoms of poisoning: headache, dizziness, sweating, insomnia, nausea, and general malaise. NAS (1977) reports that there is sufficient evidence of accidental contamination with **Endrin**, showing convulsions in humans already at levels of 0.2–0.25 mg/kg. Nevertheless, no available data were then recorded to consider positive mutagenicity and carcinogenicity by **Endrin**, but some evidence of teratogenicity was recorded. NAS consequently advised USEPA that more research was needed prior to standard setting.

In 1976 USEPA questioned the use of **Endrin** as a pesticide, basing their reasons for limitations on presumptions of acute toxicity to widelife, teratogenicity, and risk of significant population reductions of non-target organisms. In 1985 USEPA registered **Endrin** only for control of cutworms, grasshoppers, and moles.

In the standard setting (NPDWR, 1985), USEPA summarized metabolism and toxicity of **Endrin**, stating that it tends to accumulate in liver, brain, kidneys, and fat, but at lower doses **Endrin** is quickly metabolized and eliminated from the body.

*One-Day assessment*: recorded by USEPA (1985) at 0.2 mg/kg/day as the NOAEL of animal tested, with an uncertainty factor of 100, and standard water consumption (one or two L/day, child and adult) the value derived is 0.02 mg/L for a 10-kg child and 0.07 mg/L for a 70-kg adult.

*Ten-Day assessment*: recorded a NOAEL of 0.05 mg/kg/day; a 0.005 mg/L and 0.018 mg/L were derived for child and adult, respectively.

A provisional AADI of 0.002 mg/L for **Endrin** was derived from a NOAEL of 0.045 mg/kg of body weight, using 1,000 as the uncertainty factor and 2 L/day water consumption.

### Standards

USPHS Advisory Committee, in 1968, recommended a standard of 1 ppb (parts per billion) for Endrin.

In the National Interim Primary Drinking Water Regulations, an RMCL = 0.2 µg/L was established by USEPA (1975).

In 1985, USEPA did not propose a RMCL in the Federal Standards based on the fact that Endrin is rarely detected in drinking water, and the present lack of production will decrease the probability of detection.

WHO (1973) established a guideline with a maximum intake of 2 µg/kg/day or 138.2 µg/day for a 69.1-kg person. No guideline was established for drinking water evaluation.

USEPA (1992)* issued the final rules for **Endrin** with MCLG = MCL = 0.002 mg/L for this insecticide and pesticide. Target organs for potential health effects were liver, kidneys, and heart. USEPA's approved analytical methods were Method

---

*USEPA: National Primary Drinking Water Regulations; Synthetic Organic Chemicals and Inorganic Chemicals; Final Rule. *Fed. Reg.* (July 17, 1992). (Phase V).

505, 508, and 525.2. Granular Activated Carbon was the regulated BAT.

### Analytical Methods

USEPA (1985) stated that the analytical methods available for analyzing Endrin in drinking water include the Solvent-Extraction Gas Chromatography Technique.

### Notes

**Endrin** is usually evaluated with the other cyclodienes such as Aldrin, Dieldrin, Chlordane, Heptachlor, and Heptachlor epoxy. Chlordane, Heptachlor, and Heptachlor epoxy are evaluated in this chapter. Few comments are necessary here for Aldrin and Dieldrin.

### Aldrin

1,2,3,4,10,10-hexachloro-1,4,4a,5,8,8a-hexahydro-endo-1,4-exo-5,8-dimethanonaphthalene.

It is practically insoluble in water. It was detected in U.S. rivers in concentrations of 0–0.006 ppb.

In 1968, USPHS Advisory Committee recommended a 17 ppb for Aldrin.

NAS (1977) recorded the following acute toxicity in rats for cyclodienes for oral $LD_{50}$ in mg/kg:

|          | Male | Female |
|----------|------|--------|
| Aldrin   | 39   | 60     |
| Dieldrin | 46   | 46     |
| Endrin   | 17.8 | 7.5    |

The comments on the health aspects of **Endrin** can be repeated here for Aldrin as far as metabolism and target organs. NAS (1977) concluded that the potential limitations for **Endrin** can be extended to Aldrin: basically insufficient evidence is available to set drinking water standards.

### Dieldrin

6,7-epoxy Aldrin

In U.S. rivers concentrations of 0.08 to 0.122 ppb were recorded. In the grab samples from the Mississippi and Missouri rivers, 40% of treated water samples contained Dieldrin at levels up to 0.25 ppb (1969).

In evaluating proposed standards, NAS (1977) entered the following data for the cyclodienes—finished water (evidently showing more concern from 1965–1968):

| Recommended In | 1965 | 1967 | 1968 |
|---|---|---|---|
| Aldrin | 32* | 0.25 | 17 |
| Dieldrin | 18 | 0.25 | 17 |
| Endrin | 1 | 0.1 | 1 |
| Heptachlor | 78 | 1.0 | 18 |
| Heptachlor epoxy | 18 | 1.0 | 18 |
| Chlordane | 52 | 0.15 | |
| DDT-T[†] | 42 | 0.5 | |

*All concentrations measured in parts per billion.
[†]Includes DDT, DDE, and DDD.

Dieldrin is very stable in respect to Aldrin and therefore appears everywhere in the environment where Aldrin was used.

These notes on Aldrin and Dieldrin should be read together with notes on **Endrin**.

# Hexachlorobenzene (HCB)
## $C_6CL_6$

## Contaminant

A white solid of very low solubility in water (6 μg/kg). CAS #118-74-1.

## Use

The production of **Hexachlorobenzene** has been discontinued in the United States. HCB was used as a fungicide in cereal, vegetable, and other crops. HCB was used in dye manufacturing, as an intermediate in organic synthesis, and as a wood preservative.

## Occurrence in the Environment

As a waste by-product of chlorinated solvents and pesticide production, HCB may enter into the environment in very small concentrations in air, food, and water.

Concern for human exposure originated when this chemical was found in human tissue, milk, fish, and birds.

Very limited data are available from sampling drinking water. USEPA reported in 1975 raw water samples containing 0.01 ppb and finished water with concentrations as high as 0.06 ppb.

## Health Effects

NAS (1977) recorded low acute toxicity but considerably more significant toxicity in prolonged exposure to **Hexachlorobenzene (HCB)** with porphyria (defined as an inborn error in porphyrin metabolism) influencing females more than males.

NAS (1980) reported also carcinogenicity of HCB in hamsters and suggested a no-adverse-response level (SNARL) of 0.03 mg/L for a seven-day exposure, based on animal studies, uncertainty factor of 1,000, a 70-kg human, and 2 L/day consumption. Mutagenicity was not established but carcinogenicity suspected.

NAS (1983) confirmed earlier conclusions showing concern

about porphyrinogenic effect of HCB on humans and indicated a need for additional epidemiological data.

WHO, NAS, and USEPA reported the five-year evaluation of the 1955–59 epidemic in Turkey from exposure of contaminated seed wheat used for food. The epidemic was classified as HCB-induced PCT (porphyria cutaea tarda, actually causing cutaneous lesions and hyperpigmentation). A daily consumption of 0.05–0.2 grams of HCB was estimated.

*One-Day* and *Ten-Day assessments* were derived by USEPA (1985) using a NOAEL of 0.5 mg/kg and an uncertainty factor of 100, resulting in 0.05 mg/L for a child, and 0.175 mg/L for an adult.

The AADI was calculated from a NOAEL of 1.6 ppm (0.084 mg/kg/day) with an uncertainty factor of 100, water consumption of 2 L/day, resulting in an AADI = 0.029 mg/L.

### Standards

USEPA (1985) RMCL was not proposed as primary regulation because **Hexachlorobenzene (HCB)** is no longer detectable in drinking water.

IARC classification of HCB is in Group 2B—inadequate evidence of carcinogenicity in humans and sufficient evidence of carcinogenicity in animals.

USEPA (1985) listed HCB in Group B2 ("carcinogenic in rats, hamsters, mice. Liver cancer, no human evidence").

CAG and NAS recorded risk estimates of 0.02 and 0.54 µg/L, respectively, for "projected upper limit, excess lifetime cancer risk ($10^{-6}$) concentration in drinking water."

WHO guideline is 0.01 µg/L based upon a risk of one additional case of cancer per 100,000 population (2 L/day consumption of drinking water). This value was derived from the linear multistage extrapolation model for a cancer risk (conservative, hypothetical, and mathematical models that cannot be experimentally verified).

USEPA (1992) issued final rules for **Hexachlorobenzene** in Phase V with MCLG = zero and MCL = 0.001 mg/L, indicating cancer as the potential health effects and Granular Activated Carbon for the BAT. USEPA listed approved methods as 505, 508, and 525.2.

### Analytical Methods

The 1988 Supplement to the 16th edition of *Standard Methods* includes **Hexachlorobenzene (HCB)** in the method listed,

"Extractable Base/Neutrals and Acids," Liquid-Liquid Extraction GC/MS Method.

USEPA (1985) stated that "analytical methods available for analyzing hexachlorobenzene in drinking water include the solvent-extraction gas chromatography technique." (USEPA, Method 625)

### Notes

The World Health Organization also reported the following data in its 1984 *Guidelines for Drinking-Water Quality*:

- HCB in air is found near sites of high HCB concentration.
- In water HCB was detected always in very low concentrations in surface water tested. In Italy in 108 samples the average levels were 2.5 ng/L.
- The FAO/WHO conditional ADI of 0.6 µg/kg of body weight has been withdrawn and substituted with a guideline value for HCB of 0.01 µg/L on the basis of the above-mentioned $10^{-5}$ risk level.

# Simazine
## 6-Chloro-N,N-diethyl-1,3,5-triazine-2,4-diamine

## Contaminant

One of the four derivatives of **Cyanuric Chlorides** examined by NAS, 1977: CAS #122-34-9.

|  |  | Production ratio in the U.S. |
|---|---|---|
| Atrazine | See *Atrazine* on page 268 | 90% |
| Simazine | 2-Chloro-4,6-Dimethyl-Amino-S-Triazine | 5% |
| Propazine | 2-Chloro-4,6-Diisopropyl-Amino-S-Triazine | 4% |
| Cyanazine | 2-Chloro-4-(1-Cyano-1-Methylethylamino)-6-Ethylamino-S-Triazine | 1% |

## Use

Used as herbicide (corn, sorghum, sugar cane). Simazine is also used with citrus fruits, pineapples, turf grasses, and as an algaecide and weed control.

## Occurrence in the Environment

**In Air**: No data available.
**In Food**: No concentration of **Simazine** above the quantitation limits (10–100 µg/kg).
**In Water**: In surface water supply (finished water) maximum values recorded during peak period of pesticide use:
**Simazine** = 0.026–0.883 µg/L (USEPA 1985), but also in 1983 state and USEPA surveys 0.077–0.30 µg/L were recorded. In groundwater concentrations of 0.5–3.5 µg/L in 6 out of 166 wells in California (USEPA, 1985) were recorded.

## Health Effects—Standards

NAS (1977) reported inadequate toxicological data in humans and animals. USEPA (1985) adopted a LOAEL = 215 mg/kg/day and an uncertainty factor of 1,000 to derive an ADI of 0.215 mg/kg/day from NAS, but later it was considered invalid. IARC: No evaluation reported officially on carcinogenicity. USEPA (1985) classified **Simazine** in Group D (inadequate data to classify).

USEPA (1985) RMCL = not issued

N.Y.S. Health Dept. adopted in 1988 a maximum contaminant level of 50 µg/L for "Unspecified Organic Contaminant" and 100 µg/L for the total of unspecified and "Principal Organic Contaminant."

To compare the toxicity of these Triazines, **acute toxicity** can be used as reported by NAS (1977):

**Atrazine**: $LD_{50}$ = 3,080 mg/kg (rats) and 1,750 mg/kg (mice)
**Propazine**: $LD_{50}$ = 500 mg/kg (rats and mice)
**Cyanazine**: $LD_{50}$ = 334 mg/kg (rats) (NAS, 1977)
**Simazine**: $LD_{50}$ = 5 g/kg (mice) (Water Quality Criteria)

NAS (1977) calculated ADIs for these Triazines as follows:

**Atrazine** = 0.0215 mg/kg/day
**Propazine** = 0.0464 mg/kg/day
**Simazine** = 0.215 mg/kg/day

USEPA (1992) issued final rules for **Simazine** with MCLG = MCL = 0.004 mg/L for this herbicide, indicating cancer as the potential health effects. USEPA's approved methods are 505, 507, and 525. Granular Activated Carbon technology was preferred for BAT.

### Analytical Methods

Solvent-extraction gas chromatography technique is reported by USEPA (1985) as the advisable method.

### Notes

From two-year study of rats, no difference between treated and control animals in gross appearance and behavior was recorded.

# 1,1,2-Trichloroethane
## Vinyl Trichloride
### $C_2H_3Cl_3$

## Contaminant

Insoluble in water.

## Use

As a solvent for fats, oils, waxes, and resins; also in the synthesis of organic chemicals.

## Occurrence in Water

After chlorination—water treatment disinfection—detected in the finished water of New Orleans in the range of 0.1–8.5 µg/L (USEPA, 1975).

## Health Effects

Chronic oral exposure data are insufficient; for mutagenicity and teratogenicity, there are no data; for carcinogenicity, some similarity with Carbon Tetrachloride has been found (NAS, 1977).

## Standards

No specific standard has been issued by USEPA or WHO, but the contaminant is under observation in sampling and animal testing. New York State considered this contaminant a "Principal Organic Chemical" and as such a standard was issued with MCL = 5 µg/L as a total of POCs (N.Y.S. Health Department, 1988).

USEPA (1992) issued final rules for **1,1,2-Trichloroethane** with MCLG = 0.003 mg/L and MCL = 0.005 mg/L. Granular Activated Carbon and Packed Tower Aeration were listed as the BAT. Also approved analytical methods were 502.2 and 524.2.

### Analytical Methods

*Standard Methods* (1988) and USEPA (Method 524) listed the Purge and Trap Capillary Column GC/MS Method and the Purge and Trap GC Method (USEPA 601) and the Purge and Trap Capillary Column GC Method (USEPA 502.2) as recommended and approved methods.

## Bromobenzene
## Monobromobenzene

### Contaminant

Insoluble in water.

### Use

Used as an intermediate in organic synthesis and as an additive in motor oil and fuels.

### Occurrence in Water

A possible by-product of chlorination, it has been found in finished water in the lower Mississippi River area (USEPA, 1972).

### Health Effects

Bromobenzene irritates skin and is a central nervous system depressant in humans.

Chronic effects: studies limited to inhalation by rats.
Mutagenicity: weakly positive in Salmonella test.
Carcino/teratogenicity: no available data.

### Standards

No specific standard has been issued by USEPA or WHO, but the contaminant is under observation in sampling and animal testing. New York State considers this contaminant a "Principal Organic Chemical." A standard was issued with MCL = 5 $\mu$g/L as a total POCs (N.Y.S. Health Dept., 1988).

USEPA (1995) did not issue final rules for **Bromobenzene**, which remains on the list of Unregulated Contaminants.

### Analytical Methods

*Standard Methods* (1988) and USEPA listed the following methods for analysis of drinking water:

Purge and Trap GC Method (USEPA 503.1)
Purge and Trap Packed-Column GC/MS Method (USEPA 524)
Purge and Trap Capillary-Column GC/MS Method (USEPA 524)
Purge and Trap Capillary-Column GC Method (USEPA 502.2)

## Dichlorodifluoromethane
## $Cl_2CF_2$

### Contaminant

Also known as **Freon 12**. A relatively inert, nonflammable liquid, quite insoluble in water.

### Use

Mainly as refrigerant.

### Occurrence in Water

No survey data available, but detected.

### Health Effects

Chronic toxicity: low effects even at high doses (NAS, 1980).
   Mutagenicity, carcinogenicity, teratogenicity: no evidence in animals.
   SNARL: *One-Day* exposure: 350 mg L
            *Seven-Day* exposure: 150 mg L
            Chronic exposure: 5.6 mg L (u.f. = 100; 20° intake
            from water (NAS, 1980)

### Standards

No specific standards were issued by USEPA or WHO, but the contaminant is under observation in sampling and animal testing. New York State lists **Dichlorodifluoromethane** as a "Principal Organic Chemical." A standard was issued with MCL = 5 μg L as a total POCs (N.Y.S. Health Dept., 1988).

### Analytical Methods

*Standard Methods* (1988) and USEPA listed the following methods for analysis of drinking water:

   Purge and Trap Packed-Column GC/MS Method (USEPA 524)
   Purge and Trap Capillary-Column GC/MS Method (USEPA 524)
   Purge and Trap Capillary-Column GC Method (USEPA 502.2)

## n-Propylbenzene

### Contaminant

**1-Phenylpropane.**
Insoluble in water.

### Use

Used in manufacturing methylstyrene and in textile dyeing.

### Occurrence in Water

Out of 10 water supplies tested, 2 were detected positive in finished water: Miami with 0.05 µg/L and Cincinnati with 0.01 µg/L (USEPA, 1975).

### Health Effects

This contaminant is classified as an irritant of the mucous membranes, eyes, nose, throat, and skin with depression of the central nervous system.

Mutagenicity/carcino/teratogenicity: no data available.

Chronic effects: minor problems in rabbits fed at the rate of 0.25 and 2.5 mg/kg/day (NAS, 1977).

NAS (1977) also stated: "estimates of the effects of chronic oral exposure at low levels cannot be made with confidence."

### Standards

USEPA, WHO, and EEC did not issue specific standards for n-Propylbenzene, but this contaminant is under observation with water sampling and animal testing.

New York State lists this contaminant as a "Principal Organic Chemical." A standard was issued with MCL = 5 µg/L for the total POCs (N.Y.S. Health Dept., 1988).

## Analytical Methods

*Standard Methods* (1988) and USEPA listed the following methods for analysis of drinking water:

Purge and Trap Packed-Column GC/MS Method (USEPA 524)

Purge and Trap Capillary-Column GC/MS Method (USEPA 524)

Purge and Trap GC Method (USEPA 503.1)

Purge and Trap Capillary-Column GC Method (USEPA 502.2)

## 1,1,1,2-Tetrachloroethane

### Contaminant

Water soluble (1 g in 350 ml, 25°C).

### Use

Used as a solvent in preparation of insecticides, herbicides, soil fumigants, bleaches, and paints (USEPA, 1975 classification).

### Occurrence in Water

Potentially formed in chlorination of water (water treatment—disinfection). In finished drinking water concentrations of **1,1,1,2-Tetrachloroethane** were recorded as 0.11 µg/L in New Orleans (USEPA, 1974) and as 1 µg/L in the District of Columbia (NAS, 1977).

### Health Effects

Mutagenicity/carcino/teratogenicity: no available data (NAS, 1977).

Chronic effects: Data available can be classified as incomplete and insufficient. Only data on acute toxicity were reported by NAS (1977) as $LD_{50}$, where this contaminant was approximately one-half or one-third toxic when compared with **1,1,2,2-Tetrachloroethane**.

### Standards

USEPA, WHO, and EEC did not issue a specific standard or guideline for **1,1,1,2-Tetrachloroethane**, but this organic chemical is under observation with water sampling and animal testing, since SDWA lists it as future drinking water quality parameter for USEPA MCL determination. New York State lists this contaminant as a "Principal Organic Chemical;" therefore a standard was issued with MCL = 5 µg/L for the total POCs (N.Y.S. Health Dept., 1988).

USEPA (1995) did not issue final rules for **1,1,1,2-Tetrachlo-**

**roethane**, which remains on the list of Unregulated Contaminants.

### Analytical Methods

*Standard Methods* (1988) and USEPA listed the following methods for analysis of drinking water:

Purge and Trap Packed-Column GC/MS Method (USEPA 524)

Purge and Trap Capillary-Column GC/MS Method (USEPA 524)

Purge and Trap Packed-Column GC Method (USEPA 502.1) for purgeable organics or halocarbons like **1,1,1,2-Tetrachloroethane**.

## Dinoseb

### Contaminant

**2-sec-Butyl-4,6-Dinitrophenol.** CAS #88-85-7.
Dinoseb is slightly soluble (52 mg/L at 25°C).

### Use

Used as a herbicide and insecticide since 1945 (control of annual weeds in cereal and crop).

### Health Effects

Mutagenicity: Insufficient data (1 positive and 2 negative tests).
Carcinogenicity: Tendentially negative in limited tests.
Teratogenicity: Not observed after oral exposure.
Chronic exposure: NAS (1983) reported a SNARL of 39 µg/L; based on an u.f. = 1,000 and 20% intake from water, a NOAEL was recorded at 5.6 mg/kg/day.
NAS (1983) concluded: "Limited human data indicate that high exposure to **Dinoseb** can result in a variety of physical and psychological symptoms. Since there are no data on chronic lifetime exposure, this information should be generated before limits for **Dinoseb** exposure in drinking water are established."

### Standards

USEPA, WHO, and EEC did not issue a specific standard for **Dinoseb**, but this contaminant is under test and water supply systems sampling, since **Dinoseb** is listed among chemicals that must be regulated by USEPA according to SDWA (1986). The New York State Dept. of Health issued, in 1988, a MCL = 50 µg/L for "Unspecified Organic Contaminant;" this value is the N.Y.S. standard for **Dinoseb**. The State requires a limit of 100 µg/L for the combination of principal and unspecified organic contaminants.
USEPA (effective from 1/17/94) issued final rules for **Dinoseb** with MCLG = MCL = 0.007 mg/L for this herbicide, indicating thyroid and reproductive organs as potential health

effects. Analytical methods approved by USEPA were Methods 515.2 and 555. Granular Activated Carbon was the specialized process listed for the Best Available Technology.

### Analytical Methods

USEPA (1989) has not yet issued final or proposed regulations for **Dinoseb**, pending further studies and proceeding in order of public health significance and occurrence in drinking water priorities.

*Standard Methods* (1988) has not identified **Dinoseb**, but the method accepted for Phenols, the Liquid-Liquid Extraction Gas Chromatographic Method, can be used for its determination in drinking water.

USEPA (1992) issued the following final rules:

| Pesticides | MCLG (mg/L) | MCL (mg/L) |
|---|---|---|
| Dalapon | 0.02 | 0.02 |
| Diquat | 0.02 | 0.02 |
| Endothall | 0.1 | 0.1 |
| Glyphosate | 0.7 | 0.7 |
| Oxamyl (Vydate) | 0.2 | 0.2 |
| Picloram | 0.5 | 0.5 |

During the same Phase V Regulations of Pesticides, the following, already individually examined, contaminants described in this chapter were finalized: **Dinoseb**; **Endrin**; **Simazine**.

In addition, the following organic contaminants were finally regulated by USEPA, Phase V.

| Contaminants | MCLG (mg/L) | MCL (mg/L) |
|---|---|---|
| Benzo(a)pyrene | zero | 0.002 |
| Di(2-ethylhexyl)adipate | 0.5 | 0.5 |
| Di(2-ethylhexyl)phthalate | zero | 0.006 |
| Hexachlorocyclopentadiene | 0.05 | 0.05 |

Also regulated during Phase V, and already examined with other SOCs in this chapter, were **Hexachlorobenzene** and **2,3,7,8-TCDD (Dioxin)**.

USEPA (1994) issued *proposed* MCLG and MCL for the following contaminants regulated under the "Disinfectants and Disinfection By-Products" (D/DBP):*

| Contaminants | MCLG (mg/L) | MCL (mg/L) |
|---|---|---|
| Bromodichloromethane | zero | NA[†] |
| Bromoform | zero | NA |
| Chloral hydrate | 0.04 | TT[‡] |
| Chloroform | zero | NA |
| Dibromochloromethane | 0.06 | NA |
| Dichloroacetetic acid | zero | NA |
| Haloacetic acids | | |
| (HAA5; sum of 5) | | 0.060 Stage 1 |
| | | 0.030 Stage 2 |
| Trichloroacetic acid | 0.3 | NA |

†NA = not applicable
‡TT = Treatment Technique

For more information on D/DBP, see Chapters 5 and 8.

## Contaminants Classified in This Chapter.

| Organic Chemicals | Other Names | See Details Under |
|---|---|---|
| Alkyl Benzyl Sulfonates | | Detergents |
| Carbolic acid | | Phenols |
| Monochlorophenols | | Phenols |
| Dichlorophenols | | Phenols |
| Trichlorophenols | | Phenols |
| Tetrachlorophenols | | Phenols |
| 2,4,6-Trichlorophenols | | Phenols |
| Trichloromethane | Chloroform | Trihalomethanes |
| Bromodichloromethane | | Trihalomethanes |
| Dibromochloromethane | | Trihalomethanes |
| Tribromomethane | Bromoform | Trihalomethanes |
| Dichloroiodomethane | | Trihalomethanes |
| Bromochloroiodomethane | | Trihalomethanes |
| Chlorodiiodomethane | | Trihalomethanes |
| Dibromoiodomethane | | Trihalomethanes |
| Bromoiodomethane | | Trihalomethanes |
| Triiodomethane | Iodoform | Trihalomethanes |
| Benzene | | Volatile Organic Compounds |

*USEPA: National Primary Drinking Water Regulations: Disinfectants and Disinfection By-Products. Proposed Rule. *Fed. Reg.* (7/29/94).

## Contaminants Classified in This Chapter (*Continued*).

| Organic Chemicals | Other Names | See Details Under |
|---|---|---|
| Vinyl Chloride | Chloro Ethylene | VOC |
| Vinyl Chloride | Monochloroethylene | VOC |
| Carbon Tetrachloride | Tetrachloromethane | VOC |
| 1,2-Dichloroethane | Ethylene Dichloride | VOC |
| 1,2-Dichloroethane | Glycol Dichloride | VOC |
| Trichloroethylene | Trichloroethene | VOC |
| Trichloroethylene | Trichloride or Ethinyl Trichloride | VOC |
| Tetrachloroethylene | Tetrachloroethene (PCE) | VOC |
| 1,1-Dichloroethylene | 1,1-Dichloroethene (1,1-DCE) | VOC |
| 1,1-Dichloroethylene | Vinylidene Chloride | VOC |
| 1,1,1-Trichloroethane | Methyl Chloroform | VOC |
| p-Dichlorobenzene | 1,4-Dichlorobenzene (PDB) | VOC |
| Acrylamide | | Synthetic Organic Chemicals |
| Alachlor | | SOC |
| Aldicarb | Aldicarb Sulfoxide, Aldicarb Sulfone (Temik) | SOC |
| Carbofuran | | SOC |
| Chlordane | | SOC |
| Dibromochloropropane | DBCP | SOC |
| o-Dichlorobenzene | 1,2-Dichlorobenzene | SOC |
| m-Dichlorobenzene | 1,3-Dichlorobenzene | SOC |
| cis-and-trans-1,2-Dichlorethylene (acetylene dichloride) | | SOC |
| 1,2-Dichloropropane | Propylene Dichloride | SOC |
| 2,4-D | (2,4-Dichlorophenoxyacetic acid) | SOC |
| 2,4-D | Chlorophenoxys or chlorinated phenoxy acid herbicides | SOC |
| Epichlorohydrin | (1-chloro-2,3-epoxypropane) (ECH) | SOC |
| | Halogenated Alkyl Epoxide (ECH) or chloropropylene oxide or (EPI) | Epichlorohydrin |
| Ethylbenzene | | SOC |
| Ethylene Dibromide | (1,2-Dibromomethane) (EDB) | SOC |
| Ethylene Dibromide | Ethylene Bromide | SOC |
| Heptachlor | | SOC |
| Lindane | Benzene hexachloride, gamma-benzene hexachloride, or 1,2,3,4,5,6-hexachlorocyclohexane (gamma—HCH) | SOC |
| Methoxychlor | | SOC |
| Monochlorobenzene | Phenyl chloride, chlorobenzene, or MCB | SOC |
| Polychlorinated Biphenyls | PCBs | SOC |
| Pentachlorophenol | (PENTA) (PCP) | |
| Styrene | Vinyl benzene orphenylethylene, cinnamene, or styrene monomer | SOC |
| Toluene | Methyl benzene or phenyl methane | SOC |
| Toxaphene | (chlorinated camphene) | SOC |
| 2,4,5-TP | Silvex | SOC |

## Contaminants Classified in This Chapter (*Continued*).

| Organic Chemicals | Other Names | See Details Under |
|---|---|---|
| Xylene | Dimethylbenzene | SOC |
| Atrazine | | SOC |
| Dioxin | 2,3,7,8-TCDD, TCDD, or 2,3,7,8, tetrachlorodibenzo-*p*-dioxin | SOC |
| Endrin + Aldrin + Dieldrin | Endrin | SOC |
| Hexachlorobenzene | (HCB) | SOC |
| Simazine + Propazine + Cyanazine | | SOC |
| 1,1,2-Trichloroethylene | | SOC |
| Bromobenzene | | SOC |
| Dichlorodifluoromethane | | SOC |
| *n*-propylbenzene | | SOC |
| 1,1,1,2-Tetrachloroethane | | SOC |
| Dinoseb | | SOC |
| Benzo(a)pyrene | | SOC |
| Bromoform | | SOC |
| Chloral Hydrate | | SOC |
| Dalapon | | SOC |
| Dichloroacetic Acid | | SOC |
| 1,2-Dichloropropane | | SOC |
| Di(2-ethylhexyl)adipate | | SOC |
| Di(2-ethylhexyl)phthalate | | SOC |
| Diquat | | SOC |
| Endothall | | SOC |
| Glyphosate | | SOC |
| Haloacetic Acids | | SOC |
| Heptachlor Epoxide | | SOC |
| Hexachlorocyclopentadiene | | SOC |
| Oxamyl | Vydate | SOC |
| Picloram | | SOC |
| Tetrachloroethylene | | SOC |

## REFERENCES

A.W.W.A. 1981. **Analyzing Organics in Drinking Water**. Denver, CO.

A.W.W.A. 1982. **Treatment Techniques for Controlling Trihalomethanes in Drinking Water**. Denver, CO.

A.W.W.A. 1982. **Controlling Organics: Research Update**. Denver, CO.

A.W.W.A. 1983. Strategies For the Control of Trihalomethanes, *AWWA Seminar Proceedings* #20174, Denver, CO.

A.W.W.A. 1984. Chloramination for THM Control: Principles and Practice, *AWWA Seminar Proceedings*, Denver, CO.

A.W.W.A. 1984. Non-Specific Organic Analysis For Water Treatment Process Control and Monitoring, *AWWA Seminar Proceedings*, Denver, CO.

A.W.W.A. 1987. **New Dimensions in Safe Drinking Water**, Denver, CO.

A.W.W.A. 1988. **Organics**, Denver, CO.

A.W.W.A. 1990. **Journal**. No. 82. Feb. 1990. p. 32. Drinking Water Regulations. Denver, CO.

A.W.W.A. 1995. **Journal**. No. 87. Feb. 1995. p. 48. Regulatory Update. Denver, CO.

A.W.W.A. 1995. **The SDWA Advisor**. A Regulatory Update Service. Denver, CO.

APHA-AWWA-WEF. 1995. **Standard Methods for the Examination of Water and Wastewater**, 19th edition. Washington, D.C.

California Institute of Technology. 1963. **Water Quality Criteria**, Springfield, VA: NTIS.

CRC. 1985–1986. **Handbook of Chemistry and Physics**, 66th edition, Boca Raton, FL: CRC Press, Inc.

Drinking Water Research Foundation. 1985. **Safe Drinking Water**. Chelsea, MI: Lewis Publishers, Inc.

Kim and Stone, 1980 (Rev. 1981). **Organic Chemicals and Drinking Water**. Albany, NY: New York State Department of Health.

Lappenbush, W. L. 1986. **Contaminated Drinking Water and Your Health**. Alexandria, VA: Lappenbush Environmental Health, Inc.

National Academy of Sciences. 1977, 1980, 1983. **Drinking Water and Health**. The National Research Council, prepared for USEPA, Vols. 1, 3, and 5, Washington, D.C.: National Academy Press.

New York State Dept. of Health. 1988–1989. New York State Sanitary Code, Part 5, Drinking Water Supplies, Albany, NY.

USEPA. 1986. Safe Drinking Water Act, 1986 Amendments. EPA 570/9-86-002, Washington, D.C.

USEPA. 1983. **Methods for Chemical Analysis of Water and Wastewater**. Cincinnati, OH. USEPA Office of Research and Development.

USEPA. 1986. **Interim Treatment Guide For the Control of Chloroform and Other Trihalomethanes**. Cincinnati, OH: Water Supply Div., Office of Research and Development.

USEPA. Rules and Regulations published in the *Federal Register*: Vol. 44, No. 231; Nov. 29, 1979 (Trihalomethanes)/Vol. 46, No. 194; Oct. 5, 1983 Nat. Revised Primary Drinking Water Regulations/Vol. 48, No. 40; Feb. 28, 1983 (Trihalomethanes)/Vol. 49, No. 227; Nov. 23, 1984 Proposed Guidelines for the Health Assessment of Suspected Developmental Toxicants/Vol. 49, No. 277; Nov. 23, 1984. Proposed Guidelines for Exposure Assessment/Vol. 50, No. 6; Jan. 9, 1985. Proposed Guidelines for the Health Risk Assessment of Chemical Mixtures/Vol. 50, No. 219; Nov. 13, 1985 (VOCs). Vol. 50, No. 219; Nov. 13, 1985 (SOCs)/Vol. 52, No. 130; Jul. 8, 1987 National Primary Drinking Water Regulations—Synthetic Organic Chemicals; Monitoring of Unregulated Contaminants—Final Rule/Vol. 54, No. 97; May 22, 1989. National Primary and Secondary Drinking Water Regulations; Proposed Rule.

WHO. 1984. **Guidelines For Drinking Water Quality**. Vols. 1 and 2. Geneva, Switzerland.

# Microbiological Parameters

# Introduction

In water quality control technology, microbiological parameters are indicators of potential waterborne diseases and, in general, are limited to BACTERIA, VIRUSES, and PATHOGENIC PROTOZOA.* The major interest of Public Health Authorities in classifying and issuing standards is the identification, quantification, and evaluation of microorganisms connected with waterborne diseases. There are indeed microbiological contaminants, associated with undesirable tastes and odors, or generators of treatment problems in drinking water technology, such as algae and fungi, but since they are not causes of waterborne diseases, they are not regulated from a public health viewpoint.

Waterborne diseases have not been eliminated by disinfection or by this century's much improved sanitary conditions, not even in the industrialized nations.

The outbreaks of waterborne diseases registered at the present time may be caused by disease-causing agents, that are not too deadly as the slow disappearance of cholera and typhoid fever outbreaks leaves **Gastroenteritis**, **Giardiasis**, **Cryptosporidiosis**, **Shigellosis**, **Hepatitis-A**, and **Legionellosis** to provide the largest number of cases reported in the United States since 1970 or in 1996, according to the Center for Disease Control (CDC) in Atlanta, GA.

The most known, dangerous, or common waterborne diseases are briefly listed in the next section. It would be expected that the water supply will be tested for the pathogens related to the waterborne diseases. At the present time, this is impossible because of the complexity of such testing and the time and cost involved. Therefore, *routine* examination of water for pathogenic organisms is not feasible, and it is not prescribed by the Health Authorities. Even when specific pathogens are examined, there are no answers regarding the remaining pathogens. Even for the specific pathogen, a negative result may be due to the inadequacy of the test as presently described by available methodology, not sufficiently sensitive to detect a low concentration of pathogens in the sample portion examined.

A known test is still used, and was for many years in the past, for routine examination of raw and drinking water: the Coliform Test (see section on *Bacteria*) and the related coliform standards have been used with certain success in evaluating disinfection that prevents outbreaks of bacterial diseases but certainly not against the low levels of virulent pathogens. An improvement in sensitivity of the coliform test could be reached using larger volumes of water samples. Such increased laboratory work can be substituted, with more frequent tests and more sampling points of the distribution system.

---

*Turbidity can be associated with the microbiological parameters in evaluating water quality (see Chapter 2 and Chapters 8 and 9).

However, it must be taken into consideration that all laboratory work becomes meaningless if the water supply is not protected at its source, and so reaches the consumer after adequate treatment with a distribution system free from contamination. It is, therefore, clear that the coliform standards are not applicable in water reuse (for example, secondary treated liquid waste effluent used for a filtered swimming pool). Again, disinfection is not sterilization, and viruses and some pathogenic bacteria will not be destroyed by chlorination and will not appear as positive samples in routine drinking water examinations in the laboratory (Coliforms and Standard Plate Count).

## Waterborne Diseases

Waterborne diseases are certainly not microbiological parameters, but only the etiology of waterborne diseases may conduct to the selection of microbiological parameters. These indicators must be very significant to assess the microbiological condition of the water examined.

The absence of these parameters, such as pathogenic bacteria, viruses, pathogenic protozoa, would not guarantee that the water is safe. Even if it were possible to run such tests routinely, negative results would only indicate the safe water quality at that moment when the water was sampled, and it would not guarantee that the same safe water can be expected one hour before or after. Of course, a large number of samples collected in many points of the distribution system on a daily basis would give certain assurance, but the cost of such analyses would be prohibitive, if not impossible.

The prevention of waterborne diseases is, therefore, left to many precautions to be taken from the source of raw water to the ultimate consumer. With waterborne diseases in mind, the emphasis should be given to the following:

- **Disinfection**—continuous, effective, and adjustable as needed
- **Protection of the source**—also indicated as sanitary survey of watersheds to eliminate pollution
- **Protection of the distribution system**—also indicated as cross-connection survey or prevention of contamination of the distribution system
- **Water treatment**—adequate to the quality of the raw water and suitable to changes required by the variations of the raw water quality

These are the major tools to fight waterborne diseases, to be coupled with acceptable community sanitary conditions.

Treatment and disinfection of water was implemented as a consequence of clear understanding of the microbiological contamination of drinking water.

The most common and significant waterborne diseases have been reported as follows:

- cholera
- typhoid and paratyphoid
- the dysenteries
- diarrhea and enteritis

and as a secondary list of potential waterborne diseases:

- infectious hepatitis
- ascariasis
- tularemia
- ankylostomiasis (hookworm)
- dracentiasis (guinea worm)

- schistosomiasis (bilharzia)
- G.I. outbreaks of obscure etiology
- eye, ear, nose, and throat infections
- nitrate cyanosis (methemoglobinemia), see "nitrate"

The above listing of infectious and morbid conditions does not necessarily indict water as the sole means of transmission, and water is not even responsible for the majority of cases in all situations.

## CHOLERA

The striking epidemicity of Asiatic cholera or infection with *Vibro cholerae* has given well-known cause to drinking water. In conjunction with high fatality rates, cholera remains at the top of the list of potential waterborne diseases. It has been practically eliminated in the last 20 years in the industrialized countries.

While India and China may be considered endemic, the epidemic incidence is preponderant in those countries, too, as well as the rest of the world. Flies and food (especially cut fruit, eaten raw food) may be the major factors in spreading endemic cholera.

The microorganism involved in this acute infectious disease is *Vibrio comma*, group vibrio, family pseudomonadaceae, with habitat in feces of patients infected with Asiatic cholera and water. Optimum temperature for growth is 37°C (limits 14°C–40°C). It is killed in 10 minutes at 55°C.

No pandemic spread was recorded after 1910. Sanitary measures have kept the disease under control with the exception of areas where primitive conditions exist. The development of higher standards of living in primitive parts of the Far East may eventually eliminate cholera epidemics but it still is potentially very dangerous because of high fatality rates coupled with the short incubation period (less than five days) and the common means constituted by the water supply.

## TYPHOID FEVER

Typhoid is the most serious of the communicable illnesses during the last century, with still millions of cases per year at the beginning of this century throughout the world. It has been curtailed to rates lower than 0.1 per thousand inhabitants annually due to vaccination, water treatment, and improved general sanitation.

A relatively large number of chronic typhoid carriers still remains but has been progressively reduced. Typhoid and Paratyphoid groups, included in the genus *Salmonella*, are gram-negative, aerobic, motile rods bacteria which ferment dextrose and many other carbohydrates with the production of acid and gas. Gastrointestinal disturbance is caused after a short incubation period. Paratyphoid fever is a slow, continuous fever of the typhoid type.

Microorganism: **Salmonella typhosa**
Group: Typhoid or **Eberthella**
Family: Enterobacteriaceae
Habitat: Stools of patients suffering from typhoid fever or carriers of
   the organism
Optimum Temperature: 37°C (limits 4°C to 46°C). Destroyed by a tem-
   perature of 56°C in 20 minutes.
Pathogenicity: Infection transmitted also by milk, food, flies, shellfish,
   fomites, and carriers

In water or soil this organism may survive weeks or months. It is not destroyed by freezing. Of people surviving the disease, 30% become carriers excreting the organism in the stool or urine. Approximately 5% remain long-term carriers excreting continuously or intermittently. The expected means of dissemination is water, milk, food, flies, and direct contact. Major epidemics have been attributed to contaminated water supplies. As far as **paratyphoid fever** is concerned, it can be stated that dissemination, clinical cause, diagnosis, prevention, and treatment are similar to typhoid fever.

## DYSENTERY—DIARRHEA—ENTERITIS

In the transmission of enteric infection, bacillary dysentery, diarrhea and enteritis, and amoebic dysentery can be listed in that order following typhoid. More prevalent in tropical climates, the total cases of dysentery, diarrhea, and enteritis are larger in number than typhoid cases. Epidemics of typhoid are usually accompanied by outbreaks of gastrointestinal diseases, salmonellosis, and sometimes shigellosis. In 1926, in Detroit, 45,000 cases of gastroenteritis followed 8 cases of waterborne typhoid.

Group: dysentery or shigella
Family: Enterobacteriaceae
Habitat: intestinal trace of bacillary dysentery and carriers

Optimum Temperature: 37°C

Pathogenicity: gastrointestinal disease accompanied by abdominal pain and diarrhea, notable nervous systems attributable to the absorption of a toxic product.

Dissemination is expected from infected food or water and probably from houseflies.

Dysentery is determined by various microbial agents. **Bacillary Dysentery** is also disseminated by convalescent carriers and rarely by healthy carriers. The incubation period varies from 3 to 7 days. Acute cases may cause death in 2 or 3 days by poisonous products (toxins) in the blood (toxemia).

**Amoebic Dysentery**, where the protozoan endamoeba histolytica is involved, is more chronic and dangerous than the bacillary dysentery with a longer incubation period.

It must be noted that cysts of amoebic dysentery are removed by normal filtration but not destroyed by normal chlorination.

## INFECTIOUS HEPATITIS

There are several waterborne outbreaks of hepatitis clearly identified. Perhaps the most known is the epidemic that occurred in December 1955 and January 1956 in Delhi, India, with 35,000 cases or 2,000 per 100,000 in the single month of the epidemic, where at that time the disease was also highly endemic with an annual morbidity rate of about 300 cases per 100,000.

The Center for Disease Control (CDC) reported for the period 1971–1974 13 outbreaks of hepatitis-A, involving 351 cases.

Infectious Hepatitis is an inflammation of the liver caused by a virus—HAV—the etiologic agent of hepatitis type A (infectious or short-incubation hepatitis). HAV is very resistant to heating (56°C for 30 min.), acid, and disinfecting agents.

This virus can be transmitted through infected human feces (cesspool contaminating water well). Chlorination for 30 minutes at the minimum level of 0.4 mg/L is capable to inactivate the hepatitis virus in heavily contaminated water when coagulation, flocculation, settling, and filtration precede disinfection.

## GASTROINTESTINAL (GI) OUTBREAKS OF OBSCURE ETIOLOGY

These can be defined by acute gastrointestinal illness with a course of several days. The epidemic lasts a week or two, terminating abruptly.

Viruses or bacteria, their toxins and enzymes may be the causing agent. Toxins produced by algae that bloom in raw water reservoirs may also be a causative agent. Inadequacy of treatment and disinfection may be detected, preceding the outbreaks.

## RECENT OUTBREAKS OF WATERBORNE DISEASES

There are many water suppliers and community members who believe that waterborne diseases belong to the past or to Asia and Africa. Unfortunately, this is not so.

The U.S. Environmental Protection Agency (USEPA) in issuing the National Primary Drinking Water Regulations (NPDWR) in November 1985, referred to 427 reported outbreaks of waterborne diseases in the United States involving 106,000 individuals from 1971 to 1983, with 40 outbreaks and 21,000 cases in 1983 alone. Of course, these are only cases reported to the CDC. In a study sponsored by the USEPA in Colorado, it was determined that only 25% of the actual outbreaks were recognized and reported. The above-mentioned USEPA Regulations state the following:

From 1971–1980, 50% of the outbreaks reported were in noncommunity water systems, 39% in community systems, and 11% in private systems. Although most of the outbreaks were from noncommunity systems, about 75% of the illness occurred from outbreaks in community systems. Between 1971 and 1980, the major causes of outbreaks in community water systems were treatment deficiencies (49%) and in contamination in the distribution system (32%). Almost all outbreaks (83%) and illnesses (80%) in noncommunity systems were a result of using groundwater without treatment or using groundwater with inadequate treatment, primarily interrupted in inadequate disinfection. Most known agents of waterborne disease cause acute gastrointestinal disorders, especially diarrhea and cramps. During the period 1971–1983, the most commonly identified pathogen was the protozoan **Giardia lamblia**.* During these years, there were reported outbreaks of waterborne giardiasis involving nearly 23,000 cases. A number of bacteria also have recently been implicated in waterborne disease. These include **Salmonella** species, **Shigella** species, **Campylobacter jejuni, Yersinia enterocolitica** *Pseudomonas sp*, enteropathogenic *E. coli*. Viral agents implicated in recent waterborne illnesses include Norwalk and Norwalk-like agents, rotaviruses, and the hepatitis-A agent. In about half the waterborne outbreaks, the causative agent has not been found. There is growing suspicion that many of these may be due to viruses. Unfortunately, the unavailability of suitable analytical techniques have impaired efforts to resolve this issue.

These statements from USEPA are reported to stress the reality of continuous outbreaks and the necessity to provide conventional treatment and proper disinfection. Of course, proper design and maintenance of water treatment and distribution systems must be accompanied by adequate sampling and laboratory analyses. Such control will trigger alarms with consequent increased disinfection and control of potential pollutional sources influencing raw waters.

More recently, for the period 1986 to 1992, the USEPA confirmed the high number of waterborne-disease outbreaks: 47,000 people were affected in

---

*See *Giardiasis* (*Pathogenic Protozoa*—this chapter).

110 reported cases. The reported cases are expected to be well below the real number of victims, not all of whom are registered in waterborne disease outbreaks. During these more recent years, it should be assumed that stricter regulations from USEPA and from the states had caused more careful disinfection of drinking water and better technological control of conventional water treatment plant operations. Another point in favor of a drop in expected infection cases is the successes obtained or forecasted in protecting the rivers, lakes, and watersheds. Such improvements were expected due to the implementation in the 1980s and 1990s of the new regulations resulting from enforcement of the Clean Water Act, issued and updated by Congress. Also, strict enforcement of the Safe Drinking Water Act (SDWA) was expected to produce better results, not only in organic and inorganic chemicals but also on the microbiological parameters analyzed in this chapter. Nonetheless, USEPA remains pessimistic and now plans additional research on microbiological parameters.

# Bacteria

## IDENTIFICATION

The NAS (1977) in reviewing bacterial agents that cause human intestinal diseases disseminated by drinking water has listed the following:

**Salmonella typhi**—typhoid fever
**Salmonella paratyphi-A**—paratyphoid fever
**Salmonella**—salmonellosis, enteric fever
**Shigella dysenteriae, S. flexneri, S. sonnei**—bacillary dysentery
**Vibrio cholerae**—cholera
**Leptospira** sp.—leptospirosis
**Yersinia enterocolitica**—gastroenteritis
**Francisella tularensis**—tularemia
**Escherichia coli\***—gastroenteritis
**Pseudomonas aeruginosa**—various infections
**Edwardsiella, Proteus, Serratia**, and **Bacillus** are other genera of the Enterobacteriaceae associated with gastroenteritis

## INFECTING DOSE

It is difficult to determine the number of viable pathogenic cells necessary to produce infections. The NAS reported values varying from $10^3$–$10^9$ pathogenic cells per person, with subjects infected representing from 1%–95% of the total subjects tested.

---

*Specific enteropathogenic strains.

Of course, age and general health as well as previous exposure are important factors. The survival of the organism in water, the water temperature, and the presence of colloidal matter in water are also significant influencing factors.

## INDICATOR ORGANISMS

Microbiologists have researched extensively to compare the presence and the significance of specific organisms in water to the traditional coliform group and waterborne diseases to pinpoint the best indicator of contamination in water. Such studies, intensified during the last 30 years, will continue searching for a quick, economic, reliable determination to be used possibly in routine examination or at least during outbreaks of the above-mentioned diseases.

The **Multiple-Tube Fermentation Technique (with results reported in MPN) for members of the Coliform group** was almost exclusively used until 1960 and then substituted by the **Membrane Filter Technique (MF) for members of the Coliform group** because of its simpler and quicker analysis. Both methods are still recognized as sufficiently reliable with relatively simple techniques and economic equipment to be run as often as required by the monitoring activity of water quality control.

## THE COLIFORM GROUP

Originally the target of the contaminating bacteria was the *Escherichia coli*, highly present in the human intestine. It was not and is not considered a pathogen, although it may be associated with pathogenic strains.

*E. coli* are classified enteric bacteria of the family enterobacteriaceae, facultatively anaerobic and able to ferment sugars with the production of organic acid and gas.

*E. coli* differ from the "coliform group" primarily on the basis that they are well defined as sugar-fermentation reaction, motility, production of indole from tryptophan, lack of urease, inability to utilize citrate as sole carbon source, and inhibition of growth by potassium cyanide. More generality is summarized in the 1985 "Approval of the Standard Methods Committee" (*Standard Methods*, 16th edition) for the coliform group: "all aerobic and facultative anaerobic, gram-negative, nonspore forming, rod shaped bacteria that ferment lactose with gas formation within 48 hours at 35°C." This is the definition for the Multiple Tube Fermentation (MPN) technique. This grouping is practical for the purpose of measuring contamination in drinking and wastewater technology. The coliform group includes such organisms in addition to the *E. coli* as the *Klebsiella pneumoniae* and Enterobacter aerogenes. Since *Salmonella* and *S. typhi* do not produce gas from lactose, they would produce negative results for coliform count in spite of their pathogenicity. Operating with the Membrane Filter Technique (MF), the equivalent definition by *Standard Methods* is: "all aerobic and facultative anaerobic, gram-negative, nonspore-forming, rod-shaped bacteria that produce a dark colony with a metallic sheen within 24 hours at 35°C on an endo-type medium containing lactose."

In spite that the group represented by MPN is not the same as the one represented by MF, they basically have the same sanitary significance.

As stated before, the coliform test is a reliable indicator of the possible presence of fecal contamination and, consequently, a correlation with pathogen, but there are deficiencies reported by microbiologists:

- Coliforms may be suppressed by a high concentration of other organisms.
- False-positive results may be caused by aeromonas in warm weather.
- False-negative results may be caused by strains that are unable to ferment lactose.

The **Fecal Coliform** test commonly used in polluted water (organisms that develop on media incubated at 44.5°C) is not useful in drinking water testing because the number of fecal coliforms would be much lower than total coliforms and, consequently, the test would be less sensitive. A similar conclusion can be reached for **Fecal Streptococci** (organisms that develop in a sodium azide medium).

Other tentatives to standardize a better indicator have been tried for **Clostridium perfringens, Bifido bacterium, Pseudomonas,** and **Staphylococcus.**

It must be underlined here that the sanitary engineer and the public health official have to rely on microbiological determination not only for a quick determination of a new condition of contamination, but also for evaluation of long-established monitoring results giving a history of contamination of raw water. Therefore, the change in indicator is only possible after a long period of analyses performed with the old and new method in order to establish a meaningful correlation of data.

While the coliform group is certainly a reliable method, the NAS, USEPA, and Health Authorities are encouraging research for other indicators of contamination in drinking water. Engineers and public health officials are satisfied with the reliability of the coliform tests. Nevertheless, they would prefer to see a more rapid test, necessary to correlate the disinfection treatment with the rapid change of bacterial concentration. In the meantime, such correlation can be established with turbidity readings.

For the time being, the fastest standardized method is the MF technique, reduced to 18 hours.

## STANDARDS

The 1962 USPHS standards officially introduced the MF method, setting the limit at one coliform per 100 mL as the arithmetic mean coliform density of all standard samples examined per month. The standards also called for a single sample, maximum coliform colonies of 3/50 mL, 4/100 mL, 7/200 mL, or 13/500 mL in (a) 2 consecutive samples; (b) more than 1 standard sample, when less than 20 are examined per month; or (c) more than 5% of the standard samples when 20 or more are examined per month.

The standards also called for daily sampling, when standards are exceeded, from the same sampling point and for samples to be examined until the results obtained from at least 2 consecutive samples show the water to be of satisfactory quality.

For the MPN test of 10 mL portions, not more than 10% in any month should show the presence of the coliform group. The presence of the coliform group in 3 or more of the 10 mL portions is not allowable if this occurs (a) in 2 consecutive samples; (b) in more than 1 sample/month, when less than 20 are examined per month; or (c) in more than 5% of the samples when 20 or more are examined per month.

The National Interim Primary Drinking Water Regulations (NIPDWR, 1977 and 1982) issued by the USEPA, basically continue the USPHS regulations recognizing the effect of a single positive sample, when 10 or fewer samples are collected monthly, allowing, therefore, to exclude one positive routine sample from the monthly calculation with the Health Authority's consent based on adequacy of the distribution system's disinfection and prior bacteriological record.*

In 1985, in a proposed rule of the National Primary Drinking Water Regulations (NPDWR), the USEPA proposed a recommended MCL of zero for total coliform, using either MPN or MF method, suggesting that final regulations could specify that 95% of the samples examined over a given period of time have no coliform present. Turbidity limitation was coupled with the microbiological parameters setting a recommended limit of 0.1 NTU (Nephelometric Turbidity Unit).

Since 1965 the American Water Works Association (AWWA) had adopted the goal for bacteriological factors—coliform organism of "no coliform organisms." At the same time, a turbidity goal was set at <0.1 unit. The European Community standards use a maximum concentration of total coliforms of zero for the MF technique and a MPN <1 for the MPN technique for a 100 mL portion. Similar values are given for fecal coliform, fecal streptococcus, with a note for total coliforms stating that 95% of the 100 mL of a sufficient number of samples must not show coliform bacteria throughout the year.

The World Health Organization (WHO) has similar standards. Specifically, the WHO (1993) issued the guidelines contained in Table A-II, page 527, in the Appendix.

Microbiological standards are not so clearly defined as the physical/chemical parameters, presented in Chapters 2, 3, and 4. If we continue to write the microbiological standards in the same style, summarizing USEPA's present regulations, the microbiological parameters are:

---

*USEPA = *Fed. Reg.* Vol. 54, No. 124, June 29, 1989—40 CFR—Parts 141 and 142—NPDWR—Total Coliforms—Final Rule.

MCLG = zero for **total coliform bacteria**, including fecal coliforms and *E. coli*

MCL = Not more than 5% of the monthly samples may be positive. When water systems are allowed to collect less than 40 samples per month, the **Total Coliform Rule** (TCR) allows not more than one sample per month to be positive.

This immediately indicates that monitoring and compliance must be dictated, selecting the minimum total number of samples that must be collected monthly according to the population served; a sampling plan must be established by each water system and approved by USEPA or by the delegated state. Also, regulations must be specific regarding where and when resampling of the "incriminating" positive sample must take place, specifying the number of resampling samples. Then specific rules are expected to be given for reporting to the state the monthly positive and negative samples. This section of the standards is specified in Chapter 8 in regard to total coliform reporting (TCR) (see page 382).

USEPA (1991–1996) **Final Rules**:

*E. coli*: MCLG = zero; MCL = not more than 5% of samples per month may be positive. When water systems* are allowed to collect less than 40 samples per month, not more than one sample per month may be positive.

**Fecal coliforms**: MCLG = zero; MCL = same rule issued for total coliforms or *E. coli*. See Total Coliform Rule (TCR).

*Giardia lamblia*: MCLG = zero; MCL = Treatment Technique.[†]

**Heterotrophic bacteria**: MCLG = MCL = standards not issued and not expected; this final rule is related to surface water; it may be applied in the future to groundwater systems.

*Legionella*: MCLG = zero; MCL = Treatment Technique.[†]

**Turbidity**: MCLG = not issued; MCL = turbidity is usable as performance standard[†] (see Chapters 8 and 10; for description and interpretation, see Chapter 2).

**Viruses**: MCLG = zero; MCL = Treatment Technique.[†]

*Cryptosporidium*: Status of regulation is still "proposed" as MCLG = zero; MCL = Treatment Technique.[†]

---

*"Water system" is an expression that indicates the water supplier of a public water system. Sometimes the term is abbreviated to "system," for example: "large, medium, small public systems." Another term often referred to is "water purveyor" or "public water purveyor."

[†]"Treatment Technique" is a reference to the Surface Water Treatment Rule (SWTR) of 6/29/89 or to the proposed "Enhance Surface Water Treatment Rule," (ESWTR) issued 7/29/94 as an addition to regulations to protect the drinking water in regard to *Cryptosporidium*, to be also effective in the elimination of *Giardia lamblia* and Viruses (USEPA: Filtration and Disinfection; Turbidity, *Giardia lamblia*, Viruses, *Legionella*, and Heterotrophic Bacteria. Final Rule. *Fed. Reg.* 6/29/89 and National Primary Drinking Water Regulations; Enhance Surface Water Treatment Requirements; Proposed Rule. *Fed. Reg.* 7/29/94). See also this chapter for the microbiological indicators and Chapter 11 for SWTR and ESWTR.

USEPA with the Final Rule for Total Coliforms prescribed the following analytical methodology:

**Total coliform analyses** are to be conducted using the 10-tube Membrane Filter Technique (MTF), the Membrane Filter Technique (MF), the Presence-Absence (P-A) Coliform Test, or the Minimal Media ONPG-MUG (MMO-MUG) Test* (Autoanalysis Colilert System). A water supply system may also use the 5-tube MTF Technique (using 20-ml sample portions) of a single culture bottle containing the MTF medium, as long as a 100-ml water sample is used in the analysis.

**A 100-ml standard sample volume** must be used in analyzing for total coliforms, regardless of the analytical method used.

**Fecal coliform analysis** must be conducted using the method described for *E. coli*.

### Escherichia-coli (*E. coli*)

The presence of *E. coli* or fecal coliforms theoretically confirms the suspected presence of pathogenic organisms. There were no approved methods for the determination of *E. coli* when the Final Rules were issued in 1989. USEPA proposed three methods on June 1, 1990 based on the ability of *E. coli* to produce the enzyme Beta-glucuronidase.

In the Jan. 1991† Analytical Techniques Final Rule two methods were approved by USEPA: EC medium plus MUG and nutrient agar plus MUG. USEPA postponed decision on the third method (the Minimal Medial ONPG-MUG (MMO-MUG) test).‡

*Standard Methods** (1995) lists the *Escherichia-coli* Procedure as a **proposed** analytical method (#9221 F). This procedure is used for confirmation after prior enrichment in a presumptive medium for total coliform bacteria. Direct detection of *E. coli* can be obtained using the Chromo Substrate Procedure (Section #9223), approved by the Standard Methods Committee in 1994 and reported in the 19th edition of 1995. In the first procedure EC-MUG medium is used; in the second procedure (#9223) ONPG-MUG medium is used.

### Presence-Absence (P-A) Coliform Test

The presence-absence (P-A) test for the coliform group is a minor change of the multiple-tube procedure. A 100-ml test portion is used in a single culture bottle. The scope is to obtain qualitative information only. The result may indicate a negative result (absence of total coliforms) or may declare the sample

---

*"Standard Methods for the Examination of Water and Wastewater," APHA, AWWA, WEF, 1995, 19th edition.
†USEPA, National Primary Drinking Water Regulations; Analytical Techniques; Coliform Bacteria; Final Rule. *Fed. Reg.* (Jan. 8, 1991).
‡*E. coli* procedure: MUG = the fluorogenic substrate 4-methylumbelliferyl-Beta-D-glucuronide; ONPG = chromogenic substrates, such as ortho-nitrophenyl-Beta-D-galactopyranoside.

positive for total coliform. If positive, the test may be continued to determine other indicators like fecal coliform, Aeromonas, Staphylococcus, Pseudomonas, fecal streptococcus, and Clostridium on the same qualitative basis.

It is also possible and advisable to proceed to a quantitative determination in drinking water quality monitoring or surveillance investigations. *Standard Methods* lists the P-A analytical method in procedure #9221 D. (1995, 19th edition).

### Fecal Coliform Procedure

This test can be performed by one of the multiple-tube procedures or membrane filter methods. *Standard Methods* presents the Fecal Coliform Test (EC Medium) and the Fecal Coliform Direct (A-1 Medium) with approved procedures #9221 E and Section 9222 (19th edition).

The USEPA standards for fecal coliforms are MCLG = zero and MCL = same as total coliforms (see Total Coliforms Rule) (effective disinfection, treatment, operation of water plants, etc.).

### ANALYTICAL METHODS

In this chapter, in both the introduction and the section on "indicator organism," a definition and evaluation of the coliform group are presented. Here the procedures available to laboratories and the data reported by laboratories are briefly examined, since all details are well presented in *Standard Methods*.

### Multiple-Tube Fermentation Technique for Members of the Coliform Group (MPN—Most Probable Number)*

The result of this test is reported in a number based on the mathematical estimate of the mean density of coliforms in the sample. This estimate becomes more accurate when a large number of sample portions are used in tubes to be incubated. Each sample may be diluted in a selected way, when a high coliform concentration is expected. In examining drinking water from the distribution system, a zero or negative result is expected or a minimum concentration of coliform. A typical test is the incubation of five 10 mL portions, and the final result is reported as follows:

| | | | |
|---|---|---|---|
| *Positive* 0 | 5 *Negative* | *less than* | 2.2 *MPN Index* |
| *Positive* 1 | 4 *Negative* | | 2.2 *MPN Index* |
| *Positive* 2 | 3 *Negative* | | 5.1 *MPN Index* |
| *Positive* 3 | 2 *Negative* | | 9.2 *MPN Index* |
| *Positive* 4 | 1 *Negative* | | 16.0 *MPN Index* |
| *Positive* 5 | 0 *Negative* | *more than* | 16.0 *MPN Index* |

---

*Additional information on the following microbiological parameters are provided in Chapter 8, "Microbiological Determination," in the sections dealing with sampling and monitoring: *Total Coliform Group*, Giardia, *Heterotrophic Plate Count*, and *Pathogenic Viruses*.

In the case of 5 positive, the coliform concentrations could have been in the hundredths or thousands, but the results cannot be predicted higher than 16 in numbers, because no dilution was employed. In the case of 5 negative, the MPN cannot be reported 0 but less than 2.2 in the probability calculation.

Higher and lower levels can be obtained using dilution or using a larger volume of portions. For example, using five 100-mL portions (for expected low bacterial concentrations):

| | | | |
|---|---|---|---|
| Positive 0 | 5 Negative | less than | 0.22 MPN Index/100 mL |
| Positive 1 | 4 Negative | | 0.22 MPN Index/100 mL |
| Positive 2 | 3 Negative | | 0.51 MPN Index/100 mL |
| Positive 3 | 2 Negative | | 0.92 MPN Index/100 mL |
| Positive 4 | 1 Negative | | 1.60 MPN Index/100 mL |
| Positive 5 | 0 Negative | more than | 1.60 MPN Index/100 mL |

For example, using three dilutions of 10 mL, 1.0 mL, and 0.1 mL in 5 tubes (from *Standard Methods*, reporting only a few combinations of positive and negative):

| Combination of Positives | | |
|---|---|---|
| 0-0-0 | less than | 2 MPN/100 mL |
| 1-1-1 | | 6 MPN/100 mL |
| 3-2-1 | | 17 MPN/100 mL |
| 4-1-2 | | 26 MPN/100 mL |
| 4-3-0 | | 27 MPN/100 mL |
| 5-0-2 | | 40 MPN/100 mL |
| 5-2-0 | | 50 MPN/100 mL |
| 5-3-0 | | 80 MPN/100 mL |
| 5-4-0 | | 130 MPN/100 mL |
| 5-5-0 | | 240 MPN/100 mL |
| 5-5-2 | | 500 MPN/100 mL |
| 5-5-3 | | 900 MPN/100 mL |
| 5-5-4 | | 1,600 MPN/100 mL |
| 5-5-5 | more than | 1,600 MPN/100 mL |

A simple formula was devised by H. A. Thomas (American Water Works Association *Journal* No. 34, p. 572) for different combinations of tubes and dilutions:

$$\mathrm{MPN}/100\,\mathrm{mL} = \frac{\text{No. of Positive Tubes} \times 100}{\sqrt{\text{mL sample in negative tubes} \times \text{mL sample in all tubes}}}$$

Using a 95% confidence limit, the lower and upper limits of each reported value give an indication of the field values being tested in looking at results in MPN/100 mL. For example (from *Standard Methods*):

| Reported Positive | MPN/100 mL | | Lower | Higher |
|---|---|---|---|---|
| *Out of five 10 mL* | 0 *less than* | 2.2 | 0 | 6.0 |
| | 1 | 2.2 | 0.1 | 12.6 |
| | 2 | 5.1 | 0.5 | 19.2 |
| | 3 | 9.2 | 1.6 | 29.4 |
| | 4 | 16.0 | 3.3 | 52.9 |
| | 5 *more than* | 16.0 | 8.0 | Infinite |
| *10 mL; 1.0 mL; 0.1 mL dilution* | 0-0-0 *less than* | 2 | — | — |
| | 1-1-1 | 6 | 2.0 | 18 |
| | 3-2-1 | 17 | 7.0 | 40 |
| | 4-1-2 | 26 | 12 | 63 |
| | 4-3-0 | 27 | 12 | 67 |
| | 5-5-2 | 500 | 200 | 2,000 |
| | 5-5-4 | 1,600 | 600 | 5,300 |

To estimate the time required for the test and to interpret the data from the laboratory, it is useful to know the following aspects of the procedure (see also upper part of Fig. 5-1).

**First Presumptive Phase**   Tubes are incubated for 24 ± 2 hours at 35°C in lauryl tryptose broth.

**Second**   Checked after 24 hours;
If negative, incubate for additional 24 hours.
If positive, transfer to "brilliant green lactose bile broth," and incubate for 48 hours at 35°C.
**Third**   If negative after 48 total hours or 72 total hours, the sample portion is considered **negative**. This is the **confirmed case**.
If positive after 48 total hours, transfer to "brilliant green lactose bile broth."
If positive after 48 hours in "brilliant green lactose bile broth," they are classified **positive** or coliform group. **Confirmed**.

Therefore, the laboratory can report the following:

- After 24 hours, a number of presumptive positive.
- After 48 hours, a number of presumptive negative and/or a number of presumptive positive.
- After 72 hours, a number of confirmed positive and/or a number of confirmed negative.
- After 96 hours, a number of confirmed positive and/or a number of confirmed negative (these are the presumptive negatives that have been confirmed on "brilliant green lactose bile broth" for an additional 48 hours, finally turning either positive or negative).

To be considered "**positive**," the incubated portion of the sample must show formation of gas in any amount in the inverted tubes or vials at the end of the incubation period. The gas formation is accompanied by cloudiness in the broth. Gas formation in the brilliant green broth is reported as "**confirmed**" and a value is given for MPN/100 mL according to the number of positive tubes in the sample.

If this sample comes from raw water or from the water treatment plant prior to final chlorination, no further tests are necessary and the value is recorded for the monthly report and water quality evaluation.

But if such a sample has been collected after final chlorination or from the distribution system or from a faucet of consumer's complaint case, it is necessary to complete the test to demonstrate that coliform bacteria are involved and to control and confirm the procedure used for the "presumptive" and "confirmed" tests. It is again repeated here that coliform bacteria resulting either from the "confirmed" or from the "completed" tests are not a direct knowledge of the presence of pathogens but an assumption that pathogens may be present, particularly when high coliform density is reported. In all cases, a "completed" test positive result is a clear indication of contamination and, more likely, an indication of recent contamination since coliform bacterial have less survival resistance than bacteria measured with a Standard Plate Count (heterotrophic bacteria). Unfortunately, pathogenic bacteria and viruses appear to have longer survival rates than coliform bacteria.

**The Completed Test**  A positive result for the "completed test" is reported when formation of a gas in the secondary tube of lauryl tryptose broth within 48 hours is recorded with demonstration of gram-negative, nonspore-forming, rod-shaped bacteria from the agar culture (see Figure 5-1). The schematic diagram showing the completed test includes in the first line the **Presumptive** positive and in the final incubation in the brilliant green lactose bile broth shows the **Confirmed** positive. The test continues with the transfer from the gas producing tubes into the LES endo agar plate with an inoculating needle for an incubation period of 24 hours at 35°C.

The procedure continues with the recording of negative colonies, if any, and the transfer of the colonies developed on LES endo agar that will be classified as follows:

Typical = nucleated with or without metallic sheen
Atypical = opaque, unnucleated, mucoid, and pink after 24 hours
Negative = all others

Possibly using the *typical* or similar colonies, a transfer is made into a lauryl tryptose broth fermentation tube and onto a nutrient agar slant,* incubating the broth for 24–48 hours and the agar between 18–24 hours, both at 35°C.

The gram-stain portion of agar procedure that completes the test may be avoided for drinking water samples in consideration that it is very unlikely

---

*Slant culture or slope culture is a slanted, thin layer of solid medium dispensed in tubes, sloped before setting to facilitate incubation by wire loop.

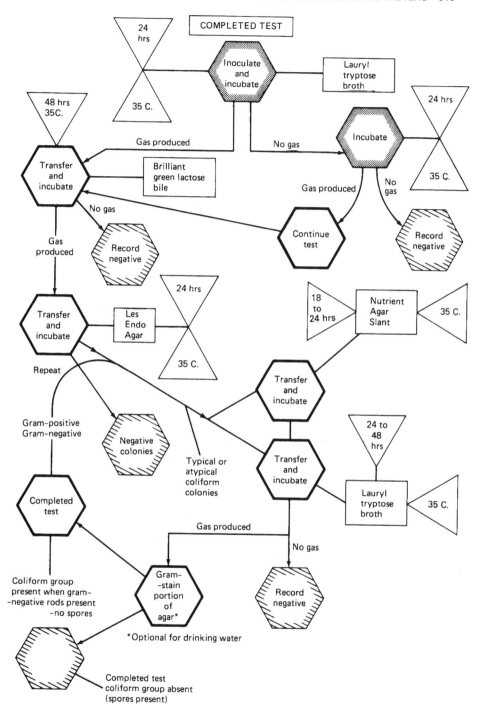

Figure 5–1. Completed Test. Multiple-Tube Fermentation Technique for Members of the Coliform Group (MPN).

that gram-positive bacterial and spore forming organisms survive such selected procedure in water of very low turbidity (<10 NTU).

Gram stain is a microscope procedure requiring a bacteriologist's skill and experience:

Gram + Organisms are blue or violet

Gram − Organisms are pink or violet

**Fecal Coliform (MPN) procedure** (EC Medium) can be used as an alternative method for raw water.

The *Standard Methods* (1985) described a "tentative" method, "Presence-Absence (P-A) Coliform Test" which is a simple modification of the multiple-tube procedure. In drinking water sampling, it is important to know as soon as possible when a sample is positive in order to investigate and resample. The P-A test requires also 24–96 hours for "presumptive" to "confirmed."

## Membrane Filter Technique (MF)

When compared with the Multiple-Tube Fermentation technique (MPN), MF has the advantage of requiring less time and giving more definite results, but MF may not be as reliable as MPN when the sample turbidity is high or whenever high concentration of noncoliform bacteria is expected.

When extensive records are available for the water supply in question analyzed with the MPN technique, it is advisable to analyze now with both techniques so as to compare data when the decision is reached to use MF only.

The standard volume to be filtered in this method is 100 mL. Larger volumes can be used, such as 1L subdivided in portions, and then the portion results combined per liter are reported. But with the goal of zero coliforms from the distribution system, numerical count of bacterial concentration has reduced importance as long as well over 95% of the samples routinely collected are negative.* Numerical values remain very important in special surveys or studies.

*Standard Methods* describes in detail the technique, equipment, materials and culture media. The laboratory reports results as total coliform density per 100 mL. If more than 200 colonies are counted, the report will be TNTC "too numerous to count," otherwise the laboratory will report

$$\text{(total) coliform colonies/100 ml} = \frac{\text{coliform colonies counted} \times 100}{\text{mL sample filtered}}$$

---

*See Standards for coliforms (USEPA, 1987 or WHO, 1984).

Using a 95% confidence limit it can be seen that, in spite of the greater "statistical" reliability of the MF versus the MPN, a reported number is to be judged within the following variations:

| Coliform Colonies Counted | Lower Limit | Upper Limit |
|:---:|:---:|:---:|
| 1 | 0.025 | 3.0 |
| 2 | 0.35 | 4.7 |
| 3 | 0.81 | 6.3 |
| 4 | 1.4 | 7.7 |
| 5 | 1.6 | 11.7 |
| 10 | 4.8 | 18.4 |

The Membrane Filter Technique requires simple equipment. Laboratory technicians are easily trained; inoculation can be achieved in the field immediately after sampling when surveys are conducted.

## STANDARD PLATE COUNT—HETEROTROPHIC PLATE COUNT

Practically all water plants and all public water supply distribution systems have a historical record of laboratory examination for Standard Plate Count (SPC), also recorded under traditional abbreviations as "total count," "plate count," "bacterial count," "water plate count," or "total bacterial count." This test provides an estimate of the total number of bacteria in a sample that will develop into colonies during a period of incubation in a nutrient.

### Standard Plate Count

Incubation time for SPC is set at 35°C for 24 hours or 20°C for 48 hours. The culture medium recommended is tryptone glucose extract agar. The 48-hour incubation period is preferable. Sample water volume is planted in such amount (usually one mL) to provide from 30–300 colonies on a Petri dish or plate. The results are reported as the average of all plates containing a number of colonies within the above-mentioned limits.

*Standard Methods* recommended (not reported in the 16th edition) to use only two significant figures to give a more realistic view to the questionable accuracy; i.e., 122 is reported as 120, 135 is reported as 140, and 35 is reported as 35.

It must be noted that only a portion of the total bacteria in the sample is represented in this count, namely, the bacteria that grow at that temperature, in the stated time and media. The SPC data never obtained a high degree of recognition in interpretation of water contamination data probably because official standards were never issued. Also, the high variation of concentrations, let us say from 100–600 colonies per mL, left the water supply still classified as potable when total coliform were negative. A certain respect for the test reappeared when it was demonstrated that high concentration of bacteria may suppress the growth of coliforms, questioning the validity of the statement that the absence of coliforms would necessarily indicate a safe supply even when bacterial count is high.

Interpreting SPC results, it is wise to assume that when plate count is over 200/mL of colony forming units, some problems may have developed in water storage or the distribution system requiring additional chlorination. It is expected, judgmentally, that when 0.2 mg/L of free chlorine residual is constantly maintained in stored water or in a distribution system, standard plate counts are expected under 50/mL densities.

Testing 923 water systems with required chlorine residual, 60% gave SPC densities of 10/mL or less as reported by the NAS (1977) and Geldreich (1973) (*Proc. Am. Water Works Assoc.*, Water Quality Technology Conference, Cincinnati, OH, Dec. 1973). It is the author's experience that the SPC is a valuable tool to investigate the variation of water quality, particularly when the turbidity is higher than 0.4 NTU or when the size of the distribution system is such that a minimum of 0.2 ppm of chlorine residual cannot be constantly maintained.

The test is not a substitute for the MPN or MF tests, but it should be used in conjunction with the total coliform counts. The NAS (1977) stated in reviewing SPC that, "the scientific information available makes it reasonable to establish the upper limit of the SPC initially at 300 mL, as developed in 35°C, 48-hr plate count procedure."

## Heterotrophic Plate Count (HPC)

The Standard Methods Committee approved in 1985 a major change to the SPC; the 16th edition (1985) of *Standard Methods* presented three alternative methods and two new media attempting to improve recovery of the heterotrophic bacterial population. It must be recognized that this improved method does not pretend to report all bacteria of the water sample examined, but it is definitely an improved procedure and a valuable test to measure water treatment plant efficiency and/or after growth typical of transmission lines (distribution system joints, valves, and dead ends).

*Standard Methods* lists the following three different procedures:

1. **Pour Plate Method**—48-hr. 35°C incubation with tryptone glucose extract agar and a 48-hr, 35°C incubation with tryptone glucose yeast agar.
2. **Spread Plate Method**—48-hr, 35°C incubation with same media as in No. 1.
3. **Membrane Filter Method**—m-HPC agar or R2A agar (72-hr at 35°C for R2A).

The laboratory is expected to report in Colony Forming Units (CFU) stating the method use, the temperature, the incubation time, and the medium: for example; CFU/mL—pour plate method—35°C/48-hr, plate count agar.

## STANDARDS

USEPA (Nov. 1985) NPDWR recognized the importance of sampling and analyzing SPC and now the Heterotrophic Plate Count (HPC). The importance is due to "Many heterotrophs in water are opportunistic pathogens* (30% in one study). There is some evidence that numerous hospital-acquired infections have been caused by waterborne opportunistic pathogens," according to USEPA. In addition, USEPA recognizes the water quality deterioration implied by high densities of HPC and the previously mentioned interference of coliform growth in MPN or MF, when HPC shows densities higher than 500 colonies/mL.

A recommended MCL[†] has not yet been proposed, but the agency "is considering incorporating a level of HPC control into the total coliform monitoring requirements since high densities of HPC interfere with coliform analysis."

It is expected that Health Authorities will encourage or issue standards establishing a certain percentage of samples collected and analyzed from the distribution system to be also analyzed for SPC or HPC and consider "potable" water with counts under 100/mL and "questionable" for values between 100 and 500/mL. It is here stressed again that relationship factors exist between bacteria concentration and the lack of disinfection of groundwater source and distribution systems as well as turbidity, not sufficiently removed by a conventional water treatment.

## HETEROTROPHIC BACTERIA (HPC)

No MCLG/MCL is proposed. Proposed regulation is based on HPC **interference** with total coliform analysis.

If the coliform sample produces a turbid culture in the absence of gas production using the multiple-tube fermentation technique, or produces a turbid culture in the absence of an acid reaction using the presence-absence (P-A) test, or produces confluent growth or a colony number that is "too numerous to count" using the membrane filter technique, the system may either accept the sample as coliform-positive or declare the sample invalid and collect and analyze another water sample. The second sample is analyzed for both total coliforms and HPC. If the HPC is greater than 500 colonies/mL, then the sample is considered coliform-positive, even if total coliform analysis is negative.

---

*Pathogenic to the sensitive population (age, illness, etc).
†USEPA. 1989. Limited MCLG-0 to *Giardia Lablia*, Viruses, and *Legionella* (*Fed. Reg.* Vol. 54, No. 124–40 CFR Parts 141 and 142, NPDWR. 6/29/89).

## Legionellaceae—Legionellae

*Standard Methods* subdivides *Legionella*, a gram-negative, aerobic, nonspore-forming bacteria, into at least seven species* with **Tatlockia** as a second and **Fluoribacter** as a third proposed genus. Most of these species have been implicated in outbreaks of pneumonia-like diseases encountered in the United States from 1947 but most prominently from the 1976 Legionnaires' Convention in Philadelphia. Transported via the airborne route, legionellae may be originated by environmental conditions of cooling towers, river water, mud, and excavation sites near hospitals and hotels. The USEPA (NPDWR, 1985) estimates between 50,000 and 100,000 cases of legionellosis to occur annually within the United States, caused by one of the 26 currently recognized species of the genus *Legionella*. The Health Authorities have issued guidelines for *Legionellae* in finished water to be zero; but it is not practical to monitor for these organisms. Substitutes for MCL are filtration, disinfection, and cross-connection control implementation. Air-conditioning, cooling towers, hot water systems in institutions and homes, suspected to allow growth of *Legionella*, should be controlled to help to remove and/or inactivate *Legionella*.

# Viruses

Since viruses are parasitic entities, they are unable to multiply or adapt to the hostile water environment and lack of living host. These particles still can infect humans through recently contaminated drinking water, the relativity of time being connected with the survival ability of these particles in the natural and man-made hostile environment.

The viruses examined by the water microbiologist are practically limited to enteric viruses (infections of the intestinal tract).

Isolation of viruses has improved considerably during the last 35 years, particularly by the fear that **Poliomyelitis** epidemics were transmitted by the water route (with limited scientific confirmation) and then by many reported episodes with conclusions that **Infectious Hepatitis** was transmitted by drinking water. Virologists made considerable progress in isolating and reducing in study groups the viruses of interest to water supply examiners. In spite of this virological progress in recovering and identification, unidentified viruses may also be the cause of waterborne gastroenteritis and diarrhea of unknown etiology.

Recovery of viruses from water to be tested is a complicated procedure, in general limited by the presence of interfering substances in water (turbidity, organic matter, salts, heavy metals, pH), by the difficulty of the laboratory host system (experimental animals and cells), and prominently by the limited concentration of viruses in drinking water. Enteric viruses are detected using primary-monkey-kidney or human embryonic kidney cell cultures.

In this procedure the virus of hepatitis-A and other viruses are not recovered in spite of prominent laboratory updated procedures. The NAS (1977) recognizes that presently available methods are adequate for recovery of enteroviruses but perhaps not of reoviruses (found in the respiratory and enteric tracts of animals and man), and adenoviruses (associated to epidemic or sporadic cases of respiratory diseases, isolated from tonsils and adenoids and from fecal specimens).

Particularly during the last 25 years, efforts were concentrated to determine certain viruses as indicator viruses. Polioviruses could be detected by a rapid method, but enteric viruses could be detected also by a similar recovery method, making unnecessary the selection of poliovirus as an indicator virus. It must be pointed out that more than 100 serotypes of enteric viruses are known and are recoverable from wastewater normally.

In 1977, the NAS, in their book *Drinking Water and Health*, reached the following conclusions:

1. The presence of infecting virus in drinking water is a potential hazard to public health, and there is no valid basis on which a no-effect concentration of viral contamination in finished drinking water might be established.
2. Continued testing for viral contamination of potable water should be carried out with the facilities and skills of a wide variety of research establishments, both inside and outside of government, and methodology for virus testing should be improved.
3. The bacteriological monitoring methods currently prescribed or recommended in this report (coliform count and standard plate count) are the best indicators available today for routine use in evaluating the presence in water of intestinal pathogens including viruses.

The above-mentioned statement no. 1 should not be interpreted to mean that viruses are not removed by conventional treatment and disinfection. On the contrary, drinking water produced by an effective conventional treatment and distributed after disinfection is expected to have significantly reduced concentrations of viruses inactivated by the treatment. The infectious dose of viruses by the oral route is not yet known, leaving questionable interpretation of a minimum viral contamination. Several studies have been conducted to evaluate the viral removal of the components of traditional water treatment plant. They can be summarized as follows:

**Coagulation/Flocculation and Settling**. Virus removal over 90% expected under carefully controlled conditions.

**Filtration**. Virus removal of over 90% was reported for sand filtration* preceded by coagulation and sedimentation. Both results were obtained with diatomaceous earth and activated carbon media. Nevertheless it may

---

*The virus itself is nonfilterable, but host organisms and adjacent colloidal matters are reduced by the sedimentation/filtration process.

not be so efficient in actual operation, because of desorption or resuspension from filters in need of backwash.

In both treatments no inactivation of viruses is reached. Such inactivation may be reached in water softening process with lime, if a high pH (near 11) is maintained under treatment.

**Disinfection.** This is the most efficient viral removal treatment at low turbidity levels, i.e., when the maximum effectiveness of disinfection can be reached.

Concentration of chemical and contact time are the important factors. A free chlorine residual of 0.4 mg/L and a 30-minute contact time was reported by the NAS (1977) as a disinfection treatment that inactivated viruses added to the drinking water, then disinfected and ingested by volunteers.

Temperature and pH are factors influencing disinfection. Bromination and iodination showed similar ability to inactivate viruses as chlorination. Ozonation is also showing equivalent ability to inactivate viruses in drinking water of low turbidity.

USEPA (NPDW—40 CFR Part 141—Nov. 13, 1985) examined available literature on pathogenic viruses and concluded with a proposed goal (MCL) for human pathogenic viruses in drinking water of zero. USEPA recognizes that between 1978 and 1982, 18 reported waterborne disease outbreaks were caused by viruses with over 5,700 cases. These outbreaks are to be considered in addition to the outbreaks of unknown origins caused by viruses. Norwalk and Norwalk-like agents, rotavirus (causing gastroenteritis), and hepatitis-A agent (causing hepatitis) are listed as the most likely to be involved in outbreaks. Less common, but potentially more dangerous (causing meningitis, paralysis, myocarditis, diarrhea) are the enteroviruses (poliovirus, echovirus, and coxsackievirus) not involved in recent outbreaks but occasionally found in drinking water.*

USEPA pointed out that some strains of waterborne viruses have a minimum infective dose at very low concentration (17 plaque-forming units [PFU] of echovirus, 12 may be considered sufficient to infect 1% of a tested population).

USEPA[†] recognized the difficulty of routine procedures for detection of the most important waterborne viruses. The WHO and the European Community standards have a zero goal for pathogenic viruses as measured by the enteroviruses.

## STANDARD METHODS

A revised section on "Detection of Enteric Viruses" was approved by the Standard Methods Committee in 1985. All proposed methods are still con-

---

*"Transmission of Viruses by the Water Route" by G. Berg., et al (1966), FWPCA—DOI. J. Wiley & Sons. New York.
†USEPA (1985) and USEPA (1987) *Fed. Reg.* Vol. 50. No 219, 11/13/85 and Vol. 52. No. 212. 11/3/87—NPDWR Proposed Rule.

sidered "Tentative." The use of such tests is proposed not for routine analyses, but for special research studies, disease outbreak investigations, or wastewater reclamation projects, performed by competent and specially trained water virologists having adequate facilities.

After collection of samples one difficult step preceding the identification is the procedure to concentrate the viruses in the sample. Concentration is necessary due to very limited presence of viruses in water and, consequently, the need for a very large volume of water to be tested. For the concentration procedure *Standard Methods* (1985) listed the following tentative methods:

- Virus Concentration by Adsorption to and Elution from Microporous Filters (usable for drinking water).
- Virus Concentration by Aluminum Hydroxide Adsorption-Precipitation.
- Hydroextraction-Dialysis with Polyethylene Glycol.
- Recovery of Viruses from Suspended Solids in Water and Wastewater.
- Assay and Identification of Viruses in Sample Concentrates.

The USEPA (1987) concluded on *Giardia* and viruses that it is not economically or technologically feasible to measure the level of *Giardia* or enteric viruses in drinking water for the following reasons: Presently acceptable methods require levels of expertise that utility personnel normally do not possess or the methods would be too expensive if analyzed by private laboratories. Validation procedures have not yet been established. Continuous monitoring would be required. Monitoring also does not provide advance notice to assure the safety of drinking water at the consumer's tap.

The maximum contaminant levels for *Giardia* and viruses remain zero as a goal or guideline.*

## Pathogenic Protozoa

As potential causes of waterborne diseases, protozoan organisms are in the minds of waterworks scientists as a connection to the *Entamoeba histolytica*, the cause of amebic dysentery and amebic hepatitis, and *Giardia lamblia*, the protozoan parasite presently causing a very large number of gastrointestinal outbreaks. Sewage contamination transports the eggs and cysts of parasitic protozoa and helminths (tapeworms, hookworms, etc.) into raw water supplies, leaving to water treatment and disinfection to diminish the danger of contaminated water to the consumer.

---

*See References, USEPA (1989). See also Chapter 8 (Microbiological Determination).

## GIARDIASIS

Symptoms: diarrhea, fatigue, abdominal cramps, and other gastrointestinal symptoms.

The NAS (1977) reported 14 outbreaks of giardiasis in the United States between 1969 and 1975 involving at least 700 cases. The Rome, NY, 1974–1975 outbreak has been well documented with 395 cases of symptomatic giardiasis and an attack rate of 10.6% influencing 4,800 persons considered affected. More cases have been recently reported to the Center for Disease Control (CDC) due to the improved understanding of *Giardia* infection.

Tourists in the former Leningrad, U.S.S.R., reported acquiring giardiasis in 23% of 1,419 travelers with 14 days elapsing between the first day in the U.S.S.R. and the appearance of symptoms with illness lasting 6 weeks. Consumers of tap water reported the highest attack rate.

USEPA (NPDWR—1985) stated that protozoan *Giardia lamblia* caused more waterborne disease outbreaks in the U.S. than any other single, identified, etiologic agent, causing gastrointestinal symptoms lasting for several days to several months. The CDC reported 23,000 cases in 77 outbreaks between 1971–1983. Most of the cases involved water supplies treated with disinfection only with limited or no free chlorine residual. Some cases involved conventional treatment but with inefficient operations. Surface supplies untreated, particularly with limited or no chlorination, are more susceptible to *Giardia* contamination danger to the consumer.

USEPA estimates that in the United States there are 65 million people presently served by surface water systems that provide only disinfection or no treatment at all. Disinfection alone is not sufficient protection for the elimination of *Giardia*. In studies on humans and animals, the assumption was confirmed that even a very low dose of *Giardia* is infective in humans and animals.

Based on the above-mentioned information, USEPA proposed in 1985 a RMCL of zero viable cysts.

*Standard Methods* first introduced *Giardia lamblia* as a flagellate protozoan shed in the feces of man and animals most often in the cyst stage, with an infective dose for humans at 10 or perhaps fewer viable cysts ingested in drinking water. Cysts, shed in excess of $10^6$ cysts/g of stool, may survive for months in drinking water at 8°C.

Special sampling techniques shall be used starting with 380 L (100 gal) of the sample collected with a *Giardia* sampling device described in the 16th edition of *Standard Methods*. Results are reported as *Giardia* cyst-like structures larger than 8 μm and less than 19 μm, possibly documenting with photomicrographs of subject cysts (μm = $10^{-6}$ meter).

USEPA concedes that the present method is not available for monitoring of the waterborne *Giardia*, because the examination is difficult in recovery and detection, beside the impossibility to distinguish between viable and nonviable cysts.

# CRYPTOSPORIDIUM

*Cryptosporidium* is a protozoan, a parasite that infects humans and a large variety of animals. Known since 1907, *Cryptosporidium* was not considered a pathogen to humans, until waterborne disease outbreaks involved not only untreated water but even drinking water distributed from a conventional water treatment plant. Recognized as a pathogen since 1976, this protozoan has been recently investigated to evaluate the infection cycle, the path of the oocysts in the metabolism of water and food. USEPA was particularly concerned about this "new" pathogen. In the reexamination of regulations on water treatment and disinfection, USEPA had to issue MCLG and MCL for *Cryptosporidium*. The similarity to *Giardia lamblia* and the necessity to provide an efficient conventional water treatment capable of eliminating viruses at the same time, forced USEPA to regulate the surface water supplies in particular. The "Enhance Surface Water Treatment Rule" (ESWTR) was proposed including regulations from watershed protection to specialized operation of treatment plants (certification of operators and state overview) and effective chlorination that beyond the bactericidal, and virucidal, effects respects the limits of TTHM. Protections against *Cryptosporidium* include control of waterborne pathogens, such as *Giardia* and viruses.

## Cryptosporidium: The Pathogen

Of the four species recognized, two are related to mammals: *C. parvum* and *C. muris*. The illness of humans and animals is primarily connected with *C. parvum*. The oocysts (defined as a stage in the development of any sporozoan in which after fertilization a zygote is produced that develops about itself an enclosing cyst wall; zygote is the developing ovum), once ingested and reaching the small intestine, split open, releasing sporozoites.* Additional oocysts are created, additional sporozoites are generated, and the infection is continued unless they are excreted in the feces.

## Cryptosporidium: The Environment

*Cryptosporidium* was confirmed in raw sewage, treated wastewater, and effluents from a cattle slaughterhouse in concentrations as high as 14,000 oocysts. Unfortunately, the oocysts were also found in rivers, lakes, reservoirs, and in treated waters. Infections were detected in many animals, including cattle, dogs, cats, deer, rabbits, etc.

---

*A *sporozoite* is defined as any of the cells resulting from the sexual union of spores during the life cycle of a sporozoan. It refers specifically to the elongated, nucleated cells produced by the multiple fission of the zygote contained in the oocyst of the female.

## Cryptosporidium: Cryptosporidiosis

The incubation varies between 2–12 days (average = 7 days). Symptoms are diarrhea, abdominal cramps, nausea, occasional vomiting, and a low fever. The condition may last 10–14 days; children are the most susceptible. Recovery is related to the patient's immune system. Diarrhea treatment is in general applicable to treatment of cryptosporidiosis. The disease is more serious for the sensitive population (infants, chronically ill, and cancer patients) but it can be fatal for the immuno-suppressed, AIDS patients, organ transplant recipients, or the sensitive population. There are nevertheless individuals who do not suffer from Cryptosporidiosis, in spite of the contamination or exposure to *Cryptosporidium*.

The infective dose is not easy to be determined, but it may vary between 10 and 500 oocysts.*

## Cryptosporidium: Sampling and Detection

Sampling requires the collection of large volumes of water (raw and treated) to be significant for "positive" and "negative" sample determination. Oocysts must be filtered, concentrated, and examined by microscopy.

The absence of oocysts cannot be interpreted as indicating that the water supply tested is free of *Cryptosporidium*, even if the laboratory results are also negative for coliform, fecal coliforms, and *Giardia lamblia*. The contamination of its oocysts may be found later during different water temperatures, or seasons, or following flooding conditions. It is useful to create a sampling background information in case of positive samples for the water system of interest, and, in general, for a first period of "orientation" and public information. The cost, the time, the difficulty of the interpretation of data are not presently suggestive to develop a routine sampling program. It all depends on the expectations that prevention and effective water treatment plant operation may be capable to eliminate coliforms, fecal coliforms, strongly reduce viruses, *Giardia*, and perhaps other pathogens. It is granted that *Cryptosporidium* is probably more resistant to disinfection than other pathogens. Disinfection and watershed control and quality and training of plant operators are essential parameters for compliance with a standard of MCLG = zero.

See also the **Surface Water Treatment Rule**, the **Enhanced Surface Water Treatment Rule**, and the **Disinfection/Disinfection By-Products Rule** in Chapters 8, 10, and 11.

---

*Pontius, F. W. (1993). "Protecting the Public Against *Cryptosporidium*." A.W.W.A. *Journal*. No. 85, Aug. 1993, p. 18. A.W.W.A.: Denver, CO.

## Cryptosporidium: The Experience

Many studies have been conducted after the first very significant outbreak, in April 1993, in Milwaukee, WI, when 400,000 people became ill with cryptosporidiosis and over 100 deaths were registered. Of course, before that other outbreaks had occurred in this country and in England.

We learned several lessons from these studies. Having a certain scientific experience is extremely important to those to whom the responsibility belongs for issuing new standards for prevention, water treatment, disinfection, protection of the sensitive population, proper identification and count of oocysts (new standard methods). In addition, it was necessary to determine the infection dose for cryptosporidiosis and what intelligent information should be given to the public after detection of *Cryptosporidium* at the water plant.

What did we learn?

*Cryptosporidium* is prevalent, or even expected, in surface waters; in streams, rivers, lakes, reservoirs, in the raw water and sometimes in treated water, especially in those not properly equipped and controlled conventional treatment plants. Some infected people may not be disturbed by the oocysts, but they become carriers for transmission of this pathogenic protozoan.

A median dose of infection can be approximately established in 130 oocysts.

Cryptosporidiosis does not have a special drug or treatment for cure; antidiarrhea drugs are advised together with the attempt to reduce the expected dehydration.

Extra precautions should be taken by people with a weakened immune system; the agency for information is the Center for Disease Control in Atlanta, GA, which provides a special service (AIDS Hotline).*

There do not as yet exist any USEPA approved standard methods; this is a serious limitation in the validity of the routine sampling for *Cryptosporidium* in the water system monitoring program.

What else can the water supplier do to protect the public? Kenneth J. Miller,† who studied the Milwaukee plant in 1993, suggested the following actions in 1994 (summarized here):

Take all the possible precautions to protect the watersheds from human and animal wastes. If possible use ozone instead of chlorine, due to the more effective results of ozone on *Cryptosporidium*. Monitor the water system for coliform, fecal coliform, *Giardia* and *Cryptosporidium*. Try to mantain a filtered water with turbidity of less than 0.1 NTU, in spite of a nonconstant relationship of turbidity with oocysts concentrations. Curtail the use of recycling backwash water. A rehabilitation

---

*Pontius, F. W. "*Cryptosporidium:* Answers to Common Questions," A.W.W.A. *Journal.* Vol. 87, Sept. 1995, p.10. A.W.W.A.: Denver, CO.
†Miller, Kenneth J. "Protecting Consumers From Cryptosporidiosis," A.W.W.A. *Journal.* Vol. 86, p. 8—"A Viewpoint," Dec. 1994. A.W.W.A.: Denver, CO.

of the existing surface water treatment plant to reduce further the turbidity value of 0.1 NTU should be examined.

## HELMINTHS

The NAS (1977) considers that drinking water in the U.S. may transmit among intestinal worms (nematodes) the following:

- **Ascaris Lumbricoides**—the stomach worm
- **Trichuris Trichiura**—the whip worm
- **Ancylostoma Duodenale** ⎫ hookworms
- **Necator Americanus** ⎭
- **Strongyloides Stercoralis**—the threadworm

Derived from sewage and wet soil, nematodes multiply in aerobic biological wastewater treatment plants (particularly trickling filters) and, therefore, appear in large concentration in treated domestic liquid waste.

Nematodes ingest bacteria, so entering the distribution system, and they may protect pathogens from disinfection of the water supply and reach the consumer.

Active motile nematode larvae can penetrate sand filters and survive chlorination, but they are not normally expected to cause parasitic nematode infections, according to NAS (1977).

Nematodes may be macroscopic ≤1 cm long) or microscopic (≤0.5 mm long), feeding on bacteria, yeast, and minute algae. Free-living nematodes have a life cycle consisting of egg, four larval stages, and one adult stage. Eggs are easily recognizable in finished water, while raw water may have excessive microfaunal forms to allow identification.

*Standard Methods* has a section for *Nematological Examination*, approved by the Standard Methods Committee in 1985, detailing sample collection and printing an "Illustrated Key to Fresh Water Nematodes," listing 100 names and graphical representations.

In general, it can be stated that pathogenic parasites may not be completely eliminated by conventional treatment and chlorination, particularly true for *Entamoeba histolytica* and *Giardia lamblia*. Ozone appears more effective than chlorination. Also, bromination appears more effective than chlorination and iodination. Filtration facilities can be adjusted in depth, prechlorination, filtration rate, and backwashing to become more effective in the removal of cysts. The pretreatment of protected watershed raw waters is a major factor in the elimination of pathogenic protozoa (see the section on *Pretreatment* in Chapter 10).

## REFERENCES

APHA, AWWA, WEF, 1995. **Standard Methods for the Examination of Water and Wastewater** (19th edition). Washington, D.C.: American Public Health Association.

American Water Works Association. 1985. *Giardia Lamblia* **in Water Supplies—Detection, Occurrence, and Removal**. Denver, CO.

American Water Works Association. 1995. **Journal**. Dedicated to *Cryptosporidium* and *Giardia*. Vol. 87, Sept. 1995. Denver, CO.

Berger, Paul. 1996. "New EPA Program Targets Pathogens in Drinking Water." Environmental Protection **Journal**. Jan. 1996. Waco, TX: U.S. Environmental Protection Agency.

Carpenter, Philip L. 1967. **Microbiology**. Philadelphia and London: W. B. Saunders Co.

Hilleboe, H.E., and Larimore, G.W. 1965. **Preventive Medicine**. Philadelphia and London: W. B. Saunders Co.

McDermott, James H. 1974. Virus Problems and Their Relation to Water Supply. AWWA **Journal** (No. 12, p. 693). Denver, CO: American Water Works Association.

McKinney, Ross E. 1962. **Microbiology For Sanitary Engineers**. New York, NY: McGraw-Hill.

National Academy of Sciences. 1977. **Drinking Water and Health**. (Vol. 1) Washington, D.C.: The National Research Council, National Academy Press.

National Academy of Sciences. 1982. **Drinking Water and Health**. (Vol. 4) Washington, D.C.: The National Research Council, National Academy Press.

Sobsey, Mark D. 1975. Enteric Viruses and Drinking Water Supplies. AWWA **Journal** (No. 8, p. 414). Denver, CO: American Water Works Association.

Taft, Robert A. Sanitary Engineering Center. 1966. **Transmission of Viruses by the Water Route**. F.W.P.C.A. Dept. of Interior. New York, NY: J. Wiley & Sons.

University of Illinois. 1971. Virus and Water Quality: Occurrence and Control. Water Quality Conference. Urbana, IL: Engineering Publication Office.

USEPA. 1989. Vol. 54, No. 124, 40 CFR Parts 141 and 142, June 29, 1989. NPDWR, Final Rule. Part II Filtration, Disinfection, Turbidity, *Giardia Lamblia*, Viruses, *Legionella*, and Heterotrophic Bacteria; and Part III Total Coliforms. *Fed. Reg.*

USEPA. 1990. National Primary Drinking Water Regulations: Analytical Techniques; Coliform Bacteria. Proposed Rule. *Fed. Reg.* June 1, 1990. The Final Rule was issued with *Fed. Reg.* Jan. 8, 1991. Another Final Rule was issued with *Fed. Reg.* of Jan. 15, 1992.

USEPA. 1994. National Primary Drinking Water Regulations: Enhance Surface Water Treatment Requirements. Proposed Rule. *Fed. Reg.* July 29, 1994.

USEPA. 1978. **Evaluation of the Microbiology Standards For Drinking Water**. Springfield, VA: NTIS.

World Health Organization. 1984. **Guidelines For Drinking Water Quality**. (Vol. 1—Recommendation; Vol. 2—Health Criteria and Other Supporting Information). Geneva, Switzerland: WHO.

# CHAPTER
# 6

# Radionuclide Parameters

RADIOACTIVITY IN DRINKING WATER

NUCLEAR CHEMISTRY DEFINITIONS

RADIOACTIVITY CONCEPTS

DRINKING WATER STANDARDS (RADIONUCLIDES)

RADIOACTIVITY OCCURRENCE IN WATER

HEALTH EFFECTS AND STANDARDS INTERPRETATION

USEPA REGULATIONS (1991–1995)

ANALYTICAL METHODS

RADON

NOTES

# Radioactivity in Drinking Water

Since the late 1950s and early 1960s, under the fear of increased fallout from nuclear testing, nuclear plant malfunction or "meltdown," increased use of nuclear material in medicine and industry, and nuclear waste disposal, it was essential and a pressing issue to evaluate raw and drinking water and its possible consequences to the consumer's health. Compared with the background exposure of human beings to natural causes, it can be safely stated that radiation effects associated with water consumption are normally negligible. There is, nevertheless, the possibility of encountering groundwaters with relatively high concentrations of nuclear parameters. Sensitivity, nevertheless, remains connected with the constant fear of potentially disastrous effects of radiation in humans due to the continual increase in the number of nuclear plants, nuclear weapon arsenals, nations possessing nuclear power, and the potential of nuclear explosives in hijackers' hands. Prior to issuing standards for maximum concentration levels or reevaluating existing standards, Health Authorities had to review and quantify the developmental and teratogenic effects; i.e., the specific genetic, somatic, and carcinogenic effects of radioactive material.

To understand the standards issued by the regulatory agencies and evaluate their significance, the following sections have been developed in this chapter:

- Nuclear Chemistry Definitions
- Radioactivity Concepts
- Drinking Water Standards
- Radioactivity Occurrence in Water
- Health Effects and Standards Interpretation
- Analytical Methods
- Notes and References

In this introduction, a few explanations are necessary: *Genetic damage* is the effect of radiation limited to genes and chromosomes (reproductive materials). *Somatic damage* is the product of radioactivity that affects an organism during its life and ranges from nuclear "burns" to cancer (leukemia, bone cancer, etc.). *Carcinogenic effects* of radiation target tissues that reproduce in the body very rapidly, specifically, bone marrow, blood-forming tissues, and lymph nodes.

The effects of acute exposure to radiation are relatively well known, understood, and classifiable because they are more biologically measurable. Radiation at low or safe levels of exposure, or potential, low-level, beneficial effects of radiation are difficult to correlate with health effects; and, unfortunately, at these levels standards must be developed with the proper factor of safety.

*Radiation doses* are measured in Curie (Ci), a unit related to nuclear disintegration in time (see *Definitions*—this chapter). RAD, RBE, and REM are units used in evaluation: RAD measures energy in kilograms of tissue; RBE is a factor related to the *Relative Biological Effectiveness* of the radiation; and REM is the effective radiation dosage (*Roentgen Equivalent Man*); the number of rems is the product of RBE × RADS (see *Definitions*—this chapter).

For orientation in short-term exposure to radiation effects, the following range of values can be presented:

| Dose (REM) | Effect |
|---|---|
| 0–25 | Nondetectable |
| 25–50 | Slight; temporary decrease in white blood cell counts |
| 100–200 | Nausea; significant decrease in white blood cell counts |
| 500 | Death of 50% of exposed population within 30 days after exposure |

For orientation in long-term exposure to ionizing radiation on the typical U.S. population, the following values may be summarized:

| Source | Contribution (Percent) | Radiation Level (Millirem) |
|---|---|---|
| Natural background | 52 | 100 |
| X-ray (medical) | 43 | 80 |
| Uranium mining (other) | 2 | 4 |
| Fallout from nuclear weapon testing | 3 | 6 |
| Nuclear power plant | 0.14 | 0.3 |
| Consumer products (color televisions, etc.) | 0.02 | 0.04 |

For the general population, Health Authorities limit the annual exposure to 500 mREM and 5 REM for occupational exposure above the background (natural) radiation.

## NUCLEAR CHEMISTRY DEFINITIONS

**Radioactive Nuclides** are defined as atoms that disintegrate by emission of corpuscular or electromagnetic radiations. Alpha, beta, or gamma are the most commonly emitted rays; classified as the following:

*Primary* have a half-life time exceeding $10^8$ years. Primary rays may be alpha-emitters or beta-emitters.

*Secondary* are formed in radioactive transformations starting with uranium-238, uranium-235, and thorium-232.

*Induced* have geologically short lifetimes and are formed by induced nuclear reactions occurring in nature. All these reactions result in transmutations.

**Radioactivity** is the spontaneous disintegration of atomic nuclei with emission of corpuscular or electromagnetic radiations or a number of spontaneous disintegrations per unit mass and per unit time of a given unstable (radioactive) element, usually measured in curies or becquerel (Bq). Curie (Ci) is the unit rate of radioactivity decay; the quantity of any radioactive nuclide which undergoes $3.7 \times 10^{10}$ disintegrations per second. This is the same number of disintegrations of one gram of radium.

**Picocurie** is the one trillionth or $10^{-12}$ of a curie (or 2.22 nuclear transformation per minute). **Millicuries** = $10^{-3}$ Ci; **Microcuries** = $10^{-6}$ Ci, and **Nanocuries** = $10^{-9}$ Ci.

**Half-life** is the average time required for one-half the atoms in the sample of a radioactive element to decay. One-half of the remainder will be lost in another half-life and the process continues until the end product is substantially inert.

**Alpha-Particle**   One of the particles emitted in radioactive decay; it consists of two protons plus two neutrons bound together. Therefore, it is similar to the helium nucleus. A positively charged moving alpha-particle with a velocity as high at $10^9$ cm/sec is strongly ionizing, consequently, losing energy rapidly in passing through matter—dangerous when ingested and deposited within the body. It produces disintegration on many elements upon which it impinges.

**Radium**   A radioactive metallic chemical element, which undergoes spontaneous atomic disintegration through several stages, emitting alpha, beta, and gamma rays and finally forming an isotope of lead. Symbol Ra; atomic weight 226; atomic no. 88; specific gravity 5?; melting point 700°C; boiling point 1,140°C; valence 2.

It decomposes in water and is a member of the alkaline-earth group of metals. One gram of Ra-226 undergoes $3.7 \times 10^{10}$ disintegration/sec. The half-life of Ra-226 is 1,620 years. One gram of radium produces approximately 0.0001 mL of emanations of radon gas/day. Radium causes well-known biological hazards. Both Ra-226 and Ra-228 primarily occur in groundwater in specific geographic areas.

**Radon**   A radioactive, colorless, gaseous, chemical element formed together with alpha rays as a first product in the atomic disintegration of radium. Twenty isotopes are known. Symbol Rn; atomic weight 222; atomic no. 86; density 9.73 g/L; melting point −71°C; boiling point −61.8°C; valence usually 9.

| Isotope | Derivation | Half-Life | Note |
| --- | --- | --- | --- |
| Radon 222 | Radium | 3.823 days | alpha emitter |
| Radon 220 | Thorium | 54.5 sec | alpha emitter called Thoron |
| Radon 219 | Actinium | 3.92 sec | alpha emitter called Actinon |

Radon is released in negligible amounts into the atmosphere by the soil, assessed to contain an average of one gram of radium per square mile to a depth of 6 inches.

Ambient air contains radon in the ratio of 1:1 sextillion. Radon-222 maximum permissible concentration in air is set at $3 \times 10^{-8}$ uCi/cc (lung) for an 8-hour day/40-hour week. (See also radon in water in this chapter.)

**Beta Particle**   One of the particles which can be emitted by a radioactive atomic nucleus. The negatively charged beta-particles are of identical properties with those of ordinary electron radiation; the positively charged type—positron—differs from the electron in having equal but opposite electrical properties. They can be generally classified as moderately damaging and moderately penetrating. They have smaller ionizing power than alpha particles.

**Gross Beta Particle Activity**   The total radioactivity due to the beta particle emission as inferred from measurements on a dry sample.

**Gross Alpha Particle Activity**   Used in drinking water standards, it can be defined as the total radioactivity due to alpha particle emission as inferred from measurements on a dry sample.

**Gamma Rays**   They are unit quantities of electromagnetic wave energy similar to, but of much higher energy than, ordinary X-rays. Gamma rays are highly penetrating, an appreciate fraction being able to traverse several centimeters of lead. Sometimes they are called "Nuclear X-Rays" or "Radiant Energy."

**Neutron**   A neutral elementary particle of mass number one. It is unstable with respect to beta-decay, with a half-life of about 12 min. It produces no detectable primary ionization in its passage through matter, but interacts with matter predominately by collision, and to a lesser extent, magnetically. It is a high-speed particle, deeply penetrating and highly damaging.

**Roentgen—R**   Unit of Radiation (energy) absorbed in air—$1R = 2.58 \times 10^{-4}$ Coulomb*/kg of air. It is the amount of gamma or X-radiation (including alpha, beta, and neutron particles) produced in 1 cc of air ions carrying one electrostatic unit of charge of either sign, at standard temperature and pressure ($1R = 83.6$ ERG†/gram of air).

**Strontium**   A silvery, rapidly turning pale yellow, metallic chemical element resembling calcium in properties but softer and found only in combination. Strontium-90 is a radioactive isotope of strontium with a half-life of 28 years.

Symbol Sr; atomic weight 87.62; atomic no. 38; specific gravity 2.54; melting point 769°C; boiling point 1,384°C; valence 2. It is used in glass for color

---

*Coulomb is the quantity of electricity transported in 1 second by a current of 1 ampere. One A = 1 Coulomb/sec.
†ERG is work performed by a force of 1 dyne acting through a distance of 1 cm. A dyne = the force that accelerates 1 g of mass of 1 cm/sec/sec.

television tubes, in producing ferrite magnets, and in refining zinc; strontium is a present and future product of the nuclear industry.

**Tritium**   A radioactive isotope of hydrogen having an atomic weight of 3, and a half-life of 12.5 years; its nucleus is made up of a proton and two neutrons; it decays by beta-particle emission and is used in hydrogen bombs and as a radioactive tracer. Tritium exists uniformly in the environment as a result of cosmic radiation and residual fallout of nuclear weapon testing. Tritium is also produced by nuclear reactors. The major part remains in the fuel and is released when the fuel is reprocessed. USEPA MCL of tritium is 20,000 pCi/L.*

**Roentgen Equivalent Man—REM**   Unit of radiation dose equivalence. An equal dose expressed in REM produces the same biological effects, independent of the type of radiation involved. Numerically, the amount of ionizing radiation of any type which produces the same biological damage to man as 1 roentgen of about 200 kv-X radiation (1 REM = 1 RAD in tissue $\times$ RBE). A millirem (mREM) is 1/1000 of a REM.

**Radiation Absorbed Dose—RAD**   An ionizing radiation unit corresponding to an absorption of energy in any medium (except air) of 100 ERG/g (or equal to the transfer of 0.01J† of energy per kilogram of material) or $10^{-2}$ gray.‡

**Relative Biological Effectiveness—RBE**   This is an adjustment factor used to convert RADS to REMS. The adopted values for RBE of different kinds of radiation are as follows:

- Gamma and X-Rays      1 RBE
- Beta Particles        1 RBE
- Alpha Particles       10 RBE
- Neutrons and Protons  10 RBE
- Heavy Recoil Nuclei   20 RBE

The exact value of RBE actually varies with dose rate, total dose, and the type of tissue affected.

**Cesium (CS-137)**   A soft, silver-white, ductile, alkali, metallic chemical element—the most electropositive of all the elements. It reacts vigorously

---

*This MCL was set in the USEPA Interim Regulations of Jul. 9, 1976. No modification was proposed by USEPA in the formal proposal of Jul. 18, 1991.
†J(Joule) = $10^7$ ERG
‡Gray is used in the International System of Units for measuring absorbed doses of ionizing radiation. Gray is equal to 1 joule per kilogram.

with water; liquid at room temperature. Symbol Cs; atomic weight 132.90; atomic no. 55; specific gravity 1.873; melting point 28.4°C; boiling point 670°C; valence 1.

Cesium-137 is a radioisotope with a half-life of 37 years, a fission product used in cancer research and radiation therapy. The USEPA—IPDWR standard is 80 pCi/L for Cs-134 and 200 pCi/L for Cs-137.

**Iodine (I-131)**   A bluish-black, nonmetallic chemical element belonging to the halogen family and consisting of grayish-black crystals that volatize into a blue-violet colored vapor with an irritating odor. Symbol l; atomic weight 126.90; atomic no. 53; specific gravity 4.93; melting point 113.5°C; boiling point 184.35°C; density of the gas 11.27 g/L; valence 1, 3, 5, or 7.

Only slightly soluble in water, iodine forms compounds very important in organic chemistry and very useful in medicine. Iodine vapor is a serious hazard to the eyes and mucous membranes. In air it should not exceed 1 mg/mc (8 hr/40 hr) for safety. The USEPA MCL is set at 3 pCi/L for I-131.

Iodine-131 is an artificial radioactive isotope of iodine with a half-life of 8 days; it has been used for treatment of the thyroid gland.

**Uranium**   A hard, heavy, ductile, silvery-white, moderately malleable, radioactive metallic chemical element. It is important in work on atomic energy, especially in the isotope uranium-235, which can undergo continuous fission, and in the more plentiful elemental uranium-237, from which plutonium is produced by nuclear conversion. Symbol U; atomic weight 238.03; atomic no. 92; specific gravity 19; melting point 1,132.3°C; boiling point 3,818°C; valence 2, 3, 4, 5, or 6. All 14 isotopes are radioactive. Uranium-238 has a half-life of $4.51 \times 10^9$ years.

Since one pound of completely fissioned uranium has an equivalent fuel value of 1,500 tons of coal, several hundred nuclear power reactors around the world use it as fuel. Uranium is also used in the production of nuclear weapons, the preparation of isotopes for peaceful purposes, and research work.*

**Man-Made Beta Particle and Photon Emitters**   As used in radioactivity, standards are all radionuclides emitting beta particles and/or photons, Th-232, U-235, and U-238, and their progeny.

**Dose Equivalent**   Term used in the drinking water standards for radioactivity, it means the product of the absorbed dose from ionizing radiation and such factors which account for differences in biological effectiveness due to the type of radiation and its distribution in the body as specified by the

---

*"Natural Uranium," as defined by USEPA, means uranium with combined U-234 plus U-235 plus U-238 which has a varying isotopic composition but typically is 0.006% U-234, 0.7% U-235, and 99.27% U-238.

International Commission on Radiological Units and Measurements (ICRU).

**Fission**   The splitting of an atomic nucleus into two more-or-less equal fragments. Fission may occur spontaneously or may be induced by capture of bombarding particles. In addition to the fission fragments, neutron and gamma rays are usually produced during fission.

**Absolute Risk**   Excess or incremental risk due to exposure to a toxic or injurious agent (e.g., radiation). The difference between the risk (or incidence) of disease or death in the exposed populations and the risk in the unexposed population. Usually expressed as the number of excess cases in a population of a given size, per unit time, per unit dose (e.g., cases per million of exposed population/year/REM).

**Relative Risk**   Ratio of the risk in the exposed population to that in the unexposed population. Usually given as a multiple of the natural risk.

**Latent Period**   Period between time of exposure to a toxic or injurious agent and appearance of a biological response.

**Plateau Period**   Period of above normal, relatively uniform, incidence of disease or death in response to a toxic or injurious agent.

**LET (Linear Energy Transfer)**   Average amount of energy lost by an ionizing particle or photon per unit length of track in matter.

**Nuclear Transmutation**   Conversion of one kind of atom to another.

**Radioisotope**   An isotope that is radioactive; i.e., it is undergoing nuclear changes with emission of nuclear radiation.

**Photon**   The elementary quantity, or quantum, of radiant energy. Photons are generated in collision between nuclei or electrons and when an electrically charged particle changes its momentum (i.e., it has no charge or mass but possesses momentum).

**Ionizing Radiation**   Any radiation consisting of directly or indirectly ionizing particles or any alpha particle, beta particle, gamma ray, X-ray, neutron, high speed electron or proton, or any other atomic particle producing ionization but does not include any sound or radio wave or visible, infrared, or ultraviolet light (nonionizing radiation).

**Becquerel (Bq)**   A special unit of radioactivity in the international system of units (SI). One Becquerel is equal to one disintegration per second or 27 pCi approximately; one Curie = $3.7 \times 10^{10}$ Bq.

**Sievert (SV)**   The unit of dose equivalent in the international system of units (SI) from ionizing radiation to the total body or any internal organ or organ system; one Sievert = 100 REM.

**Effective Dose Equivalent**   The sum of the products of the dose equivalents in individual organs and the organ weighting factor.

**Organ Weighting Factor**   The ratio of the stochastic risk for that organ to the total risk, when the whole body is irradiated uniformly.

**Activity**   The nuclear transformations of a radioactive substance which occur in a specific time interval.

## RADIOACTIVITY CONCEPTS

Radioactivity in water is more significant in relation to the potential deleterious biological effects and public fear connected to nuclear fallout than the actual concentration experienced and the frequency of contamination detected in drinking water.

Many springs and deep well waters, nevertheless, may have high levels of radioactivity. Surface water may be the subject of atmospheric nuclear detonation or nuclear industry accidents with resulting fallout of Sr-90, Cs-137, and I-131. Nuclear industry mining, manufacturing, and transporting accidents may be potential pollutional sources. Also, more frequent use of radioisotopes in health care and related industries may be a potential pollutional source, particularly connected to the disposal of waste.

The major problem of radioactivity is that it cannot be neutralized or nullified by known chemical or physical treatment. Natural decay is a very slow reduction process. This is particularly true for radium, plutonium, and uranium.

Radioactively polluted water may be caused by four general categories of radiation: alpha and beta particles, gamma rays, and neutrons. The biological effects of radiation have been studied on animals as well as humans as a consequence of war, nuclear accidents, and medical treatment. Humans have been exposed to radiation in nature (cosmic rays and radioactive minerals) at levels estimated in the order of 80–170 millirems per year, leaving doubts of the possible natural tolerance of a minimum level of radiation. In some cases man has been exposed to significant levels of radioactivity from mineral waters. In some radioactive springs, 100,000 pCi/L has been registered. Natural waters indicate radioactivity in the range of 1–1,000 pCi/L. Radon content of mineral springs has been detected as high as 750,000 pCi/L in New York State, frequently exceeding 4,000 pCi/L in France, and as high as 1,130,000 pCi/L in New England.

Following nuclear weapon testing in the air, a reading of 300,000 pCi/L (rainfall) was noted over a pre-testing background of 50 pCi/L. It was assumed by sampling experience that 1%–10% of rainfall levels can be converted into predictable levels of surface water concentrations. The effects of acute radiation, well known from a biological viewpoint, are of no concern in the drinking of polluted water. Chronic effects, not well identified, are of interest in drinking water evaluation or setting standards.

Target conditions may be listed as eye cataracts, induction of leukemia from irradiation of the bone marrow, bone cancer, skin cancer, and the suspected reduction of the life span related to an imperceptible impairment of several body functions. Threshold levels of radioactivity initiating these biological impairments are not presently known. Pending further scientific research, Health Authorities have followed the basic principle of keeping radiation exposure to the lowest level. Unfortunately, keeping the threshold for radiation damage undefined produces a negative view of the benefits that radiation is providing to humans in medicine, research, and industry. Consequently, some biological risk must be taken and standards should be developed to balance risk and benefit.

## DRINKING WATER STANDARDS (RADIONUCLIDES)

### USEPA—NIPDWR 1976 and 1982

The NIPDWR for radionuclides issued in 1976 (41 CFR-28404) were basically derived from the USPHS Drinking Water Standards of 1962. Uranium and radon were excluded because of uncertainties about their occurrence, toxicity, and route of exposure. Radium standards were calculated to produce a radiation dose to the bone of 150 millirem per year or an excess cancer risk rate of approximately one case per ten thousand per lifetime.

To measure compliance with the radium drinking water standards, the gross alpha-particle activity was adopted as a screening device. If the level was found greater than 5 pCi/L then the level of radium had to be determined. The regulations also require testing for gross beta-particle activity, tritium, and strontium as a screening device to measure man-made radionuclides. Strontium was selected for its toxicity and tritium because it will not appear in the gross beta screening procedure.

If gross beta-particle activity is greater than 50 pCi/L then the sampled water must be again analyzed to determine which other radionuclides are present. All man-made radionuclides cannot result in a dose that exceeds 4 millirem/yr in the standards; such value was selected with the judgment that such a low level can be reached and set well below the 170 millirem/yr maximum dose recommended by the Federal Radiation Council (FRC, 1961) for the general public.

Monitoring for man-made radionuclides was required for surface water systems serving more than 100,000 persons. Frequency of sampling was selected at quarterly intervals for one year every four years. If gross alpha-particle activity exceeds 5 pCi/L then the sample must be analyzed for radium-228.

**Natural** Maximum Contaminant Level is as follows:

- Combined Radium-226 and Radium-228—5 pCi/Liter
- Gross Alpha-Particle Activity*—15 pCi/Liter

If gross alpha activity does not exceed 5 pCi/L, measure of combined radium may be omitted (including radium-226 but excluding radon and uranium).*

Man-made*—beta particle and photon radioactivity from man-made radionuclides:

- The average annual concentration shall not produce an annual dose equivalent to the total body or any internal organ greater than *four millirems per year.*
- Average annual concentration assumed to produce a total body of organ dose of 4 millirems/yr shall be calculated on the basis of 2 liters per day drinking water intake using the 168-hour data listed in "Maximum Permissible Body Burden and Maximum Permissible Concentration of Radionuclides in Air or Water for Occupational Exposure," *NBS Handbook 69* (amended Aug. 1963).

| Radionuclide | Critical Organ | Concentration (annual average assumed to produce a dose of 4 mREM) |
|---|---|---|
| Tritium | total body | 20,000 pCi/L |
| Strontium-90 | bone marrow | 8 pCi/L |

## WHO—International Standards for Drinking Water (1963)[†]

The following is a guide to the maximum acceptable limits of radioactivity in drinking water used by consumers in large populations over a lifetime:

- Strontium-90            30   pCi/L
- Radium-226            10   pCi/L
- Gross beta concentration   1,000[‡] pCi/L

In the 1964 WHO publication by Charles R. Cox—"Operation and Control of Water Treatment Processes," it was recorded that 70%–90% of radioactive materials from surface waters are removed by conventional water treatment plants. Radioactive material percolating into the soil with rainwater is removed to an extent exceeding 99%.

---

*Man-made radionuclides produced by nuclear fission may enter drinking water supplies through activities such as nuclear fallout, discharges from plants or medical facilities, leaching from radionuclear waste storage, or accidents.
†Now superseded—see WHO (1984) same section.
‡In the absence of strontium-90 and alpha-emitters.

## USEPA—ANPR (1986)

The USEPA issued on September 30, 1986, an Advance Notice of Proposed Rulemaking (ANPR of NPDWR—40 CFR Part 141) for **Radionuclides**. This is a proposal for issuing goals (MCLGs) and drinking water enforceable standards (MCLs) for the following parameters: radium-226, radium-228, natural uranium, radon, gross alpha-particle activity, and gross beta and photon activity.

With the 1986 Amendments of the Safe Drinking Water Act, MCLGs and MCLs must be developed simultaneously with MCLs as close to the MCLGs as feasible, this adjective being defined as using the best technology for efficacy available, taking cost into consideration.

USEPA then adopted, in 1986, an MCL goal set at zero for **Radium-226, Radium-228**, natural uranium, radon, and gross alpha particle activity.

## RADIOACTIVITY OCCURRENCE IN WATER

All drinking water is expected to show minute traces of radioactivity with constituents related to soil and rock strata. Most abundant radionuclides expected are:

| LOW LET GROUP | HIGH LET GROUP (only traces expected) |
| --- | --- |
| potassium-40 | radium-226 |
| tritium | the daughters of radium-228 |
| carbon-14 | polonium-210 |
| rubidium-87 | uranium |
| | thorium |
| | radon-220 |
| | radon-222 |

Potassium-40 is the most significant radionuclide; but from a 2 L diet, only 8 pCi per day are ingested versus a food intake of 2,300 pCi. It must be noted that potassium-40 occurs at 0.012% of total potassium (NAS, 1977).

The most significant levels of Ra-226 and Ra-228 in groundwater in the United States are in the Midwest where a concentration of 5 pCi/L is estimated for a population of 1 million, with localized population of 120,000 people using well water containing 9–80 pCi/L of Ra-226.

Radon-222, a radioactive gas formed by the decay of Ra-226, is generally 1 pCi/L in surface water but a few thousand times greater in groundwater where it may reach 500,000 pCi/L or 0.5 μCi/L in mineral water. Drinking water is expected to contribute not more than 2% of the alpha-emitting radionuclides ingested daily by the food diet.

Surveys of potable water produced radium concentrations of 0–5.79 × $10^{-9}$ mg/L of radium* with a medium of 0.028 × $10^{-9}$ mg/L. Raw water

---

*$10^{-9}$ mg/L of radium = 1 pCi/L = 0.037 Bq/L

radium concentration ranged from $0.02 \times 10^{-9}$ to $6.54 \times 10^{-9}$ mg/L with a median of $0.049 \times 10^{-9}$ mg/L.* Groundwater sources contained higher concentrations than surface water supplies in surveys in the United States. Other surveys recorded variations in raw waters from $0.01 \times 10^{-9*} - 7.1 \times 10^{-4}$ mg/L (spring water in Japan) with an average of $0.07 \times 10^{-9}$ mg/L for river waters to $0.08 \times 10^{-9}$ mg/L in sea waters.

Pathological examination of humans deceased after long residence in areas with average radioactivity concentrations indicated radium in the order of $120 \times 10^{-9}$ mg: this is one-third of a safe level used in the radium industry. Consequently, it was concluded ("Hursh-Radium Content in Water Supplies" AWWA *Journal*, 1954) that water and food do not constitute a health hazard for the population.

The USEPA (1986) reported nationwide monitoring data collected from 50,000 public water supply systems with the following comments:

- Radium, uranium, and radon are seldom found in high concentrations together.
- Radon in drinking water shows concentrations ranging from less than 10 pCi/L to a reported high of 2,000,000 pCi/L (groundwater, public well).
- Uranium is universally found in the United States at low levels (1–10 pCi/L).
- Radium also has a widespread occurrence, but the low level of detection is not much below the current standards (5 pCi/L).
- Population-weighted average concentrations of natural radionuclides in the U.S. community water supplies in pCi/L were reported as follows:

| | | | |
|---|---|---|---|
| Radium-226 | 0.3–0.8 | Lead-210 | less than 0.11 |
| Radium-228 | 0.4–1.0 | Polonium-220 | less than 0.13 |
| Natural uranium | 0.3–2.0 | Thorium-230 | less than 0.04 |
| Radon-222 | 50–300 | Thorium-232 | less than 0.01 |

## HEALTH EFFECTS AND STANDARDS INTERPRETATION

It can be unequivocably stated that toxicology related to radioactivity is more complex than that of organic toxicology. The complexity is due to many factors, such as the following:

- Variation in cosmogenic radionuclide concentrations due to the variable interaction of cosmic rays with the earth's atmosphere and stratosphere;
- Variation in biological effects among components of radioactive elements examined;
- Difficulty in measuring dose rate and absorbed rate prior to evaluating biological effects;

---

*$10^{-9}$ mg/L of radium = 1 pCi/L = 0.037 Bq/L

- The plurality of target organs involved and the complexity of metabolic distribution and retention patterns;
- The potential genetic damage and its related complexity in short-term determination of toxic effects; and
- Low-level radiation, or even relatively large radiation doses, do not produce evidence of somatic or genetic effects; nevertheless, the interdependence of cells, tissues, and organs may result in damage of others.

Based on the above-mentioned complexities alone, one may justify the adopted principles of many health authorities that any exposure to radiation should be avoided; but background radiation cannot be avoided, and medical use of radiation is necessary or helpful at times. It is nevertheless possible to predict that in the future better understanding of the physical phenomena of radiation combined with better knowledge of biological effects may produce a scientific interpretation of low-level radiation that may lead to threshold values (determination of safe or perfectly acceptable low levels).

Epidemiological data can be briefly summarized as follows:

- *RADIUM*—Carcinogenic to humans and animals. Ra-226 may cause bone sarcomas and head carcinomas, and Ra-226 and Ra-228 are more deleterious than Ra-224.
- *URANIUM*—It causes noncarcinogenic bioeffects to kidneys (chemical toxicity) beside carcinogenic risk. Radiotoxicity of natural uranium is evaluated at one-half that of radium.
- *RADON*—Rn-222 may be linked with lung cancer. See detailed information in "Radon" at the end of this Chapter.

In general it can be stated that radioactivity in drinking water is essentially a natural radionuclide consequence, and only rarely can it be attributed to man-made radioactivity.

## Standards

In the United States drinking water standards were issued by the USEPA based on recommendations received by the NAS. The NAS reviewed (and concluded in 1977, Vol. 1—see Appendix A-VIII, "General References") the report submitted by the Subcommittee on Radioactivity in Drinking Water. The material reviewed by the Subcommittee included the National Academy of Sciences Advisory Committee on the Biological Effects of Ionizing Radiation (BEIR), the United Nations Scientific Committee on the Effects of Atomic Radiation (UNSCEAR), the International Commission on Radiological Protection (ICRP), and the National Council on Radiation Protection and Measurements (NCRP).

In 1972, BEIR estimated that the annual dose in millirem from natural radiation in the United States is 102 total or 44 from cosmic rays and 58 from terrestrial radiation.

The NAS (1977), reviewing the developmental and teratogenic effects, concluded, "We anticipate that no measurable developmental and terato-genic effects of radionuclides in drinking water will be found at the levels studied." Moreover, in estimating genetic risk in man, NAS concluded: "The wide fluctuation in bone dose caused by fluctuations in the radium concentration of drinking water could not have any sensible effect on the genetically significant dose, because radium is predominately a bone seeker and will deliver very little radiation to the gonads," and in further conclusions, "The radiation associated with most water supplies is such a small proportion of the normal background to which all human beings are exposed, that it is difficult, if not impossible, to measure any adverse health effects with certainty. In a few water supplies, however, radium can reach concentrations that pose a higher risk of bone cancer for the people exposed."

Recent drinking water standards (1962, 1976, 1982) for natural radioactivity in the United States set the maximum contaminant level for combined Ra-226 and Ra-228 at 5 pCi/L. This level corresponds to $5 \times 10^{-9}$ mg/L of radium, or using stricter standards issued by WHO International and European drinking water standards, this level is 1.0 pCi/L or $1.0 \times 10^{-9}$ mg/L of radium, revised at 0.1 Bq/L for gross alpha activity in 1984.

In 1984, the WHO published the following guidelines for radioactive constituents:

Gross alpha activity 0.1 Bq/L
Gross beta activity   1 Bq/L

with the remarks: "If the levels are exceeded more detailed radionuclide analysis may be necessary. Higher levels do not necessarily imply that the water is unsuitable for human consumption."

The above-mentioned values are applied to the mean of all radioactivity measurements obtained during a "sampling period that is appropriate for the source water." These guidelines were issued assuming that the most toxic radionuclides are expected to be present in significant quantities, Sr-90 and/or Ra-226, contributing to the gross radioactivity of the drinking water.

The WHO (1984) also stated that Rn-222 and Rn-220 should be eliminated before starting the analysis of alpha activity. Also reported was a list of naturally occurring alpha-emitting radionuclides of known high toxicity: Ra-226, Ra-224, Po-210, Th-232, U-234, and U-238. Also associated with these radionuclides are beta-emitters, such as Ra-228 and Pb-210, because of the alpha-emitting daughters.

The beta-emitting radionuclides of high toxicity have been recorded with Sr-90, Sr-89, Cs-134, Cs-137, I-131, and Co-60. The WHO (1984) also recommended a careful selection of the sampling points of the water distribution system and supplies to make the sample representative since many radionuclides are readily absorbed by surfaces and solid particles.

The International Commission on Radiological Protection (ICRP, 1977) recommended that all exposure be "as low as reasonably achievable, economic and social factors being taken into account." To reduce excessive exposure to radium, WHO recommended lime or lime-soda softening, cation exchange softening, or reverse osmosis as typical water treatment processes. Anion exchange was listed as a promising treatment, since uranium is often expected in the negatively charged uranyl ion.

Aeration may be used for radon, an inert gas. The WHO (1984) in issuing guidelines noted that the gross activity screening procedures adopted had to meet two criteria; namely, the associated exposure will be low enough to avoid further detailed analyses first, and second, guideline values must be high enough to ensure that the great majority of drinking water supplies satisfies such levels, therefore avoiding additional sampling and detailed analyses.

In 1984, WHO also confirmed the statement on the negligible effects of drinking water in general on total radiation effects, stating: "Drinking water is a relatively minor constituent of total radiation exposure." Such a conclusion is the summary of the WHO review of exposure levels from natural and man-made radioactivity reported by the UNSCEAR.

In evaluating **Radon** activity, WHO (1984) concluded that some groundwater sources recorded values of radon as high as 1,000 Bq/L, but due to the volatility of this noble gas, it is impossible to determine how much is ingested. It is also important to note that the risk of inhalation of atmosphere in the room where radon is accumulated is greater than the risk due to ingestion of water, when radon has a high concentration.

In the United States, the selection of the 5 pCi/L standard could have been based on no adverse health effects observed in a relatively large population that consumed groundwater containing less than 5 pCi/L and in a smaller population consuming drinking water with 9–80 pCi/L.

USEPA (1986), submitting new goals and proposals for standard promulgation, reviewed BEIR III Committee's report and the USEPA Health Criteria Documents. The following observations may summarize the latest data on potential biological damage:

- *RADIUM-226 and 228*—Cancer arising from skeletal tissue. Bone sarcomas and head carcinomas.
- *URANIUM*—The primary chemically toxic effect of natural uranium is the kidneys. The AADI—60 µg/L (70 kg adult; 2 L/day) and the NOAEL—1 mg/kg/day (based on this value, the Canadian Government set a guideline at 20 µg/L). See Chapter 3 for AADI and NOAEL explanations.
- *RADON*—Lung cancer in humans (wide range of histological types).
- *MAN-MADE RADIONUCLIDES*—Approximately 200 man-made radionuclides have been determined to be carcinogenic. They are listed in Appendix A-VI (Radionuclide Data), reporting the concentration in drinking water yielding a risk equal to that from a dose rate of 4 millirem/yr.

## USEPA REGULATIONS (1991–1995)

In July 1991,* USEPA issued a **Proposed Rule** for the following radionuclides:

| Contaminant | MCLG | MCL |
|---|---|---|
| Radium-226 | zero | 20 pCi/L |
| Radium-228 | zero | 20 pCi/L |
| Radon-222 | zero | 300 pCi/L |
| Uranium | zero | 20 ug/L |
| Adjusted gross alpha emitters | zero | 15 pCi/L |
| Beta and photon emitters | zero | 4 mREM/year |

In consideration of the USEPA classification of these radionuclides as carcinogens (Group A), the selection of MCLG was apparently easy with the value of "zero." Some question marks were left for the classification of uranium MCL, in consideration of the toxicity to the kidneys. Many comments are expected to be evaluated prior to the issuance of the Final Rule for uranium. A similar statement can be recorded for radon due to the toxicity of radon in the air and the cost of remediation to contain radon below the 300 pCi/L level in the effected water systems.

Radon is examined, separately, at the end of this chapter. The 1991 Proposed Rule also included regulations for monitoring, reporting, and public notification for the above-mentioned contaminants.

### Analytical Methods

USEPA recognizes the following Proposed Methods for the identification and measuring of these radionuclides:

**Key:** PQL = Practical Quantitation Level; BAT = Best Available Technology
- *RADIUM-226*—Radon emanation; radiochemical (PQL = 5 pCi/L).
  As BAT: Ion Exchange; Lime Softening; Reverse Osmosis.
- *RADIUM-228*—Radiochemical; liquid scintillation (PQL = 5 pCi/L).
  As BAT: Ion Exchange; Lime Softening; Reverse Osmosis.
- *URANIUM*—Radiochemical; fluorometric; alpha spectrometry (PQL = 5 pCi/L).
  As BAT: Coagulation + Filtration; Anion Exchange; Lime Softening; Reverse Osmosis.

---

*USEPA: National Primary Drinking Water Regulations; Radionuclides; Proposed Rule, *Fed. Reg.* July 18, 1991. Corrected. *Fed. Reg.*, Sep. 3, 1991. Notice of Corrections to Proposed Rule and Extension of Public Comment Period, *Fed. Reg.*, Oct. 18, 1991.

A Court-ordered deadline was originally set for Apr. 30, 1995, for the final rule of radionuclides. Final rule for radon will be delayed and regulated separately.

- *RADON-222*—Liquid scintillation; Lucas cell (PQL = 300 pCi/L). As BAT: Aeration.
- *GROSS ALPHA EMITTERS*—Gross alpha activities (PQL = 15 pCi/L). As BAT: Reverse Osmosis.
- *GROSS BETA EMITTERS*—Gross beta activities (PQL = 30 pCi/L). As BAT: Ion Exchange; Reverse Osmosis.
- *GAMMA and PHOTON EMITTERS*—Gamma Ray Spectrometry. As BAT for photon emitters: Ion Exchange; Reverse Osmosis.
- *RADIOACTIVE CESIUM-134 and -137*—Precipitation (PQL = 10 pCi/L).
- *RADIOACTIVE IODINE*—Precipitation (PQL = 20 pCi/L).
- *RADIOACTIVE STRONTIUM-89 and -90*—Precipitation; for strontium-90: Radiochemical (PQL = 5 pCi/L).
- *TRITIUM*—Gamma Ray Spectrometry.

Monitoring is expected to begin January 1996. See Chapter 8 for monitoring requirements for radionuclides.

## ANALYTICAL METHODS

Analysis of radioactive material in water is a rather complex operation due to the necessity of eliminating background radiation (properly designed and built laboratory physical facility), to use and maintain very sensitive instrumentation and complications due to the slow and complex identification of radionuclides when required.

Standards first require the determination of **gross alpha** and **gross beta** measurements; these readings require simple instrumentation and limited laboratory work and, therefore, can be considered inexpensive analyses. These results will not provide information about radionuclides.

Evaporation of the sample may cause radionuclide losses by volatilization. It is very important in the process of identifying radionuclides that selected chemical/physical knowledge be available in the laboratory.

Sampling is also a delicate operation—in selecting a representative sample and in acidifying the sampling bottle, which should be plastic or glass. For tritium sampling, a glass bottle only should be used. The size of the container should be from 1/2–18 L according to the analyses scheduled.

*Standard Methods* describes or prescribes counting room, counting instruments, laboratory reagents and apparatus, the statistic involved, and the quality assurance.

In reporting results, all three units related to curie can be used:

| | | |
|---|---|---|
| picocuries | per liter at 20°C pCi/L | 1 pCi = $10^{-12}$ Ci |
| nanocuries | per liter at 20°C nCi/L | 1 nCi = $10^{-9}$ Ci |
| microcuries | per liter at 20°C µCi/L | 1 µCi = $10^{-6}$ Ci |

"Gross Alpha" reported means unknown alpha sources. "Gross Beta" reported means unknown sources of beta including some gamma radiation.

USEPA standards may be satisfied by the simple examination of gross alpha and gross beta radioactivity, when the average annual concentration of gross beta activity is less than 50 pCi/L and if the average annual concentration of tritium and strontium-90 is less than 20,000 pCi/L and 8 pCi/L, respectively.

When gross beta activity is over 50 pCi/L, identification of the major radioactive contaminants must follow so that the "Body Doses" may be calculated to search for the upper limit that shall not exceed 4 millirem per year.

So the examination for gross alpha and gross beta activity can be used for monitoring and for early detection of radioactive contamination.

USEPA (NIPDWR, 1982) revised regulations call for compliance based on the analysis of an annual composite for four consecutive quarterly samples or the average of the analysis of four samples obtained at quarterly intervals. Frequency of sampling requires at least once every four years monitoring procedure (quarterly samples). Annual monitoring is required when the Ra-226 concentration exceeds 3 pCi/L. *Standard Methods* lists the following examination procedures approved by the 1981 and 1985 Standard Methods Committee:

- Gross Alpha and Gross Beta Radioactivity (Total, Suspended, and Dissolved)
- Total Radioactive Strontium and Strontium-90 in Water
- Radium in Water by Precipitation
- Radium-226 by Radon in Water (Soluble, Suspended, and Total)
- Radium-228 (Soluble—Tentative)
- Tritium
- Radioactive Iodine (Precipitation, Ion Exchange, Distillation Methods)
- Uranium (Radiochemical—Tentative, Fluorometric Methods—tentative)

## RADON

Radon has been defined in the *Nuclear Chemistry Definitions* section of this chapter. Radon is an air contaminant, a water contaminant, and, again, an air contaminant as a consequence of water spraying in daily kitchen, bathroom, or laundry uses, when the source of water supply is groundwater with high radon concentrations.

**Radon** is a colorless, odorless, tasteless, gaseous, inert element, originating in rocks or soil containing uranium or radium. A radioactive decay chain from uranium and radium produces radon, which, as a gas, seeps mainly from soil into the air.

Many epidemiologic studies were available to the Health Authorities who evaluated the health effects of radon and translated them into drinking water standards for radon. Unfortunately, these studies were only related to mine workers, notoriously exposed to high concentrations of gaseous radon.

In Oct. 1992 Congress mandated USEPA to prepare a report with findings regarding all the risks of human exposure to radon, including the cost associated with the control and mitigation, and including the risks related to treatment and disposal of wastes produced by the water treatment.

The USEPA report to Congress stated that, "The cancer risks from radon in both air and water are high. While radon risk in air typically far exceeds that in water, the cancer risk from radon in water is higher than the cancer risks estimated to result from any other drinking contaminant."

In summary of USEPA's report,* it was stated: "When ingested, **radon** is distributed throughout the body, which increases the cancer risk to many organs." Also, "EPA estimates that an individual's combined risk during a lifetime from constant use of drinking water with one picocurie of radon per liter is close to 7 chances in 10 million of contracting fatal cancer."

Since only groundwater sources are involved, USEPA was concerned with the 81 million Americans served by community groundwater supplies. The average value per user was calculated in 246 picocuries per liter of water. Moreover, radon in water exceeds 100 pCi/L in 72% of the groundwater sources surveyed. Another of USEPA's conclusions from this study is that, "Approximately 2 of every 10,000 individuals exposed at 300 pCi/L would develop a fatal case of cancer as a result of exposure to radon at this level."†

USEPA presented to Congress the following table as a summary of Risk, Fatal Cancer Cases, Cancer Cases Avoided, and Cost for Mitigating **Radon** in Drinking Water:

| | |
|---|---|
| Number of Fatal Cancer Cases per year: | 192 |
| Proposed Radon Level: | 300 pCi/L |
| Individual Lifetime Risk of Fatal Cancer at Target Level: | 2 in 10,000 |
| Total Number of People Above Target Level: | 19 million |
| Number of Fatal Cancer Cases Avoided<br>Annually by Meeting the Target Level: | 84 |
| Total Annual Cost for Mitigating Radon: | $272 million |
| Average Cost for Fatal Cancer Case Avoided: | $3.2 million |

---

*USEPA. March 1994. "Report to the United States Congress on Radon in Drinking Water—Multimedia Risk and Cost Assessment of Radon," EPA 811-R-94-001.
†The USEPA Proposed Maximum Contaminant Level (MCL) for **Radon** is 300 pCi/L. This value is under evaluation by USEPA (600 comments submitted) due to the high number of water systems in violation of this proposed MCL. The estimated cost of compliance varies from $180 million per year to $2.5 billion for the national compliance cost. The water systems expected to exceed the proposed MCL vary from 25,900 (USEPA) to 32,750 (AWWA publications). Congress prohibited USEPA from spending fiscal year 1995 money to promulgate a radon standard.

At the same time USEPA presented its estimate for Indoor Air. The number of fatal cases of cancer per year was 13,600; the proposed target level was 4 pCi/L$_{air}$*; the individual lifetime risk of cancer at target level was 1:100; the number of fatal cancer cases avoided was 100 (perhaps 2,200 cases in the future after implementation of a very successful air pollution control program); and the total annual cost for mitigating **radon** was estimated at $1.5 billion.

## NOTES

*RADON.* In preparing another groundwater sampling for radon, New York State reported that radon in water has been found to contribute about 1 pCi/L radon to indoor air for each 10,000 pCi/L in water. The first survey was conducted between 1978 and 1988 by the N.Y.S. Health Department. New York State temporarily adopted the USEPA concern regarding radon pending issuance of regulatory standards:

| | |
|---|---|
| Less than 1,000 pCi/L | low level; no follow-up |
| 1,000–10,000 pCi/L | possible follow-up |
| More than 10,000 pCi/L | follow-up, air monitoring, and possible remedial work |

The 1978–88 N.Y.S. survey recorded the following results for radon in pCi/L from 235 well water supplies:

| Region | Min. | Max. | Median | No. of Sampled Water Supplies |
|---|---|---|---|---|
| Albany | 100 | 3,160 | 191 | 23 |
| Buffalo | N.D. | 1,810 | 370 | 31 |
| Syracuse | N.D. | 952 | 306 | 45 |
| Dutchess | 34 | 1,437 | 558 | 11 |
| Nassau | 4 | 314 | 82 | 38 |
| Orange | 8 | 135,000 | 112 | 26 |
| Putnam | 318 | 3,240 | — | 2 |
| Rochester | N.D. | 880 | 181 | 12 |
| Rockland | 193 | 638 | 398 | 5 |
| Suffolk | 12 | 460 | 90 | 26 |
| Sullivan | 505 | 791 | 647 | 6 |
| Ulster | 12 | 728 | 100 | 7 |
| Westchester | 193 | 17,500 | 395 | 3 |

N.D. = not detected
— = decimal point omitted

*The target level of 4 pCi/L$_{air}$ is an acceptable level for radon concentration, called *action level* in buildings, to initiate a radon abatement program. Radon is tested with charcoal canisters or with the alpha track detector.

# REFERENCES

American Public Health Associates, Inc. 1966. **Ionizing Radiation**. Washington, D.C.: A.P.H.A., Inc.

American Water Works Association. 1988. July **Journal**, Issue dedicated to Radionuclides. Denver, CO: AWWA.

American Water Works Association. 1991. April **Journal**. Issue dedicated to Radionuclides. Denver, CO: AWWA.

American Water Works Association. 1994. April **Journal**. Issue dedicated to Removing Radionuclides. Denver, CO: AWWA.

Brown and LeMay. 1977. **Chemistry**. Englewood Cliffs, NJ: Prentice-Hall.

Bunger, Cook, and Barrick. 1981. **Health Physics** 40:439–455.

California Institute of Technology. 1963. **Water Quality Criteria**. Pasadena, CA: NTIS, Springfield, VA.

CRC. 1985–86. **Handbook of Chemistry and Physics**. (66th edition) Boca Raton, FL: CRC Press, Inc.

Federal Radiation Council. 1961. **Background Material For the Development of Radiation Protection Standards**, Report No. 2. Washington, D.C.: U.S. Department of Health, Education and Welfare, USPHS.

International Commission on Radiological Protection. Publication #2. New York, NY: Pergamon Press.

National Academy of Sciences. 1977 and 1984. **Drinking Water and Health**. (Vols. I and V)—Safe Drinking Water Committee. Washington, D.C.: The National Research Council.

National Council on Radiation Protection. Subcommittee on Permissible Internal Dose 1959—NBS #69. Washington, D.C.: National Bureau of Standards.

National Council on Radiation Protection and Measurements. **NCRP Report No. 50**. Washington, D.C.

Royal Society of Health. 1964. **Radiation Levels: Air, Water, and Food**. London, England.

Salvato, J. A. 1982. **Environmental Engineering and Sanitation**. (Third edition). New York, NY: J. Wiley & Sons.

USEPA. 1976. Drinking Water Regulations. Radionuclides—*Fed. Reg.* 41.28402. Washington, D.C.

USEPA. 1991. National Primary Drinking Water Regulations; Radionuclides; Proposed Rule. *Fed. Reg.* (July 18, 1991). Notice of Correction to Proposed Rule. *Fed. Reg.* (Sep. 3, 1991). Notice of Extension. *Fed. Reg.* (Oct. 18, 1991).

USEPA. 1993. Report to the U.S. Congress on **Radionuclides in Drinking Water**. Multimedia Risk and Cost Assessment of Radon in Drinking Water. Office of Water, Washington, D.C. (7/9/93).

USEPA (March 1994). Report to the U.S. Congress on **Radon in Drinking Water**. EPA 811-R-94-001: USEPA.

World Health Organization. 1984. **Guidelines For Drinking Water Quality**. Vols. 1 and 2, Geneva, Switzerland: WHO.

# CHAPTER
# 7

# Carcinogens

INTRODUCTION

RESEARCH CONCEPTS

CARCINOGENS AND DRINKING WATER STANDARDS

# Introduction

Carcinogens can be defined as substances or agents that stimulate the formation of cancer.

Carcinogenicity is studied to search for formation of both benign and malignant tumors. Cancer is a new growth of tissue possessing an atypical structure of the body tissues or organs. It has an unlimited and uncontrolled power of growth with the tendency to spread to distant locations where it may assume a renewal of growth.

Malignant tumors are characterized by the ability of cancer cells to grow when deposited elsewhere, while benign tumors are characterized by local growth without the ability to spread. There are as many varieties of cancer as there are organs and tissues within the body.

Predominantly, cancer occurs in humans but also in domestic animals and in mice where the majority of experimentation takes place.

Biology, chemistry, and physics are the fields of cancer research: Since growth is related to chemical reactions within the cell, chemistry has an important role. To produce successful cell division, physics (biophysics and physical chemistry) is involved. Biology is the base of cancer research because the cell is the active part of tissue, and a similar relationship exists between the tissue and organ, and organs are coordinated with other organs.

One theory of cancer formation is the **mutation** of one or more cells. Such mutations are listed as changes in the total number of **chromosomes**, changes in size or structure of individual chromosomes, or changes in the chemical organization of the genes.

The genetic risk associated with human exposure to existing industrial chemicals and pesticides must be evaluated by the Health Authorities. The scope here is to provide a judgment concerning the weight of evidence that a specific agent is a potential human mutagen with respect to transmitted genetic changes. When confirmed by evidence, the impact on public health must also be judged. The major component of regulatory decision is **risk assessment** (see this chapter). **Mutation** can be defined as a change in the characteristics of an organism as a result of changes in genes or other hereditary factors. Mutations can be considered spontaneous or induced, but both are random. Ultraviolet and X-irradiations, as well as some chemicals and temperature increase, may cause mutations (induced). Agents or substances capable to induce mutations are called **mutagens**.

A **chromosome** is the microscopic body within the cell that contains the genes (the units responsible for hereditary characteristics; sperm cells of the male and the egg cells of the female).

Mutagens are, therefore, studied because they increase the frequency of gene mutations and are correlated with **carcinogenicity** and **teratogenicity** (the cause of malformation of a fetus as an agent, chemical or disease).

## Research Concepts

Particularly during the last 20 years, the scientific community, working on evaluation and standard setting of chemicals in drinking water, has been under great public pressure to reach "final" conclusions on the carcinogenicity and toxicity of perhaps several thousand chemicals that could have found a way to pollute raw and treated water. Such public pressure was originated by information on domestic and industrial wastes discharged upstream of raw water river intakes or potential contamination of aquifers that feed groundwater used for human consumption.

The Health Authorities had to decide, indeed in limited time, on at least several hundred chemicals in order to set standards. It appeared then by preliminary studies that statistically higher rates of cancer were found in populations subjected to industrially contaminated drinking waters. Consequently, based on advisory committees and research groups, the Health Authorities in the United States did their best to advise and regulate as much as possible with the information available. Unfortunately, sometimes informational material was very limited (incomplete studies or research work based on limited animal studies only).

When these proposed regulations reached the public and water works administrators, engineers, and operators, it appeared that very accurate determinations were made by the Health Authorities, particularly when extremely low concentrations were selected for standards (parts per billion or even fractions of microgram per liter). Little emphasis was made to underline that a factor of safety of 1,000 was often used and the fact that straight mathematical proportions were made necessarily from a few small, disturbed, irritated, laboratory animals to compare with human adults or children.

It is to be expected that now and in the future carcinogen risk assessment and standard proposals will be based on a more accurate identification of agent hazard. The research work quantitative determination is expected to be more standardized and evaluated in the exposure assessment and in the proper risk characterization.

The National Research Council (NRC) of the National Academy of Sciences, the International Agency for Research on Cancer (IARC), and the USEPA have provided guidelines for research and evaluation of data for Health Authority's standard setting. It must be noted, however, that carcinogenesis investigation is undergoing a technological revolution with rapid changes, and perhaps a future breakthrough in cancer research may rapidly leave present guidelines and studies in a state of obsolescence.

The Health Authorities, in standard setting, must properly measure the overall risk evaluation and the administration of risk as related to potential legislation based on factual examination from a biological, engineering, socioeconomic, and perhaps psychological viewpoint.

USEPA, in its November 23, 1984, "Proposal Guidelines for Carcinogen Risk Assessment; Request for Comments" advised that "risk assessment" include one or more of the following components:

- Hazard Identification (qualitative risk assessment)
- Dose-Response Assessment
- Exposure Assessment
- Risk Characterization

These testing procedures are certainly not of concern to water works engineers/administrators/operators, but are summarized here to indicate the complexity of present-day standard setting.

## Hazard Identification

**Review of biological and physical-chemical properties**   Related to the body metabolism and to degradation products, including associated contaminants.

**Identification of parameters relevant to carcinogenesis**   These include metabolic and pharmacokinetic properties from (macromolecular interaction to excretion from the body) toxicologic effects—prechronic and chronic toxicity, target organ effects, pathophysiologic reactions, preneoplastic lesions, and dose-response and time-to-respond analyses.

Short-term tests—these include tests for point mutations, numerical and structural chromosome aberrations, DNA damage/repair, tests in tubes and within a living organism.

Long-term animal studies—increase the number of tissue sites affected; the increase in number of animal species, the occurrence of clear-cut dose-response relationship; statistical significance vs. the control group; the time-to-tumor and time to death with tumor; and dose-related changes in the incidence of preneoplastic lesions.

Human studies—epidemiological studies usually provide excellent data and have known factors with proper selection and characterization of exposed and control groups; adequacy of duration and quality of the follow-up; and identification of confounding factors and bias, etc.

**Weight of Evidence**   The IARC classification method lists: "sufficient," "limited," and "inadequate" for evidence that an agent is producing cancer in humans and animals. Similarly, USEPA uses the following classification:

- GROUP A—Carcinogenic to humans
- GROUP B—Probably carcinogenic to humans
- GROUP C—Possibly carcinogenic to humans
- GROUP D—Not classifiable as to human carcinogenicity
- GROUP E—No evidence of carcinogenicity for humans

## Dose-Response Assessment

The quantitative assessment done by the statistician must be guided by the individuals working on the qualitative assessment (toxicologists, pathologists, pharmacologists, etc.) for proper selection of data.

The choice of mathematical extrapolation model to be worked from high dose to low dose in animal studies must be carefully prepared for each agent. Linearity of tumor initiation is assumed pending better understanding of the mechanism of the carcinogenesis process.

Equivalent exposure units among species; the correlation of data from laboratory animals to humans is complicated by a variety of factors, such as life span, body size, genetic variability, population homogeneity, existence of concurrent disease, pharmacokinetic effects, such as metabolism and excretion patterns and exposure regimen.

Standardized dosage scales normally used in studies include:

- mg/kg body weight per day
- ppm in the diet of water
- $mg/m^2$ body surface per day
- mg/kg body weight for lifetime

## Exposure Assessment

The exposure of the population of interest must be evaluated with the dose-response assessment. This must be done on a case-by-case basis, since the magnitude, duration, and frequency of exposure are variables. The level of uncertainty in this evaluation should be included in the risk characterization, so as to have an impact on the final risk estimate.

## Risk Characterization

Numerical risk estimates are dealt in reviewing synthetic organics (see SOCs). Whenever numerical estimates are used, the qualitative weight-of-evidence classification should also be included. For example, $2 \times 10^{-4}$ (C) represents a lifetime of individual risk resulting from exposure to a "possible human carcinogen—Group C." (USEPA *Guidelines*, Part VII–Nov. 23, 1984).

# Carcinogens and Drinking Water Standards

In evaluating evidence of carcinogenicity inhalation versus ingestion in setting drinking water standards, data must be separated. Unless there is data confirming no carcinogenicity by ingestion, the use of inhalation data may be applied on a case-by-case basis in standard evaluations.

## Table 7-1. Preliminary Classification of SOCs and Inorganics into Three Categories Approach.

| Chemical | Category |||
|---|:---:|:---:|:---:|
| | I | II | III |
| DBCP | X | | |
| Dioxin | X | | |
| Epichlorohydrin | X | | |
| Hexachlorobenzene | X | | |
| Alachlor* | X | | |
| Toxaphene | X | | |
| Acrylamide | X | | |
| EDB | X | | |
| Chlordane* | X | | |
| Heptachlor* | X | | |
| PCBs | X | | |
| Heptachlor epoxide | X | | |
| Lindane | | X | |
| Styrene | | X | |
| 1,2-Dicloropropane | | X | |
| Monochlorobenzene | | X | |
| Pentachlorophenol | | | X |
| Aldicarb | | | X |
| cis-1,2-Dicloroethylene | | | X |
| trans-1,2-Dicloroethylene | | | X |
| o-Diclorobenzene | | | X |
| m-Dichlorobenzene | | | X |
| 2,4-D | | | X |
| Endrin | | | X |
| Ethylbenzene | | | X |
| Methoxychlor | | | X |
| Toluene | | | X |
| 2,4,5-TP | | | X |
| Xylene | | | X |
| Atrazine | | | X |
| Simazine | | | X |
| Carbofuran | | | X |
| All the inorganics except arsenic and asbestos | | | X |
| Asbestos (ingestion only fibers > 10 μm) | | X | |
| Arsenic | | X | |

*Classified by IARC in Group 3. However, reexamination has resulted in classification in USEPA Group B2, and thus these chemicals have been placed in the "probable" carcinogen category.

USEPA, in the proposed ruling of November 1985 for NPDWR—Synthetic Organic Chemicals (SOCs), Inorganic Chemicals (IOCs), and Microorganisms—selected a three-category approach for setting the recommended maximum contaminant levels (RMCLs):

- **Category I**—*Strong evidence of carcinogenicity*—including USEPA Group A or Group B and IARC Group 1, 2A, or 2B.
- **Category II**—*Equivocal evidence of carcinogenicity*—USEPA Group C or IARC Group 3.
- **Category III**—*Inadequate or no evidence of carcinogenicity in animals*—USEPA Group D or E or IARC Group 3.

From the USEPA above-mentioned proposed rule (40 CFR Part 141), Table 7-1 is reproduced where the preliminary classification of synthetic organic chemicals and inorganics into the three categories is listed. Table 7.2 and Table 7.3 list the classification of Synthetic Organic and Inorganic Chemicals based on proposed USEPA guidelines. Table 7-4 compares adjusted acceptable daily intake (AADI) versus lifetime cancer risk concentration at 100,000 ($10^{-5}$) or one million ($10^{-6}$) population.

To take into account the possible evidence of carcinogenicity, AADI is reduced further by USEPA (1985), dividing by an arbitrary factor of 10. A lifetime risk calculation in the range of $10^{-5}$ to $10^{-6}$, using a conservative method should be considered to be protective, at least with the data presently available.

Table 7-2. Classification of Synthetic Organic Compounds Based on Proposed USEPA Guidelines.*

| Contaminant | EPA Classification | Research Conclusions |
|---|---|---|
| DBCP | B2 | Carcinogenic in rats/mice. Carcinomas in forestomach in rats and mice of both sexes, positive in short tests. |
| 2,3,7,8-TCDD | B2 | Carcinogenic in rats/mice. Cancer in liver, thyroid, tongue, etc. Inadequate evidence in humans. |
| Epichlorohydrin | B2 | Carcinogenic in rats. Cancer in forestomach (oral), subcutaneous injection sites, and nasal turbinates (inhalation). Mutagen. No human evidence. |
| Hexachlorobenzene | B2 | Carcinogenic in rats/hamsters/mice. Liver cancer. No human evidence. |
| Alachlor | B2 | Carcinogenic in rats/mice. Both sexes and dose responsive. |
| Toxaphene | B2 | Carcinogenic in rats/mice. Both sexes and dose responsive. |
| Acrylamide | B2 | Carcinogenic in two species, tumors at multiple sites. |
| EDB | B2 | Carcinogenic in rats/mice by gavage, inhalation. Dermal in mice only. |

## Table 7-2. (Continued)

| Contaminant | EPA Classification | Research Conclusions |
|---|---|---|
| Chlordane | B2 | Carcinogenic in both sexes of mice. Liver hepatocellular carcinoma. |
| Heptachlor | B2 | Carcinogenic in mice. Liver hepatocellular carcinoma. |
| Heptachlor epoxide | B2 | Carcinogenic in mice and rats. Hepatocellular carcinoma. |
| PCBs | B2 | Certain PCBs carcinogenic in mice and rats (oral). Produces benign and malignant neoplasms. |
| Lindane | C | Marginal tumors of the liver of both sexes in mice. Carcinogenic metabolite. |
| 1,2-Dichloropropane | C | Limited evidence of carcinogenicity in mice. Evidence in rats equivocal (based on NTP draft report). |
| Styrene | C | Carcinogenic in rats. Alveolar/bronchiolar adenomas and carcinomas in both sexes of rats (oral). However, these studies are not conclusive. |
| Monochlorobenzene | C | Increased occurrence of neoplastic nodules of the liver in high dose male rats. (Based on NTP draft report.) |
| Pentachlorophenol | D | Negative in studies in rats and mice; mouse data not really solid. Negative for mutagenicity. No human evidence. |
| cis and trans-1,2-Dicloroethylene | D | Not tested. |
| o-Dichlorobenzene | D | Negative results in both rats and mice NTP (draft report) gavage studies. Some caution should be used since high dose may have been below MTD. |
| m-Dichlorobenzene | D | Not tested. |
| 2,4-D | D | Inadequate animal data to classify for carcinogenicity. |
| 2,4,5-TP | D | Inadequate animal data to classify for carcinogenicity. |
| Ethylbenzene | D | Not tested. |
| Methoxychlor | D | Inadequate animal evidence. Inconclusive results. |
| Toluene | D | Negative in one CHT bioassay (inhalation) up to 300 ppm. MTD was not reached. |
| Xylene | D | Insufficient information to determine whether or not xylene itself is carcinogenic. |
| Atrazine | D | Inadequate data to classify. |
| Simazine | D | Inadequate data to classify. |
| Endrin | E | Negative results in studies, including NCI 1979 bioassay. |
| Carbofuran | E | Negative in two species and negative in short-term tests. |
| Aldicarb | E | Negative results in several studies including the NCI bioassay. |

*Reprinted courtesy of USEPA.

## Table 7-3. Inorganic Chemicals.*

| Chemical | EPA Classification | Research Conclusions |
|---|---|---|
| Asbestos | A | By inhalation, carcinogenic in humans and animals. By ingestion of intermediate (> 10 μm length) range chrysotile asbestos, limited evidence in animals—benign polyps in male rats. However, the available epidemiologic/experimental data are inadequate to conclude that the chemical is carcinogenic via ingestion. |
| Arsenic | A | Carcinogenic in humans by inhalation and ingestion. However, this chemical has potential essential nutrient value. |
| Chromium | A (based on data for Cr$^{+6}$) | Carcinogenic in humans by inhalation and rodents by intratracheal instillation. However, regulating as "D" since there is inadequate evidence to conclude that the chemical is carcinogenic via ingestion. |
| Cadmium | B1 | Limited evidence in humans exposed to cadmium fumes, cancer in rats exposed to cadmium chloride aerosol, injection site tumors in animals given cadmium salts. However, regulating as "D" since there is inadequate evidence to conclude that the chemical is carcinogenic via ingestion. |
| Nickel | B1 (based on subsulfide and carbonyl) | Limited evidence in humans by inhalation, sufficient evidence in animals by inhalation and injection. However, regulating as "D" since there is inadequate evidence to conclude that the chemical is a carcinogenic via ingestion. |
| Lead | B2 | Sufficient evidence in animals. Kidney tumors by oral route in rats. However, insufficient basis to regulate as human carcinogen via ingestion. |
| Barium | D | Inadequate data to classify. |
| Nitrate/Nitrite | D | Inadequate data to classify. |
| Sodium | D | Inadequate data to classify. |
| Cyanide | D | Inadequate data to classify. |
| Copper | D | Inadequate data to classify. |
| Selenium | D | Inadequate data to classify. |
| Silver | D | Inadequate data to classify. |
| Molybdenum | D | Inadequate data to classify. |
| Sulfates | D | Inadequate data to classify. |
| Mercury | D | Inadequate data to classify. |

*Reprinted courtesy of USEPA.

## Table 7-4. RMCL Options for Category II Contaminants.

| Contaminant | AADI With Added Factor of 10 ($\mu$/L) Considering 20% Drinking Water Contribution | $10^{-5}$ Cancer Risk (CAG) ($\mu$/L) | $10^{-6}$ Cancer Risk (CAG) ($\mu$/L) |
|---|---|---|---|
| 1,2-Dichloropropane | NA | 6 | 0.6 |
| Lindane | 0.2 | 0.28 | 0.026 |
| Monochlorobenzene | 60 | 24 (NAS) | 2.4 (NAS) |
| Styrene | 140 | NA | NA |
| Asbestos (medium and long fibers) | NA | 71,000,000 f/L | 7,100,000 f/L |

# REFERENCES

Conway, R. A. 1981. **Environmental Risk Analysis of Chemicals**. New York: Van Nostrand Reinhold.

IARC/NAS and Commission of the European Communities, Directorate General For Research Science and Education. 1986. **Screening Tests in Carcinogenesis**. Lyon, France: IARC Scientific Publications.

National Academy of Sciences. 1977. **Drinking Water and Health**. Vol. 1 (pages 793–856) and Vol. 3 (pages 5–66). Washington, D.C.: National Academy Press.

National Academy of Sciences. 1987. **Pharmacokinetics in Risk Assessment—Drinking Water and Health** (Vol. 8). Washington, D.C.: National Research Council, National Academy Press.

USEPA. 1980 and 1982. National Primary Drinking Water Regulations (45*FR*57346 and 47*FR*10999).

USEPA. 1984. Proposed Guidelines For Carcinogen Risk Assessment; Request For Comments. *Fed. Reg.* (Vol. 49, No. 7).

USEPA. 1984. Proposed Guidelines For Mutagenicity Risk Assessment; Request For Comments. *Fed. Reg.* (Vol. 49, No. 227).

WHO, IARC (International Agency for Research on Cancer) & Institut National de la Santé et de la Recherche Médicale. 1979. **Carcinogenic Risks—Strategies for Intervention**. Lyon, France: IARC Scientific Publications.

CHAPTER

# 8

# Water Analyses

# Sampling and Monitoring

## PARAMETERS AND STANDARDS

Any water supply system must be adequately sampled to meet the Health Authority's regulations and to determine on a continuous basis the potability of the finished water. In addition, it is necessary to maintain an acceptable record to document the water quality supplied, and to evaluate changes in water quality in time, in season, in drought condition, in sudden deterioration of quality, or during periods of maximum and minimum demands.

The sampling program can be subdivided as follows:

- Raw water sampling (streams, rivers, natural or artificial reservoir outlets)
- Sampling points indicative of pretreatment, intermediate, and post-treatment (plant effluent)
- Distribution system sampling in conjunction with the Health Authority* control program (mandatory minimum sampling at selected points for routine sampling)
- Watershed annual sampling; special survey sampling; citizens' complaints sampling.

When a water treatment plant is also operated, a selected sampling program is compulsory (as provided in the specific "Operation Manual" supplied by the consulting engineers who designed and supervised construction). The responsibility here is to monitor to maintain the set standards for the plant effluent.

From a laboratory and evaluation viewpoint, sampling can be subdivided as follows:

- Physical parameters
- Inorganic chemicals
- Organic chemicals
- Turbidity
- Microbiological determinations
- Radionuclide surveys
- Watershed sanitary surveys

---

*"Health Authority" is defined in Chapter 1 in the section on *The Role of Health Authorities*, page 8. In this case it may represent the County or City Health Department, the State Health and Environmental Conservation agencies, who also represent the federal agency; in the United States, USEPA with its specific rules. Regulations must be understood and promptly observed; but there is no substitute for competent operators with continuous training in the field of water quality control. Operators and supervisory technical staff are expected to utilize primarily their experience and a dedicated professionalism.

## Physical Parameters (Maximum Contaminant Level Goal)

| | |
|---|---|
| Color: | 15 CU |
| Odor: | 3 TON |
| Total Dissolved Solids: | 500 mg/L |
| pH: | not less than 6.5 or not more than 8.5 |

Descriptions of the above-mentioned parameters are provided in Chapter 2. Also, taste, temperature, alkalinity, hardness, specific conductance are described in Chapter 2. Alkalinity, hardness, and pH are also water quality parameters that may influence water treatment plant operation, and, therefore, are also treated in Chapter 10.

## Inorganic Chemicals

Health Authorities issue Maximum Contaminant Levels (MCL) for inorganic chemicals when the toxicity or carcinogenicity requires specific maximum level values not to be exceeded. The MCLs are enforceable standards.

MCLGs are recommended standards. A recommended standard is a suggested or advisable standard based on the goal to acquire a lower level of the contaminant concentration than the level previously accepted or practiced. An enforceable standard is the standard related to public health safety; aesthetics or taste and odor considerations are not included. A guideline or recommended standard is a goal toward better drinking water quality. An MCL* is the law, i.e. it is mandatory, and, if violated, remedial action must be immediately taken, the Health Authority must be informed, and, according to the Safe Water Drinking Act, consumers and the public must be notified.

USEPA also issues Secondary Drinking Water Standards (SDWS).† This classification is for contaminants that may adversely affect the odor or appearance of the water. As a consequence there is an adverse effect on the public welfare and in that relation it may partially or psychologically influence public health.

USEPA listed the following contaminants as secondary standards: *Color, Odor, pH, Total Dissolved Solids* as listed here under Physical Parameters. USEPA also listed *Aluminum, Chloride, Fluoride, Iron, Manganese, Silver, Sulfate,* and *Zinc* in the secondary standards together with *Foaming Agents* (SMCL = 0.5 mg/L); *Corrosivity* (SMCL = noncorrosive). All 14 above-mentioned parameters are considered secondary standards.

---

*The 1986 Revision of the Safe Drinking Water Act requires USEPA to issue Maximum Contaminants Level as close as possible to the Maximum Contaminant Level Goal. This is why many MCLG = MCL or the difference is minimal.

†USEPA (1979) National Secondary Drinking Water Regulations; Final Rule. *Fed. Reg.* (July 19, 1979). Aluminum and Silver were listed in the National Secondary Drinking Water Regulations. Final Rule. *Fed. Reg.* (Jan. 30, 1991). Fluoride was listed in the National Primary and Secondary Drinking Water Regulations; Fluoride; Final Rule. (April 2, 1986).

The current MCL or MCLG or SDWS for the inorganic chemicals are contained in the Summary Tables (A, B, C, D) for the four groups considered in Chapter 3.

## Organic Compounds—Chemical Parameters

The Organic Compounds are described in Chapter 4. Separately are treated the Volatile Synthetic Organic Chemicals (VOCs), the Synthetic Chemicals (SOCs), and the Trihalomethanes (THM). The Summary Tables of Chapter 4 list the MCLG and the MCL for these contaminants, including the Analytical Methods and the Best Available Technology for reduction of the contaminant concentrations when the MCLs are/or could be violated. This chapter must now indicate the necessary monitoring requirements for both Inorganic and Organic Chemical parameters.

USEPA issued a Proposed Rule in May 1989 and a Final Rule in January and July of 1991. These rules contain specific requirements for sampling and monitoring for the evaluation of compliance with the standards issued at that time (known as Phase II). In addition, 30 unregulated contaminants are listed in this Phase II requiring monitoring, depending on the vulnerability assessment conducted by a water supply system in accordance with the state criteria.

The contaminants included in Phase II are:

| Inorganics | Volatile organics (solvents) |
|---|---|
| Asbestos | cis-1,2-Dichloroethylene |
| Barium | 1,2-Dichloropropane |
| Cadmium | Ethylbenzene |
| Chromium | Monochlorobenzene |
| Mercury | o-Dichlorobenzene |
| Nitrate | Styrene |
| Nitrite | Tetrachloroethylene |
| Total nitrate-nitrite | Toluene |
| Selenium | trans-1,2-Dichloroenthylene |
|  | Xylenes (total) |

In addition to these nine inorganic chemical parameters and the 10 volatile organics, USEPA listed in the Phase II the following 20 synthetic chemical parameters:

| Pesticides, herbicides, PCBs | |
|---|---|
| Acrylamide* | Dibromochloropropane (DBCP) |
| Alachlor | 2,4-D (Formula 40) |
| Aldicarb (Temik)† | Epichlorohydrin |
| Aldicarb Sulfone | Ethylenedibromide (EDB)* |
| Aldicarb Sulfoxide | Heptachlor |
| Atrazine‡ | Heptachlor Epoxide |
| Carbofuran | Lindane |
| Chlordane | Methoxychlor |

Pesticides, herbicides, PCBs (*continued*)

| | |
|---|---|
| PCBs | Toxaphine |
| Pentachlorophenol | 2,4,5-TP (Silvex) |

*Acrylamide and Epichlorohydrin are listed without an MCL, but with an MCLG = zero. Acrylamide is used as a flocculant in water treatment plants; Epichlorohydrin is considered a contaminant of polymers in water treatment. For these contaminants the MCL is controlled at the water treatment plant with the limitation of the percentage used.

†The effective date of regulation of Aldicarb, Aldicarb Sulfoxide, and Aldicarb Sulfone, originally set for Jan. 1, 1993, was postponed by USEPA with a Notice of Postponement of Certain Provisions of Final Rule. *Fed. Reg.* May 27, 1992.

‡Atrazine, Simazine, and Cyanazine received a Notice of Initiation of Special Review with *Fed. Reg.* of Nov. 23, 1994.

In issuing the Phase II contaminants Final Rule, USEPA, as expected, issued the *monitoring requirements* for the 38 contaminants listed and for 30 additional Unregulated Contaminants. In consideration of the increased number of contaminants regulated, the complexity of the requirements increased because it is not possible to standardize the frequency for all of them. Trying to simplify, the periods of sampling schedules have been divided by USEPA into nine-year compliance cycles, divided into three-year compliance periods. Specifically, 1993/94/95 are in the first compliance period, followed by the second period, 1996/97/98, and the third will be 1999/2000/2001. States may stagger the initial period to avoid congestion in the laboratories involved. Frequency of sampling is also determined by the states when there are available prior sampling data that may reduce the increased monitoring requirements or when values greater than the MCL limits may not be expected. The "increased monitoring," to demonstrate compliance, requires analyses at quarterly intervals until values inferior to MCL are well established.

The 30 unregulated contaminants must also be monitored and the state must determine the vulnerability of the specific water supply system based on prior surveys or operational records.

The following tables give a specific schedule, as expected by these USEPA final rules of Phase II.

## USEPA—PHASE II—COMPLIANCE MONITORING REQUIREMENTS

| Contaminants | Minimum Requirements |
|---|---|
| Asbestos | One sample first year only |
| Inorganics | One sample each year for Surface Water; One sample every 3 years for Groundwater |
| VOCs (10) | Quarterly sample for the first year; Annually after first year; If no detection, reduce sampling to every 3 years |
| SOCs (18) pesticides, herbicides, and PCBs | Four samples per year; Reduce to one per year after one round of no detection |
| Unregulated Contaminants (30) 6 TOCs + 24 SOCs | One sample for 4 consecutive quarters |

*Notes:* For all water supply systems, the base monitoring may be reduced by the states after 3 sample analyses show lower concentrations than the MCL. The Phase II went into effect Jan. 1, 1993. The above-indicated schedule has also been standardized by USEPA for Phase V and for future regulations. *Detection* is defined as exceeding the method detection limit (MDL) for the contaminant.

### USEPA—General Monitoring Requirements for Organic and Inorganic Chemicals*

The schedules regulated by USEPA Phase II include public community water supply systems and nontransient, noncommunity water supply systems.

When a water supply system detects a contamination for any regulated parameter, the schedule for sampling is increased with intervals of three months. The return to a normal schedule depends on the analysis results. The state must determine that the values of concentrations are now "reliably and dependably" below the MCL for inorganics, while the VOCs concentrations must remain under 0.0005 mg/L. Groundwater systems must increase the monitoring for at least two additional quarterly samples; surface water systems must collect a minimum of four additional quarterly samples. States determine the reliability and dependability. Investigation of the possible source of contamination, and related actions taken, may guide to the determination of the expectation that the MCLs will not be violated.

The Phase II rule allows waivers that may eliminate some of the prescribed monitoring. Waivers may be requested by a specific water supply system to reduce the number of specific contaminant sampling (asbestos, pesticides, and unregulated contaminants). The process is called a *vulnerability assessment*; it may be conducted by the state, the water supply system or by a third-party organization.

### USEPA—PHASE II—Unregulated Contaminants

**Inorganics**   Antimony; Beryllium; Cyanide; Nickel; Sulfate; Thallium

**Organics**

| | |
|---|---|
| Aldrin | Dicamba |
| Benzo(a)-pyrene | Dieldrin |
| Butachlor | Dinoseb |
| Carbaryl | Diquat |
| Dalapon | Endothall |
| Di(2-ethylhexyl)adipate | Glyphosate |
| Di(2-ethylhexyl)phthalates | Hexachlorobenzene |

---

*Valid for USEPA Phases I, II, V, VIb, and future rules.

| | |
|---|---|
| Hexachlorocyclopentadiene | Oxamyl (Vydate) |
| 3-Hydroxycarbofuran | Picloram |
| Methomyl | Propachlor |
| Metolachlor | Simazine |
| Metribuzin | 2,3,7,8-TCDD (Dioxin) |

The above-mentioned contaminants are not subject to compliance rules. States determine which specific contaminants must be analyzed based on the vulnerability assessment.

### Analytical Methods

Only state-certified laboratories are accepted when *compliance* samples are analyzed. Laboratories may select an approved analytical method that measures the largest number of contaminants. Composting must be done only in the laboratory.

### Compliance with Final Rules*

For quarterly sampling, compliance is based on an annual running average for each sampling point. For annual sampling, compliance is based on a single sample unless the state requires a confirmation sample.

For **nitrate** and **nitrite**, another sample must be taken within 24 hours, if analyses are above the MCL limits.

### USEPA—PHASE V—INORGANIC, SYNTHETIC, AND VOLATILE ORGANIC CHEMICALS

USEPA, after the Proposed Rule of July 25, 1990, issued the Final Rule, released in May 1992 and published July 17, 1992. Six Inorganics, 3 VOCs, 9 SOCs (pesticides, herbicides), and 6 SOCs (non-pesticides) were regulated.

As expected from USEPA, MCLG = zero was set when evidence of carcinogenicity via ingestion had been detected. In those cases, the MCLs were set at the Practical Quantitation Levels (PQL). **Phase V** is the identification name for these Synthetic Organic Chemicals and Inorganic Chemicals.

The compliance determination effective date was Jan. 17, 1994, with exception of **endrin** (effective Aug. 17, 1992).

As usual, water supply systems must report the results of all monitoring to the state for documentation of compliance.

Violations of the Final Rules must be reported to the general public with specified mandatory language (see Chapter 4) for all phases regulating Inorganics and SOCs and VOCs.

---

*USEPA—National Primary Drinking Water Regulations. Synthetic Organic Chemicals and Inorganic Chemicals; Monitoring for Unregulated Contaminants. N.P.D.W.R. Implementation; N.S.D.W.R. Final Rule. *Fed. Reg.* (Jan. 30, 1991).

Variances and exceptions are accepted as mentioned in Phase II.

Final rules for **sulfate** have been postponed, due to the high cost of treatment for removal (BAT). On Dec. 20, 1994, USEPA issued a *Proposed Rule* for sulfate with MCLG = MCL = 500 mg/L.

## USEPA—PHASE V—List of Regulated Contaminants

**Inorganics**    Antimony; Beryllium; Cyanide; Sulfate (deferred); Thallium

### Organics (VOCs)

> Dichloromethane (methylene chloride)
> MCLG = zero; MCL = 0.005 mg/L
> 1,2,4-Trichlorobenzene
> 1,1,2-Trichloroethane

**Synthetic Organics (pesticides, herbicides)**    Dalapon; Dinoseb; Diquat; Endothall; Endrin; Glyphosate; Oxamyl (Vydate); Picloram; Simazine

**Synthetic Organics (non-pesticides)**    Benzo(a)-pyrene; Di(2-ethylhexyl)adipate; Di(2-ethylhexyl)phthalate; Hexachlorobenzene; Hexachlorocyclopentadiene (HEX); 2,3,7,8-TCDD (Dioxin).

*Notes:*    The MCLG and MCL for the inorganics are recorded on the Summary Tables of Chapter 3. The MCLG and MCL for synthetic organics are recorded in Chapter 4 (see index).

### Phase V—Monitoring

The same monitoring requirements applicable to Phase II are applicable to Phase V.* The first three-year monitoring period is selected by the states. Also, the states have jurisdiction on resampling for compliance, determination of vulnerability, and approval of laboratories.

For VOCs, detection is defined as any analytical result greater than 0.0005 mg/L.

Composting may be allowed up to five samples.

Analytical Methods and Best Available Technology are described for each contaminant in Chapters 3 and 4.

### USEPA—PHASE I—Volatile Organic Chemicals Rule

USEPA issued a Final Rule under Phase I for 8 VOCs and monitoring requirements for 51 additional contaminants; published on July 8, 1987 and

---

*USEPA—National Primary Drinking Water Regulations. Synthetic Organic Chemicals and Inorganic Chemicals. Final Rule. *Fed. Reg.* July 17, 1992.

corrected on July 1, 1988. Five of the listed VOCs are known or suspected human carcinogens; consequently, the MCLG was set at zero.

## Summary of Volatile Organic Chemicals

| Compound | MCLG (mg/L) | MCL (mg/L) |
|---|---|---|
| Benzene | zero | 0.005 |
| Carbon Tetrachloride | zero | 0.005 |
| para-Dichlorobenzene | 0.075 | 0.075 |
| 1,2-Dichloroethane | zero | 0.005 |
| 1,1-Dichloroethylene | 0.007 | 0.007 |
| 1,1,1-Trichloroethane | 0.20 | 0.20 |
| Trichloroethylene | zero | 0.005 |
| Vinyl chloride | zero | 0.002 |

## Analytical Methods

The five approved methods are based on gas chromatography or GC-mass spectrometry technique. See Chapter 4 for details for each contaminant.

## Best Available Technology

Packed-tower stripping is the only BAT for vinyl chloride. The other compounds may use packed-tower stripping or granular activated carbon adsorption.

## Compliance Requirements

See Phase II and Phase V for compliance determination.

In addition, **Phase I** regulated that the water supply systems were expected to have completed the **Initial Monitoring** by Dec. 31, 1991, if a population of less than 3,300 was served, and, by Dec. 31, 1988, if the population served was more than 10,000. Therefore, compliance was expected for the Initial Monitoring before 1992, and **Repeat Monitoring** has been of interest.

The following sampling schedule was ruled by USEPA:

| Results | Groundwater | Surface Water |
|---|---|---|
| VOCs not detected | Every 5 years | States decide |
| VOCs not detected, but source is vulnerable* | Every 3 years | Every 3 years |
| VOCs detected | Quarterly | Quarterly |

*Water supply systems with less than 500 connections may reduce sampling to "every 5 years."

*Note*: Basic monitoring for Phase I requires one sample per quarter at each entry point of the distribution system for groundwater and surface water. Composite samples of up to 5 sampling points are permitted. Sampling points are approved by the states. Also, the states must decide on the necessity to sample or not for **vinyl chloride** for each water supply system.

**Monitoring requirements for 51 additional contaminants** divide them in three groups: list 1: compounds to be monitored by all water supply systems; list 2: to be monitored only when "vulnerability" was established; list 3: monitoring at the discretion of the state.

## USEPA—PHASE I—VOCs—MONITORING REQUIREMENTS

**LIST 1** (34 VOCs)

| | |
|---|---|
| Bromobenzene | 1,1-Dichloroethane |
| Bromodichloromethane | 1,2-Dichloropropane |
| Bromoform | 1,3-Dichloropropane |
| Bromomethane | 2,2-Dichloropropane |
| Chlorobenzene | 1,1-Dichloropropene |
| Chlorodibromomethane | 1,3-Dichloropropene |
| Chloroethane | Ethylbenzene |
| Chloroform | Styrene |
| Chloromethane | 1,1,1,2-Tetrachloroethane |
| *o*-Chlorotoluene | 1,1,2,2-Tetrachloroethane |
| *p*-Chlorotoluene | Toluene |
| Dibromomethane | 1,2,3-Trichloropropane |
| *m*-Dichlorobenzene | Tetrachloroethylene |
| *o*-Dichlorobenzene | 1,1,2-Trichloroethane |
| *trans*-1,2-Dichloroethylene | *m*-Xylene |
| cis-1,2-Dichloroethylene | *o*-Xylene |
| Dichloromethane | *p*-Xylene |

**LIST 2** (2 VOCs of Phase I)

1,2-Dibromo-3-chloropropane (DBCP)
Ethylene dibromide (EDB)

**LIST 3** (15 VOCs of Phase 1)

| | |
|---|---|
| Bromochloromethane | Fluorotrichloromethane |
| *n*-Butylbenzene | Hexachlorobutadiene |
| Dichlorodifluoromethane | Isopropylbenzene |

| | |
|---|---|
| *p*-Isopropyltoluene | 1,2,3-Trichlorobenzene |
| Naphthalene | 1,2,4-Trichlorobenzene |
| *n*-Propylbenzene | 1,2,4-Trimethylbenzene |
| *sec*-Butylbenzene | 1,3,5-Trimethylbenzene |
| *tert*-Butylbenzene | |

*Notes:* To document compliance, water supply systems must report the analytical results to the state. Violations of final rules for regulated contaminants are expected to be reported to the general public. Variances and exceptions are applicable for this rule as for rules of Phases II and V.

In all phases of USEPA Final Rules for VOCs and SOCs, the term **vulnerability** is used. Vulnerability is determined by the states for each water supply system. The criteria for vulnerability are based on:

- Evaluation of previously recorded monitoring results
- Proximity of the plant infrastructures to commercial or industrial use, disposal, or storage of contaminants
- Evaluation of the level of protection provided for the water source, such as regulations and surveys for protection of the watershed or wellhead and aquifer surveillance

The state may act under its legal and/or public health jurisdiction or in representation of USEPA. Such USEPA/State relationship is defined as **primacy** in accordance with the Safe Drinking Water Act (SDWA). **Primacy** is mainly a delegation of authority for enforcement with federal financial help. SDWA gave to the states "with primacy" the authorization to exercise primary enforcement responsibility for all the National Primary Drinking Water Regulations, as long as the state adopts no less stringent regulations.

## USEPA—PHASE VIb—PRELIMINARY PROPOSAL

USEPA is still working on a proposal that will be finalized later regarding the next group of contaminants. The MCLGs and the MCLs will be proposed with the other information requested by the SDWA. The Final Rule may be issued in 1997 and become effective for enforcement in 1998. The USEPA Drinking Water Regulatory Schedule is under revision and subjected to a court-ordered deadline to be accepted by the parties.*

---

*Pontius, F.W. AWWA Regulatory Update, A.W.W.A. **Journal** No. 87, Feb. 1995 and Mar. 1996, AWWA, Denver, CO.

The following data are presented for informational purposes:

## USEPA—Phase VIb—Inorganic and Organic Contaminants

| Contaminant | Expected MCLG (mg/L) | Expected MCL unit (mg/L) |
|---|---|---|
| **Inorganics** | | |
| Boron | 0.6–1 | 0.6–1 |
| Manganese | 0.2 | 0.2 |
| Molybdenum | 0.04 | 0.04 |
| Zinc | 2 | 2 |
| | | |
| **Pesticides** | | |
| Acifluorfen | zero | 0.002 |
| Bromomethane | 0.01 | 0.01 |
| Cyanazine | 0.001 | 0.001 |
| Dicamba | 0.2 | 0.2 |
| 1,3-Dichloropropene | zero | 0.0006 |
| Ethylene thiourea | zero | 0.025 |
| Methomyl | 0.2 | 0.2 |
| Metolachlor | 0.1 | 0.1 |
| Metribuzin | 0.2 | 0.2 |
| Trifluralin | 0.005 | 0.005 |
| | | |
| **Other SOCs** | | |
| Acrylonitrile | zero | 0.003 |
| 2,4- and 2,6-Dinitrotoluene (mixture) | zero | 0.003 |
| Hexachlorobutadiene | 0.001 | 0.001 (HCBD) |
| 1,1,1,2-Tetrachloroethane | 0.07 | 0.07 (TCE) |
| 1,2,3-Trichloropropane | zero | 0.0008 (TCP) |

**Unregulated Monitoring**
Bromacil; Prometon; Methyl-*t*-butyl-ether

*Notes:*   Final Rules must be issued by USEPA for at least 25 contaminants under VIb and the Disinfectants/Disinfection By-Products (D/DBP).

## TRIHALOMETHANES—MONITORING REQUIREMENTS

Information on THMs was given in Chapter 4. No major change was regulated by USEPA after the Final Rule was issued on Nov. 29, 1979, and a clarification on the Best Available Technology issued on Feb. 28, 1983. Also, an update on Analytical Techniques was issued on Aug. 3, 1993. (see *THMs*, page 165).

The MCL = 0.10 mg/L has remained the Final Rule for TTHMs. This upper limit includes the total concentrations for chloroform, bromodichloromethane, dibromochloromethane, and bromoform.

**Base Monitoring** includes the following schedules:

- Four samples per quarter at treatment plants as a minimum.
- In the distribution system: a minimum of 25% of samples taken at the maximum residence time; 75% of samples to be taken at representative locations.
- State approval is necessary to accept multiple wells from the same aquifer to be equivalent to a treatment plant.

## Reduced TTHM monitoring

The state may accept reduced monitoring, basing decisions on monitoring data, quality and stability of the water source, and type of water treatment.

**Compliance** requires that the running annual average be less or equal to the TTHM of 0.1 mg/L.

## Turbidity*

The Health Authority in general requires frequent sampling for **turbidity**, particularly when the water supply is from surface supply. Turbidity, originally considered only a physical parameter, has assumed during the last 30 years a "microbiological parameter" concept in the disinfection procedures. Turbidity measures the fine suspended matter in water, mostly caused by colloidal matter. This matter has the tendency to protect bacteria, viruses, and perhaps pathogenic protozoans from the disinfecting action provided by the chlorination or ozonation of water under treatment or of potable water delivered by the distribution system.

Conventional water treatments have the technological possibility to deliver water to the distribution system with very low values of turbidity (under 0.5 NTU and, very rarely, under one NTU). The major problem is to regulate the surface water supply systems *without filtration*. In these cases the standard must be raised to one NTU, allowing exceptions only in very rare and justifiable occasions. Exceptions may be granted to water supply systems that have an extensive record of successful disinfection results of the distribution system (bacteriological data to prove it; protected watershed; surveys performed routinely for cross-connection controls; multiple disinfection stations in series; qualified personnel; state approval of the water quality).

The most important parameter is the maintenance of a free chlorine residual of at least 0.2 mg/L or equivalent bactericidal treatment for other disinfectants in the distribution system.

---

*Turbidity is described as a physical parameter in Chapter 2 and a microbiological parameter in Chapter 5. Here turbidity is described as a parameter to evaluate water quality through sampling and monitoring.

An isolated, relatively high **turbidity** reading cannot be called an indicator of contaminated water. It all depends on the type of colloidal matter encountered in that particular test.

USEPA has tried to transform scientific knowledge and the experience of consulting engineers and water supply system operators into intelligent regulations that may assure consumers of potable water with a final product without pathogenic bacteria, viruses, or protozoans. Not an easy task!

USEPA issued final rules for **turbidity** with the Total Coliform Rule of 1989, contemporary with the Surface Water Treatment Rule, also final and followed by the proposed Disinfectants/Disinfection By-Products Rule (D/DBP) of 1991, and the proposed Enhanced Surface Water Treatment Rule (ESWTR) of 1994. These last two proposals have an effective date of implementation with final rules in the year 2000 or sometime early in the 21st century: this proves the difficulty of the above-mentioned task.

More information is provided in Chapter 10 on evaluating filtration requirements and disinfection.

In November 1987, USEPA concurred that an MCL is not significant and should not be issued. Turbidity levels should be used in correlation with bacteriological data and filtration/disinfection to evaluate when filtration treatment is required. In 1987, USEPA issued the *Turbidity Performance Criteria.*
These new proposed standard can be summarized as follows:

1. *For conventional treatment or direct filtration.* **Turbidity** equal or less than 0.5 NTU in 95% of the monthly measurements. States are allowed to raise this value when disinfection performance achieves 99.9% inactivation of *Giardia* cysts. In all cases turbidity should be maintained under 1 NTU in 95% of the monthly measurements and never exceed 5 NTU. Water treatment, improved to reach this standard if necessary, is expected to achieve optimal removal of *Giardia* cysts and other pathogens. It is clear that under these conditions the effectiveness of disinfection is expected.
2. *Slow sand filtration.* **Turbidity** less or equal to 1 NTU in 95% of monthly readings with a maximum of 5 NTU. Similar latitude (see No. 1) is given to the States.
3. *Diatomaceous earth filtration.* **Turbidity** less or equal to 1 NTU in 95% of monthly readings with a maximum of 5 NTU.
4. *Monitoring and reporting.*

   - **Systems not using filtration**—USEPA regulations require the following reports to be submitted by the State: monthly reports on total and fecal coliforms; number of samples collected; number of samples exceeding 100/100 mL or 20/100 mL, respectively; the cumulative total of the samples collected since the beginning of the running 6-month compliance period and the percentage of these which are below the respective criteria; values above 5 NTU and dates, and when the water supply informed its customers to boil their water.

In addition monthly disinfection conditions should be reported to demonstrate:

- the 99.9% inactivation performance criteria are met for *Giardia* cysts and enteric virus;
- a disinfectant residual of at least 0.2 mg/L is maintained all times in the distribution systems and in 95% of the monthly readings;
- temperature, pH, and contact time at peak hourly flow conditions and the actual CT values;
- annual report of the watershed control program;
- suspected occurrence of any waterborne disease outbreak within 48 hr; and
- copy of any public notice issued.
- **Systems using filtration**—Reports to the State include: filtered water turbidity; disinfectant residual concentration in the water entering the distribution system; maintenance of disinfectant residuals in the distribution system, and public notification. In addition statistical documentation is necessary, when standards are not maintained.

## Microbiological Determination*

**Total Coliform Bacteria**    Coliform densities examination samples should be collected routinely and at regular time intervals in a number proportional to the distribution system. Traditionally the minimum number of samples has been referred to the population served. Table 8-1 can be used as a guide. Table 8-2 provides sampling requirements based upon population, as recommended by USEPA.

The minimum number of monthly samples could be reduced, substituting the bacteriological samples with chlorine residual samples at representative points of the distribution system, with increased frequency or with a larger number of chlorine residual determination. The goal for MCL for total coliform should be zero. A realistic MCL could be adopted as 95%–98% of all samples examined monthly to be negative for coliforms.

## TOTAL COLIFORM GROUP—MONITORING

Fecal coliforms and *Escherichia coli (E. coli)* are part of the total coliform group. A positive sample in this group is an indication of possible presence of pathogens.

---

*See Chapter 5 for more detailed information on bacteria, viruses, and pathogenic protozoa evaluated from a viewpoint leading to identification, infecting dose, indicator organisms, current standards, laboratory analyses, and their reference to waterborne diseases.

Table 8-1. Minimum Number of
Bacteriological Samples Collected for
the Population Served.

| Population Served | Monthly Number (Minimum) |
|---|---|
| Under 1,000 | 2 |
| 1,000– 5,000 | 5 |
| 5,000– 10,000 | 12 |
| 10,000– 20,000 | 24 |
| 20,000– 40,000 | 40 |
| 40,000– 70,000 | 75 |
| 70,000– 110,000 | 100 |
| 110,000– 160,000 | 120 |
| 160,000– 220,000 | 140 |
| 220,000– 300,000 | 170 |
| 300,000– 500,000 | 200 |
| 500,000– 800,000 | 250 |
| 800,000–1,200,000 | 300 |
| 1,200,000–1,600,000 | 350 |
| 1,600,000–2,000,000 | 400 |
| 2,000,000–2,500,000 | 450 |
| 2,500,000–4,000,000 | 500 |
| Over 4,000,000 | 600 |

If a water supply system does not meet the minimum requirements of the negative sample percentage for a month, the state and the public must be notified. Moreover, the same notification is expected when a routine sample tests positive for total coliforms and for fecal coliforms or *E. coli*, and any repeat sample tests positive for total coliforms. Also, when a routine sample tests positive for total coliforms and negative for fecal coliforms or *E. coli*, and any repeat sample is positive for fecal coliforms or *E. coli*, the state and the public must be notified.

All water supply systems must prepare a sampling schedule to be submitted to the state in writing and with graphic representation for approval. The total number of the monthly samples must be proportional in number to the population served, as a minimum (see Table 8-2).

### Resampling Rules

Any routine sample that results positive for total coliforms must be followed by 3 or 4 repeat samples taken as soon as possible (within 24 hours of notification). Of these repeat samples, one must be from the same tap of the original sample, while the other 2 or 3 must be taken from within five service connections of the original sample, one upstream and the other downstream. If these results indicate that the original problem was of local origin, the state can invalidate the positive routine sample.

In the evaluation of total coliform results it is necessary to consider that

Table 8.2.  USEPA—Recommended Final Coliform
Rule (6/29/1989) Drinking Water Regulations
Sampling Requirements Based Upon Population.

| Population Served | Minimum Number of Samples per Month |
|---|---|
| 25 to 1,000 | 1 |
| 1,001 to 2,500 | 2 |
| 2,501 to 3,300 | 3 |
| 3,301 to 4,100 | 4 |
| 4,101 to 4,900 | 5 |
| 4,901 to 5,800 | 6 |
| 5,801 to 6,700 | 7 |
| 6,701 to 7,600 | 8 |
| 7,601 to 8,500 | 9 |
| 8,501 to 12,900 | 10 |
| 12,901 to 17,200 | 15 |
| 17,201 to 21,500 | 20 |
| 21,501 to 25,000 | 25 |
| 25,001 to 33,000 | 30 |
| 33,001 to 41,000 | 40 |
| 41,001 to 50,000 | 50 |
| 50,001 to 59,000 | 60 |
| 59,001 to 70,000 | 70 |
| 70,001 to 83,000 | 80 |
| 83,001 to 96,000 | 90 |
| 96,001 to 130,000 | 100 |
| 130,001 to 220,000 | 120 |
| 220,001 to 320,000 | 150 |
| 320,001 to 450,000 | 180 |
| 450,001 to 600,000 | 210 |
| 600,001 to 780,000 | 240 |
| 780,001 to 970,000 | 270 |
| 970,001 to 1,230,000 | 300 |
| 1,230,001 to 1,520,000 | 330 |
| 1,520,001 to 1,850,000 | 360 |
| 1,850,001 to 2,270,000 | 390 |
| 2,270,001 to 3,020,000 | 420 |
| 3,020,001 to 3,960,000 | 450 |
| 3,960,001 or more | 480 |

heterotrophic bacteria of high count may interfere in the test of total coliforms giving false negative results. In these suspected cases the samples
involved may be declared invalid.*

*USEPA Final Rules of June 29, 1989 stated the following in regard to the interference of
Heterotrophic bacteria:
Heterotrophic bacteria can interfere with total coliform analysis. Therefore, if the total
coliform sample produces: (1) A turbid culture in the absence of gas production using the
Multiple-Tube Fermentation (MTF) Technique; (2) a turbid culture in the absence of an acid
reaction using the Presence-Absence (P-A) Coliform Test; or (3) confluent growth or a colony
number that is "too numerous to count" using the Membrane Filter (MF) Technique, *the sample
is invalid.*

Unfiltered surface water systems and systems using unfiltered groundwater under the direct influence of surface water must analyze one coliform sample taken every day in the proximity of the first consumer when the **turbidity** of the source water indicates a value that exceeds **one NTU**. MCL violations must be reported to the state no later than the end of the next business day after the water supply system learns of the violation. (USEPA, Total Coliform Rule effective since Dec. 31, 1990.) Violations of these rules must be reported to the general public with mandatory language issued by USEPA to read as follows:

The United States Environmental Protection Agency (EPA) sets drinking water standards and has determined that the presence of total coliforms is a possible health concern. Total coliforms are common in the environment and are generally not harmful in themselves. The presence of these bacteria in drinking water, however, generally is a result of a problem with water treatment or the pipes which distribute the water, and indicates that the water may be contaminated with organisms that can cause disease. Disease symptoms may include diarrhea, cramps, nausea, and possibly jaundice, and any associated headaches and fatigue. These symptoms, however, are not just associated with disease-causing organisms in drinking water, but also may be caused by a number of factors other than your drinking water. EPA has set an enforceable drinking water standard for total coliforms to reduce the risk of these adverse health effects. Under this standard, no more than 5.0 percent of the samples collected during a month can contain these bacteria, except that systems collecting fewer than 40 samples/month that have one total coliform positive sample per month are not violating the standard. Drinking water which meets this standard is usually not associated with a health risk from disease-causing bacteria and should be considered safe.

The difficulties in the determination with reliable analytical methods for **viruses**, *Giardia lamblia*, and *Cryptosporidium* should not limit our confidence to judge water quality from a microbiological viewpoint, based on careful examination and control of the following important aspects of water quality:

- water treatment performance
- turbidity data and variations in turbidity data
- free chlorine residuals or other disinfectant residuals
- total coliform organisms and Heterotrophic Bacteria data; resampling of points where positive samples were collected; confirmation or not with fecal coliforms or *E. coli.*

USEPA has tried to promulgate rules that, if complied with, can assure that waterborne microbial diseases are eliminated or reduced to an absolute minimum. USEPA's efforts are contained in its **Total Coliform Rule** of June 29, 1989; the **Surface Water Treatment Rule** of June 29, 1989; and the **Enhanced Surface Water Treatment Rule** proposed on July 29, 1994, expected to be effective for all water supply systems by June 2000.

Maximum Contaminant Level Goals (MCLG) and MCLs for Total Coliforms, *E. coli*, Fecal Coliforms, *Giardia, Legionella*, and *Cryptosporidium* are described in Chapter 5.

**Giardia**   *Giardia lamblia* may cause Giardiasis, as described in Chapter 5. In the same chapter it is mentioned in the USEPA Final Rules (1991–1996) that MCLG = zero for *Giardia lamblia* and MCL = Treatment Technique (implementation of the Surface Water Treatment Rule).

USEPA's goal remains zero viable cysts, leading to the conclusion that all surface water supply systems need a conventional or modified filtration treatment.

The 19th edition of *Standard Methods* (1995) considers *Giardia lamblia, Entamoeba histolytica*, and *Cryptosporidium parvum* the pathogenic protozoans of primary concern in drinking water. These pathogenic organisms cause diarrhea or gastroenteritis and waterborne outbreaks. In this edition, *Standard Methods* proposed for analytical method the Immunofluorescence Method for *Giardia* and *Cryptosporidium* (#9711) and Direct Microscopic Examination for *Entamoeba histolytica*.

**Pathogenic Viruses**   While a virology department (directed by an experienced virologist) of a specialized water laboratory can isolate and identify pathogenic viruses in a satisfactory reporting method, nevertheless, routine determination cannot yet be required by the Health Authority in the microbiological evaluation of maximum contaminant levels for lack of a standardized reliable method. The goal of all standards is to have no viruses in drinking water.

Not only the most significant pathogenic viruses such as hepatitis A agent, rotaviruses, and Norwalk-like agents, but also the traditional enteroviruses (poliovirus, echovirus, and coxsackievirus) occasionally found in drinking water cannot be used for routine analysis (see Chapter 5, *Viruses*). Consequently, the only reliable assurance in water quality at present is treatment, such as effective filtration and disinfection for surface water and disinfection for groundwater, in addition to watershed monitoring and pretreatment (see Chapters 1 and 10).

*Standard Methods* (1995) reports the detection of enteric viruses with analytical methods approved in 1990 by the Standard Methods Committee. *Standard Methods*, nevertheless, did not recommend the routine examination of water for enteric viruses; exceptions are the microbiological evaluation of outbreaks (very rare in water supply systems, and, in those cases, confined to the distribution system).

**Legionellae**   These are bacteria causing legionellosis, either in the form of pneumonia or in non-pneumonia (milder cases of Legionnaires' Disease), also known as Pontiac fever. Health Authorities are not yet ready to issue a MCL for *Legionellae* and cannot recommend a specific treatment due to inefficiency of bactericidal effect of both typical filtration and low dosage of chlorination. Routine sampling, therefore, is not mandatory in spite of the abundance of *Legionellae* in ambient water. Because of that pathogenicity of the *Legionellae*, a zero recommended standard is desirable and expected from the Health Authorities (see Chapter 5).

**Heterotrophic Plate Count**   Derived from the old and traditional **standard plate count**, the **Heterotrophic Plate Count Method** has been improved and standardized. In 1994 the Standard Methods Committee approved three different methods and four different media were described, namely:

- Pour Plate Method
- Spread Plate Method
- Membrane Filter Method

To compare data, unfortunately, it is necessary to always use the same procedure and the same medium.

The purpose is to identify **heterotrophic bacteria**, a group that includes pathogenic as well as innocuous bacteria.

When high bacteria concentration is reported, especially when higher than the usually expected numbers (previously recorded) it is an indication of deterioration of the water quality. These test data can be used for evaluation of water treatment, as well as for monitoring the distribution system. These tests should be used in conjunction with coliform determination.

Health Authorities have traditionally avoided setting standards for plate counts; possibly for lack of pathogenicity to be associated with heterotrophic bacteria and due to a great variation in density encountered. Density under 100 colonies/mL is expected in the distribution system with maximum concentration of 500 (certainly a limit for potability). It is advisable to maintain a sampling record for bacteriological evaluation for both coliforms and heterotrophic bacteria (see Chapter 5).

In laboratory reports of the heterotrophic bacteria concentrations, the term **colony-forming units (CFU)** is descriptive of the method used, but the report must include also the specific method used, the incubation temperature and time, and the medium. For example: CFU/mL, Pour Plate Method, 35°C/48 h, plate count agar.

## Collection of Samples

**Volume**   In general, a sample volume of two liters is sufficient for physical and chemical examination of water. Sterilized bottles of 100 mL may be sufficient for a single bacteriological examination. Separate samples are collected for chemical and microscopic examination. A two-liter volume can be used for collection of radioactive samples.

Much larger volumes are required for pathogenic microorganisms; for example, 380 L (100 gal) for pathogenic protozoa. Special equipment is needed for pathogen recovery.

In organizing the sampling activity, the public health sanitarian or engineering technician must obtain from the laboratory, where the samples will eventually be examined, sampling bottles specifically prepared and pretreated for each parameter to be tested. A sampling field program is always initiated at the laboratory. Moreover, personnel at the laboratory must understand the purpose of sampling and approve and confirm the volume of each sample constituent necessary for analysis and repeat analysis whenever possible. Once the field activity is completed, the samples must be immediately taken to the laboratory.

It is highly advisable that in case of doubts, the public health program organizer check the sampling procedure in *Standard Methods* and train the technician accordingly. In case of disagreement with laboratory personnel, in terms of sampling technique, the last word is with the Laboratory Director.

**Time**   The sample collected should be examined immediately to provide the maximum reliability of results. Whenever possible, field work (portable lab) should be done with the proper instrumentation and equipment. Chlorine residual, temperature, pH, and specific conductance can be easily determined in the field.

Bacteriological examination should be initiated within one hour after collection. Maximum time between collection and examination should be 30 hours.* Field examination of samples (incubation) should be contemplated whenever routine delays of over 24 hours between collection and examination are expected.

### Handling of Samples

*Routine Bacteriological Examination.*   Bottles resistant to action of water and repeated sterilization should be used. Ground-glass stoppered bottles, preferably widemouth, are recommended. When shipment is necessary, special plastic bottles may be used. A metal foil should be used prior to sterilization to protect tops and necks of sampling bottles.

Laboratory procedures for washing and sterilization are prescribed in the *Standard Methods for the Examination of Water and Wastewater* published by the American Public Health Association, the American Water Works Association, and the Water Pollution Control Federation.† Dechlorination should be practiced when sampled water is expected to contain chlorine residual to stop the bactericidal effects of chlorine. Sodium thiosulfate is an acceptable dechlorinating agent, added before sterilization of bottles. An approximate sodium thiosulfate concentration of 100 mg/L in the sample should be obtained. Air space should be left in the bottle to allow for mixing of the sample.

---

*This time is not valid for research work and probably for legal enforcement action as well.
†Now the "Water Pollution Control Federation" has been renamed as the "Water Environment Federation."

The bottle, transported free of potential contamination, should be opened only at the time of filling. The stopper and the protective foil should be removed as a unit. The bottle, held at the base, should be filled without rinsing; stopper and foil should be replaced and secured, avoiding any accidental contamination by human and transportation means. The bottle containing the sample may not be refrigerated, but care should be used to avoid storage at temperature higher than the temperature of the water at the moment of sampling. In case of doubt, refrigeration of the sample is preferable.

The tap used as a sampling point should be without attachment and should not be too distant from the main: it should be opened fully to allow water to run for 2 minutes or more to obtain a representative sample of the main or major feeding branch. The flow from the outlet should then be restricted to allow filling without splashing.

When raw water from the surface supply or treated water from the reservoir is collected, a representative sample should be collected (as close as possible from the level of drawoff) avoiding points of stagnation.

Minimum volume should be 100 mL. Accurate identification of the sample with location, time, name of agency and sampler, chlorine residual, temperature, and remarks related to the survey or monitoring program should be stressed.

*Routine Chemical Samples.   Standard Methods* recommends the use of acidified bottles to avoid loss by adsorption on or ion exchange with the wall of glass containers, particularly in samples used for examination of aluminum, cadmium, chromium, copper, iron, lead, manganese, silver, and zinc.

Significant changes can take place in a matter of minutes at the time of collection of the samples in temperature, pH, and dissolved gases (loss of chlorine, oxygen, carbon dioxide, hydrogen sulfide, or a gain of oxygen and carbon dioxide). These factors may affect precipitation of calcium carbonate, influencing data for calcium and total hardness.

It is strongly advised that the laboratory technician, prior to collecting chemical samples, contact both the responsible professional in the laboratory and the engineer who ordered the sampling so as to utilize the best technique for accurate collection and analysis preparation.

To avoid contamination in sampling and transportation, the general principle (described in the preceding section) should be used here, but filling of the bottle is preceded with rinsing 2 or 3 times with the water to be sampled unless bottles have been pre-acidified.

Table 8-3 lists the recommended handling of water samples according to expected analysis and reporting volume of the sample required, acceptable container (plastic or glass), expected preservation, and maximum holding time.

## Watershed Sampling

Watershed sampling is part of the general inspection of the watershed, prescribed by the Health Authority. Such inspection is required at least once

## Table 8-3. Recommended Container Handling of Water Samples (P = plastic, G = glass).

| Parameter | Sample Size (mL) | Sampling Bottle | Preservation | Max. Holding Time |
|---|---|---|---|---|
| *Physical Parameters* | | | | |
| Color | 50 | P,G | Cool, 4°C | 48 hours |
| Conductance | 100 | P,G | Cool, 4°C | 28 days |
| Hardness | 100 | P,G | $HNO_3$ to pH<2 | 6 months |
| Odor | 200 | G only | Cool, 4°C | 24 hours |
| pH | 25 | P,G | None Req. | Analyze immediately |
| Residue | | | | |
| Filterable | 100 | P,G | Cool, 4°C | 7 days |
| Nonfilterable | 100 | P,G | Cool, 4°C | 7 days |
| Total | 100 | P,G | Cool, 4°C | 7 days |
| Volatile | 100 | P,G | Cool, 4°C | 7 days |
| Settleable Matter | 1,000 | P,G | Cool, 4°C | 48 hours |
| Temperature | 1,000 | P,G | None Req. | Analyze immediately |
| Turbidity | 100 | P,G | Cool, 4°C | 48 hours |
| *Metals* | | | | |
| Dissolved | 200 | P,G | Filter on site $HNO_3$ to pH<2 | 6 months |
| Suspended | 200 | | Filter on site | 6 months |
| Total | 100 | P,G | $HNO_3$ to pH<2 | 6 months |
| Chromium$^{+6}$ | 200 | P,G | Cool, 4°C | 24 hours |
| Mercury Dissolved | 100 | P,G | Filter $HNO_3$ to pH<2 | 28 days |
| Total | 100 | P,G | $HNO_3$ to pH<2 | 28 days |
| *Inorganics, Non-metallics* | | | | |
| Acidity | 100 | P,G | Cool, 4°C | 14 days |
| Alkalinity | 100 | P,G | Cool, 4°C | 14 days |
| Bromide | 100 | P,G | None Req. | 28 days |
| Chloride | 50 | P,G | None Req. | 28 days |
| Chlorine | 200 | P,G | None Req. | Analyze immediately |
| Cyanides | 500 | P,G | Cool, 4°C NaOH to pH>12; 0.6 g ascorbic acid | 14 days |
| Fluoride | 300 | P,G | None Req. | 28 days |
| Iodide | 100 | P,G | Cool, 4°C | 24 hours |
| Nitrogen | | | | |
| Ammonia | 400 | P,G | Cool, 4°C $H_2SO_4$ to pH<2 | 28 days |
| Kjeldahl, Total | 500 | P,G | Cool, 4°C $H_2SO_4$ to pH<2 | 28 days |
| Nitrate plus Nitrite | 100 | P,G | Cool, 4°C $H_2SO_4$ to pH<2 | 28 days |
| Nitrate | 100 | P,G | Cool, 4°C | 48 hours |
| Nitrite | 50 | P,G | Cool, 4°C | 48 hours |

Table 8-3. (*Continued*)

| Parameter | Sample Size (mL) | Sampling Bottle | Preservation | Max. Holding Time |
|---|---|---|---|---|
| Dissolved Oxygen | | | | |
| Probe | 300 | G bottle and top | None Req. | Analyze immediately |
| Winkler | 300 | G bottle and top | Fix on site and store in dark | 8 hours |
| Phosphorus | | | | |
| Ortho-phosphate, Dissolved | 50 | P,G | Filter on site Cool, 4°C | 48 hours |
| Hydrolyzable | 50 | P,G | Cool, 4°C H$_2$SO$_4$ to pH<2 | 28 days |
| Total | 50 | P,G | Cool, 4°C H$_2$SO$_4$ to pH<2 | 28 days |
| Total, Dissolved | 50 | P,G | Filter on site Cool, 4°C H$_2$SO$_4$ to pH<2 | 24 hours |
| Silica | 50 | P only | Cool, 4°C | 28 days |
| Sulfate | 50 | P,G | Cool, 4°C | 28 days |
| Sulfide | 500 | P,G | Cool, 4°C add 2 mL zinc acetate plus NaOH to pH>9 | 7 days |
| Sulfite | 50 | P,G | None Req. | Analyze immediately |
| *Organics* | | | | |
| BOD | 1,000 | P,G | Cool, 4°C | 48 hours |
| COD | 50 | P,G | Cool, 4°C H$_2$SO$_4$ to pH<2 | 28 days |
| Oil and Grease | 1,000 | G only | Cool, 4°C H$_2$SO$_4$ to pH<2 | 28 days |
| Organic Carbon | 25 | P,G | Cool, 4°C H$_2$SO$_4$ or HCl to pH<2 | 28 days |
| Phenolics | 500 | G only | Cool, 4°C H$_2$SO$_4$ to pH<2 | 28 days |
| MBAS | 250 | P,G | Cool, 4°C | 48 hours |
| NTA | 50 | P,G | Cool, 4°C | 24 hours |

per year, but in less sensitive or vulnerable cases once every three years is acceptable. Of course, sampling is much more frequent. The sampling involved is related to the potential pollution sources within the boundary of the watershed. Sampling of effluents, streams, ponds, lakes, rivers, recharging basins, agricultural runoff, wells, and reservoirs will be conducted with similar sampling technique used for raw water.

It is essential to maintain good records of all results for evaluation of changes in water quality. Past records of pollutional problems may lead to concentration of efforts on those critical points and, consequently, to planning more frequent inspections.

It is unquestionable that from a viewpoint of water quality control, the inspection of watersheds is an activity really neglected in the past. Of course, lack of jurisdiction on the part of the water purveyors and the undefined limit of the watershed have caused the reluctance to implement this necessary program.

Routinely organized joint inspections of water purveyors with representatives of the regulatory agencies will detect problems and expedite solutions.

### Radionuclides—Sampling and Monitoring

For USEPA regulations with regard to sampling and monitoring of radionuclides, see Chapter 6.

See Table 8-4 for sampling and monitoring according to the Sanitary Code, enforced by the New York State Department of Health, Bureau of Public Water Supply Protection.

## Laboratory Analyses

Laboratory analyses are performed either by certified private or municipal laboratories under the direction and responsibility of a certified professional or by an on-site certified laboratory that is part of the water treatment plant.

Where there is no water plant for conventional treatment of water supply but the required minimum treatment for disinfection and/or pH control, and fluoridation is provided, then certified operators run field tests and operational control tests at the pumping stations and/or chemical feeding stations. All other laboratory work is performed in city/town/county environmental or health departments or contracted out to a private certified laboratory.

All above-mentioned certifications are issued by the Health Authority having territorial responsibility, based on public health laws issued by the legislature having territorial jurisdiction.

### LABORATORY PERSONNEL

Health Authorities require daily sampling of public water supplies in a number proportional to the population served (see *Sampling*, pages 387–390). Analyses of such Sampling Programs must be conducted by certified

## Table 8-4. Radiological Minimum Monitoring Requirements.

| Contaminant | Type of water system | Source type All | Source type |
|---|---|---|---|
| Combined radium-226 and radium-228 and gross alpha particle activity | Community | Once every four years, an annual composite of quarterly samples; or four quarterly samples must be obtained[a,b,c,d] | |
| | Noncommunity | Not applicable | |
| | | Groundwater only | Surface only or surface and groundwater |
| Beta particle and photon radioactivity from man-made radionuclides | Community serving over 100,000 people | State discretion[e] | Once every four years, an annual composite of quarterly samples; or four quarterly samples must be obtained[f,g] |
| | Community serving 100,000 or fewer people | State discretion[e] | State discretion[e] |
| | Noncommunity | Not applicable | Not applicable |

*Note:* From Sanitary Code: Subpart 5-1, Public Water Systems; issued by the New York State Department of Health, Bureau of Public Water Supply Protection, effective Jan. 6, 1993.

[a]Gross alpha particle activity measurement may be substituted for the required radium-226 and radium-228 analysis, if the measured gross alpha particle activity does not exceed five picocuries per liter at a confidence level of 95 percent (1.65 sigma, where sigma is the standard deviation of the net counting rate of the sample). When the gross alpha particle activity exceeds five picocuries per liter, the same or an equivalent sample shall be analyzed for radium-226. If the concentration of radium-226 exceeds three picocuries per liter, the same or an equivalent sample shall be analyzed for radium-228.

[b]The State may permit the substitution of the analysis of a single sample for quarterly sampling when the average annual concentration is less than one half of the MCL.

[c]The State may require suppliers of water to conduct annual monitoring when the radium-226 concentrations exceeds three picocuries.

[d]If the average annual MCL for gross alpha particle activity or total radium is exceeded, monitoring at quarterly intervals shall be continued until the annual average concentration no longer exceeds the MCL, or until a monitoring schedule as a condition to a variance, exemption, or enforcement action is effective.

[e]When the State determines that a community water system is using water contaminated by effluents from nuclear facilities, the supplier of water shall initiate quarterly monitoring for gross beta particle and iodine-131 radioactivity and annual monitoring for strontium-90 and tritium.

[f]Monitoring compliance may be assumed without further analysis if the average annual concentration of gross beta particle activity is less than 50 picocuries per liter and if the average annual concentration of tritium is less than 20,000 picocuries per liter and the average annual concentration of strontium-90 is less than 8 picocuries per liter if both radionuclides are present, the sum of their annual dose equivalents to bone marrow shall not exceed four millirems per year.

[g]If the gross beta particle activity exceeds 50 picocuries per liter, an analysis of the sample must be performed to identify the major radioactive constituents present, and the appropriate organ and total body doses shall be calculated to determine compliance.

personnel in Chemical or Microbiological Laboratories since certification should be separated for the two branches of quality control.

The purpose of certification of laboratories is to assure that analyses of samples are performed *in an adequately equipped laboratory, staffed with competent personnel using acceptable methodology to produce accurate results.*

The first certification in order of importance is issued to the Laboratory Director. He or she will be certified for chemical and/or microbiological analysis based on education and experience. The educational requirements will specify the minimum college credits, usually a baccalaureate degree in chemical or biological sciences, sanitary/public health engineering or medicine, or environmental sciences, with a minimum of 24 college credit-hours in chemistry, and three courses in microbiology.

Experience in a properly supervised subordinate role of one or more years must be demonstrated with length of time inversely proportional to the education based on academic degree. Health Authorities will review credentials of applicants and then will interview the candidate on site during evaluation of the physical condition of the laboratory and adequacy of equipment (see Fig. 8-1).

Laboratory personnel reporting to a certified director will meet qualifica-

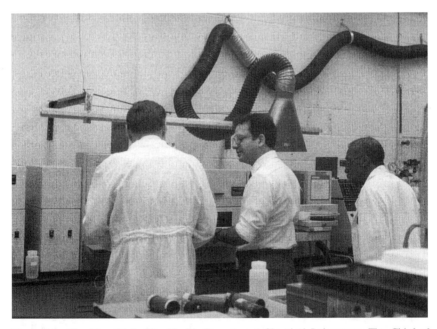

Figure 8-1. The New York City Health Department Chemical Laboratory. The Chief of Toxicology supervises a procedure on a special sample determined by atomic absorption spectroscopy. The sample is analyzed at a very low concentration range with a spectrophotometer equipped with a furnace accessory. (Courtesy of the N.Y.C. Department of Health.)

tions related to the size of the laboratory and the population served. Such personnel when required to staff a medium to large size laboratory may be classified as a Junior Chemist, Assistant Chemist, Chemist, Senior Chemist or Junior Bacteriologist, Assistant Bacteriologist, Bacteriologist, and Senior Bacteriologist. Their classification is based on education, experience, ability, and reliability, and in public administration also related to civil service entrance or promotional examinations.

## PHYSICAL FACILITIES

The New York State Health Department, for example, requires laboratory space of at least 150 to 200 sq.ft/person (14 to 19 sq. m/person) with an approximate 15 linear feet of usable bench space (4.6 m) for a chemical laboratory with electricity, a sink with hot and cold running water, a source of distilled and/or deionized water, and an air exhaust hood for analysis of organic chemicals and trace metals.

For the Bacteriological Laboratory the State requires adequate space (i.e., 200 sq. ft [19 sq. m] and 6 linear feet [1.83 m] of bench space per analyst). Possibly a separate area for cleaning glassware and sterilizing materials (a must for medium and large laboratories) may be required.

## EQUIPMENT

### Chemical Laboratory

For a Chemical Laboratory the following equipment should be part of the standard equipment or be easily available:

- analytical balance
- photometer (spectrophotometer—filter photometer)
- magnetic stirrer
- pH meter
- specific ion meter
- atomic absorption spectrophotometer
- recorder of atomic absorption
- gas chromatograph
- recorder for gas chromatograph
- conductivity meter
- drying oven
- dessicator
- hot plate
- refrigerator
- glassware
- stirred boiling water bath

## Bacteriological Laboratory

For analyzing samples for the total coliform membrane filter or most probable number procedures, the bacteriological laboratory should have available the following:

- pH meter
- balances
- thermometers or continuous recording devices
- air (or water jacketed) incubators/water baths/aluminum block incubators
- autoclave
- refrigerator
- optical/counting/lighting equipment
- inoculation equipment (mandatory)
- membrane filtration equipment
- membrane filters
- laboratory glassware, plasticware, and metal utensils
- measuring equipment
- culture dishes
- culture tubes and closures

## METHODOLOGY

Reliable results are the products of adoption and application of proper methodology performed by competent personnel in a properly equipped and maintained laboratory.

In many of the laboratories in the United States and most of those in other countries around the world, the analytical methods used come primarily by the advice and recommendation provided by a publication jointly prepared and published by the American Public Health Association, the American Water Works Association, and the Water Environment Federation. The references indicate the *Standard Methods*, by its full title: *Standard Methods for the Examination of Water and Wastewater*. The first edition was published in 1917; the 16th edition was published in 1985; the 19th edition in 1995.

All methods are presented as standard unless marked "**proposed**"; that means, not yet approved by the Standard Methods Committee.

Another source of analytical methods is the publication prepared by the American Society of Testing and Materials. In relation to water, the publication is the *Annual Book of ASTM Standards*, Part 31, "Water." USEPA may or may not approve methods of analysis that are not contained in their official list of approved methods or contained in their publication, known as the *Methods for Chemical Analysis of Water and Wastes* (see Appendix A-VIII, "General References").

USEPA also recognizes only analyses performed in **approved laboratories**; that means, laboratories specifically approved by USEPA or by the states connected to USEPA by **primacy**, as provided by the Safe Drinking Water Act.

*Standard Methods* also describes in its general introduction: apparatus, reagents, and techniques; expression of results (units); precision, accuracy, and correctness of analyses; collection and preservation of samples; ion exchange resins; and reagent water, laboratory hazards, and safety.

## Interpretation of Laboratory Reports

Analyses performed by the laboratory are reviewed by the laboratory director and released for use of sanitary and public health engineers in their evaluation of water treatment performance and water quality for certification of potability at the consumer's tap. These data are then utilized to prepare the statistics of the annual, quarterly, monthly, weekly, or daily reporting systems.* In other instances the data received from the laboratories are to be used for a specific case; investigation of a suspected waterborne disease; complaints regarding discolored water, bad taste, excessive chlorine in the distribution system; suspicion or evidence of contamination of raw water; violation of MCLs; lead or copper survey; corrosivity, etc.

Sometimes a result obtained from laboratories may show, in a single sample, a totally unexpected high concentration. These results are of much concern because they are real and may be above or well above the maximum contaminant level. Such a sudden unexpected reading surprises both the laboratory director and the public health engineer. What went wrong? Laboratory error, sampling error, improper handling, and labeling of sample, or actual true value?

Anyone with experience in monitoring water supplies has seen many cases of unexplainably high results. Accusations may start during interpretation of these data, when team effort is necessary to resolve the problem quickly. The best way to resolve this problem is to avoid occurrence, taking some or all of the following precautions:

1. Try to collect more than one sample.
2. Collect sufficient volume of sample material to be able to run another analysis to confirm.
3. Make sure that laboratory personnel participate in the survey organization to maintain their alertness on awkward readings.
4. Make sure that the laboratory director or his/her assistant is alert at all times as to what bottles are used by the sampler for the specific

---

*See Appendixes A-III and A-VI for sampling examples.

test and what pretreatment is recommended and available in properly marked sampling bottles (only bottles prepared by the laboratory and properly labeled should be used and not stored for convenience by the sampler).

5. Make sure that the engineer and the field assistant have a complete understanding of where the sample is to be taken, with what precautions, and with the clear understanding that the sample or samples must be *representative* of the supply at the point of sampling.

The first important step is to get the "wrong" result immediately known (let us say by telephone) so that a resampling visit may be immediately scheduled. Definitely more than one sample will be taken for better understanding of the eventual problem. Of course, a "smaller" water supply is more susceptible to surprises than a "large" water supply; in other words, a shallow well is more susceptible to a sudden contamination than a deep well, and also a small stream is more affected by sudden, limited contamination than a very large natural or artificial reservoir.

Taking the above-mentioned precautions and operating with an intelligent team of cooperating professionals (engineering—sampling—analyzing), the "false alarm" of peculiar readings can be minimized.

Maximum emphasis should be given to the laboratory reported unit. The one milligram per liter concentration is 1,000 times the one microgram per liter reading; the one MPN per 100 mL is certainly one-tenth of one MPN per 10 mL and one hundredth of one MPN per mL. One nanocurie is 1,000 times one picocurie; one gram of solute is one thousand times one milligram of solute; while these are apparently difficult errors to make, in reality they are mentioned here to emphasize the importance of accurately recording the unit together with each number with which it is associated all the time. It is also important to understand the difference in laboratory work between **precision** and **accuracy**. *Standard Methods* defines **precision** by stating that it "refers to the reproducibility of a *method* when it is repeated on a homogeneous sample under control conditions, regardless of whether or not the observed values are widely displaced from the true values as a result of systematic or constant errors present throughout the measurements. Precision can be expressed by the standard deviation."

**Accuracy** refers to the agreement between the amount of a component measured by the test method and the amount actually present. It is, therefore, understandable that many parameters can be evaluated for precision due to simple repetition of tests of a homogeneous sample, since only the **reproducibility** is in question; while it is not always possible to evaluate accuracy, since it is difficult to create a **known** sample with standard substances to test the **percent recovery** which represents accuracy (typical example here is a test for biochemical oxygen demand with its unknown natural components). Of course laboratory work can be evaluated for accuracy when a standard solution (artificial sample) is prepared and properly distributed to several laboratories.

## REFERENCES

APHA, AWWA, WEF. 1995. **Standard Methods for the Examination of Water and Wastewater.** 19th edition. Publication Office: 1015 Fifteenth Street, NW, DC, 20005: American Public Health Association.

ASTM. 1994. **Annual Book of ASTM Standards**. Vols. 11.01 and 11.02. 1916 Race St., Philadelphia, PA 19103: American Society of Testing and Materials.

AWWA. 1995. The SDWA Advisor. Regulatory Update Service. Analytical Methods. F. W. Pontius. Denver, CO.: AWWA.

AWWA. 1993. **Journal**, Sept. 1993. Vol. No. 85, p. 63–69. USEPA Drinking Water Laboratory Certification Program. Denver, CO: American Water Works Association.

USEPA. 1993. Analytical Methods; Trihalomethanes. *Fed. Reg.* (Aug. 3, 1993.)

USEPA. 1994. Analytical Methods for Regulated Drinking Water Contaminants. Final Rule. *Fed. Reg.* (Dec. 5, 1994).

USEPA. 1983. "Methods for Chemical Analysis of Water and Wastes." EMSL (EPA 600/4-79-020, March 1979, revised March 1983), Cincinnati, OH 45268: U.S. Environmental Protection Agency.

# Public Health Regulations

# Public Health Engineering

## PROGRAM MANAGEMENT

Constructive concern for the environment was first manifested by civil engineers. Working to provide adequate treatment for water and wastewater or evaluating pollutional effects in streams and lakes, civil engineers reinforced and expanded that branch of civil engineering known as "sanitary engineering" to further specialize, particularly in water quality control. Drinking water, bathing in polluted waters, air pollution effects on the environment, contamination of oil, food, sanitary housing, and management of solid wastes control programs were areas of daily concern to civil/sanitary engineers.

When these problems were analyzed from a specific viewpoint of public health, and when improvement of our environment was initiated with the cooperation of other public health workers (biologists, chemists, virologists, physicists, physicians, and lawyers), the sanitary engineers practicing in this field began to be identified as "public health engineers" (to emphasize the viewpoint of public health concern and protection and the engineering specialization developed in graduate education during the last 30 years).

Prevention of further deterioration of the environment, coupled with the betterment of the existing environment overstressed by the Industrial Revolution, was a goal of both the public health physician and the sanitary engineer, aided by the technicians who had been assisting in monitoring the general sanitation of urban and rural environments with the laboratory professionals.

This need of improved environmental sanitation and technologically upgraded sanitary facilities originated in preventive medicine, the school of public health officials and scholars, and in civil engineering in specialization in hydraulic applications or sanitary engineering, the school of public health engineers, and in research. Chemists, microbiologists, and engineering technicians joined the public health sanitarians to perform the necessary inspections, sampling, monitoring, and in evaluating the licensing of facilities closely related to public health.

This development of a progressing new field closely related to science, preventive medicine, and engineering facilities design and construction originated, in principle more than in results, during the end of the last century. Only during the first half of this century had applicable technology sufficiently developed with research well underway to consider that a new science, "environmental protection," was born. Actually, between 1930 and 1960, the field of sanitary engineering had produced sufficient research work in the most industrially advanced nations of the world to determine the pollutional damage to our water, air, soil, and housing. Engineers were seriously alarmed by the level of pollution. They easily convinced the related professionals (toxicologists, chemists, microbiologists, and public health administrators) to help them. The battle, however, could not be fought without the understanding and cooperation of the public and political representatives.

It should be noted that mainly from 1960 on, news of environmental concern did reach the public not only through conscientious reporting of the press, radio, and television, but also, unfortunately, through panic-producing, exaggerated publicity and Doomsday predictions.

Now the serious scientist wonders about the benefit and the problem of public support that sometimes changes into public fury and misunderstanding. Moreover, unquestionably environmental protection has become, unfortunately, a field of confrontation in politics.

The "green" political parties of Europe are confirmation of confusion when a country goes from total neglect of the environment to the panic and dreams of illusory panaceas.

How can public health workers now hope to regain control so as to allow planning of future life on this planet with the realism that is a product of true scientific knowledge? Extremism has never produced an intelligent evaluation of facts that are needed now and in the future to assess with professional knowledge and experience the demand of a richer consumer, a chemical industry in booming years, a population explosion, and the existing and limited environment that must be protected.

It can be done! Reasonably, patiently, and scientifically, the professional must continue to work at his/her best.

In the engineering field, public health engineers will undoubtedly continue to progress and begin the 21st century, perhaps with not the same confidence but with definitely more knowledge than civil engineers began the 20th century.

*Water Quality* has been scrutinized in Chapters 2, 3, 4, 5, and 6, looking closely at the internal components and contaminants, such as an examination of a water sample under the microscope. On the contrary, water quality can now be looked at from a very external viewpoint. Notice the influence of many environmental factors, all connected with urbanism, consumer demand, and high standards of living of the industrialized nations. These factors are the environmental problems that demand controls originated by preventive medicine and implemented by engineering technology. Guided by these concerns and concepts, public health engineers evaluate and control the engineering technology of urban sanitation.

Some examples follow here to relate water quality control to other administrative/regulatory environmental controls.

## SPECIFIC PROGRAMS

Public Health Engineering in serving the community is usually subdivided into the following sections:

1. Drinking Water Control (the subject of this Handbook)
2. Wastewater Control

3. Air Pollution Control
4. Solid Waste Control
5. Radiation Control
6. Recreation Area Sanitation
7. Housing Sanitation
8. Environmental Planning

The seven remaining branches of public health engineering are only briefly outlined here:

*Wastewater Control.*   The review of design of wastewater treatment facilities; the monitoring of a safe and effective operation; the evaluation of pollutional load on bodies of receiving waters; and the control of industrial wastewater discharges.

*Air Pollution Control.*   Includes the examination of air pollutants, sampling, measurement, emission control (design, operation, and maintenance of equipment); ambient air quality evaluation, standard setting, and compliance; and noise control.

*Solid Waste Control.*   Includes an overview of planning, designing, and operation of treatment and disposal of solid waste; sanitary landfills; transfer stations; incineration stations; resource recovery; composting; pyrolysis; high-temperature incineration; wet oxidation; high-density compaction; inventory, storage, treatment, and disposal of hazardous waste; and recycling.

*Radiation Control.*   Evaluation of biological effects of radiation; monitoring and regulating the proper use of X-ray equipment; radiation potential in industrial and commercial establishments; implementation of guidelines to reduce radiation exposure; and collection, treatment, storage, and disposal of radioactive material.

*Recreation Area Sanitation.*   The examination of plans, construction, operation, and maintenance of swimming pools with licensing, sampling, and monitoring of all public pools and bathing establishments; sampling, monitoring, and classification of beach areas, including surveys to determine pollutional sources; evaluation of sanitary aspects of temporary residences; i.e., camps, migrant labor camps, mass gatherings; and marinas and boat pollution.

*Housing Sanitation.*   Includes evaluation of citizens' complaints or reported cases of nuisance, unsanitary conditions; rat control program; animal control in or near housing; ragweed and noxious weed control near housing; lead paint poison control; safety and accident prevention in housing; substandard housing; and ventilation hazards.

*Environmental Planning.* The preliminary approval of community developments for compliance with public health standards listed in the above-mentioned activities of public health engineering with particular emphasis given to individual well water and private sewage disposal systems; standard setting; scheduling of priorities re sanitary surveys; evaluation of program performance; record keeping and data processing; in-service training; education (media releases, meeting with communities, teaching of short courses or seminars); preparation of data for legal enforcement of health standards.


## THE PROFESSIONAL TEAM

To reach the goal of an improved environment, it is unquestionably teamwork (effort and research) that is necessary. The unknowns of the environment are still all around us, but they become more understandable when they are studied by several professionals together and discussed in meetings, seminars, and symposiums. The question becomes: Who will be the promoter or team leader?

Team members have to slightly sacrifice their professional individuality to participate in co-workers' knowledge and efforts. Also, unfortunately, professional jealousy may develop and discourage the professional who may feel alienated or distracted by an activity apart from his or her own field. Compulsory cooperation is not only a potential risk (reluctant team of experts), but a widespread difficulty that has actually alienated some professionals with abstentions and consequent damage to research projects. An example of this problem is clearly identified when analyzing practically all scientific papers and qualifications of writers: the "complete team" is always missing. This is also evident in textbooks prepared by several writers who are experts in different disciplines. They provide excellent materials but the materials are poorly coordinated. Analyzing this problem further: the physiologist, toxicologist, and oncologist have great input because he/she can assess the consequence of pollutants and, therefore, has major input in standard settings. The chemist is the scientist who, in creating most of "the chaos," is more prepared to substitute or eliminate the hazardous substances from production and also has major input in standard setting. The virologist and the bacteriologist can identify the real cause of diseases that public health programs attempt to eliminate. The physicist is the main partner in radiation control and the philosopher in the long-range planning of our planet's problems. The public health sanitarian is a trained biologist with field experience, and is, therefore, the practical public health worker with daily human contact. The physician in public health is the motivator and administrative leader of all public health programs, priorities, and budgeting. However, it is the engineer, the theoretical and practical coordinator, who is trained to evaluate and organize ideas, sampling, statistics, structures, and site selections. With input from science and technology, the engineer can purify the water, filter the air, shield the radiation, make wastewater reusable, sterilize garbage or cremate it, improve habitation, and realistically plan the total environment. Hence, the engineer should not be alienated from the teamwork, but maintain leadership. His/her success is guaranteed only by professionalism (an oath to public health and safety, a duty to specialization and continuing education).

Indeed all public health workers must fight the natural introversion, generated by scientific upbringing and related work, with enthusiasm, humility, and dedication to public welfare.

## Drinking Water Regulations

The local Health Authority has a duty to the community to promulgate, update, explain, publish, advertise, and enforce drinking water standards. The local Health Authority may have sufficient knowledge and particular reasons *to issue more restrictive regulations* than the region, the state or interstate, or federal/national regulations. This should not be interpreted to mean that the local Health Authority must promulgate its own standards; it must primarily enforce and publish/explain/advertise water standards published at regional, state, or national levels.

In smaller countries or smaller states, it would be advisable that national or state standards be locally enforced; and when dictated by local circumstances, "interim standards" or "recommended standards" be issued as a reachable goal requiring specific and reasonable water treatment and/or disinfection. After publication of these standards, ample time should be given to the community for comments; and revision of the interim/recommended status should be preceded by the latest literature research review.

Local standards must be the same or more restrictive than the national standards. Whenever the local authority disagrees with the national standards by finding some parameter too restrictive, it should gather strong scientific evidence to request modification of the national standards.

While water standards should be strictly enforced, nevertheless, they *should always be considered ready to be revised.* This is particularly true in the field of carcinogens where negative or positive evidence is continuously surfacing as new studies are completed, mainly due to the unreliability of tests on small animals.

If new rules are issued by the Health Authority or complete revision is contemplated, regulations should contain the following items:

1. *Statutory requirements*—reference to national/state regulations.
2. Clear *definitions* of terms and jurisdiction.

3. *Requirements for filing* of construction plans, alteration plans, and preliminary plans. Suggestions should be encouraged for request of opinions on general siting or general proposal of development of new water supplies.

4. Requirements on professional *responsibility of the filing engineer* (state license requirements).

5. *Inorganic chemicals*—maximum contaminant levels (MCLs) should be set for primary parameters (Arsenic, Barium, Cadmium, Chromium, Fluoride, Lead, Mercury, Nitrate, Selenium, and Silver). At the very least, guidelines should be stated for the secondary parameters (Aluminum, Asbestos, Copper, Cyanide, Iron and Manganese, Molybdenum, Nickel, Sodium, Sulfate, Zinc, Hardness, Alkalinity, and Corrosivity). Range of acceptable values should be issued for the other compounds listed in Chapter 3.

6. *Physical characteristics and radioactivity*—MCLs should be set for radium, gross alpha activity—gross beta radioactivity, with guidelines for pH, color, odor, and total dissolved solids conductivity.

7. *Organic chemicals*—MCLs should be set for very toxic and/or carcinogenic compounds. Guidelines should be issued for the synthetic organic compounds described in Chapter 4.

8. *Microbiological parameters*—MCLs should be established for total coliforms with guidelines for heterotrophic bacteria. Pathogenic bacteria, viruses, and protozoa MCL should be zero with no requirements for testing, pending approvals of reliable, simple, and economic methods of analysis.

9. *Frequency of sampling* should be stated, and best location for sampling (in principle) should be advised.

10. *Certification of laboratories* and qualifications for laboratory directors regarding the analyses specified in the regulations.

11. *Resampling procedures* should be specifically stated whenever MCLs are violated.

12. Routine and immediate *reporting system* from the water supplier to the Health Authority in regard to sampling program and resampling in the case of MCL (primary or secondary) violations. Reports to the public and/or consumer should be specified in relation to key parameters.

13. *Certification of water treatment plant operators* is expected, stating qualifications (minimum experience and education) with possible requirements for examinations and continuing education. (See Figure 9-1.)

14. Regulations in regard to *cross-connection controls* should be stated with specific reference to a minimum number of surveys and potential sources of cross-connection.

15. Regulations for private *well water* sources for drinking and non-drinking purposes should be listed.

Figure 9–1. Plant operators' certification and continuing education certificates, in the entrance of a treatment plant, show not only compliance with the Health Authority's requirements but also management interest in encouraging specialization and theoretical knowledge. (Courtesy of the City of Poughkeepsie, NY.)

16. *Fluoridation*—acceptable fluorine compounds should be stated with requirements for storage, application, sampling, and reporting.
17. *Watershed protection* regulations, including equalizing and distribution reservoirs, should be specific, at least as guidelines.
18. *Bottled water*—regulations on approved source, bottling, disinfection, sampling, approval, and licensing of operation and distribution.
19. *Request for variance* or exception procedures should be stated, including enforceability of regulations.

Chapter 8 contains sampling and monitoring procedures and laboratory requirements that can be used as examples and guides for general regulations on drinking water.

## Turbidity Regulations

Turbidity and its significance as a physical and microbiological parameter have been analyzed in prior chapters. It is analyzed here from a viewpoint of Health Agencies' regulations.

Sampling and analysis are expected to be required daily at a sampling station located after treatment but before the distribution system, and the first consumer. The monthly average is expected to be below 1 turbidity unit, allowing a two-consecutive-day average high limit of 5 turbidity units (for surface water, high turbidity levels can be caused by exceptional rainstorm, seasonal overturn of reservoirs, aqueduct/main repair). Whenever such limits are exceeded, the Health Authority requires notification. After notification, the water supplier is expected to collaborate with the Health Authority, having the investigating team monitor the following critical aspects:

- Determine the cause of change in turbidity and forecast an end of the problem, scheduling a control sampling program.
- Repeat samples should be taken as soon as possible and analyzed by the investigating team.
- Immediately evaluate free chlorine residuals in the distribution system to ascertain that a minimum of 0.2 mg/L of free chlorine residual is maintained throughout the distribution system. Whenever possible, a slight increase of chlorine feeding is suggested, particularly when low levels of chlorine residuals are detected, coupled with turbidity units above set limits.
- Immediately examine bacteriological samples. Increased sampling is suggested as a safety measure and to determine for future use a record of effects of high turbidity readings on the distribution system water quality.
- If high turbidity readings are generated by sudden inefficiency of a water treatment plant, the chief plant operator should be included in the

investigative team; and the plant engineering consultant should make a joint report with the operator to analyze the causes of poor performance, upgrading procedures, temporary measures, and time of corrective measure implementation.

Regulations may allow, at least temporarily, a water supplier to operate with a monthly turbidity average of more than one and less than five as a MCL. This is expected in the case of an applicant who has had an established record of reliable performance (bacteriological results) under the management and operation of a qualified, professional staff. Moreover, it is important that the free chlorine residual has been maintained throughout the distribution system. In this case, the Health Agency will examine more carefully the overall performance and the microbiological parameters with emphasis on disinfection. If bacteriological results are not satisfactory, filtration will be required.

The following specific aspects of the operation will be under scrutiny:

- Chlorine residual maintained throughout the distribution system. Monthly operational reports must be compared with prior record.
- Microbiological record will be equally reviewed and compared (for meeting of microbiological standards, possibly confirmed by satisfactory heterotrophic bacteria concentration).
- Reevaluation of source protection measures with strict enforcement of watershed rules or issuance of new or additional controls to protect the source.
- Turbidity data will be reviewed with additional sampling from sources in the distribution system.
- Health Authority will request feasibility studies for development of complete or partial water treatment plants, or compulsory filtration for surface supplies.

After review of a water supplier's application for an entry point variance of five turbidity units, a temporary approval can be expected from the Health Authority once it has been established that:

- Disinfection has been determined to be effectively implemented throughout the distribution system and, therefore, the relatively high turbidity did not interfere with the disinfection or the microbiological data.
- The supplier has increased the sampling program of the distribution system (above the minimum requirement for sampling) and the data collected confirms the effectiveness of disinfection.

The 1987 stricter regulations proposed by USEPA are contained in *Turbidity*, in Chapter 8, page 380.

# Fluoridation

## THE PROGRAM

It is advisable that the reader first become familiar with **fluoride** as a chemical compound (see *Fluoride* in Chapter 3, page 75, where the chemistry of fluorine, the environmental exposure, health effects, and the drinking water standards are reported).

**Fluoridation** is dealt with separately in this chapter, in addition to the review of the compound (in Chapter 3) as a potential contaminant, for the following reasons:

- The fluoridation program has been very successful in the United States.
- Initiation and approval of fluoridation is still left to the community (nonmandatory program).
- Fluoridation at the very beginning did receive a great deal of criticism (mainly during the 1950s) and, therefore, has a historical background of accusations that have sometimes prevailed, negating the health benefits in certain communities.

**Fluoridation** is a drinking water treatment intended to reduce the incidence of caries in the teeth. It is part of preventive medicine, it is a community-requested and approved public health program implemented with the addition of fluoride to the drinking water supply at an optimum and safe level.

After statistical studies and scientific research had practically demonstrated the advantages and the safe water treatment of fluoridation, many questions were raised by the opponents of this program, concerned mainly with the potential side effects of fluoride ingestion, such as toxicity, carcinogenicity, mutagenicity, fear of technical ability to feed the chemical at a constant rate, fears of potential sedimentation of the chemical in the water distribution system, and consequent unequal concentration at the consumer's tap, accidental "overdose" at the treatment plant, and sabotage.

The Health Authorities did carefully investigate all these aspects (in the United States mainly between 1945 and 1955) and found them unsupported.

Research and studies conducted in recent years can be summarized as follows:

It has been established beyond any scientific doubt that the natural presence or the addition of fluoride to the drinking water at a level of at least 0.8 mg/L (or between 0.80–1.20 mg/L) has reduced dental cavities or tooth decay by 60%. It has been determined also that slightly higher concentration—up to 3.0 mg/L—has an even higher percentage of reduced tooth decay, but the formation of mottling (staining of teeth) may affect children's teeth when concentration is higher than 2 mg/L. Fluoridation, at the practiced optimum level of 1.0 mg/L, is a safe operation approved and encouraged by all the American Health Authorities.

## RECENT REGULATORY REVIEWS—STANDARDS

USEPA issued on May 14, 1985, the National Primary Drinking Water Regulations (NPDWR) related only to Fluoride as "Proposed Rule." In 1975, USEPA had promulgated the National Interim Primary Drinking Water Regulations (NIPDWR) under Section 1412 of the Safe Drinking Water Act, setting a MCL for fluoride of 1.4–2.4 mg/L, depending upon annual average ambient air temperatures. This RMCL was the same enforceable standard issued in 1962 by the U.S. Public Health Service. USEPA set these limits based on the consideration that higher concentrations would produce adverse health effects by increasing the occurrence of mottling (dental fluorosis) of the dental enamel manifested by staining (dark spots).

In 1980, the NAS* reached the following conclusions:

- Fluoride did not show itself unequivocally to be an essential element for human nutrition.
- Dental mottling and changes in tooth structure may develop in children at fluoride concentrations higher than 0.7–1.3 mg/L, depending on ambient temperature and diet.
- A daily ingestion of 20–80 mg of fluoride for a 10–20 year period could result in crippling skeletal fluorosis.
- Pending further studies, the optimal levels for anticariogenic benefits should not be exceeded. The Surgeon General summarized in 1982 the findings of his researchers as follows:

  a) No sound evidence exists which shows that drinking water within the various concentrations of fluoride found *naturally* in public water supplies in the United States has an adverse effect on general health or on dental health as measured by loss of function and tooth mortality.
  b) Public water supplies should not cause undesirable cosmetic effects to teeth.
  c) Public water supplies should have the optimum concentration of fluoride for protection against dental caries.

In January 1984, the Surgeon General stressed the following points:

- The fluoride content of drinking water should not be greater than four times the optimal level of any community water supply.

---

*NAS—National Research Council—Safe Drinking Water Committee—Prepared for USEPA, "Drinking Water and Health," Vol. 3, National Academy Press, Washington, DC.

- There exists no direct applicable scientific documentation of adverse medical effects of levels of fluoride below 8 mg/L.
- Four times the optimum level in U.S. drinking water supplies would provide no known or anticipated adverse effect with a margin of safety.

In December 1984, the National Drinking Water Advisory Council (NDWAC) recommended that the MCL of fluoride be set at 2 mg/L.

In 1984, the WHO set the fluoride guidelines in the category of "inorganic constituents of health significance" at 1.5 mg/L, based on mottling of teeth. It noted that at concentrations of 3.0–6.0 mg/L, skeletal fluorosis may be observed; at concentrations higher than 10 mg/L, crippling fluorosis can ensue.

The European Community established in 1980 an MCL of 1.5 mg/L or 0.7 mg/L when the average air temperature of the zone considered falls in between 8°C–12°C or 25°C–30°C, respectively.

In May 1985, the "Fluoride; National Primary Drinking Water Regulations: Proposed Rule (40 CFR Part 141)" based the Primary Drinking Water Regulation upon protection from crippling skeletal fluorosis (defined as an adverse health effect caused by excessive amounts of fluoride in drinking water consumed over a long period of time). Therefore, USEPA proposed an RMCL of 4 mg/L ("no known or anticipated adverse effects on health of persons occur and which allow an adequate margin of safety"). USEPA conclusions can be summarized as follows:

1. Crippling fluorosis should be considered an adverse health effect.
2. Osteosclerosis (abnormal hardening of the bones) is not viewed by USEPA as an adverse health effect within the meaning of the Safe Drinking Water Act.
3. The objectionable dental fluorosis is associated with excess fluoride in drinking water above 1–2 mg/L, and it should be the base for secondary regulations intended to protect the public welfare.
4. A secondary MCL was recommended at 2 mg/L.
5. The temperature dependency for fluoride maximum level determination was eliminated because there were insufficient data to predict quantitatively the role of temperature in drinking water consumption (also the most extensive use of air-conditioners in southern climates or during summer weather may contribute to minimizing the relationship of water consumption and temperature).

## COMMENTS AND SUMMARY

Dental fluorosis is not an adverse health effect requiring regulation under the *primary* drinking water standards. (Opinion supported by USEPA November 1985, American Medical Association, American Dental Association, the National Institute for Dental Research, et al.)

A maximum contaminant level of 4 mg/L has an adequate factor of safety to prevent crippling skeletal fluorosis even for a person of higher water intake (USEPA, et al.).

To relate the fluoride MCL to air temperature is not necessary. No other contaminant is related to air temperature or expected higher water intake during warmer weather. There is insufficient evidence to sustain that larger water consumption is automatically associated to the annual average temperature. (USEPA, et al.) There is insufficient epidemiologic evidence to relate fluoride to oncogenicity (USEPA, et al.). Fluoride has not been demonstrated to be a mutagenic hazard to humans (USEPA, et al.). There is no sensitive subgroup of the population (such as cancer patients or individuals with arthritis, thyroid impairment, or fetuses, infants, the elderly, the sick or malnourished) that is at risk due to fluoridation of the water supply (USEPA, et al.). There is insufficient evidence to conclude that sensitivity and/or allergic effects are produced by fluoride in drinking water (WHO, NAS, USEPA, et al.).

## THE NEW YORK EXPERIENCE WITH FLUORIDATION

The New York City Health Code prescribes an optimum level of 1.0 mg/L and a maximum of 1.5 mg/L of fluoride. These standards were adopted in 1965 when fluoridation started. After many years of experience during implementation of this program, these standards remained confirmed.

The N.Y.C. Department of Health has monitored the fluoridation of 5 billion liters (1.35 billion gal) per day serving over eight million persons during the last 24 years. The monitoring consisted in evaluation of chemical analysis of over 100,000 samples, in addition to similar sampling programs undertaken by the City Department of Environmental Protection; this City Agency equally confirmed the safe and effective implementation of **Fluoridation**.

The N.Y.S. Health Department issued a MCL of 2.2 mg/L in the 1977/1989 revisions of the State Sanitary Code.

The State monitoring program has confirmed the safety and effective implementation of the fluoridation program during the last 45 years.

The addition of fluoride chemical compounds to public water supplies in the optimum amount has been experienced now for over 40 years. The compounds used are hydrofluosilicic acid ($H_2SiF_6$), sodium fluoride (NaF), sodium silicofluoride ($NaSiF_6$), and rarely ammonium silicofluoride ($NH_4SiF_6$). These compounds have been selected after safety and cost considerations. Fluoride is added at the end of the water conventional treatment or before the point of entry of the distribution system. Solution and gravimetric or volumetric dry feeders are designed, installed, and operated to provide the maximum control and allow a minimum margin of variation so that the optimum level, usually 1.0 mg/L, be maintained with $\pm$ 10%.

Conventional treatment has little or no effect on the fluoride concentration when natural fluoride concentration is high (for removal of fluoride, see Chapter 3). The feeding of fluoride compounds to the water supply does not require special technology. The cost of equipment and the annual operational costs are negligible in the total operation. For example, the New York City fluoridation program was implemented in 1965 at the cost of under twenty cents per capita for installation and under ten cents per capita for operational cost. Presently* the cost for operation is at fifty cents per capita per year.

The concentration of fluoride at the feeding point is then maintained throughout the distribution system. In other words, there is no "dropout" of fluoride or higher values to register since there is no settling. This was confirmed in over 20 years of testing in New York City where a distribution system of over 9,000 kilometers (5,600 miles) conveys fluoride in drinking water to over eight million consumers. New York City selected hydrofluosilicic acid as the compound that requires the least storage space and offers more safety and simplicity in handling.

In 1955 the New York City Department of Health and the Board of Health recommended to the Mayor the adoption of fluoridation.† The report is summarized in few statements, reported in Appendix A-VII together with the summary and conclusion of the National Academy of Science (1977) and references (bibliography). USEPA recently issued the following Regulations for explanation and justification of fluoride MCL and fluoridation:

- May 14, 1985—Fluoride; National Primary Drinking Water Regulations; Proposed Rule. 40 CFR Part 141.
- November 14, 1985—National Primary Drinking Water Regulations; Fluoride; Final Rule and Proposed Rule. 40 CFR Parts 141, 142, and 143.
- April 2, 1986—National Primary and Secondary Drinking Water Regulations; Fluoride; Final Rule. 40 CFR Parts 141, 142, and 143.

USEPA, April 1986 (Fluoride) Final Rules became effective Oct. 2, 1987 and contained the following regulatory decisions:

- The revised MCL is set at 4.0 mg/L.
- The SMCL (secondary standard) is set at 2 mg/L.
- Public notice is required at levels above 2.0 mg/L.
- Analytical methods are revised (see *Fluoride*, in Chapter 3) and labora-

---

*1996–1997 New York City, Department of Environmental Protection, Bureau of Water Supply, Budget Indication.
†"Report to the Mayor on Fluoridation for New York City" submitted by the Board of Health, Leona Baumgartner, M.D. Chairman—1955 (NYC Department of Health, 125 Worth Street, New York, NY 10013)

tory performance requirements of $\pm$ 10% of the reference value are established.

The secondary MCL is based on potential cosmetically objectionable dental fluorosis so the 2 mg/L guideline is established for protection of public welfare. Since the 2 mg/L is also a very efficient way to provide significant protection from dental caries, it could be interpreted as advice to raise the fluoridation rate to that level. This is not recommended by USEPA, now leaving the current recommendations of the Center for Disease Control at 0.7–1.2 mg/L optimum level.

## USEPA (1986–1996)

The USEPA Final Rule for **fluoride** was promulgated on Apr. 2, 1986; monitoring began Oct. 2, 1997 when the regulation became effective. At that time and in 1996 MCLG = 4.0 mg/L; MCL = 4 mg/L; and the secondary standard SMCL = 2.0 mg/L.

No changes were intended by USEPA in 1990 but the Safe Drinking Water Act requires reevaluation every three years. Consequently, USEPA in 1990 requested all information available in regard to reevaluating this potential contaminant. The National Research Council (NRC) reviewed health effects and risk assessment of **fluoride**.

Also, in 1991, an Ad Hoc Subcommittee on fluoride published a report, the *Review of Fluoride Benefits and Risks*, presented by the USPHS of the Department of Health and Human Services. This report reconsidered the treatment of **fluoridation** and health, and reevaluated this program.

The following summary of the study was recommended by the U.S. Public Health Service with full support of fluoridation:

- Use of fluoride should be continued to prevent dental caries and improve dental health.
- Water supply systems should continue the effort to maintain the optimal level (between 0.7–1.2 mg/L) used in their treatment of fluoridation.
- USEPA should evaluate future studies of the health effects concerning *naturally occurring fluoride* in drinking water.
- Health professionals and the public should avoid excessive and inappropriate fluoride exposure.
- State health agencies should be encouraged and supported to inform physicians, dentists, pharmacists, and local communities about the fluoridation status of drinking water.

# Water Treatment

# The Basic Essentials

After examining the water quality parameters, divided in different groups such as physical, inorganic chemicals, organic chemicals, microbiological, and radionuclide parameters (from Chapter 2 to Chapter 7), it was necessary to introduce the sampling and monitoring of these parameters (Chapter 8) and the environmental engineering aspects from a public health viewpoint (Chapter 9).

Now it is essential to investigate the **makings of potability** from the raw water at the source and the maintenance of potability in the distribution system. At this end is the most important person, the consumer of potable water.

This chapter deals with the makings of potability: the public water supply systems that manage and operate the water treatment plants, the pumps for transmission of this precious product, the storage tanks, the disinfection, and the safety of a complex distribution system. Not secondary for the plant managers and operators are the sampling and monitoring of water quality, the prompt handling of emergencies at any time of the day or night or on holidays, the maintenance of delicate instruments and machinery of this essential service which is provided 24 hours a day, 365 days a year.

The duty of this manual (not a design manual for professional engineers) is to indicate the components of this infrastructure, and the chemicals or physics used in the various treatments as far as they are related to **water quality**.

# Water Works Components

The major functional components of a water supply system can be listed as follows:

- Watersheds or Aquifers
- Water Works Intakes
- Pretreatment (when required)
- Water Treatment Plants
- Distribution System

A definition of these components is presented first.

## Watersheds or Aquifers

Watershed is a reserved, protected area, usually distant from the treatment plant where natural or artificial lakes are utilized for water storage, natural sedimentation, and seasonal pretreatment with or without disinfection. Watershed is also defined as a collecting area into which water drains. Watersheds are related with surface water (usually fed by gravity) to distinguish them from groundwater (usually fed by pumping). **Aquifers** are defined as geological formations through which water can percolate, sometimes very slowly, for long distances. Springs and wells are charged from aquifers. They feed the groundwater to water wells.

There is another term often used in the federal regulations: **groundwater under the direct influence of surface water**. Its legal definition is

any water beneath the surface of the ground with (1) significant occurrence of insects or other macroorganisms, algae, or large diameter pathogens such as *Giardia lamblia*, or (2) significant and relatively rapid shifts in water characteristics such as turbidity, temperature, conductivity, or pH which closely correlate to climatological or surface water conditions. Direct influence must be determined for individual sources in accordance with criteria established by the state. The state determination of direct influence may be based on site-specific measurements of water quality and/or documentation of well construction characteristics and geology with field evaluation.

When groundwater is so defined, the monitoring regulations for surface water are applied to this type of groundwater.

The legal definition of **surface water** is "all water open to the atmosphere and subject to surface runoff."

### Water Works Intake

These are the connecting structures where the raw water is captured and conveyed by aqueducts to the water treatment plant. *Surface supplies* include intakes from lakes, rivers, streams, and reservoirs (natural or artificial). For *groundwater supplies* intakes are the aquifers where the wellpoints are penetrated as a test or as final wellpoints with screens from where the groundwater is pumped to the surface.

### Pretreatment

Usually, pretreatments are obtained with large tanks or small artificial lakes utilized to obtain the presedimentation advantages. It was considered an ideal place for prechlorination, but normally, at this stage, the *precursors* are in high concentrations, and, consequently, it would increase the concentrations of the trihalomethanes (see Chapter 4, page 165). If the pretreatment takes place in the vicinity of the treatment plant, pretreatments become part of the conventional water treatment plant.

### Water Treatment Plants

These are the hearts and brains of the process to improve the water quality through chemical mixing, coagulation, flocculation, sedimentation, filtration, post-filtration treatment, including disinfection, pH control, fluoridation, and final storage at the plant.

The filtration process is so important that sometimes the conventional (or modified) water treatment plants are called by the type of filtration used, such as "slow sand filtration," "gravity rapid sand filtration," "direct filtration," or "pressure filters" (for small plants), and "diatomaceous earth filtration" plants.

## Distribution System

From the treatment plant effluent line to the ultimate consumer, a small, medium, or large distribution system may include storage tanks, equalizing tanks, additional chlorination stations with or without corrosion control facilities. The excellent water quality leaving the treatment plant is potentially threatened by the water quality problems characteristic of the distribution system: cross-connections, loss of disinfecting power, pipeline materials impact, corrosion and sediment, regrowth of bacteria in dead ends of the distribution system, etc.

To underline areas that influence water quality, it is necessary to review each component of the water supply system from the source to the consumer (see Figure 10-1). To keep in mind the interest of public health

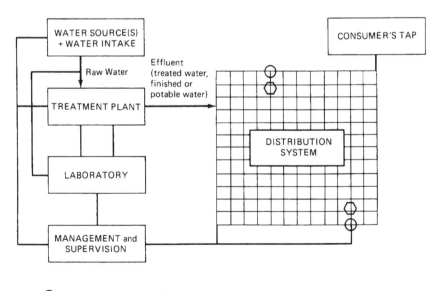

Figure 10–1.   Schematic of Water Works Components.

engineering, water quality controls are reviewed as they should be planned and enforced by the Health Authority or as they are presently enforced by the Health Authority, having jurisdiction in the territory where the water works components are located.

The task is to follow the water flowing from the intake structure to the consumer's faucet. Potential contamination is always threatening the source (watershed or aquifer): contamination threatens the treatment plant, the storage facilities, and the distribution system.

There are very few cases when a completely new plant is proposed, including new intake structures. More likely a new plant is proposed where previously water was supplied with limited treatment, such as chlorination and pH control. For a new plant or alteration of an existing facility, the Health Authority is expected to examine preliminary projects with descriptive reports that indicate the protection of the watershed, existing or proposed, to assure the elimination of pollutional sources. The engineering project must include an evaluation of raw water quality, a sanitary survey of the reservoirs and watersheds, and a classification of treatment requirements related to the raw water condition. The goals of water quality improvements must be attempted, analyzing the existing conditions and the improvements suggested.

The report will evidence the sanitary engineering experience of the designer, the depth of the field and laboratory investigation, and the concern to stress quality in all components of proposed water works. Regarding Health Authority expectations, the submission of preliminary or final approval should contain not only general information but also periodic reports. Operational records of preceding years will constitute the basis of the documentation submitted, and, if those records appear insufficient, the proposal should specify the additional laboratory work expected.

Construction is not expected to start until the final project has been approved by the Health Authority. In the United States, this review and approval is usually granted by the State Department of Health.

In the review, the states presently use published sanitary engineering standards. For example, the New York State Department of Health presently uses the publication subscribed by 10 states, and briefly referred to as the "10 State Standards." Specifically, they are titled *Recommended Standards for Water Works* which contains "Policies for the Review and Approval of Plans and Specifications for the Public Water Supplies—A Report of the Committee of the Great Lakes—Upper Mississippi River Board of State Sanitary Engineers."

Besides state approval of design and construction of new water treatment plants, as well as state approval of major alterations, the states monitor plant performance through the documentation to be submitted by the water supply engineer or operators or administrators. This documentation is listed below under *General Information* and *Periodic Reports*.

## General Information

A **public water supply system** must have readily accessible engineering design plans and specifications for the water treatment facilities' equipment and instrumentation. Such documents must be kept current. Modern water treatment plants are expected to receive a Manual of Maintenance from the design engineer that describes mainly the required ordinary maintenance, possible operational problems, handling of emergencies, data collection for daily operation, safety precautions, and watershed water quality control. This information is particularly important for large water supply systems (those serving more than 50,000 people). The new regulations, issued by USEPA with the final rule (May 1996) of the Information Collection Rule (see Chapter 11), require data on water quality performance and available treatment capacity.

Other necessary data and information are listed below.

The description and the summary of results of a sanitary survey of the watershed, reporting all sources of pollution, real or potential, including implemented or proposed pollution control action.

A long-range plan is advisable to take into consideration future development of recreational activities and, unfortunately, in some cases, residential use of the watershed.

Laboratory reports on raw water quality of reservoirs, tributary lakes, and streams and field tests collected and recorded during the preceding five years or longer period of time when such data are available.

Documentation on certification of laboratory services (see Chapter 8, *Laboratory Analyses*). Documentation on issuance or proposal for watershed rules and regulations.

A description of water intakes with emphasis to demonstrate that they are effectively protected from contamination.

A description of water storage, be it pre- or post-treatment or part of the distribution system, to demonstrate that protection from contamination has been assured during design and construction. Recommendations for proper maintenance to preserve potability should be included. All water storage tanks must be covered in urban areas.

A report indicating that the water supplier has adopted an acceptable Cross-connection Control Program (see *Cross-Connection Control*, page 468, and *Distribution System*, page 466).

Documentation on the certification of water treatment plant operators, including training and continuing education.

## Periodic Reports

The operation of a public water supply system shall be recorded daily. A summary shall be submitted monthly to the Health Authority. Such record shall consist at least of the following data:

BACTERIOLOGICAL RECORD. Date, place, time of sampling, name of sampler, type of sample (routine, check sample, raw water, survey, etc.), date of analysis, laboratory and person responsible, method used and result). Sufficient number of bacteriological samples shall be collected monthly from selected points of the distribution system. The minimum number of samples is usually determined by the population served (see Chapter 8, *Sampling*, page 387).

CHEMICAL SAMPLING RECORD. Most chemical samples are not collected daily, and synthetic organic chemicals and radionuclides are collected at several month intervals or even years according to the vulnerability or likelihood to detect contaminants of high toxicity or carcinogenicity. Their frequency of sampling, the type of analysis, the problems with samples showing concentrations above the MCLs are items of public health concern spelled out in the drinking water standards (see Chapters 3 and 8).

OPERATIONAL RECORD. Details of operational key data are entered in the log by the operator. Data include quantity of chemicals used in treatment, rate of feeding, time of application, operational laboratory data, control samples, operational malfunction, units not in service, control panel data, personnel on duty, and important observations during the operator's shift.

## Water Works Intake

Water intake or intake structures are defined as the components of the water supply related to the capture of raw water, be it an intake crib or intake tower, or gate house for surface supplies, or a groundwater well for groundwater supplies. The intake structure may include a pumping station, a tunnel or intake pipeline, or storage facility for chemicals.

The design of an intake structure is preceded by the collection of data on water demand (population served and per capita consumption), the knowledge of the hydraulic profile with an estimate of head losses from intake structures to tunnel, treatment, storage, and distribution system. Topographical knowledge of the watershed, drainage characteristics of the catchment area, and bacteriological and chemical data of the raw water provide essential information for design together with data on rainfall, evapo-transpiration, and runoff. Pollution sources surveys and controls are described in the preceding section.

Intake structures are permanent constructions, but the water quality changes with time due to weather conditions (drought or flooding) and due to the activities in the watershed (level of pollution). The design of the intake structure must take into consideration these above-mentioned factors, allowing the maximum flexibility in water withdrawing as far as water level is concerned. Water quality must be taken into consideration before quantity, leaving to the operator the choice to withdraw the best water available according to season, water temperature, rainfalls, and climate. Usually the worst quality is near the surface and near the bottom.

Surface supplies are subject to air pollution, sudden contamination due to changing weather conditions or involuntary spillage connected to transportation, potential sabotage, and residential/commercial activities in the watershed area.

Groundwater supplies are more protected than surface supplies and particularly safer for sudden contamination, but contact with the subsoil and rocks provides a variable concentration of chemicals.

Shallow wells, normally less than 100 feet, are more subject to surface contaminations listed for the surface supplies. Normally several wells must be used for a public water supply multiplying the problems related to the single intake.

While the pollution problems of surface waters are more numerous than those in the groundwater supplies, a survey of a surface watershed is easier than a groundwater survey.

Whenever possible, artificial reservoirs should be adopted. This makes possible dam construction to develop deeper water storage, less influenced by surface contamination problems, such as algae bloom, high turbidity related to excessive rainfalls, freezing in cold climate, contamination from traffic and eutrophication, sudden temperature changes, and excessive evaporation.

Deeper reservoirs allow installation of intake towers with many ports on the vertical line so that water can be drafted at different levels and in all cases away from surface contamination and bottom sediments. Deep wells in protected aquifers are in general producing more constant water quality of almost potable grade.

There are unfortunately cases in which both surface and groundwater must be used to supply the demand of an increased population. The blending of the two supplies should take place before treatment is provided or at least near the plant effluent prior to storage. This will help reach an almost constant quality in mineral contents essential for industry and consumers' tolerance. Of course the ratio of flow from the different supplies should be maintained as much as possible for uniformity of physical and chemical parameters.

Radioactivity is normally much higher in groundwater.

## Surface Supplies

Considering that surface supplies are used more extensively in the world than groundwater supplies, additional comments are presented for water supplies derived from surface sources.

Surface supply is usually the only alternative for medium and large cities or when large demand of water is requested by the community and local industry. Surface supply could be a lake, a river, or a stream as well as an artificial lake fed by a watershed and its streams that collect rainfalls or melted snow. More difficult than groundwater to protect from pollution, surface water is normally more reliable in the estimation of raw water availability. Reliable data on runoff or stream flow are more likely available than for underground water sources. Nevertheless, surface water is definitely more susceptible to periods of drought.

Searching for sources, the engineer must make an inventory of all possible sources of water supplies and analyze all potential sources from the following viewpoints:

- Raw water mining (lakes, rivers, streams).
- Future reliability of flow. Probable watershed yield estimation.
- Ease for protection of the source—watershed control.
- Economic factors.
- Existing water quality analysis, including consistency of quality.
- Treatability of raw water.
- Transportability of supply to the treatment plant location and/or storage.
- Percolation and absorption characteristic of the watershed examined. Determination of evaporation.
- Adaptability of watershed to changes necessary to meet future demands.
- Evaluation of water rights.
- Study of present and future land development into the watershed.
- Gathering of information and evaluation of the geological history of the watershed and/or aquifers.

## Water Treatment Plants

The major components of the water treatment plants are:

| | |
|---|---|
| **PRETREATMENT** | **Prechlorination** |
| | **Sedimentation** |
| | **Aeration** |
| | **Multiple Chlorination** |
| | **Eutrophication Control** |
| **PREFILTRATION** | **Coagulation and Flocculation** |
| **PROCESS** | |
| **SEDIMENTATION** | **Direct Filtration** |
| **FILTRATION** | **Slow Sand Filtration** |

|                       | Rapid Sand Filtration |
|                       | Diatomaceous Sand Filtration |
|                       | Activated Carbon Treatment |
| CHEMICAL TREATMENT    | Water Softening |
|                       | Iron/Manganese Control |
| FINAL TREATMENT       | Fluoridation |
|                       | Disinfection |
|                       | pH Control |

## PRETREATMENT

Following the raw water path through the various treatment processes, the parameters influencing water quality are reviewed, neglecting necessarily hydraulic, structural, and construction problems.

In a plant the pretreatment may involve the following processes: prechlorination, sedimentation, aeration, multiple chlorination, and control of eutrophication problems.

### Prechlorination

*Prechlorination* (or predisinfection, if chlorine or chlorine compounds are not used) can be provided with or without on-site water storage. The advantage to having initial storage capacity using tanks or open reservoirs is a successful step in water quality control, utilizing preliminary sedimentation with consequent decrease of turbidity; but, above all, what counts is the practicality to provide more contact time for the disinfectant to reach successful bactericidal effects. *Prechlorination* also allows the reduction of total chlorine dosage, if chlorination is applied as disinfectant, but trihalomethane formation must be controlled with values under 100 µg/L.

*Ozone* is a preferable disinfectant due to its oxidation process and the consequent power to be also virucidal and more effective in eliminating pathogenic protozoans like *Giardia lamblia* and *Cryptosporidium*. Ozone disinfection by-products are still subject to research. *Bromate* is a by-product of ozonation when *bromide* is present; the MCL for bromate is presently proposed at 0.010 mg/L.

The major inconvenience of prolonged storage (sedimentation) is the accumulation of sludge at the bottom of artificial lakes or tanks that sooner or later may have to be removed from the bottom, bypassing the storage facility when out of service for cleaning. This problem is also encountered with very large distribution system storage (balancing reservoirs).

The construction of pretreatment tanks is expensive and, therefore, pretreatment storage is not provided in medium-size plants. Prechlorination with sedimentation has been used effectively in large plants where the tanks are actually artificial lakes because of the huge water supply handled. Artificial lakes provide a sizable distance between the upstream gatehouse where the disinfectant is injected and the downstream gatehouse where the aqueduct starts with or without additional chlorination.

The City of New York, in providing over 5.3 million cubic meters (1.4 billion gallons) per day to a population of well over 8 million residents, commuters, and visitors, has been using storage reservoirs as "pretreatment." A treatment plant was considered necessary only during the last 40 years. Drinking water standards were met. Using chlorination in small dosages of 1–1.5 mg/L of free chlorine residual and obtaining satisfactory bacteriological record, multiple artificial lakes in series were built provided with chlorination facilities and no other treatment besides sedimentation, pH control, and fluoridation. In one instance New York City operated a storage reservoir, Jerome Park Reservoir in the Bronx, for approximately 100 years before taking it out of service for sludge removal, while the other reservoirs have been in operation for 60–100 years without bottom sludge removal. Problems in water quality resulted particularly during summer and during sudden changes of temperature, but the origin of the actual problem was difficult to define due to the complexity of the system (multiple reservoirs and over 320 km [200 miles] of tunnels).

Pretreatment storage would be particularly effective when the raw water is derived from a river that consequently has a wide variation of turbidity. Storage would be acting as an "equalization chamber" and would contribute to a more steady process in coagulation effectiveness.

In case of storage located in proximity to an urban development or industrial zone, the storage surface is expected to be covered whenever feasible to avoid air pollution fallout, potential sabotage, illegal swimming, and bird infestation. In all cases the storage area should be protected by a high fence, runoff contamination, facilities for flow bypassing and bottom sludge cleaning and flushing, and equipped with sampling pumps for easy access to water strata for sampling. In case the surface supply withdraws water containing floating materials, fish, large and fine solids, beside coarse screens, it will be advisable to use fine, stationary, or movable screens according to the material that can be intercepted. This protection, besides being effective to protect pumps, will improve the efficiency of the coagulation process, particularly when direct filtration is used (see *Filtration*, page 445). When filtration is not provided, microstrainers can be used to remove small objectionable particles. The effectiveness of microstrainers can be cataloged as an aesthetic improvement since bacteriologic improvements are questionable.

### Sedimentation

The sedimentation process in large storage areas, mainly artificial reservoirs, is a settling of suspended matter without the aid of chemical coagulation. Classic parameters that influence the settling process, namely surface area and depth or distance from inlet to outlet or protection of turbulence near the entrance and outlet, are not important in presedimentation. The size is always large, avoiding bypassing of the process with channeled flow through the storage. The depth is regulated by local topography. Sludge is not removed, so deep reservoirs are recommended.

These artificial lakes are particularly beneficial during and following heavy rains, to permit uniformity in turbidity prior to treatment. Prechlorination is recommended at a low dosage to avoid THMs; this is possible since detention time is significant here in its effectiveness of disinfection. In very large reservoirs chlorination is applied at the outlet; in medium-size reservoirs chlorination can be provided at both ends; in small reservoirs chlorination can be adopted at the inlet only, since prechlorination is being applied here. When treatment is not provided these applications of chlorination can be applied at several stages so as to provide the benefit of multiple chlorination.

## Aeration

This is the process to remove gases, usually to eliminate or reduce offensive taste and odor. Aeration is not frequently used as a treatment; it is actually avoided because of the hydraulic loss and because it is a suspected cause of excessive oxygen, a negative aspect in corrosion control. When topographic conditions allow head losses and the artificial lake is far from urban air problems, aeration is recommended, but the engineer must take into consideration cost-effectiveness since the public health benefits of aeration are normally small. When power is not required, this process can be successful, since the operation requires little maintenance. It is necessary to stress that the above statements apply to aeration as part of the pretreatment process.

## Multiple Chlorination

When chlorination is applied at several points from the first intake structure to the distribution system, this treatment can be considered both pretreatment and perhaps the only treatment necessary to make the raw water potable. The advantages are: economy and simplicity of construction and maintenance, effective bactericidal treatment, and use of low dosage, since the chlorine demand is minimal and the contact time is extensive. Extensive bacteriologic and turbidity data must show satisfactory results, and watershed sanitary surveys and controls must be highly reliable. A major disadvantage is the inability to control a sudden deterioration of water quality (industrial pollution, illegal dumping, domestic liquid waste contamination, sabotage, algae bloom). The City of New York has made use of multiple chlorination with economic success for over 100 years. The distance of the reservoirs from the city distribution system, 65–200 km (40–120 miles), as well as the necessity to develop large balancing reservoirs—one for a 30-day and one for a one-day supply of over 3.4 million cubic meters (0.9 billion gallons)—gave New York the possibility to install several chlorination stations where the effectiveness of treatment is also evaluated through the daily collection of bacteriological samples.

In general, multiple chlorination is effective when the *raw water* quality is good (in general coliform concentration of less than 10 MPN per 100 mL in a monthly average, turbidity under 5 NTU and chemical parameters that meet standards). But, *filtration* is the factor of safety.*

Protection of watersheds is a must in these circumstances. If residential areas are within the limits of the watersheds and pollutional sources cannot be eliminated, a conventional treatment plant should be requested. Multiple chlorination should still be developed whenever feasible in the pretreatment process.

## Control of Eutrophication Problems

Natural reservoirs, when shallow and exposed to residential and commercial/industrial pollution, have the tendency to become rich in plant nutrient minerals and poor in dissolved oxygen concentrations, particularly during summertime.

Nitrogen and phosphorus are the major nutrients. Algae and aquatic plants grow slowly and progressively while sediments make the water shallower. Reservoirs become smaller and difficulty in maintenance is aggravated by the progressive worsening of the raw water available. This phenomenon is accelerated also by domestic liquid waste pollution, even when secondary treatment is provided with chlorination. Pretreatment here involves an extended algae control treatment with copper sulfate evenly distributed at the surface and chlorination in limited dosage so as not to cause the fish to be killed. Dosage of copper sulfate between 120–320 g per one thousand cubic meters (1.0–2.7 lb per million gallons) and free chlorine residual under 0.35 mg/L is normally applied.

Aquatic plants may be removed by dredging, aquatic harvesters, or by wire or chain drags. Organic material should be removed since this is the major cause of bad taste and odor, change in color and turbidity, as well as higher bacterial concentration and the related potential contamination by viruses and pathogenic bacteria. Reservoirs with a high degree of eutrophication are the cause of citizens' complaints because of taste and odor, particularly aggravated by algae bloom. Blue-green algae may produce a toxin deleterious to fish and animals. Strict implementation of watershed control rules and regulations are essential with constant monitoring of critical parameters.

---

*In 1951, a Mayor's Committee (Engineering Panel on Water Supply) strongly recommended the construction of a modern filtration plant for the New York City water supply.

## PREFILTRATION PROCESS

### Coagulation and Flocculation

The development of rapid sand filters has emphasized the need to eliminate, as much as possible, the particles of suspended matter through processes known as coagulation and flocculation. *Coagulation* is the step to treat raw water with chemicals to coagulate the particles, and *flocculation* is the growth of the coagulated particles to unite the colloidal and larger particles of suspended matter leading to an easier settlement and more effective filtration to follow.

In recent years a theoretical review of these processes has reached a better understanding with less reliability on plant laboratory tests. To implement this process, facilities for chemical storage, dosing, mixing, and flocculating must be designed to precede filtration units (see Fig. 10-2).

The theory of coagulation and flocculation is quite complex. It involves the colloidal dispersion in liquids, the electrokinetic properties (hydrophobic, hydrophilic, and zeta potential), the effects of electrolytes, the influence of valence, the exchange capacity of the removable colloids, the velocity gradients imparted by stirring mechanism in flocculation, and the multiple additional factors that influence the process, such as degree of agitation, detention period, temperature, color, turbidity, alkalinity, hardness, pH, free carbon dioxide, and synthetic detergents.

The scope of this section in this handbook, as well as the following sections dealing with water treatment, is to summarize the factors influencing water quality control only. To implement these quality factors, it is necessary to review the major units of the process, taking also into consideration the possibility of operational control.

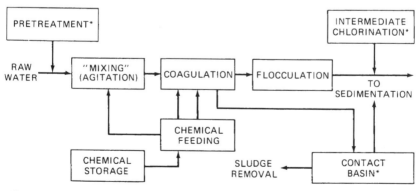

*Facoltative Treatment

Figure 10–2. Prefiltration Process—Schematic Flow Diagram.

## Process Summary

The very fine material of suspended solids will not settle quickly unless a coagulant is used. Chemicals used are expected to be safe for drinking water, when used according to standards.*

The coagulant must be added to the raw water and perfectly distributed into the liquid; such uniformity of chemical treatment is reached through rapid agitation or mixing.

A precipitate of an insoluble, almost gelatinous particle is formed in complex physical and chemical reactions. The next step is to reach the agglomeration of particles, obtained in slow agitation of the water tested, leading to formation of a settleable floc, through the absorption of suspended and colloidal matter, and, consequently, substantial removal of organic and inorganic matter, bacteria, and color.

The floc will go to a sedimentation basin or first to a solid contact basin for a more complete flocculation and preliminary sedimentation. The floc, not removed by sedimentation, will be removed by filtration. Water treatment plants, once designed, built, or already in operation are equipped with properly engineered mixing, coagulation, and flocculation structures which leave ample operational adjustment to control chemical feeding, chemical changes, coagulant aids (polyelectrolytes) action, agitation speed, and sludge withdrawal volume control with the related laboratory assistance. Modification of existing facilities are many times necessary to maximize the goals of providing the best water to reach the filtration beds. The following factors must be scrutinized to obtain the above-mentioned goals:

## Colloids

Colloidal particles range in size, normally between 1 and 100 millimicrons.

Colloidal matters have electrical charges; consequently, the natural tendency to repel each other and remain in suspension must be eliminated through pH control. The binding of colloids is preceded by destabilization and aggregation actions of coagulants like alum.

## Alum

Alum ionizes in water:

$$Al_2(SO_4)_3 \cdot 18H_2O \rightleftarrows 2Al^{+++} + 3SO_4^{--} + 18H_2O$$

Presence in water of sulfate ions ($SO_4$) or calcium ($Ca$) influences the pH range in relation to the capacity to precipitate.

---

*AWWA Standards (Coagulation—No. 42402 to 42407) AWWA, Denver, CO.

## Constituents in Water

Added chemicals are interrelated with the constituents of water treated and additionally influenced by temperature, time, and intensity of agitation.

## Jar Test

The jar or coagulation or flocculation test in the plant laboratory is the method to determine the cost-effective dose of a coagulant for the time and intensity of agitation selected. Maximum reduction of turbidity and color, coupled with an acceptable floc formation in size, shape, and actual ability to settle is the trial-and-error purpose of the test.

## Coagulants

Coagulants can be evaluated by subdividing them in the following groups:
- Alum Coagulants (primary flocculant)
- Ferric Coagulants (primary flocculant)
- Clay
- Calcite or Whiting
- Coagulant Aids (polyelectrolytes)
- Alkali (Soda Ash—Quicklime—Hydrated Lime)

## Alum Coagulants

*Filter Alum.*   In granular/powdered/lump form, filter alum or aluminum sulfate is readily soluble in water. It reacts with natural alkalinity of water or with the added alkali. Theoretically, 1 grain/gal. of commercial alum requires 7.7 mg/L of alkalinity as $CaCO_3$. If sulfuric acid is added to lower the pH, then the alkali will be added after filtration to avoid corrosivity in the distribution system.

*Liquid Alum.*   For the practicality of transporting and feeding, it is used extensively in medium and large plants.

*Sodium Aluminate* ($Na_2Al_2O_4$).   Consisting of aluminum oxide stabilized with caustic soda, it is used with alum in special instances, such as very cold water, or highly colored water, or in the lime-soda softening of water. It is used in concentration of approximately 0.2 grain/gal.

*Activated Silica.*   Used with alum it helps to produce large and heavy flocs. It is formed by the reactions between dilute solution of sodium silicate and sulfuric acid or alum solution or ammonium sulfate, or chlorine, or sodium bicarbonate or carbon dioxide (also by the reaction between dilute solutions of alum and sodium silicate). It speeds up the production of coagulation. Silica in the distribution system may be objectionable because of steam boiler problem.

*Sulfuric Acid.*   It assists alum in the coagulation of colored water where low pH values are required to reach effective flocculation and relative low cost of chemicals.

## Ferric Coagulants

*Copperas.*   Ferrous sulfate ($FeSO_4 \cdot 9H_2O$) is a granular acid compound effective at high pH, used, therefore, in conjunction with lime. Effective in lime-soda water softening and in coagulation of iron and manganese; not effective in colored waters because of the required high pH.

*Chlorinated Copperas.*   Addition of chlorine to copperas produces ferric sulfate $Fe(SO_4)_3$ and ferric chloride ($FeCl_3$). Useful when prechlorination is required and always advantageous where a variation of pH values is expected. It produces a floc that settles efficiently; also efficient in color removal.

*Ferric Sulfate.*   Ferric sulfate reacts with the natural alkalinity of water or the alkalizing lime added to form ferric hydroxide floc.

*Ferric Chloride.*   Ferric chloride reacts with the natural alkalinity of water or with the added lime to form ferric hydroxide floc. Ferric coagulants floc is heavier than alum floc.

**Clay**   In bentonite, Fuller's earth, or other absorptive clays, this alum coagulant acid has been proven successful in reducing alum dosage when applied to relatively clear waters. It has also been proven useful in absorbing taste and odor substances.

**Calcite or Whiting**   As a substitute of clays, powdered calcium carbonate (calcite or whiting) has proved effective with alum in treating cold, soft waters of low turbidity.

**Coagulant Aids (Polyelectrolytes)**   Coagulant aids are in general classified patented products marketed by experienced and reliable manufacturers. They have been proven effective with alum even in doses in the range of 0.5–4.0 ppm.

The use of highly ionized coagulant aids has increased in recent years. The main function of all coagulant aids is the addition in the water to be treated of substances that provide nuclei for floc formation.

## Alkali

*Soda Ash.*   Soda ash is practically pure sodium carbonate. It is very effective when sufficient alkalinity does not exist in the water to be treated. It also reduces the noncarbonate hardness of water (see *Hardness/Alkalinity* in Chapter 2 and in this Chapter pages 29, 31 and 451) and reacting with alum reduces the production of carbon dioxide when compared with an alum reaction with natural alkalinity; with sufficient soda ash, carbon dioxide formation can be eliminated.

*Quicklime.* Quicklime or calcium oxide is effective with alum or copperas to increase alkalinity not sufficient in natural water. Under controlled time and temperature, the slaking of the lime will produce the calcium hydroxide that is only slightly soluble. Use of excess lime can create deposit problems in the filter grains, clear well, and distribution system.

*Hydrated Lime.* Also known as slaked or calcium hydroxide, it can be used directly in mixing tanks or in dry-feeding equipment. It is more economic than quicklime, but only in small plants.

## Summary of Flocculation Control Factors

**Degree of Agitation** Coagulant must be distributed throughout the raw water with rapid agitation (called "flash mixing"). Initial agitation must be sizeable, followed by controlled agitation for proper flocculation, 0.2–0.6 m/sec (0.6–2.0 ft/sec). Low velocities may produce premature sedimentation, while high velocity may prevent the formation of proper floc.

**Detention Period** A 10–30 minute treatment period is required for flocculation. The longer the period, the lower the dosage of coagulant required, and a lower degree of agitation is required for mixing.

**Temperature** With lower temperatures, the demand for larger coagulant doses or longer flocculation period is expected; but colder waters have higher viscosity* with consequent greater agitation required at any given velocity (flow or paddles).

---

*Values of viscosity (see Chapter 2) of water used for drinking purposes, with temperatures that may vary between a minimum of 1°C and 25°C, are reported as follows:

| °C | $\eta$ (cp) | °C | $\eta$ (cp) |
|----|------|----|------|
| 1  | 1.728 | 16 | 1.109 |
| 4  | 1.567 | 19 | 1.027 |
| 7  | 1.428 | 22 | 0.9548 |
| 10 | 1.307 | 25 | 0.8904 |
| 13 | 1.202 |    |      |

cp = centipoise = 0.01 poise
$\eta$ = coefficient of viscosity (modifying factor in use in viscometers calibrated with water at 20° C and 1 atm. based on use of accepted value for viscosity of water at 20° C of 1.002 cp)
 Data contributed by the National Bureau of Standards. The above values indicate the increase of viscosity of water at lower temperature.

**Chemicals**    The selection of chemicals and coagulant aids, the dosage per optimum degree of agitation, and their relationship with turbidity and temperature of the raw water is summarized in the chemical characteristics and related tables for operator's guidance.

**Specific Factors**    The control of coagulation is affected also by color, turbidity, alkalinity, hardness, pH, free carbon dioxide, and synthetic detergents in the raw water.

**Zeta-Potential**    The lowering of the zeta-potential of the colloid is an essential force to be considered. The zeta-potential (Z) is a definition of the potential magnitude of the electric charge. It is mathematically defined as

$$Z = \frac{4\pi\delta q}{D}$$

where

$q$ = charge of particle
$\delta$ = thickness of the zone of influence of the charge
$D$ = dielectric constant of the liquid

Colloid stability fails and, consequently, coagulation begins when alleviation occurs in the double layer ions so as to reduce the zeta potential below the respective critical level.

**Objectives**    The main objective of flocculation is the formation of a clear water in which floc are visible and in suspension. Plant observations are the necessary tools of controls, but laboratory work supplements the evaluating factors and the background statistical record through daily readings of turbidity, color, pH, and alkalinity. Jar tests are necessary before implementing any change due to variations in water qualities or in the above-mentioned influencing parameters.

Special patented solids contact clarifiers, where more efficient floc formation takes place, avoiding short circuiting, are designed (see Fig. 10-3).

**Water Quality Control**    The water quality control factors of this process—whether at the stage of plant design, or construction, alteration, and operation—are briefly analyzed here, keeping in mind the theory and influencing factors previously noted in this section.

**Engineer-Operator Cooperation**    It is essential that a good relationship be established between the engineer responsible for the plant and the chief operator. Every problem should be reviewed from these two viewpoints; information must be exchanged and operating/laboratory records should be regularly reviewed together. The laboratory director, the assistant chief

Figure 10–3.  Conventional water treatment plant at Poughkeepsie, NY. Seen from the roof of the main building are two of the three solids-contact units. The metal circular tanks confine the contact process: raw water from the rapid mixing chamber enters a center well where mixing takes place with the previously formed floc. The mixture is then forced upward by an impeller; the small floc now grows to larger and heavier particles. About 85%–90% of the solids are removed by a sludge scraper and bottom drain. The clarified water rises outside the metal tank and enters a loop of perforated pipe leading the flow to the rectangular settling basins. (Courtesy of the City of Poughkeepsie, NY.)

operator, the chief of maintenance may be called for providing their input toward the solution of problems. When the responsible engineer is not the design (project) engineer, the design engineer may be contacted, particularly when operational changes are to be implemented to improve or solve the problem under consideration.

**Raw Water Quality**   Since the coagulation process is mainly used for surface waters, changes in water quality are expected not only due to seasonal effects but also due to industrial waste; prolonged storms, with stream/river flow variations; sudden changes in temperature; prechlorination treatment changes; and macro/micro biological changes. Consequently, the original setting of coagulant and coagulant feeders must be altered to suit the variation in raw water quality. The setting of agitation control mechanism should be altered to suit the variation in raw waters. The contact time may be reduced or increased by using out-of-service or stand-by tanks.

Careful observations of pH values of raw and treated water should be kept, and pH may have to be changed with addition or reduction of pH influencing chemicals. Accurate records must be maintained registering successes and failures.

**Limits of Operational Parameters**   The plant has limitations in operational influencing factors. Nevertheless certain broad limits were allowed in plant design. Specifically:

- *Alum dosage* may vary between 10–50 ppm.
- *Detention time* for coagulation and flocculation following mixing is possible between 15–45 min. Periods between 10–30 min have been used for flocculation alone.
- *pH range* strictly from a viewpoint of best coagulation results vary between 5.8–6.4 for soft waters and 6.8–7.8 for harder waters using alum. Raw waters high in color (more than 30 ppm) require a pH range between 5.0–6.0.
- *Flow velocity* designed for gentle agitation essential in the flocculation is expected to obtain an optimum degree of turbulence with values between 5.5–18 m/min (0.3–1.0 f/sec).

**Alkalinity**   The natural alkalinity of water being treated, essential in the coagulation process, may fall below the acceptable minimum concentration (approx. 20 mg/L of alkalinity as $CaCO_3$). Therefore, addition of alkali is necessary, if not previously used, or the alkali treatment has to be slightly increased (normal dose 0.35 kg or grain of lime or 0.5 kg or grain of soda ash per kg or grain of alum).

**Virus Removal**   In the prefiltration process (including sedimentation) a removal from 90%–99% of viruses has been noted in studies conducted in recent years. Alum and ferric chlorides were proven effective. Prechlorination* or preozonation may be necessary when excessive organic matter is encountered. It must be noted that the virus removed is not inactivated, so the settled floc is potentially infectious.

**Organic Removal**   Humic and fulvic acids are coagulated by iron and aluminum salts. Therefore, this prefiltration treatment becomes effective in eliminating the humic substances that are precursors for the formation of chloroform and other halomethanes. Prechlorination is, therefore, to be reduced in favor of intermediate (after flocculation or before filtration) or final chlorination (plant effluent).


## SEDIMENTATION

In a conventional water treatment plant, the sedimentation process is preceded by coagulation/flocculation for better results and improved utilization

---

*See THMs formation under *Trihalomethanes* in Chapter 4, page 165.

of the settling basins and followed by the filtration process. In the past, but not necessarily confined to the past, filtration may be preceded only by coagulation: here filtration is provided only after a few minutes of contact with consequent additional stress to the filters. Lack of sedimentation means less reliability on operation of filters, when water quality suddenly changes characteristics (see *Filtration* this Chapter).

The adoption of patented flocculation basins where the sludge is used to improve flocculation and allow sedimentation of excessive sludge may result in diminished importance of settling basins. In this case when the settling basins are adopted, continuous removal of sludge is normally not necessary, since the continuous removal is adopted in the preceding basins (see Figure 10-3).

When coagulation is eliminated (raw water quality permitting) the sedimentation process is called **plain sedimentation**. In the section *Pretreatment*, under sedimentation, a type of plain sedimentation is described. Conventional treatment plants comprehend at least the four processes; namely, coagulation, flocculation, sedimentation, and filtration.

The principle of operation is the same. Concrete or steel tanks, called sedimentation basins or settling basins (see Figure 10-4), that allow continuous flow are used, replacing the old fill-and-draw method. The efficiency of the basins is measured by the removal of turbidity and bacteria. In spite of the simplicity of the operation—giving time to settle and removing sludge at the bottom without causing problems—the results are good (see Figure 10-5). A removal of over 30% is expected in plain sedimentation, but following coagulation the expected removal is above 50%. These numbers are very approximate because efficiency is usually evaluated combining the process of

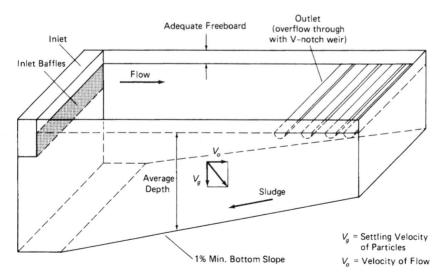

Figure 10–4.  Rectangular Settling Basin (sludge hopper and mechanism not shown).

Figure 10–5.  Three settling basins following coagulation and flocculation in the preceding contact tanks where sludge is mechanically removed. The settling basins generate a minimum quantity of sludge, removed manually twice per year. The concrete tanks are equipped with intermediate chlorine feeding devices to reduce THMs formation. Sedimentation is followed by aeration and filtration. The open basins allow formation of ice in the winter, when temperature fluctuates between $-10°C$ ($-18°F$) average minimum and $-3°C$ ($+27°F$) of average maximum during cold periods. (Courtesy of the City of Poughkeepsie, N.Y.)

filtration; anyway the figures have significance for a specific plant or raw water, since they are influenced by many factors connected with raw water quality. The major factors influencing the efficiency of the settling process are as follows:

- AREA and DEPTH of the basin
- DISTANCE between inlet and outlet
- VELOCITY of flow at the inlet and outlet
- SIZE and SPECIFIC GRAVITY of the settling particles
- TEMPERATURE of water influencing viscosity
- ABILITY to operate the basins

**Flocculation** efficiency and electrical phenomena are related. Protection from wind may also be an important factor, but good design is always an important factor.

The following factors are key parameters for the project engineer:

- *Period of detention*
- *Number of basins* to maintain the proper low value of velocity of flow
- *Number and shape of inlet and outlet devices*; particularly important is dissipation of energy at inlet

- *Sludge storage capacity*
- *Depth and surface area*
- *Overflow rate at outlet*

Theoretical formulas have been derived to guide design, but the best results were obtained testing existing basins, particularly considering the relationship of depth with area. Experience has demonstrated more reliability than formulas. Summarizing present experience and standards recently issued, the following limits should be used to handle the above-mentioned influencing factors.

**Detention Factor**   Health Authority standards require a minimum of 4 hr. Under special conditions of raw water and pretreatment, detention can be reduced between 2–4 hr. Detention time is the tank volume divided by flow considered.

**Size of Basins**   Multiplying flow by detention times the volume of the single basin or the total volume of all basins is obtained. Normally at least two basins should be designed for flexibility of operation and shut down for maintenance and repair (particularly important for the moving parts of sludge removal mechanism).

**Depth**   A *depth* between 2.5–6 m (8–20 ft) has been used, with an average of 4 m (13 ft) with a recent tendency in reduction to 3 m (10 ft). A cross-section of the tank should be designed to provide a velocity of flow not to exceed 0.3 m/min (1 ft/min).

**Surface Area**   This is one of the major influencing factors related to volume and flow. This parameter, called surface loading, originally recommended between 12–160 mc/mq/day* (300–4,000 gal/sq.ft/day) for granular solids but 30–80 mc/mq/day (800–2,000 gal/sq.ft/day) for slow settling solids; for flocculent material 40–50 mc/mq/day (1,000–1,200 gal/sq.ft/day) has recently been used.

**Size**   Rectangular size is normally used (ratio to vary between 1:3–1:5) with the exception of smaller plants where circular basins may be more economical. It must be noted that large surface areas are more susceptible to winds; wind breakers may be used, but several smaller tanks are advisable. The depth of the basins is more necessary near the inlet where sludge volume is deeper; the first one-third of the basin gets more sludge. Consequently, the bottom slopes from a minimum depth at the outlet to a maximum depth at the inlet. A minimum bottom slope of 1% is necessary for maintenance.

**Overflow Rate**   To prevent formation of currents and breaking of floc, the overflow rate should not exceed 170 L/min/m of weir length (20,000

---

*Or 12–160 m/day.

gal/day/ft of weir length). Whenever possible a smaller amount should be designed (according to "Recommended Standards For Water Works—Great Lakes-Upper Mississippi River Board of State Sanitary Engineers"). Usually outlet weirs are designed as throughs equipped with V-notch weir; throughs or weir should be adjustable for operational flexibility.

Covered tanks are desirable (for wind, ice, and algae control) but not essential. There are also disadvantages with covered tanks (more difficult access for maintenance and repair beside higher cost).

**Velocity of Flow**  The flow through settling basins shall be maintained under 9 m/h (0.5 ft/min)

The above "Standards" (recommended) issued by the Health Authority require sludge collection equipment and disposal facilities, provision for basin drainage, and safety equipment (for safe operation and maintenance).

**Inlet**  At the **inlet** a diffuser wall should be provided equipped with small openings (100–200 mm [4–8 in.] diameter) for dissipation of energy and elimination of potential currents.

**Important Notes—Summary of Parameters**

Water depth: 3–5 m min.-max. (10–16 ft)
Detention time: 2–4 hours min.-max.
Weir loading (outlet): 140–270 mc/day/linear m (8–15 gal/min/linear ft)
Velocity at inlet: 0.15–0.60 m/sec (0.5–2 ft/sec)

**Filtration Without Sedimentation**  Complete elimination of sedimentation has been experimented, based on better performance of present filtration media and controlled operation. Good results have been obtained when low turbidity and uniformity of raw water quality was expected. It is difficult to prescribe a dividing line in evaluating water quality only. It has been suggested that the settling process is desirable when turbidity exceeds 50 ppm (*Manual of British Water Engineering Practice*, Volume III).

**Pre- and Post-Sedimentation**  Sedimentation can be designed before and after the coagulation/flocculation process. The initial sedimentation can be part of the pretreatment process. A typical need for this double use of sedimentation is the case of river water transporting a heavy load of silt or high turbidity raw water. It is clear that this type of process reduces stress and reliability on filters, giving the operator the ability to handle a great variation in raw water (typical of flooding or persistent drought).

**Observations**  The major interest in removing settleable particles from the raw water is the lowering of high concentrations of turbidity, hardness, iron, and color. Such removal is primarily related to the efficiency of the

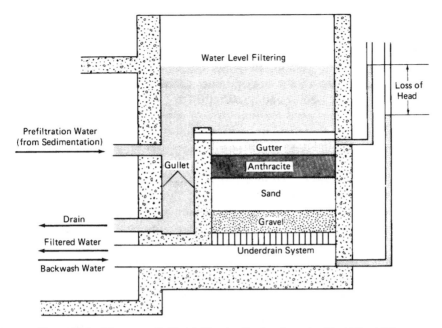

Figure 10–6. Diagrammatic Sketch Showing Section through a Rapid Sand Filter.

coagulation process. Settling efficiency should be evaluated mostly in the removal of suspended solids (particularly true when plain sedimentation only is provided). In a raw water with concentration of 100–300 mg/L of suspended solids, a 50% removal is obtained with a detention time of 1.0–1.5 hr; with the recommended detention time of 4 hr, such percentage of removal reaches values between 60%–70%.* It must be noted that with turbid waters the volume to be available for sludge should be as high as 25%; but with the present technology it is possible to operate a treatment plant almost completely with automatic control devices, consequently, the availability of personnel to perform manual removal of sludge at intervals is no longer a determining factor.

## FILTRATION

The most commonly used filtration type, GRAVITY RAPID SAND FILTERS have been in existence for almost a century, therefore parameters for design and operation have been tested and standardized (see Figure 10-6). PRESSURE FILTERS have also been standardized during the last 50 years but limited to small plants. Experience is still lacking for plants larger than

---

*Water Supply and Wastewater Disposal.* Fair and Geyer. New York: J. Wiley & Sons.

10 million gallons per day. SLOW SAND (GRAVITY) FILTERS have practically disappeared in the United States, due to the large filter area they require and the manual operation of backwashing. Effective use of SLOW SAND FILTERS is still possible when economically feasible. FILTER MEDIA is officially standardized when the traditional sand layer over gravel is used or finely crushed anthracite coal, mixed media of the above, or diatomaceous earth is used.

In conventional treatment plants, filtration is preceded by coagulation, flocculation, and sedimentation, with or without pretreatment preceding the chemical mixing leading to coagulation.

DIRECT FILTRATION is the process where no chemical treatment is preceded, with exception of pretreatment. This is feasible only with non-turbid water of good bacteriological quality; for example, well-protected artificial reservoirs for surface water or groundwater.

The mechanics of filtration are very simple: settled water introduced at low velocity to replace filtered water. Unfiltered water passes through a filtered media, where near the surface the fine media eliminates or reduces incoming turbidity. The filtered water is conveyed through the supporting media, like gravel and an underdrainage system, to a storage area or for additional treatment (pH control, fluoridation, and disinfection). When filters experience a loss of head due to the difficulty of unfiltered water to pass through the fine media, the filter beds are drained, and water pressure is provided under the filter media to backwash the fine media and settled material on top. Backwashed water is normally considered liquid waste and drained to sewers or to a wastewater treatment plant. Filter beds are then refilled and put back in operation.

There are structural, mechanical, hydraulic, electrical, topographical, foundation, and environmental problems connected with this relatively simple process, but WATER QUALITY CONTROL is the major concern.

Practically the final product of the filters is drinking water that is initially stored prior to distribution in the plant itself—in a tank or storage area defined as the CLEAR WELL. Such water will be used for backwashing also and for plant operational service.

Assuming that the civil, mechanical, and electrical engineering features were properly handled, the WATER QUALITY problems are influenced by the following parameters:

1. Quality of settled water reaching the filter beds.
2. Equal distribution of unfiltered water to the entire surface of all filter beds at proper feeding rate.
3. Effective layer of properly selected, fine media.
4. Equal distribution through the beds of wash water.
5. Controlled velocity of backwash water through the media.
6. Capacity to gently remove all the settled particles accumulated over the filter media during a controlled backwashing time.

7. Flexibility to handle raw water quality changes to maintain constant quality of the final product (drinking water with turbidity lower than 1 NTU and a goal of 0.1 NTU).
8. Facilities designed for simplicity of operation so as to allow relatively new personnel to backwash filters.
9. Ability to maintain sufficiently long periods of time between backwashes (long filter runs) to handle increased water demand or sudden changes in temperatures and viscosity of raw water or malfunctions in the coagulation process.
10. Operator's capacity to understand the filtration process.

## Filtration-Operation

Testing in pilot plants or in actual operational plants has been extensive in this century. The removal of particles during filtration is well above what would be expected from simple mechanical straining and sedimentation on top of the fine layer. Evidently other actions of a chemical, physical, and biological nature take place at the top (*Schmutzdecke*) and inside the filter media (gelatinous coating) but almost exclusively in the first layer of the media, the coarse material, being practically only a supporting layer between the underdrain and the fine layer. Mechanical straining and sedimentation do take place, followed by adsorption, electrical and chemical effects, and biological changes particularly significant at a low rate of feeding (slow sand filters).

The removal of turbidity and production of filtered water, particularly adaptable to effective chlorination, is successfully achieved in filtration. At the same time it must be noted that a change in water quality and temperature and consequent filtration rate may immediately create serious operational problems like "mud balls" on top of the filter media making backwash slowly and progressively ineffective.

Any water treatment plant operator has clear and visual memories of unexpected problems of difficult solutions after perhaps a long period of almost perfect operation. Turbidity of the filtered water may suddenly increase above set parameters and, unfortunately, there is no additional treatment on line to remove turbidity after filtration.

Sedimentation resulting in the clear well is another problem of a very difficult solution (plant shutdown and manual removal are necessary after several problems in filtration operation).

It is clear that the 10 points previously selected and listed are of primary importance from a water quality viewpoint, but they are also parameters used for final design. So it can be stated that a project, developed by a competent and experienced sanitary engineer, sustained by a team of competent civil, mechanical, and electrical engineers, is expected to provide a filtration unit that meets those requirements.

When the plant is in operation for many years, the media is not performing at its best (loss of material, uneven settling after backwash, etc.), underdrain at the bottom collects sand, so backwash is not distributed uniformly, filtration rates have changed, and the seasonal variations or the malfunctions at prefiltration treatment are no longer dealt with as easily. Preferably, four times per year and once per year, inevitably, the chief operator should meet the engineer (possibly the designer or his/her representative) and reevaluate all factors influencing water quality and review the performance of each filter bed.

## SUMMARY OF DESIGN DATA FOR RAPID SAND FILTERS

Presently a typical filtration design is expected to be a gravity down-flow sand and anthracite dual media filter with rotary surface agitators for quick removal of floc accumulation and reduction of backwash water (volume or time).

*Media*   Gravel      25–45 cm   (9–18 in.)
            Sand        45 cm        (18 in.)
            Anthracite  30 cm        (12 in.)
*Filtration Rate*  Minimum   80 L/min/mq   (2 gal/sq.ft/min)
            Maximum   200 L/min/mq   (5 gal/sq.ft/min)

Two gpm/sq. ft is also equivalent to 125 MGD/acre.

*Bed Size*   Depends on many factors such as, hours of operation (8, 16, or 24), size of plant, and minimum number of filters. Large-size filter beds can be limited between 4,000–20,000 mc/day (1–5 MGD/unit).

*Gravel Size*   2.5–5 cm at the bottom (1–2 in.) and 0.3 cm at the top (1/8 in.). Porosity: 35%–45%.

*Sand Size*   0.52–0.55 mm
            Uniformity coefficient 1.50 or less
            Specific gravity 2.65
*Anthracite Size*   0.80–0.85 mm
            Uniformity coefficient 1.50
            Specific gravity 1.50–1.65
            Hardness > 2.7 on MOHS scale
*Underdrain*   Usually special precast concrete or glazed tile blocks, porous plate bottoms (trade names), or manifolds and perforated-pipe laterals.

Water depth over the media is maintained usually between 1.2–1.5 m (4–5 ft) with a total of 8–10 ft or 3 m from water surface to underdrain.

*Backwash*   50% expansion of filtered bed is required
            Flow rate: 600–800 L/min/sq.m (15–20 gal/min/sq.ft)
            Upward flow rate: 60 cm/min (24 in./min)

Recently rates of 90 cm/min (36 in./min) have been used; also with high water temperatures up to 120 cm/min (48 in./min).

## ACTIVATED CARBON (PAC AND GAC) TREATMENT

Activated carbon in the powdered (PAC) or granular (GAC) forms has been used in this century almost exclusively for taste and odor control. The theoretical process involved is ADSORPTION—a physical/chemical activity leading to accumulation of water impurities at the solid-liquid interface. ACTIVATED CARBON is the solid ADSORBENT, while the predominant organic contaminants are the ADSORBATES.

The physicochemical adsorption phenomena are complex; the laboratory pilot-plant tests under different conditions of flow, contact time, adsorbent materials, water content of contaminants, sequence of treatment, and regeneration requirements are producing essential data necessary for the preliminary design. The U.S. Safe Drinking Water Act (1986 Amendments) officially sanctioned the feasibility and reliability of GAC for the control of synthetic organic chemicals (SOCs).

USEPA research work on the bench and at the pilot-scale as well as at the actual GAC contactor confirmed the adsorption ability of the majority of SOCs scheduled for regulation. Unfortunately, each supply must be evaluated separately with pilot columns designed, installed, and operated with a method leading to prediction of full-scale operation.

For taste and odor control, PAC in a slurry form is normally introduced before coagulation at the rate of 3–15 mg/L, but 2–5 mg/L is usually sufficient.

In the United States PAC has been used for control of taste and odor intermittently. For this exclusive use, GAC has also demonstrated efficiency in treatment and a long life prior to replacement or regeneration. The two- or three-year life makes the operation economically feasible.

During the last 20 years, GAC has been particularly under scrutiny, following concern over the THMs formation and the carcinogenicity of volatile and synthetic organic chemicals (VOCs and SOCs). Many pilot plant studies have been Conducted but very few attempts to experiment with the treatment in full-scale and large flow. Few new plants have been tested in Europe, but in sufficient number to draw conclusions in present and future treatment design with GAC. Present treatment technology leaves many options in design and the need for important data prior to final design.

The parameters and concepts, necessary in the preparation of reasonable options leading to treatment design, are summarized as follows:

**Treatment Goal**—Current and expected MCLs for synthetic organic chemicals and standards for THMs must be considered first for goals vs. cost.

**Total Organic Carbon**—TOCs are easy to determine, and they include VOCs and SOCs. It is important to connect TOC concentrations with replacement of GAC prior to SOCs exhaustion.* Also nature and concentration of organics present are necessary data for comparison with other studies.

**EBCT**—Empty Bed Contact Time—Has an impact on carbon usage rates and GAC performance. Carbon life is directly proportional to EBCT, and the THM formation potential (THMFP) is significantly related to EBCT. So EBCT and carbon usage significantly affect the cost of contactors, GAC handling, and regeneration costs (85% of capital cost and 75% of O. and M. cost estimate). Initially assumable rates for EBCT are 15–25 min.

**GAC Type**—Type and size of activated carbon have not yet been standardized; different types and diameters can be studied in pilot plants. For gravity filters in deep beds, a 8 × 30 mesh granular carbon can be tried using a smaller mesh size for pressure filters.

**Bed Depth**—Also adsorber volume is part of the GAC contactors design. Deep beds of 9 m (29 ft) have been built and shallow beds of 0.6 m (24 in.).

**GAC Usage Rate**—Related to factors under examination.

**Regeneration Cycle**—Also related to other parameters. Periods between 12–24 months have been experienced.

**Surface Loading Rate** (SLR) or **Superficial Velocity**—Rates of 5–12 m/hr (2–5 gpm/sq.ft) are expected, but higher velocities have been used.

**Location of Adsorber in the Flow Diagram**—Prefiltration, post-filtration, and filtration/adsorption are GAC design options. GAC configuration is upflow, downflow in parallel, or in series.

**Design Flow—Organic Compounds and Carbon Usage, Rate**—Useful in predicting carbon usage rates are the **Langmuir** or **Freundlich isotherms** and **Equations**.[†]

**Availability of GAC Reactivation Technology**—Including storage, transportation, and regeneration alternatives.

**Pilot Study**—Where variables of design options will be studied.

## Notes

In the 1989 National Primary and Secondary Drinking Water Regulations[‡] issued by USEPA, the GAC was indicated and proposed as the BAT for the reduction of all regulated SOCs with the exception of Acrylamide and Epichlorohydrin (Polymer Addition Practices were proposed). GAC was also proposed for reduction of Mercury (see Chapters 3 and 4 for listings of BAT for each chemical or contaminant).

---

*SOCs exhaustion is reached when TOCs of effluent is equal to TOCs of influent.
†"Kornegay, Billy H. Determining Granular Activated Carbon Process Design Parameters." AWWA Seminar Proceedings, June 1987, #20018, AWWA.
‡*Fed. Reg.* Vol. 54, No. 97. (May 22, 1989.)

## WATER SOFTENING

Water softening, or removal of hardness, is not a frequently expected or conventional treatment plant process for public water supplies. Rarely does raw water reach such high levels of hardness to require removal.

When no other source of raw water is available and hardness is above 150 ppm, treatment is considered. When a conventional treatment plant is under design, softening can be considered at 120 ppm.

Hardness (property, reporting units, laboratory determination, and public health significance) is described in Chapter 2: Hardness components are described in Chapter 3. Hardness is not limited by public health standards; but the closely related limitation is the total dissolved solids (TDS) recommended at 500 mg/L as an upper limit both by the United States and by the WHO as a guideline (constituent of no health significance). When necessary, treatment for municipal water supplies can be limited, leaving a hardness of approximately 50 mg/L to 90 mg/L from original values of groundwaters with several hundreds of mg/L of hardness.

Water softening can be accomplished with two conventional methods:

1. **Lime-Soda Process**: lime $Ca(OH_2)$ + soda ash $(Na_2CO_3)$
2. **Cation-Exchange or Zeolite Process**

Expected results of water softening can be summarized as follows:

- Reduction of soap and detergents consumption in homes and laundries.
- Reduce or prevent scale in water heaters and industrial equipment.
- Satisfy aesthetic requirements and avoid tastes.

### Lime-Soda Process

This process consists of chemical treatment of raw water with lime and soda ash. The addition of chemicals is followed by mixing, flocculation, settling, preceding filtration, using similar design and operations applied in chemical coagulation (see *Coagulation* on pages 434–438). Consequences of this process are an increased pH, decreased alkalinity, and practically no residual bicarbonate alkalinity, and removal of $CO_2$. The unwanted by-product is the large volume of sludge produced by the chemical reactions (2 ppm for 1 ppm of hardness removed). Handling and feeding of chemicals presents no specific problem, but more space for storage and handling and feeding equipment is necessary.

The lime-soda process can remove carbonate hardness down to a theoretical amount of 17 mg/L with pH = 9.4, but a practical value of carbonate hardness of 35 mg/L is achievable.

### Lime-Soda Process—Major Chemical Equations    Under the assumption that all the reactions are carried to completion:

$$CO_2 + Ca(OH)_2 \rightarrow CaCO_3 \downarrow + H_2O$$
$$Ca(HCO_3)_2 + Ca(OH)_2 \rightarrow 2CaCO_3 \downarrow + 2H_2O$$
$$CaCl_2 + Na_2CO_3 \rightarrow CaCO_3 \downarrow + 2NaCl$$
$$Mg(HCO_3)_2 + 2Ca(OH)_2 \rightarrow 2CaCO_3 \downarrow + Mg(OH)_2 + 2H_2O$$
$$MgCl_2 + Ca(OH)_2 + Na_2CO_3 \rightarrow CaCO_3 \downarrow + Mg(OH)_2 + 2NaCl$$

*Notes:* Precipitates formed and removed by settling as sludge. Solubility constant at 25°C $CaCO_3 = 4.82 \times 10^{-9}$

where

$$Ca(OH)_2 = 7.9 \times 10^{-6} \quad MgCo_3 \cdot 3H_2O = 1 \times 10^{-5}$$
$$Mg(OH)_2 = 5.5 \times 10^{-12}$$

Using the **Excess-Lime Method**, (10–50 ppm producing a pH = 10.6), the chemical reaction results in the production of calcium carbonate that can be stabilized with **Recarbonation** to avoid the tendency of colloidal carbonates of calcium and hydroxides of magnesium to precipitate on filters in the distribution system.

$$CaCO_3 \quad + \quad CO_2 \quad + H_2O \rightarrow \quad Ca(HCO_3)_2$$
calcium carbonate + carbon dioxide + water    calcium bicarbonate

Recarbonation is the process to introduce carbon dioxide through diffusers into a recarbonation chamber; a detention period of 15–30 min is expected. An alternate to recarbonation is the treatment with sodium hexametaphosphate: pH is lowered and crystals of $CaCO_3$ are dissolved.

## Split Treatment

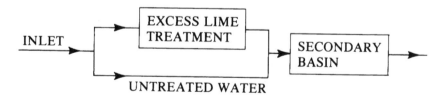

In the secondary basin, carbon dioxide and bicarbonate of the untested water react with excess of lime and precipitate the calcium at pH = 9.4.

Water quality aspects of the lime-soda softening process are the following:

- Prolonged flocculation is necessary (detention of 30–60 min).
- Mechanical sludge removal equipment is expected.

- Sufficient velocity of feeding usually prevents the settling of lime.
- Coagulation chemicals should be tested for best performance.
- Sedimentation time of 4–6 hr is calculated.
- Disinfection by chlorination of softened water must be increased because of the negative effect of a high pH.
- Associated laboratory work should be adequate for routine determination of alkalinites, pH, total hardness, magnesium, and flocculation (modified jar test).

## Cation-Exchange Process

The unit, called a **Zeolite Softener**, resembles a pressure filter, but open gravity units are satisfactory.

The calcium and magnesium of the hard water are exchanged for sodium. When all the sodium in the zeolite is consumed, regeneration is necessary, usually accomplished with brine made with salt. The sodium is retained in this "backwashing" or regeneration process, and calcium and magnesium are discharged as chlorides.

Natural greensand or synthetic zeolites are used in this zeolite softening treatment. Zeolites are hydrous silicates found naturally in the cavities of lavas (greensand); glauconite zeolites; or synthetic, porous zeolites.

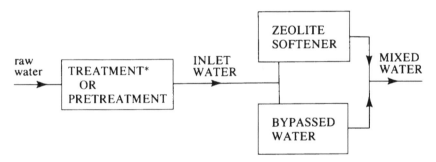

*Inlet water is expected to be clear—turbidity less than 5 NTU.

Pretreatment with lime is used to reduce high bicarbonate hardness to soften water containing approximately 85 mg/L (5 grains/gal) of bicarbonate hardness. Zero hardness could be reached, but it would leave corrosive water (see *Corrosion Control* on page 464), so mixing with untreated water is advisable.

In the **Sodium Zeolite Reactions**, the sodium zeolite with calcium/magnesium bicarbonates/sulfates/chlorides forms insoluble calcium/magnesium zeolite and soluble sodium bicarbonate/sulfate/chloride. Using the symbol Z as the zeolite radical, the following formulas may represent the treatment in the *softening* cycle:

$$Na_2Z + Ca(HCO_3)_2 = CaZ + 2NaHCO_3$$
$$Na_2Z + Mg(HCO_3)_2 = MgZ + 2NaHCO_3$$
$$Na_2Z + CaSO_4 \quad = CaZ + Na_2SO_4$$
$$Na_2Z + MgSO_4 \quad = MgZ + Na_2SO_4$$
$$Na_2Z + CaCl_2 \quad = CaZ + 2NaCl$$
$$Na_2Z + MgCl_2 \quad = MgZ + 2NaCl$$

The *regeneration* cycle can be expressed as the following:

$$CaZ + 2NaCl = Na_2Z + CaCl_2$$
$$MgZ + 2NaCl = Na_2Z + MgCl_2$$

A **Hydrogen Exchange for Metallic Ions** may be used with a *process cycle* as follows, with R = hydrogen cation exchange radical:

$$H_2R + Ca(HCO_3)_2 = CaR + 2H_2O + 2CO_2 \text{ and } Mg = Ca$$

With *regeneration reactions*:

$$Ca/Mg/Na_2R + H_2SO_4R = H_2R + CaSO_4 \text{ (or } MgSO_4 \text{ or } Na_2SO_{4)}$$

## WATER QUALITY OBSERVATIONS

In the zeolite softener with sodium, the outlet may contain sodium at concentrations higher than 100 mg/L. When the tested water is not diluted—typical of homes or small community water supplies—the high concentration of sodium is objectionable to consumers on low-sodium diets.

Hardness should not be reduced to levels too low; there are suspected beneficial effects linked to hardness (inverse correlation between the incidence of cardiovascular diseases and the amount of hardness). Water with hardness values under 50 ppm is expected to be corrosive (lead, copper, iron, zinc, etc.)

*Other Processes* potentially usable in **softening and demineralization** of water are the following:

- ION-EXCHANGE DEMINERALIZATION—Both cations and anions are removed;
- ELECTRODIALYSIS—Direct electric potential causing cations and anions to flow in opposite directions; special membranes "filter" ions from the saline water;
- DISTILLATION—Evaporation of brines or sea water produces mineral free water; and
- FREEZING—Ice crystals are separated from freezing brines.

The above-mentioned processes, not used in the past in drinking water supplies because of high costs of installation and operation, are still rarely considered in public water supplies design. Present research work will undoubtedly produce more economic equipment with simpler operation. There are no major, specific problems with water quality with the exception of cross-connection problems, potential corrosivity of treated water, and errors in operation.

## Iron and Manganese Control

**Deferrization** and **Demanganization** are chemical precipitation treatments; processes to be adopted for the removal of **iron** and **manganese**. The usual process is **Aeration**; dissolved oxygen is the chemical causing precipitation; chlorine or potassium permanganate may be required.

With Aeration, $CO_2$, $H_2S$, $CH_4$ + odor-producing substances in water are naturally removed by what can be equally defined as "physical precipitation." Groundwaters are the source of high iron concentrations usually under 10 mg/L, but higher values may be detected. Manganese concentrations may be expected not higher than 3 mg/L.

Iron is insoluble once oxidized in a pH = 7–8.5. The chemical and health characteristics of iron and manganese were examined in Chapter 3, including standards, notes, and removal methods.

# Unconventional Water Treatment

The use of Unconventional Treatments as the Best Available Technology (BAT) has been specifically stated by USEPA during the promulgation of the Final Rules issued for the Maximum Contaminant Levels (MCLs) of organic and inorganic parameters. Selection of BAT was requested by the Safe Drinking Water Act of 1986 as a must for USEPA standards.

The following treatment methods have been quoted several times in the descriptions of contaminant properties in Chapters 3, 4, and 8:

- Granular Activated Carbon
- Packed-Tower Aeration
- Reverse Osmosis
- Ion Exchange
- Desalination (Desalting or Desalinization)
- Activated Alumina
- Lime Softening
- Chlorination
- Coagulation-Filtration

Granular Activated Carbon (GAC) and Packed-Tower Aeration are normally selected for treatment of organic contaminants, while for inorganics BATs are listed with reverse osmosis, ion exchange, and the remaining treatments listed above.

Coagulation-filtration is part of the conventional treatment; effective in removal of few contaminants, economically convenient for large plants (existing or under upgrading), but not for small water systems called upon to provide specific elimination of a contaminant. In that case it would be more convenient to use an **ion exchange** or **reverse osmosis** process to provide fresh water that meets the MCLs required.

Activated alumina, lime softening, and chlorination are water treatment processes that have already been tested for many years. In conclusion, only the first four processes listed above can be considered "new technology" for large water supply systems because of the large water volume treated. Commercial and industrial establishments have experimented with these four processes in their procedures requiring water purification or demineralization, in order to produce a constant quality of water for use in their final products.

**Desalination** is a process that can be considered part of the last half of the 20th century that uses old as well as competitive new processes. Desalination or desalting should be treated separately, since it does not have a BAT issued by USEPA, but it is also not prohibited as a unique "surface water supply," namely sea water or brackish water as raw water.

## DESALINATION

Desalination or desalinization is the process used to recover fresh water from the sea. In water supply systems the fresh water is drinking water that meets potability standards.

The raw saline water to be treated in this process may be sea water or brackish water. The seawater is expected to possess about 35,000 ppm of total dissolved solids (TDS), while brackish water may have a TDS = 2,000 ppm or a higher concentration. Current drinking water standards limit TDS to 500 ppm.

Available and tested technologies are sufficient to produce fresh water with salination, but in very large plants economic factors are tied to the vital need for water in arid land; in those cases the contemporary production of electricity and the availability of low-cost energy may become essential factors. For these reasons very large desalination plants have been built in Saudi Arabia and in the United Arab Emirates.*

The physical-chemical processes available are based on:

- change of phase reachable with **distillation** or **freezing**

---

*The city of Abu Dhabi, the capital of the United Arab Emirates, has a population of over 800,000 inhabitants, and is located on the Persian Gulf in a large desert area that receives only 100 millimeters of rain per year (4 in.). Without competition from groundwater or other surface water, the local government had to finance 1.7 billion dollars to build a desalination plant capable of providing 17.5 million kWh per day and of producing 346,000 mc/day (91.4 MGD) of drinking water at full load. Besides the population consumption, the excess water will be conveyed 140 km away to an oasis where a large area of farmland may be improved.

- using osmotic principles with **membranes** with the **electrodialysis** or **reverse osmosis**
- chemical bonding: **ion exchange**

Theory and experience suggest distillation and freezing for sea water, while electrodialysis and ion exchange are more suitable for brackish water.

Reverse osmosis can be used for desalting of both sea water and brackish water.

### Desalting Plant Schematic

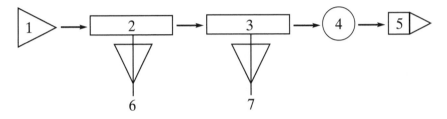

1. Raw Saline Water
2. Pretreatment
3. Desalting Plant
4. Chemical Post-Treatment
5. Treated Water to Distribution System
6. Solids Waste
7. Brine Waste

# Final Water Treatment

After filtration, water of practically potable quality is stored at the plant mainly for the following:

- Storage volume and reserve for the distribution system. Normally the clear well or part of it constitutes a wet well for the pumping facility designed to raise water at a higher elevation where additional storage and adequate pressure is provided.
- Water storage for backwash water and processed water as required.

Before leaving the plant, water usually receives additional treatment, such as the following:

1. *Fluoridation*: to implement a community or Health Authority requested public health program.
2. *Disinfection*: usually synonymous with chlorination; to disinfect water where low turbidity will make bactericidal effects very predictable. In

addition, chlorination will generate a free chlorine residual that is expected to be maintained at low, but effective doses, all the way to the end of the distribution system where a 0.2 mg/L of free chlorine residual is expected. In large distribution systems, booster chlorination stations, usually coupled with balancing reservoirs or pumping stations, will produce the additional disinfection and the ability to maintain the required free chlorine residual.*

3. *pH and Corrosion Control* may be required at the end of the treatment plant to assure a pH sufficiently high to prevent corrosion but not too high to interfere with the effectiveness of chlorination.

## FLUORIDATION

The program and problems of fluoridation have been described in Chapter 9 because of the peculiarities of this activity: it is a water treatment process not intended to improve raw water but purely to benefit the community as strongly recommended by the Health Authorities. Water supply administrators and operators are sometimes in opposition to this program (artificial fluoridation) because of potential complaints from "anti-fluoridationists" and fear of future expansion of other preventive medicine programs through water supplies.

The composition of fluoride, its occurrence in the environment, health effects, standards, analytical methods, removal of high natural fluoride concentrations are contained mainly in Chapter 3 under *Fluoride* and also in Chapter 9 under *Fluoridation*.

Equipment for liquid or dry feeding of fluoride does not require large space. It is ordinarily located in existing treatment plants where pH control chemicals are stored or in pumping stations where well water is utilized. Additional storage for chemicals must be provided inside a building, but the supply of fluoride compounds is maintained to a minimum, ordinarily up to a three-month supply to avoid potential sabotage and natural disasters.

The same precautions used in chemical treatment applications should be followed for fluoridation from storage of chemicals to cross-connection control, from accurate and detailed specifications for dry or liquid chemical feeders to solution tanks and operational safety, from protective equipment to ventilation.

Approved fluoridation chemicals are expected to meet American Water Works Association standards. Approved chemicals are hydrofluosilicic acid ($H_2SiF_6$), sodium silicofluoride ($Na_2SiF_6$), and sodium fluoride (NaF). The "Ten-State Recommended Standards for Water Works" required for fluoride feed equipment are: accuracy of feeders within 5%; dry feeders equipped with scales or loss-of-weight recorders; the fluoride compound

---

*See also *Distribution System—Disinfection*, pages 466–468.

should not be added before lime-soda softening or ion exchange softening to avoid precipitation of fluoride; if hydrofluosilicic acid is applied in a horizontal pipe, the point of application shall be in the lower half of the pipe, and the fluoride solution shall be applied by a positive displacement pump having a stroke rate of not less than 20 strokes per minute; and antisiphon devices are to be provided for all fluoride feed lines.

## DISINFECTION

### Water Chlorination

Chlorine, chlorine dioxide, chlorite, chlorate, chloramines, and chloramino acids were examined in Chapter 3 as inorganic contaminants.

Chlorine can be applied as: chlorine gas, chlorine hypochlorite, chlorine dioxide, and chloramines.

Most commonly used is chlorine gas (produced, dried, liquified, and pressurized in steel tanks). In applications, liquid chlorine immediately becomes a gas at room temperature and pressure, and the gas is mixed with water (aqueous solutions are very corrosive) in special equipment or chlorinators for safe injection of the chlorine dose. Chlorine is very toxic.

In water, chlorine is expected to react chemically as follows:

$$Cl_2 + H_2O \rightarrow \quad \underset{\substack{\text{hypochlorous}\\\text{acid}}}{HOCl} \quad + \quad \underset{\substack{\text{hydrochloric}\\\text{acid}}}{H^+ + Cl^-}$$

$$(HOCl) \rightleftarrows H^+ + \quad \underset{\substack{\text{hypochlorite}\\\text{ion}}}{OCl^-}$$

The constant for hydrolysis and ionization respectively are:

$$K_h = \frac{[HOCl][H^+]Cl^-]}{[Cl_2(aq)]} \qquad K_a = \frac{[H^+][OCl^-]}{[HOCl]}$$

*Notes:*

- pH strongly influences the ratio [OCl]/[HOCl].
- temperature also influences the same ratio to a lesser degree.
- hypochlorous acid has more bactericidal results than the hypochlorite ion.
- the lower pH and temperature values, the more efficient the disinfection.

In pure distilled water the chlorine added would result in equivalent concentration of chlorine in water or the free chlorine residual (45° line) (see Figure 10-7).

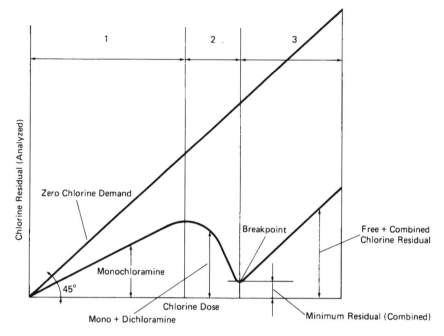

Figure 10–7.   Theoretical Breakpoint—Chlorination Curve.

In raw or potable water in the initial stage, extra chlorine is used to satisfy the "combined chlorine" (formation of chloramines), but at a certain dosage—called the BREAKPOINT—the addition of chlorine corresponds to the chlorine concentration in water (parallel to 45° degree line).

It is therefore clear that to obtain free chlorine residual, as required by standards, it is necessary to pass the breakpoint or, as it is called, use the breakpoint chlorination. It must be made clear that any substantial chlorine dosage above the chlorine demand will produce a large percentage of free chlorine residual. Therefore, referring to breakpoint chlorination does not mean that a "special chlorine" is to be used but simply pinpoints the need to feed chlorine sufficiently above the water chlorine demand (variable with raw water and with temperature, pH, time, treatment process, pipeline sediments, and bacterial growths). What happened to the chlorine "combined"? The major user of chlorine in the first stage is ammonia. Chloramines are formed until additional chlorine oxidizes the ammonia almost completely; when nitrogen is finally formed and disappears, the breakpoint is reached. Actually a small concentration of nitrogen chloride is formed. Nitrogen chloride can cause objectionable taste and odor, unless eliminated by aeration or exposure to sunlight or dechlorination. Monochloramine and dichloramine are formed and an additional chlorine breakpoint reached.

Theoretically, the chlorine/ammonia-nitrogen ratio is 7.6:1 at the breakpoint (the actual ratio is 10:1).

**Chlorine Demand** is the amount of chlorine consumed in the reactions with various water contaminants before producing the free chlorine residual at which point reliable disinfection starts. Chlorine demand can also be defined as the difference between chlorine added and available after a specified contact period. For **Chlorine Residual**, see Chapter 3, *Chlorine*, on page 134.

*Chlorination—Operational Regulations.*    Handling of chlorine means handling a hazardous material in particularly large containers (1 ton). Specific standards prescribe safety measures from the factory to transportation, storage, pipeline connections, emptying containers, disconnecting and detecting gas leaks, as well as personnel protective equipment. Above all, personnel training and preparing procedures for handling emergencies are stressed and detailed.

Health Authorities issue standard procedures for a safe operation, and engineering societies prepare training manuals and courses, but the major responsibility rests with the chief operator for proper replacement of repair kits, equipment, and continuous inspection of the chlorination facilities; these include safety of buildings, doors, chlorine feeders, analyzers, chlorine detectors, color coding, ventilation and cleanliness, respiratory protection equipment, rubber gloves, dust respirator, protective clothing, goggles or face masks, deluge shower and eyewashing devices, and scheduled training and retraining of operators and maintenance personnel (see Figure 10-8).

*Hypochlorites.*    Applied in chlorination from dry or liquid status containers, hypochlorites are salts of hypochlorous acid.

SODIUM HYPOCHLORITE, also known as liquid bleach or soda bleach solutions, is safer to handle than calcium hypochlorite because of lower chlorine concentration, but it requires more storage space. It is prepared treating chlorine with caustic soda.

CALCIUM HYPOCHLORITE is stored dry in concentrations of chlorine of 65% or 35%, according to the manufacturer. It has caused serious problems of autocombustion in storage. Storage requirement is less than sodium or LITHIUM HYPOCHLORITE, when the 65% of available chlorine concentration is used.

*Chlorine Dioxide.*    Storage of chlorine dioxide has been a serious problem, actually not yet solved. This is due to the explosive tendency at temperatures above $-4°C$. Chlorine dioxide vapors are also explosive above 40 kPa (5.8 psi). To solve this problem in the meantime, chlorine dioxide is prepared at the premises. It is usually prepared mixing chlorine with sodium chlorite. (See Chapter 3 for chemical composition.)

Figure 10–8. Safe operation of handling of chlorine is stressed in operational regulations. Visible here are doors, color coding, respiratory protection equipment, deluge showers, and eyewashing devices.

*Sodium Chlorite.* Another chlorine compound usable in chlorination. It has been applied dry or in liquid solution; not frequently used because of the unstable chlorine concentration due to storage, particularly at high temperatures and due to the presence of metals in water.

*Chloramines.* Chloramines are obtained when combining chlorine with ammonia. When disinfection is intended, then the amount of ammonia required is well above the amount commonly expected in water, raw or potable.

*Chloramination.* The process used for disinfection of water supplies. Originally it was introduced to obtain combined chlorine, more resistant to time and oxidation than free chlorine. However, when the breakpoint for chlorination was better understood and the effectiveness of hydrochlorous acid in disinfection was confirmed, particularly at low pH values, the use of chloramination was curtailed by the Health Authority, actually prohibiting the use of chloramination, unless the chloramination follows safe chlorination at the plant. In this case, chloramination in the distribution system has been tolerated. Chloramination was resuscitated when the THMs formation was detected, but, after a short period of experimentation, the Health Authority did not encourage the use of chloramination. (See also Chapter 3, *Chloramines*, on page 136.)

## OZONATION

Ozone has a strong bactericidal action and can be used effectively in disinfecting water supplies. In spite of its excellent disinfecting capacity, ozone has not substituted for chlorination due to lack of ability to produce a persistent ozone residual as it is experienced with chlorine. This is particularly true in the distribution system where additional chlorine demand is originated along the pipelines, and chlorine residual protects with residual concentration and, above all, a measurable concentration.

Ozone has other properties, being a strong oxidant, summarized as follows:

- effective in control of taste and odor,
- provides control of excessive color,
- provides oxidation of manganese and iron,
- has a positive effect on oxidation of organics,
- is an aid in flocculation and in the various phases of conventional treatment, due to reduction of macrobiological and microbiological nuisance,
- does not form THMs; on the contrary it has the ability to remove, at least partially, the precursor of THMs (see Chapter 4, *Trihalomethanes*).

It is expected in the water chemistry of **ozone** that the decomposition takes place as:

$$2O_3 \rightarrow 3O_2$$

The presence of chemicals in water considerably affects the decomposition of ozone during and after the oxidation process.

**Ozone By-Products** have been evaluated for a long time, but specifically during the last 30 years, following the concern over THMs formation in chlorination. Recently they have been cataloged into three groups; namely, organic peroxides, unsaturated aldehydes, and epoxides. Research work continues. The first results indicate no toxicity in the by-products, but additional research is underway to confirm the preliminary results.*

During the last 20 years, new technology has been developed to produce more economic and effective ozonation equipment. Real improvements have been registered in the 1980s. Large applications of ozone treatment have been installed in the United States, also in the 1980s. More applications of **Ozone** are expected in the near future, particularly in new water treatment plants or during major rehabilitation work, when polluted river water is the only available raw water for cities.

**Ozone** is not stored or made transportable, at least at this time. It must, therefore, be produced at the plant.

To produce ozone, air is utilized for cost savings and oxygen can be used in pilot plants. Ozonators, like chlorinators, are patented equipment so that the manufacturer's experience can be fully utilized but they are less subject to engineering input in design. Following the ozonator, gas is injected into a water basin of 3–5 m depth (approximately 12 ft) where contact time between 10–20 min is provided for completion of the three-stage process; i.e., the satisfaction of demand; adequate residual values obtained; and decay observation stage. The water basin should be covered; the excess oxygen or ozone should be captured and treated.

The air volume is designed approximately in the ratio of 50 liters per gram of ozone, and the air rate applied is designed in the ratio of 0.6 liter per second, per milligram per liter of ozone, per one million liters per day (approximately 5 cfm/mg/L ozone/MGD).

The state-of-the-art ozonation is **Multiple Stage Ozonation**, a product of French and German sanitary engineers' research and experimentation, which resulted in efficient and economical application. Pre-ozonation helps biological processes that lower organics and ammonia content, beside elimination of THM problems. Pre-ozonation is followed by sand filtration; GAC adsorption; post-ozonation for disinfection, and post-ozonation or post-chlorination for residual. In European treatment plants, this flow sequence was designed: pre-ozonation (1 mg/L) chemical addition, flocculation/sedimentation, ozonation (2 mg/L), filtration/(nitrification), GAC adsorption (biological active GAC), ground storage, and chlorination (0.2–0.4 mg/L).

## pH AND CORROSION CONTROL

The effects of corrosion are mainly water quality problems of the distribution system. Consequently, corrosion and corrosion control are discussed in the last part of this chapter, *Distribution System.* Corrosion control, however,

---

*"Ozonation: Recent Advances and Research Needs," *AWWA Conference Proceedings*, Jun. 1986. AWWA Publication 20005, AWWA. Denver, CO.

is practiced in water treatment with pH adjustments, that is, a chemical feeding process very similar to other treatment, fluoridation, water softening, chemical pretreatment, as far as mechanical and control equipment are concerned. In drinking water supplies, lime, caustic soda, and soda ash are more commonly used for pH adjustment, while chromates and sodium silicates are used for nonpotable water treatment.

Lime [Ca(OH$_2$)] is added with a feeding rate expected to vary between 2–15 mg/L (the increase in alkalinity will vary with 1 mg/L of lime = 1.35 mg/L of added hydroxide alkalinity).

Caustic Soda (NaOH) with a 50% solution, is fed between 3–25 mg/L (1 mg/L of caustic soda = 1.25 of added alkalinity in the hydroxide form).

Soda Ash (Na$_2$CO$_3$) with a typical feed rate of 5–30 mg/L and the corresponding increase in alkalinity is 1 mg/L versus 0.94 of carbonate alkalinity (see also *pH* in Chapter 2 and *Corrosion Control*, discussed later in this chapter).

## Zebra Mussels

Zebra Mussels are **not** water quality parameters and, therefore, were not cataloged in Chapter 5 with the microbiological parameters. Mussels belong to the class of Mollusk, that is, oysters, clams, scallops, and mussels, well recognized as food for man. The "zebra" nickname came from the exterior appearance of the adult mussels. Mollusks may be of interest in water pollution control and, consequently, may be analyzed for their potential toxicity.

The zebra mussel originated from the Black and Caspian seas, imported from Asian/European waterways around 1985. It was assumed that the major contribution was from the ballast water of ships loading cargo in the Laurentian Great Lakes. Zebra mussels are cataloged as *Dreissena polymorpha*. The size of the adult striped mollusk is approximately 25 millimeters (1 inch).

This mussel creates serious problems in raw water intakes or discharge lines because of its high proliferation (estimated around 35,000 eggs per season, per female). After the external fertilization of these eggs, the larvae that develop are free-swimmers that remain in the "column" for several days and then are transported by the water currents. These fresh water mollusks can attach to any hard solid material as used by the common mussel of salt water (*Mytilus edulis*).

The clogging near waterwork intakes will reduce water flow and intensify corrosion. Typical cases are those intakes located in shallow, eutrophic waters of the western sections of Lake Erie. When traveling screens are used, problems with zebra mussels may easily develop.

The zebra mussels have not created these problems in the Caspian Sea, probably because of the deep, cold water that diminishes their growth, now registered in the south. Another reason could be that some predators existed in the Caspian Sea that are not present in Lake Erie, where the bottom now seems to be "paved" with this mollusk.

Attached to the bottom of boats, zebra mussels then traveled south, reaching the Mississippi and Ohio Rivers and spreading also over the bottom of fresh waters of the Atlantic coast, open to navigation. The Susquehanna River may become the passageway of this invader to the Chesapeake Bay. The damage created by *Dreissena polymorpha* also influences power utilities, steam condensers, heat exchangers, boat owners, mainly by decreased hydraulic capacity and boat speed.

### Controls of Zebra Mussels

Several alternatives have been tried to minimize the increased operational cost to eliminate or reduce the above-mentioned problems:

- Use of scuba divers for the evaluation of damage and cleaning and pigging was attempted with some success.
- Use of disinfectants like ozone, potassium permanganate, chlorine, chlorine dioxide, and ultraviolet light;
- Application of nonoxidizing biocides and cathodic protection.

More recent studies on zebra mussels' mortality with chlorine* were conducted with the intention to develop a kinetic model to predict mortality of zebra mussels as a function of chlorine concentration, temperature, and contact time and a diagram useful to industries in planning chlorination treatments for controlling zebra mussels. Once it has been determined that these mussels close their shells even at the low level of 0.04 mg/L of total residual oxidant, the use of dosage above 0.5 mg/L of total residual oxidant assures that the shells remain closed all the time.

The study concluded that mortality of zebra mussels is a function of contact time, chlorine concentration, and temperature (threefold increase for a temperature variation from 10° to 20°C).

## Distribution System

### DISINFECTION

The distribution system consists of aqueducts, reservoirs, main lines, equalizing or distribution underground or elevated tanks, pumping stations, booster chlorination stations, appurtenances (hydrants, valves, flow measuring devices, etc.), street mains, and service lines.

The distribution system may be defined as the final major component of

---

*Van Benschoten, J. E. et al., "Zebra Mussels Mortality with Chlorine". A.W.W.A. Journal, Vol. 87, p. 101, May 1995, Denver, CO.

Klerks, P. L. and Fraleigh, P. C., "Controlling Adult Zebra Mussels with Oxidants," A.W.W.A. Journal, Vol. 83, p. 92, Dec. 1991, Denver, CO.

A.P.H.A., AWWA, W.E.F., *Standard Methods*, 19th Edition, 1995, Part 8000, Toxicity, 8610, Mollusks, American Public Health Association, 1015 Fifteenth St., NW, Washington, DC 20005.

the water supply, originated by the water lines leaving the treatment plant in smaller cities or by aqueducts entering city lines in large cities. The finished water transported through aqueducts or main lines is generally of low turbidity, with an almost constant pH (it commonly may vary between 6.8–7.8), chemically meeting the drinking water standards, microbiologically safe, and meeting bacteriological standards. Water may be disinfected additionally so as to meet standards before reaching the first consumer's outlet. Water leaving the treatment plants or disinfection facilities may contain concentrations between 1.0–0.20 mg/L of free chlorine residual, according to the distance between the chlorination station and the first consumer. Under normal circumstances, the first consumer should not receive more than 0.5 mg/L of free chlorine residual during hours of minimum consumption by the community (more consumer activity, more consumption, more chlorine demand, less chlorine residual). Also, the consumer at the end of the last main must be protected, receiving water with a free chlorine residual of not less than 0.20 mg/L. In this case, such a minimum should be maintained during maximum consumption. "Dead-ends" or not properly looped parts of the distribution system may end up with no chlorine residual and higher turbidities leading to bacterial growth and sources of high heterotrophic plate counts and, less frequently, sources of positive samples for coliform organisms. These weak points of the distribution system require periodic flushing of water lines with the opening of hydrants. This maintenance operation is helpful since it introduces fresh water with higher chlorine residual but causes citizens' complaints due to the stirring of sediments caused by reverse flow in pipes originated by the opening of hydrants in nonproper sequence.

Disinfection facilities should be designed to be able to supply dosages of disinfectant to the distribution system at least 100% higher than normal doses, as previously stated to provide disinfection in case of emergencies, such as waterborne disease outbreaks, major fires, water main breaks, construction for expansion, or repair of the distribution system.

When the distribution system is too large in the service area to maintain these limits of disinfection then potential booster chlorination stations should be designed for new water systems or integrated in existing distribution systems. This decision is reached after evaluation of sampling results and the necessity to locate balancing reservoirs in combination with the chlorination station. The capacity of such booster stations is, in all cases, limited since the chlorine demand of water is minimum, and the level of free chlorine residual must be maintained low since the first consumer here is expected to be in close proximity of the station.

Chlorine residual in the distribution system has the following positive aspects:

- Destruction of microorganisms as a primary function.
- Indication that all chlorine demand of water has been satisfied by the dosage imparted at the treatment or disinfection facility.

- Unexpected chlorine demands in the distribution system are met by the available chlorine residual. Such demands may be originated by cross-connections, improper disinfection after construction or after repair works (main breaks); sizable fire-fighting operation requiring heavy demand of water that forces flow reversal in mains with consequent stirring up of sediments.
- Free chlorine residual at the consumer tap is also an effective help in the disinfection of food and housewares washed in the kitchen sink or commercial/industrial equivalent activities.

The negative aspects of chlorination are the following:

- In the presence of taste- and odor-producing compounds, such as phenols, chlorine intensifies the problem.
- Chlorine is associated with the formation of trihalomethanes, suspected carcinogens (see Chapter 4).
- Chlorine is unquestionably a toxic substance at very high concentration, therefore, psychologically, it has always been looked upon with suspicion even at the safe level. It is accepted by Health Authorities and water engineers with enthusiasm not only for the safe level used, but mainly for the extraordinary practical elimination of typhoids and waterborne disease outbreaks (paratyphoids, dysentery, cholera, infectious hepatitis, etc.—see Chapter 5).

The technology of disinfection was discussed in the preceding sections— *Water Treatment Plants, Final Water Treatment*, and *Disinfection*.

## CROSS-CONNECTION CONTROL

Cross-connection may be considered an unauthorized and improper plumbing connection normally on private property and may result in actual or potential contamination of the potable water supply. Resulting contamination sometimes is limited to private property but on some occasions may also affect the public supply.

Potentially deleterious plumbing connections are commonly the product of ignorance and/or expediency. They are certainly in violation of building codes, plumbing codes, water supply regulations, and sanitary engineering design and practice.

Serious health hazards have resulted in the past as a consequence of cross-connections, requiring the Health Authority to issue regulations, organize seminars and training courses, and implement cross-connection control programs to evaluate conditions through surveys of facilities where cross-connections more likely would occur. Moreover, the Health Authorities have exercised pressure toward the water purveyors so that routine inspections be organized by their trained personnel, knowledgeable of the local plumbing code.

During the last 30 years, public health engineers employed by the Health

Authorities have developed a new field of specialization to prevent the occurrence of contaminated water supplies.

It is possible to visualize this problem, looking into the large "internal distribution system" of skyscrapers or large industrial buildings. Whether the public water supply system ends at the Corporation Valve (i.e., the connection of the service line to the street water main) or ends at the water meter of the building or first private facility, the physical (maintenance) and perhaps legal responsibility of the water purveyor ends; but, after the meter, a private water system begins serving, in some skyscrapers, over 10,000 persons exposed to potential contamination.

The probability of cross-connection problems is certainly not determined by the length of drinking water pipelines but by the complexity of "other" water lines or nonpotable lines, such as air-conditioned cooling systems, roof or intermediate water storage tanks feeding nonpotable lines, and complex hot and cool water lines. The anticorrosion chemicals used to treat cooling towers are the most likely to produce contamination from cross-connections. Less population exposure is expected in industrial or large commercial buildings, but the industrial process is more conducive to chemical contamination and hazardous backflow into the public water supply.

A clear example of potential contamination is a large sewage treatment plant that commonly is expected to use considerable volumes of drinking water for the various processes. One wrong connection between the two systems—potable line and waste line—and serious contamination will exist inside the plant and in the street main, if the water supply is not protected by an airgap or backflow prevention device, properly installed after the meter.

These examples are used to briefly illustrate the importance of cross-connection control programs from theory to training to inspections. Related literature reports long lists of accidents in the United States during the last 40 years: in most of them the sequence of events was totally unexpected, but it happened several times.

The complexity of the demands of a sophisticated society in an urban community has continuously increased the per capita consumption of water at the minimum level of between 3%–5% per year. This increasing water demand is due to industrial, commercial, and municipal activities; and domestic water consuming machines, such as dishwashers, washing machines, garbage grinders, air-conditioners, swimming pools, sprinkler systems for lawns, etc.

There is a direct proportion of these installations with cross-connection occurrence. The water purveyor may discharge his/her responsibility requiring backflow preventers in all industrial, commercial, or large water consumption customers.

The Health Authority is very much concerned with protecting the public water supply, but it is also concerned with protecting the potable water in large buildings where populations larger than a village can be served in a single building or connected residential complex. The Health Authority must protect the consumer of potable water in large or small buildings.

It is necessary to first clarify the previously mentioned contamination potentials. Considering the potable system wrongly connected to a nonpotable system, of course no hazard is created as long as the pressure in the potable system is higher than the nonpotable water pressure even during maximum consumption of potable water. But during repairs of potable lines low pressure or vacuum is created and a **Backpressure** or **Backsiphonage** will result. It is sometimes felt that cases of backsiphonage are extremely rare and perhaps negligible in a community with an extensive and updated plumbing code and an alert, competent, and experienced water purveyor. Unfortunately, this is not the record of the Health Authorities.

After having investigated hundreds of cases of suspected cross-connected contamination, the New York City Dept. of Health, in 1970, organized a special unit in the Bureau of Public Health Engineering, and during the following 10 years implemented surveys of expected causes of potential cross-connections, such as the metal plating industry, car-wash installations served by private wells, funeral homes with mortuaries equipped with embalming facilities, food processing plants, large commercial laundries, and a representative number of old roof water storage tanks, cooling towers, hospitals, and private sewage treatment plants. Such experience originated advanced checklists for inspection, materials to be used for training field sanitarians, reviewing existing city plumbing codes, and coordination with engineers of municipal and private water suppliers. The experience, of course, was mainly essential in the organization of complaint investigations when cross-connections were suspected. Moreover, the N.Y.C. Dept. of Health established a "Health Academy" which is a training center for food handlers, health inspectors, and public health sanitarians assigned to special surveys. During these training sessions, cross-connection control procedures are explained and emphasized.

The N.Y.C. Dept. of Health has investigated many cases of actual or suspected contamination of building water supplies (mostly chemical poisoning cases) to develop an attitude that in spite of code sophistication, engineering design, competent builders, or trained plumbing inspectors, the potentiality of disaster is there and awareness, preparation, education, and extensive use and maintenance of vacuum breakers, backflow preventers, or airgap installations must be continuously developed and enforced in water quality control.

## Preventive Devices

**Airgap Separation*** From any kitchen sink, washbasin, etc. to feeding of the roof tank, surge tank, chemical/food processing plant, **Airgap Separation** is the safest device.

---

*Minimum airspace between the feeding potable water line outlet and the maximum water level (flood level rim) should be twice the diameter of the potable line.

**Vacuum Breakers**   A special valve that allows entrance of atmospheric pressure and consequent blocking of the supply line to prevent a back-siphonage. Normally sufficient to prevent a backflow but insufficient to eliminate backpressure. Proper selection of the point of installation of a vacuum breaker is essential. Problems may develop with time if the pipe and, consequently, the device is subject to corrosion.

**Backflow Preventers**   This device can control backpressure also. The **Reduced Pressure Backflow Preventer** consists of two hydraulically or mechanically loaded, pressure-reducing check valves with a pressure-regulated relief valve located between the two check valves. Malfunctioning of the check valves or relief valves is indicated by a discharge of water from the relief port.

During the last 30 years, manufacturers have improved the technology and life of these devices. Health Authorities normally issue lists of "approved" devices in consultation with the technical engineering societies. An assembly of double check valves and double gate valves allows the testing of the check valves.

### Administration of Cross-Connection Control Programs

The goals of such programs should be best implemented at state and county levels. Health Codes should indicate the following:

- *Approval of Plans* for installation of acceptable devices should be required from the water purveyor and local building/plumbing inspector. Plans are to be prepared by a registered professional engineer.
- Water purveyor should list *Acceptable Protection Devices* and require annual tests by certified inspectors.
- *Certification of Testers/Inspectors* done by building/plumbing senior inspectors with guidelines prepared by the State/Local Health Departments.
- *Color Coding* should be strongly recommended for clearer understanding of piping system layouts, together with elimination of abandoned, obsolete lines prior to replacement with new pipes. Labyrinths of abandoned lines, lack of color coding practice, frequent change of maintenance personnel, unqualified stationary technicians in charge, lack of updated piping drawings are indeed leading to cross-connections.

### Notes

Backflow preventers are definitely expensive when the diameter of the feeding water main is over 50 mm (2 in.). Other negative aspects are the hydraulic head loss in the device and the maintenance required consequent of or in preparation for annual inspections.

Airgap is a very economic solution, but the loss of water pressure is not frequently tolerable.

## CORROSION CONTROL

Drinking water standards list **Corrosivity** as a secondary parameter of physiochemical characteristics. ("Secondary" means not directly related to health.) The MCL for **Corrosivity** is "**Noncorrosive**" while corrosivity shall be determined by the calcium carbonate saturation method or by an equally acceptable method.

Health Authorities issue standards related to corrosivity for the following health related reasons:

- Contaminants like **Lead** and **Cadmium** (see *Type A Inorganic Chemicals*, Chapter 3) of known toxicity are expected in finished water to be proportionally related to the aggressiveness of water (lead and galvanized pipes).
- Similarly (**Copper, Iron**, and **Zinc** [see Types A, C, and B, respectively] *Inorganic Chemicals*, Chapter 3) concentrations are expected directly proportional to the aggressiveness of water in copper, **ductile and cast iron** (unlined), and galvanized pipes.
- Microorganisms may find protection in corrosion products, a common problem in the distribution system of corrosive water.

To evaluate corrosivity the Health Authorities require sampling and analysis of corrosion factors, such as Alkalinity, pH, Total Dissolved Solids (TDS), Hardness, Temperature, and Langelier Saturation Index (LSI or SI), and the Aggressive Index (AI). The water supply system is also expected to report the pipeline materials of the distribution system as well as service lines and house plumbing.

When two different metals are in contact in the "water environment," a similar electrochemical reaction takes place, particularly sensitive when a *more active* metal (like zinc, mild steel) is associated to a *less active* metal (like copper).

Concrete pipes or asbestos-cement pipes or cement-lined pipes are subject to corrosion at low pH. At that level, the water has a potential to dissolve calcium carbonate.

To analyze the important parameters of water quality involved in the corrosion process, the first analysis is of the physical characteristics followed by other influencing chemicals.

### Velocity

**High Velocity** brings more dissolved oxygen in contact with pipe surfaces and originates frictional erosion/corrosion, also exposing new pipe surface to Dissolved Oxygen (DO). Dissolved oxygen is one of the major causes of corrosion, but dissolved oxygen has other advantages (see *DO*, Chapter 3). **Very Low Velocity** may cause stagnation, characterized by dead-end pipelines where major problems with microbiological growth and tuberculation are recorded.

### Temperature

Higher temperature means normally higher rates of chemical reactions creating more problems in corrosion; but higher temperature dissolves more calcium carbonate—the protective coating that reduces corrosion.

### Alkalinity

Properties of alkalinity were examined in Chapter 2. In corrosivity evaluation, the "buffer capacity" of the water examined is important to stabilize pH, as well as bicarbonate and carbonates involved in chemical reactions and in the potential ability to provide the protective, inert coating.

### Dissolved Oxygen

This very negative factor in corrosion control can be visualized as

$$O_2 + 2H_2O + 4e^- \rightarrow 4OH^-$$

The hydroxide ion formed helps to generate more corrosion. In addition, the free oxygen neutralizes the hydrogen gas that protects the cathode in the ionization process, slowing down the reaction (polarization of the cathode); so $O_2$ is considered a depolarization factor, with consequent increments in corrosion. Free oxygen also acts in converting soluble ferrous ion ($Fe^{+2}$) into the insoluble ferric hydroxide (see also Chapter 3, *Iron*).

### Hardness

Expressed in $CaCO_3$ equivalent quantity, hardness is a positive factor in corrosion control; normally hard waters have contributed to the protective coating (for "hard water" definition and characteristics, see Chapter 2).

### Total Dissolved Solids (TDS)

High TDS means high conductivity (see *Solids*, Chapter 2) with consequent faster ionization or a negative factor in corrosion control. High TDS, however, means more likelihood of a protective coating, a very positive factor in corrosion control.

## Chlorides and Sulfates

Concentrations of $Cl^-$ or $SO_4^-$ generate problems adding corrosion activities since they have a negative effect in forming a protective coat due to their chemical reactions with the metals in solution. Sea water is notoriously very corrosive.

## Chlorine

Chlorine residual is an oxidizing agent and by the formation of acids lowers the pH, particularly when low alkalinity exists in the water.

## Hydrogen Sulfide

$H_2S$, a gas commonly found in certain aquifers and a definite problem in related well waters, reacts with the metallic ions in a complex way, forming insoluble sulfides and making corrosivity a problem for all metal pipes.

## Bacteria

Corrosion is intensified by the presence of both aerobic and anaerobic bacteria. Protected by the tubercules in water pipes, they create a serious problem in bactericidal treatment. Particularly in dead-ends of the distribution system, they are the cause and effects of corrosion.

## HEALTH AUTHORITY REGULATIONS

The National Interim Primary Drinking Water Regulations* issued regulations for special monitoring for corrosivity characteristics. These regulations required the determination of: pH, calcium, hardness, alkalinity, temperature, Total Dissolved Solids (Total Filterable Residue), and the **Calculation of the Langelier Index**. USEPA left it to the states to determine identification of additional corrosivity characteristics such as sulfates and chlorides. When asbestos-cement pipes are involved, the **Aggressive Index** is expected to substitute for the Langelier Index. With the same regulation, USEPA requested the community water supplier to report to the State, if required, the presence of vinyl-lined asbestos-cement pipe and coal-tar lined pipes and tanks and to report in all cases the presence in the distribution system of the following:

- lead from piping, solder, caulking, interior lining of distribution mains, alloys, and home plumbing;

---

*USEPA (August 1980) Art. 141.42—45 *Fed. Reg.* 57346 (8/27/80) and 47 *Fed. Reg.* 10999 (3/12/82).

- copper from piping and alloys, service lines, and home plumbing;
- galvanized piping, service lines, and home plumbing;
- ferrous piping materials, such as cast iron and steel; and
- asbestos-cement pipe.

Water causes corrosion in pipes mainly by physical and electrochemical action; unfortunately, a certain amount of corrosion is unavoidable. The phenomena involved are rather complex, and some aspects are not scientifically understood.

**Ionization** or **Dissociation** was dealt with in Chapter 2—see *Solubility*. Ionization is represented by

$$H_2O \rightleftarrows (H^+) + (OH^-)$$

causing the presence of electrical charges involved in the electrochemical action of corrosion. When ferrous material is involved, this action can be represented by

$$\text{hydrogen} + \text{electrons} + \text{oxygen} \rightarrow \text{water}$$

or

$$4H^+ + 4e + O_2 \rightarrow H_2O$$

and

$$4Fe(OH)_2 + 2H_2O + O_2 \rightarrow 4Fe(OH)_3$$

or

$$\text{ferrous hydroxide} + \text{water} + \text{free oxygen} \rightarrow \text{ferric hydroxide}$$

and

$$4Fe^{++} + 10H_2O + O_2 \rightarrow 4Fe(OH)_3 + 8H^+$$

or

$$\text{ferrous ion} + \text{water} + \text{free oxygen} \rightarrow \text{ferric hydroxide} + \text{hydrogen}$$

## Langelier Saturation Index (SI or LSI)

Once the previously mentioned essential parameters of water quality have been recorded, the potential aggressivity of water can be calculated by formulas leading to determination of the point where **Saturation** with calcium carbonate is reached. Such a point is considered a theoretical target when the deposition of a thin protective coating is expected to take place. This pipe coating is optimized only when a thin layer covers metal or pipe internal surfaces without creating a continuous deposition leading, consequently, to internal incrustation that eventually limits the flow. This saturation is reached at a set pH value called $pH_s$. One $pH_s$ is determined by

$$SI = pH - pH_s$$

A positive value indicates supersaturation; a negative value indicates potential corrosivity. Formulas for $pH_s$ were derived by W. F. Langelier[†] and reported and calculated by several authors during the last 50 years.

The *Standard Methods* (16th ed.) 1985 reports formulas, calculations of equilibrium constants, and tabulation of values for given temperatures, total dissolved solids, and logarithms of calcium ion and alkalinity concentrations.

In simple and practical terms that allow quick use of tables, employing a factor $T$ as function of temperature and a factor $S$ as related to total dissolved solids

$$pH_s = T + S - \log[Ca^{2+}] + \log(\text{total ALK})$$

For example, with water of the following parameters: TDS = 200; Temp. = 20°C; Calcium = 30 mg/L of $CaCO_3$; pH = 7.7; ALK = 43 mg/L

$$pH_s = 2.10^* + 9.83^* - 1.48 - 1.63 = 8.82$$
$$LI = pH - pH_s = 7.7 - 8.82 = -1.1 \text{ (slightly corrosive)}$$

For example: TDS = 50; temp = 4°C $\rightarrow$ 20°C; Calcium = 15 mg/L of $CaCO_3$; pH = 6.8; ALK = 12 mg/L. At temp = 4°C

$$pH_s = 2.50^* + 9.74^* - \log(15) - \log 12$$
$$= 2.50 + 9.74 - 1.18 - 1.08 = 9.98$$

$LI = pH - pH_s = 6.8 - 9.98 = -3.2$ (very corrosive). At temp = 20°C

$$pH_s = 2.10^* + 9.74^* - 1.18 - 1.08 = 9.58$$
$$LI = -2.78 \text{ (very corrosive)}$$

In summary and with caution the LSI values can be interpreted as follows:

LSI > 0   Water analyzed is expected to be supersaturated and tends to precipitate $CaCO_3$.

---

*From tables in USEPA, *Fed. Reg.* (8/27/80) or *Standard Methods* (16th ed.) 1985.
†Langelier, W.F. AWWA. Journal. Oct. 1936.

LSI = 0  Water analyzed is considered saturated with $CaCO_3$; no precipitation or dissolution of $CaCO_3$ is expected.

LSI < 0  Water analyzed is considered undersaturated; dissolution of solid $CaCO_3$ is expected.

The Langelier Index (SI) cannot be taken as a reliable method for identification of corrosion problems within narrow margin of values because practical tests did not confirm corrosion results in the range of $+0.40-1.10$ (LI values).

The adoption of asbestos-cement pipes (AC pipe) required a corrosivity index so the Aggressive Index (AI) has been devised. AWWA Standard C-400 defined AI as follows:

$AI = pH + log[(A)(H)]$
A = total ALK as $CaCO_3$ in mg/L
H = calcium hardness as $CaCO_3$ in mg/L
AI < 10 very corrosive
AI = 10–12 moderately aggressive
AI = >12 nonaggressive
With pH = 7.6; ALK = 43 mg/L; calcium hardness = 85 mg/L;
$AI = 7.6 + log43 + log85 = 7.6 + 1.63 + 1.93 = 11.16$ (moderately corrosive)

To determine saturation with $CaCO_3$, there is also a practical test, known as the *Marble Test*, in which one portion of a split sample $CaCO_3$ is added, and then both are analyzed for alkalinity or pH. Supersaturation exists when alkalinity or pH of the untreated sample is greater than the parameters registered for the treated samples. If the contrary is true, the water is undersaturated with $CaCO_3$.

Other suggested indexes for corrosion control are the following:

- *Ryznar Stability Index* (RSI) $RSI = 2pH_s - pH$ ($pH_s$ is the same as per SI Index).
- *Riddick's Corrosion Index* (CI)—Based on field experience, using other factors such as DO, CI, silica, and noncarbonate hardness.
- *McCauley's Driving Force Index*—Based on $CaCO_3$ solubility; helpful in predicting potential weight of precipitate.

## IMPLEMENTATION OF A CORROSION CONTROL PROGRAM

As previously stated, water quality can be controlled. Increasing pH with chemical addition is a relatively inexpensive water treatment requiring an almost negligible increase of space, equipment, and personnel.

The lining of old metal pipes with prior elimination of tubercules and excessive depositions is an annual program that should be planned, budgeted, and implemented according to a capital construction distribution system rehabilitation program, integrated with replacement of water mains.

Selection of new pipe materials must be done with corrosion in mind. Unfortunately, there is no great selection of pipes because of standardization of construction and availability, economics, structural demands, and lack of experience in maintenance personnel with new materials.

The most frequently used materials are: cast iron, ductile iron (lined and unlined), cement, asbestos-cement,* galvanized iron, copper, lead, plastic, and mild steel. Plastic, asbestos-cement, and cement pipes are less subject to corrosion than lead or unlined water pipes.

When metal pipes are used, the same material or at least compatible material should be used throughout the system to avoid formation of galvanic cells and consequent corrosion.

## MAINTAINING POTABILITY

In a water quality control program, the objective is to maintain the potability achieved and measured at the outlet of the treatment plant all the way throughout the distribution system and beyond until the faucet of the ultimate consumer. Of course, an assumption is used here: the water entering the distribution system has met the physical, chemical, bacteriological, and radionuclide standards, specified by the Health Authority and described in Chapters 2, 3, 4, 5, and 6. In addition, the turbidity readings and the disinfection results are expected to meet the standards described in Chapters 8, 9, and 10.

In a small community the simplicity of the distribution system is such that very few problems are expected, particularly when residential buildings only are connected to the water system. The opposite is true for the big cities where the water system is expected to serve consumers through thousands of kilometers (or miles) of street mains interconnected with storage tanks, pumping stations, thousands of hydrants, valves, service lines, and many dead ends of the distribution system.

In a city the complex distribution system creates many potential problems, such as the following:

- Most of the water mains, or at least a large portion, may be 100 years old.
- Pipelines have been built in different times and with different standards and availability of materials.

---

*Worldwide asbestos-cement (AC) pipes are used extensively because of lower cost, light weight, and simplicity of installation. Fabrications of AC pipes continue with more safety precautions. In the U.S., the AWWA has supplied several technical publications defending the use of AC pipes for drinking water. Lately USEPA submitted proposals recommending discontinuance of use (for new or replacement of AC pipes). Pending final decision from USEPA, some States banned the use of AC pipes for drinking water.

- Old pipelines are synonymous with corrosion, incrustation and pitting, sediments, and related bacterial growth.
- Water mains are expected to feed many large commercial establishments and diversified industries where cross-connection and backflow contamination is a constant potential danger.
- The street mains are usually located not too far below the street pavement where heavy traffic and construction equipment may stress the pipe material. When in conjunction with temperature stresses, the overstressed pipe develops a small leak with the consequent beginning of additional overstressing caused by soil settlement and the following disaster of large-size water-main breaks.
- Water mains are "respected" (primary consideration) during construction of other utility lines, but at times there is not sufficient distance between the water line and the effects of excavations for storm sewers, sanitary lines, telephone lines, and sometimes steam pipelines, subways and tunnels.
- The usual lack of city maintenance funds cannot assure a proper replacement of old pipes with the end result that the city sanitary engineer (usually classified as "distribution engineer") has to cope with a system that gets older every year.

While the city "distribution engineer" has many sleepless nights and incredible emergencies, the public health engineer has less pressure while investigating an outbreak, mostly due to the slow process of bacterial examination of samples in the laboratory, but the culprit is not as visible as a watermain break.

The distribution engineer has one public health concern—the disinfection of mains prior to putting in use a new line or relined, replaced, or repaired pipings. Disinfection of mains is usually accomplished with very high chlorine dose residuals and, whenever possible, long contact time. Codes or AWWA specifications prescribe procedures for disinfection, adequate flushing, and bacteriological testing.

At the inlet of the distribution system, there is a responsible party, normally very knowledgeable about the water quality going through and leaving the plant: the chief operator. Any change in water quality going through the plant will become known to him/her through at least one of the parameters used for control. Of these parameters, some are very sensitive: turbidity, chlorine residual, filter runs, specific conductance. Of the equipment used for water purification, the filters are normally the most sensitive to changes in raw water quality.

The operator has also tabulated many parameters that he/she scrutinizes on a daily basis. The seasonal changes are also expected and have been experimented in the past. At the other end of the distribution system is another controller: the consumer. The consumer is not an expert and does not know what is in the water but is aware of changes. In all neighborhoods, there is the individual very sensitive to odor and taste.

Chlorine complaints are the most common; of course, chlorine residual is expected to vary, but sometimes the variation is more than tolerable to the consumer. It is a known fact that consumers get accustomed with time to unusual water taste, such as very hard water or very soft water, or water containing putrid hydrogen sulphide, for example. But changes in water quality are immediately detected by the very sensitive first.

While many citizens' complaints are reluctantly recorded by the Health Authority and the water purveyor, when no corrective actions can be taken by the Agency and public health is not involved (for example, high iron content—rusty water—is very visible and objectionable in old buildings where metal pipes have lost the protective galvanization), there are also extremely useful complaints brought up by the consumer, identifying a change in quality leading to the discovery of a cross-connection or the initial development of an outbreak caused by pathogenic bacteria in the water.

By examining the nature of citizens' complaints, it is possible to establish programs for *corrective measures*; such as anticorrosion treatment, increased chlorination, upgrading of the treatment plant, replacing of water mains, adding new mains to avoid dead ends, providing additional pumping stations to increase pressure, or requiring roof tanks in that zone or more likely private booster pumps in buildings, or cleaning and lining of mains or flushing hydrants at scheduled intervals.

Investigating complaints, the engineer and the technician must have available many instruments and tools. The instruments will include chlorine residual kits, thermometers, sampling bottles, pressure gauges, pH kits, proper containers for bottles, usually sufficient for the first inspection. The necessary tools are: an accurate checklist, adequate training, public relations attitude, patience, ability to listen, and courtesy.

## REFERENCES

American Water Works Association. 1974. **Journal**. Dedicated to Disinfection. Denver, CO: AWWA.

———. 1986. *Maintaining Distribution System Water Quality*. Denver, CO: AWWA.

———. 1986. *Seminar Proceedings*. Ozonation. Denver, CO: AWWA.

———. 1986. *Seminar Proceedings*. Water Quality Concerns in the Distribution System. Denver, CO: AWWA.

———. 1987. *Seminar Proceedings*. Assurance of Adequate Disinfection, or C-T or Not C-T. Denver, CO: AWWA.

———. 1987. *Seminar Proceedings*. Granular Activate Carbon Installations: Conception to Operation. Denver, CO: AWWA.

———. 1991. **Journal**. Issue dedicated to **Activated Carbon**. Vol. 83, Jan. 1991. Denver, CO: AWWA.

———. 1994. **Journal**. Issue dedicated to **Disinfection By-Products**, Vol. 86, June 1994. Denver, CO: AWWA.

———. 1994. **Journal**. Issue dedicated to **Innovations in Treatment**. Vol. 86, Aug. 1994. Denver, CO: AWWA.

————. 1994. **Journal.** Issue dedicated to **Membrane Filtration**, Vol. 86, Dec. 1984. Denver, CO: AWWA.

————. 1995. **Journal.** Issue dedicated to the **Distribution Systems.** Vol. 87, July 1995. Denver, CO: AWWA.

————. 1996. **Journal.** Issue dedicated to **Membrane Processes**, Vol. 88, May 1996. Denver, CO: AWWA.

Babbitt, H. E., Doland, J. J., and Cleasby, J. I. 1969. **Water Supply Engineering.** New York, NY: McGraw-Hill.

Boggess, W. R. and B. G. Wixson. **Lead In The Environment.** Austin, TX: Castle House Publications Ltd.

City University of New York. 1965. Water Resources and the New York Metropolitan Region, Seminar. New York, NY: The City University of New York.

Cox, C. R. 1964. **Operation and Control of Water Treatment Processes.** Geneva, Switzerland: WHO.

Culp, Wesner, and Culp. 1986. **Handbook of Public Water Systems.** New York, NY: Van Nostrand Reinhold.

Drinking Water Research Foundation. 1985. **Safe Drinking Water.** Chelsea, MI: Lewis Publishers, Inc.

Fair, G. M. and J. C. Geyer. 1965. **Water Supply and Wastewater Disposal.** New York, NY: J. Wiley & Sons.

Montgomery, James M. 1985. **Water Treatment Principles and Design.** New York, NY: J. Wiley & Sons.

Nalco Chemical Company. 1988. **The Nalco Water Handbook**, 2nd edition. New York, NY: McGraw-Hill.

National Academy of Sciences. 1980 and 1982. **Drinking Water and Health.** (Vols. 2 and 4), Safe Drinking Water Committee. Washington, D.C.: National Academy Press.

New York State Department of Health 1952. **Protection and Chlorination of Public Water Supplies.** (Bulletin No. 21). Albany, NY: Bureau of Environmental Sanitation.

New York State Department of Health. 1976. **Recommended Standards For Water Works.** (Ten-State Standards) Bulletin No. 42, Albany, NY.

Skeat, W. O. 1969. **Manual of British Water Engineering Practice.** (Vols. I, II, and III), Water Quality and Treatment. London: The Institution of Water Engineers.

USEPA. 1987. **Technologies and Costs For the Treatment of Microbial Contaminants in Potable Water Supplies.** Washington, D.C.: Office of Drinking Water.

USEPA. 1987. **Guidance Manual For Compliance With the Filtration and Disinfection Requirements For Public Water Systems Using Surface Water Sources.** Washington, D.C.

U.S. Public Health Service 1963. **Water Supply and Plumbing-Cross-Connections.** Washington, D.C.

White, George Clifford. 1992. **Handbook of Chlorination** (3d edition). New York, NY: Van Nostrand Reinhold Co.

# CHAPTER 11

# Federal Regulations

SURFACE WATER TREATMENT RULE

LEAD AND COPPER RULE

ENHANCED SURFACE WATER TREATMENT RULE

GROUNDWATER DISINFECTION

INFORMATION COLLECTION RULE (ICR)

# Surface Water Treatment Requirements

## EXISTING PARAMETERS

It is not necessary to repeat here the Maximum Contaminant Levels (MCLs) and the MCLGs in regard to chemical, microbiological, and radionuclide parameters and the rules related to monitoring to control, confirm, predict, record, and report the data that summarize the quality of the raw water and the quality of the potable water delivered to the consumer.

Up to this point, the conventional water treatment has been examined. This treatment will certainly improve the water quality and maintain a certain uniformity in the values of the water quality parameters.

Should the Health Authority therefore compel the surface water source to be subjected to a **conventional** or **modified water treatment plant,** beyond any pretreatment and efficient disinfection? Certainly yes, from a viewpoint of quality. This action will make surface water quality equivalent to groundwater supplied by reliable aquifers and treated with disinfection in addition to aeration, pH control, and perhaps iron removal. After all, the great majority of consumers receive water from surface supplies versus well water. Moreover, large cities must use surface supplies just to provide the necessary *quantity* for large populations. Large cities have water distribution systems very extended and, therefore, more subjected to contamination (see *Distribution System,* page 466). Unfortunately, the Health Authorities have not yet prescribed a mandatory construction and operation of conventional water treatment plants. Not the World Health Organization, nor the European Community, nor the U.S. Public Health Service, and not USEPA, at least until 1989, and not yet **with final rules or without exceptions.**

Economic reasons are the major factors that allow the small water system suppliers to utilize well water, as usually is the case, or reliable spring water from the mountains (very rare). The cost of construction or operation of a conventional plant is in that case a high figure per consumer for a small community. Equally, the opposite is also true: Large cities require large plants, complex operations; the capital cost is burdened by interest charges and annual operating costs that influence the annual consumer's cost of water.

New York City was practically obliged by USEPA to build a water treatment plant that would cost six billion dollars. Some will immediately say that this construction cost and related interests could better be spent helping public welfare by improving direct health care for the individual, while there are still very few and limited outbreaks of waterborne diseases in the city. Also, large cities have serious difficulty in balancing their annual and construction budgets and possess very old distribution systems notorious for water main breaks.

The *Cryptosporidium* outbreaks have improved the chances to see more water treatment plant constructions in the future.

## USEPA: THE SURFACE WATER TREATMENT REQUIREMENTS

As general requirements, USEPA (1989) stated that all public water systems, using any surface water or groundwater under the direct influence of surface water, **must disinfect,** and may be required by the state to filter, unless certain water quality source requirements and site specific conditions are met. *Treatment technique* requirements are established in lieu of MCLs for *Giardia,* viruses, heterotrophic plate count bacteria, *Legionella,* and turbidity. It was also required that *treatment* must achieve at least 99.9 percent removal and/or inactivation of *Giardia lamblia* cysts and 99.9 percent removal and/or inactivation of viruses.

All systems must be *operated by qualified operators* as determined by the state. Criteria to be met *to avoid filtration* were listed as:

- Fecal coliform concentration must not exceed 20/100 mL or the total coliform concentration must not exceed 100/100 mL before disinfection in more than ten percent of the measurements for the previous six months, calculated each month.
- *Turbidity* levels must be measured every four hours by grab samples or continuous monitoring. The turbidity level may not exceed **5 NTU.** If the turbidity exceeds 5 NTU, the water supply system must install **filtration** unless the state determines that the event is unusual or unpredictable, and the event does not occur more than twice in any one year, or five times in any consecutive ten years.

### Disinfection

Disinfection must achieve at least a 99.9 and 99.99 percent inactivation of *Giardia* cysts and viruses, respectively.

This must be demonstrated by the system meeting "CT" values in the rule ("CT" is the product of *residual concentration* (mg/L) and *contact time* (minutes) measured at peak hourly flow). See *CT Tables,* pages 488–492. Failure to meet this requirement on more than one day in a month is a violation. *Filtration* is required if a system has two or more violations in a year unless the state determines that the violation(s) were caused by unusual and unpredictable circumstances; regardless of such determinations by the state, the system must filter if there are three or more violations in a year.

Disinfection equipment must have redundant components or, if approved by the state, automatic water delivery shutoff.

Disinfectant residuals in the distribution system cannot be undetectable or HPC levels cannot be greater than 500/mL in more than five percent of the samples, each month, for any two consecutive months. Samples must be taken at the same frequency as total coliforms under the revised Coliform Rule. Systems in violation of this requirement must install *filtration* unless the state determines that the violation is not caused by a deficiency of treatment of the source water.

Systems must maintain a disinfectant residual concentration of at least 0.2 mg/L in the water entering the system, demonstrated by continuous monitoring. If there is a failure in the continuous monitoring, the system may substitute grab sample monitoring every four hours for up to five days. If the disinfectant residual falls below 0.2 mg/L, the system must notify the state as soon as possible but no later than the end of the next business day. If the residual is not restored to at least 0.2 mg/L within four hours, it is a violation and the system must filter, unless the state determines that the violation was caused by unusual and unpredictable circumstances.

## Other Conditions

Systems must maintain **an adequate watershed control program,** as determined by the state, which will minimize the potential for contamination by human enteric viruses and *Giardia lamblia* cysts.

Systems must not have had any waterborne disease outbreaks or, if they have, such systems must have been modified to prevent another such occurrence, as determined by the state.

Systems must not be out of compliance with the monthly MCL for total coliforms for any two months in any consecutive 12-month period, unless the state determines that the violations are not due to treatment deficiency of the source water.

Systems serving more than 10,000 people must be in compliance with MCL requirements for total trihalomethanes.

## CRITERIA FOR WATER SUPPLY SYSTEMS WITH FILTRATION

### Turbidity Monitoring

See Chapter 8.

### Turbidity Removal

*Conventional filtration or direct filtration* must achieve a turbidity level in the filtered water at all times less than 5 NTU and not more than 0.5 NTU in more than five percent of the measurements taken each month. The state may increase the 0.5 NTU limit up to less than 1 NTU in greater than or equal to 95 percent of the measurements, without any demonstration by the system, if it determines that overall treatment with disinfection achieves at least 99.9 percent and 99.99 percent removal/inactivation of *Giardia* cysts and viruses, respectively.

*Slow sand filtration* must achieve a turbidity level in the filtered water at all times less than 5 NTU and not more than 1 NTU in more than five percent of the samples taken each month. The turbidity limit of 1 NTU may

(Text continues page 492)

## USEPA: SURFACE WATER TREATMENT RULE

### Disinfection Requirements for Unfiltered Systems

*Table 11-1* to *Table 11-7* indicating *CT* values for 99.99 percent inactivation of Giardia lamblia cysts by free chlorine residual listed in the tables with various concentrations and pH values varying from less or equal to 6.0 to less or equal to 9.0.

The numbers in each table indicate the product of *CT*99.9 percent inactivation required in the six tables according to the water temperature that may vary from 0.5°C in Table 11-1. The other tables are prepared for temperatures in centigrade degrees of 5; 10; 15; 20; and 25.

USEPA published these values in the Final Rule of June 29, 1989 with the following note (the same for each table for clarification and interpolation):

These CT values achieve greater than a 99.99 percent inactivation of viruses. CT values between the indicated pH values may be determined by linear interpolation. CT values between the indicated temperatures of different tables may be determined by linear interpolation. If no interpolation is used, use the CT99.9 value at the lower temperature and at the higher pH.

Two additional tables were proposed by USEPA for alternative disinfectants: Table 11-7 for chlorine dioxide and ozone and Table 11-8 when chloramines are used as disinfectants, when approved by the state.

Water supply systems that use filtration do not use the CT criteria, but they are requested to monitor for turbidity (see Chapter 8) so as to achieve a 0.5-log of inactivation, once the state has approved the filtration and disinfection process, giving credit to the system, for example to conventional treatment plant, of at least 99.9 percent (3 log) removal or inactivation of Giardia cysts and 99.99 percent (4 log) for removal or inactivation of viruses, as stated in this section of Chapter 11.

## Table 11-1. CT Values.[a]

| Temperature at ≤ 0.5°C Free Chlorine Residual (mg/L) | <6.0 | 6.5 | pH 7.0 | 7.5 | 8.0 | 8.5 | <9.0 |
|---|---|---|---|---|---|---|---|
| <0.4 | 137 | 163 | 195 | 237 | 277 | 329 | 390 |
| 0.6 | 141 | 168 | 200 | 239 | 286 | 342 | 407 |
| 0.8 | 145 | 172 | 205 | 246 | 295 | 354 | 422 |
| 1.0 | 148 | 176 | 210 | 253 | 304 | 365 | 437 |
| 1.2 | 152 | 180 | 215 | 259 | 313 | 376 | 451 |
| 1.4 | 155 | 184 | 221 | 266 | 321 | 387 | 464 |
| 1.6 | 157 | 189 | 226 | 273 | 329 | 397 | 477 |
| 1.8 | 162 | 193 | 231 | 279 | 338 | 407 | 489 |
| 2.0 | 165 | 197 | 236 | 286 | 346 | 417 | 500 |
| 2.2 | 169 | 201 | 242 | 297 | 353 | 426 | 511 |
| 2.4 | 172 | 205 | 247 | 298 | 361 | 435 | 522 |
| 2.6 | 175 | 209 | 252 | 304 | 368 | 444 | 533 |
| 2.8 | 178 | 213 | 257 | 310 | 375 | 452 | 543 |
| 3.0 | 181 | 217 | 261 | 316 | 382 | 460 | 552 |

[a]See introductory notes to Tables 11-1 to 11-6 before calculating CT values.

## Table 11-2. CT Values.[a]

| Temperature at 5.0°C Free Chlorine Residual (mg/L) | <6.0 | 6.5 | pH 7.0 | 7.5 | 8.0 | 8.5 | <9.0 |
|---|---|---|---|---|---|---|---|
| <0.4 | 97 | 117 | 139 | 166 | 198 | 236 | 279 |
| 0.6 | 100 | 120 | 143 | 171 | 204 | 244 | 291 |
| 0.8 | 103 | 122 | 146 | 175 | 210 | 252 | 301 |
| 1.0 | 105 | 125 | 149 | 179 | 216 | 260 | 312 |
| 1.2 | 107 | 127 | 152 | 183 | 221 | 267 | 320 |
| 1.4 | 109 | 130 | 155 | 187 | 227 | 274 | 329 |
| 1.6 | 111 | 132 | 158 | 192 | 232 | 281 | 337 |
| 1.8 | 114 | 135 | 162 | 196 | 238 | 287 | 345 |
| 2.0 | 116 | 138 | 165 | 200 | 243 | 294 | 353 |
| 2.2 | 118 | 140 | 169 | 204 | 248 | 300 | 361 |
| 2.4 | 120 | 143 | 172 | 209 | 253 | 306 | 368 |
| 2.6 | 122 | 146 | 175 | 213 | 258 | 312 | 375 |
| 2.8 | 124 | 148 | 178 | 217 | 263 | 318 | 382 |
| 3.0 | 126 | 151 | 182 | 221 | 268 | 324 | 389 |

[a]See introductory notes to Tables 11-1 to 11-6 before calculating CT values.

## Table 11-3.  CT Values.[a]

| Temperature at 10.0°C Free Chlorine Residual (mg/L) | pH | | | | | | |
|---|---|---|---|---|---|---|---|
| | <6.0 | 6.5 | 7.0 | 7.5 | 8.0 | 8.5 | <9.0 |
| <0.4 | 73 | 88 | 104 | 125 | 149 | 177 | 209 |
| 0.6 | 75 | 90 | 107 | 128 | 153 | 183 | 218 |
| 0.8 | 78 | 92 | 110 | 131 | 158 | 189 | 226 |
| 1.0 | 79 | 94 | 112 | 134 | 162 | 195 | 234 |
| 1.2 | 80 | 95 | 114 | 137 | 166 | 200 | 240 |
| 1.4 | 82 | 98 | 116 | 140 | 170 | 206 | 247 |
| 1.6 | 83 | 99 | 119 | 144 | 174 | 211 | 253 |
| 1.8 | 86 | 101 | 122 | 147 | 179 | 215 | 259 |
| 2.0 | 87 | 104 | 124 | 150 | 182 | 221 | 265 |
| 2.2 | 89 | 105 | 127 | 153 | 186 | 225 | 271 |
| 2.4 | 90 | 107 | 129 | 157 | 190 | 230 | 276 |
| 2.6 | 92 | 110 | 131 | 160 | 194 | 234 | 281 |
| 2.8 | 93 | 111 | 134 | 163 | 197 | 239 | 287 |
| 3.0 | 95 | 113 | 137 | 166 | 201 | 243 | 292 |

[a]See introductory notes to Tables 11-1 to 11-6 before calculating CT values.

## Table 11-4.  CT Values.[a]

| Temperature at 15.0°C Free Chlorine Residual (mg/L) | pH | | | | | | |
|---|---|---|---|---|---|---|---|
| | <6.0 | 6.5 | 7.0 | 7.5 | 8.0 | 8.5 | <9.0 |
| <0.4 | 49 | 59 | 70 | 83 | 99 | 118 | 140 |
| 0.6 | 50 | 60 | 72 | 86 | 102 | 122 | 146 |
| 0.8 | 52 | 61 | 73 | 88 | 105 | 126 | 151 |
| 1.0 | 53 | 63 | 75 | 90 | 108 | 130 | 156 |
| 1.2 | 54 | 64 | 76 | 92 | 111 | 134 | 160 |
| 1.4 | 55 | 65 | 78 | 94 | 114 | 137 | 165 |
| 1.6 | 56 | 66 | 79 | 96 | 116 | 141 | 169 |
| 1.8 | 57 | 68 | 81 | 98 | 119 | 144 | 173 |
| 2.0 | 58 | 69 | 83 | 100 | 122 | 147 | 177 |
| 2.2 | 59 | 70 | 85 | 102 | 124 | 150 | 181 |
| 2.4 | 60 | 72 | 86 | 105 | 127 | 153 | 184 |
| 2.6 | 61 | 73 | 88 | 107 | 129 | 156 | 188 |
| 2.8 | 62 | 74 | 89 | 109 | 132 | 159 | 191 |
| 3.0 | 63 | 76 | 91 | 111 | 134 | 162 | 195 |

[a]See introductory notes to Tables 11-1 to 11-6 before calculating CT values.

## Table 11-5. CT Values.[a]

| Temperature at 20.0°C Free Chlorine Residual (mg/L) | pH | | | | | | |
|---|---|---|---|---|---|---|---|
| | <6.0 | 6.5 | 7.0 | 7.5 | 8.0 | 8.5 | <9.0 |
| <0.4 | 36 | 44 | 52 | 62 | 74 | 89 | 105 |
| 0.6 | 38 | 45 | 54 | 64 | 77 | 92 | 109 |
| 0.8 | 39 | 46 | 55 | 66 | 79 | 95 | 113 |
| 1.0 | 39 | 47 | 56 | 67 | 81 | 98 | 117 |
| 1.2 | 40 | 48 | 57 | 69 | 83 | 100 | 120 |
| 1.4 | 41 | 49 | 58 | 70 | 85 | 103 | 123 |
| 1.6 | 42 | 50 | 59 | 72 | 87 | 105 | 126 |
| 1.8 | 43 | 51 | 61 | 74 | 89 | 108 | 129 |
| 2.0 | 44 | 52 | 62 | 75 | 91 | 110 | 132 |
| 2.2 | 44 | 53 | 63 | 77 | 93 | 113 | 135 |
| 2.4 | 45 | 54 | 65 | 78 | 95 | 115 | 138 |
| 2.6 | 46 | 55 | 66 | 80 | 97 | 117 | 141 |
| 2.8 | 47 | 56 | 67 | 81 | 99 | 119 | 143 |
| 3.0 | 47 | 57 | 68 | 83 | 101 | 122 | 146 |

[a]See introductory notes to Tables 11-1 to 11-6 before calculating CT values.

## Table 11-6. CT Values.[a]

| Temperature at 25.0°C Free Chlorine Residual (mg/L) | pH | | | | | | |
|---|---|---|---|---|---|---|---|
| | <6.0 | 6.5 | 7.0 | 7.5 | 8.0 | 8.5 | <9.0 |
| <0.4 | 24 | 29 | 35 | 42 | 50 | 59 | 70 |
| 0.6 | 25 | 30 | 36 | 43 | 51 | 61 | 73 |
| 0.8 | 26 | 31 | 37 | 44 | 53 | 63 | 75 |
| 1.0 | 26 | 31 | 37 | 45 | 54 | 65 | 78 |
| 1.2 | 27 | 32 | 38 | 46 | 55 | 67 | 80 |
| 1.4 | 27 | 33 | 39 | 47 | 57 | 69 | 82 |
| 1.6 | 28 | 33 | 40 | 48 | 58 | 70 | 84 |
| 1.8 | 29 | 34 | 41 | 49 | 60 | 72 | 86 |
| 2.0 | 29 | 35 | 41 | 50 | 61 | 74 | 88 |
| 2.2 | 30 | 35 | 42 | 51 | 62 | 75 | 90 |
| 2.4 | 30 | 36 | 43 | 52 | 63 | 77 | 92 |
| 2.6 | 31 | 37 | 44 | 53 | 65 | 78 | 94 |
| 2.8 | 31 | 37 | 45 | 54 | 66 | 80 | 96 |
| 3.0 | 32 | 38 | 46 | 55 | 67 | 81 | 97 |

[a]See introductory notes to Tables 11-1 to 11-6 before calculating CT values.

Table 11-7.   Values (CT$_{99.9}$) for 99.9 Percent Inactivation of *Giardia lamblia* Cysts by Chlorine Dioxide and Ozone.[a,b]

| | Temperature | | | | | |
|---|---|---|---|---|---|---|
| | <1°C | 5°C | 10°C | 15°C | 20°C | ≥25°C |
| Chlorine dioxide | 63 | 26 | 23 | 19 | 15 | 11 |
| Ozone | 2.9 | 1.9 | 1.4 | 0.95 | 0.72 | 0.48 |

[a]These CT values achieve greater than a 99.99 percent inactivation of viruses. CT values between the indicated temperatures may be determined by linear interpolation. If no interpolation is used, use the CT$_{99.9}$ value at the lower temperature for determining CT$_{99.9}$ values between indicated temperatures.
[b]The use of these alternative disinfectants shall be approved by the state.

be increased by the state (but at no time exceed 5 NTU) if it determines that there is no significant interference with disinfection (see *Turbidity* in Chapter 8, page 380).

*Diatomaceous earth filtration* must achieve a turbidity level in the filtered water, at all times, less than 5 NTU and of not more than 1 NTU in more than five percent of the samples taken each month.

*Other filtration technologies* may be used if the system demonstrates to the state that they achieve at least 99.9 and 99.99 percent removal/inactivation of *Giardia lamblia* cysts and viruses, respectively, and are approved by the state. Turbidity limits for these technologies are the same as those for slow sand filtration, including the allowance of increasing the turbidity limit of 1 NTU up to 5 NTU, but at no time exceeding 5 NTU upon approval by the state.

## Disinfection Requirements

*Disinfection with filtration* must achieve at least 99.9 and 99.99 percent removal/inactivation of *Giardia* cysts and viruses, respectively. The states define the level of disinfection required, depending on technology and source water quality. Disinfection requirements for point of entry to the distribution system and within the distribution system are the same as for unfiltered systems.

Table 11-8.   CT Values (CT$_{99.9}$) for 99.9 Percent Inactivation of *Giardia lamblia* Cysts by Chloramines.*

| | Temperature | | | | |
|---|---|---|---|---|---|
| <1°C | 5°C | 10°C | 15°C | 20°C | 25°C |
| 3,800 | 2,200 | 1,850 | 1,500 | 1,100 | 750 |

*These values are for pH values of 6 to 9. These CT values may be assumed to achieve greater than 99.99 percent inactivation of viruses only if chlorine is added and mixed in the water prior to the addition of ammonia. If this condition is not met, the system must demonstrate, based on on-site studies or other information, as approved by the state, that the system is achieving at least 99.99 percent inactivation of viruses. CT values between the indicated temperatures may be determined by linear interpolation. If no interpolation is used, select the CT$_{99.9}$ value at the lower temperature for determining CT$_{99.9}$ values between indicated temperatures.

## USEPA: NATIONAL PRIMARY DRINKING WATER REGULATIONS FOR LEAD AND COPPER

It was the 1986 Safe Drinking Water Act amendments that directed USEPA to issue new drinking water regulations to reduce the risk of lead, a notoriously toxic chemical that has been blamed for causing impairment of mental abilities in small children. Of course, the water is not the only agent that causes high lead blood levels in children. (Chapter 3 presents essential information on the relationship between lead and health.) The problem with lead in the environment was known since 1925, when USPHS. set the MCL at 0.1 mg/L. With increased concern, that level was reduced to 0.05 mg/L by USPHS. The same MCL was used by USEPA in 1975 as an interim value and, later, proposed by USEPA, a value of 0.005 mg/L for lead was issued. At the same time (8/18/88) USEPA proposed an MCLG of zero in consideration of probable carcinogenicity of lead (Group B2 Class).

There were very few signs of toxicity for copper, which is still classified as a metal with nutritional values. The MCLG for copper is still 1.3 mg/L, but now the *action level* has been also set at 1.3 mg/L. The reality exists in the fact that corrosion control regulations will have more beneficial effects in reducing lead and copper than lowering the MCLs for these two metals.

Corrosion control can be improved in any water supply system without any complex technology; moreover, this treatment will eliminate problems typical of the distribution system such as nickel from galvanized pipes, possibly asbestos from asbestos-cement pipes, and aerobic or anaerobic bacterial growth, particularly associated with iron.

USEPA issued the Lead and Copper Rule with the publication of the Final Rule of June 7, 1991, followed by minor changes in 1992, 1993, and 1994. The harmful effects of copper already at 1.3 mg/L and higher values have been detected in the sensitive population.

USEPA recognized that a series of new actions was necessary to reduce notably the problems of lead and copper; specifically:

- an extensive *monitoring* of lead and copper concentrations, particularly in the initial period and for as long as the corrosion control program has been completed and confirmed by success;
- the issuance of treatment technique technology requirements for an *optimal corrosion control;*
- a *source water treatment* when requested by the reaching of the *action levels* set at 0.015 mg/L for lead and 1.3 mg/L for copper;
- *public education* to obtain the cooperation of the isolated cases in homes where the action levels are reached under specific testing of the water, still in pipes for at least six hours, possibly at the kitchen faucet;
- planning and effective *replacement of lead service lines* both by the water supply company and by the cooperation of property owners who have sole jurisdiction over their service line.

An *action level* is not an *MCL;* it is an alarm for the water supply company, alerting it to take action on its corrosion control program or to take corrective steps to identify the problems and implement, or help to implement, the other five above-mentioned steps or components of the Lead and Copper Rule.

## Monitoring

Lead and copper monitoring must be scheduled in alternate six-month periods before the installation of the corrosion control treatment and also after treatment to evaluate its effectiveness. Reduced monitoring—once a year or once every three years, according to analytical results—may follow. The sampling points are the targeted high-risk interior water taps. Water must be allowed to remain still in the pipes at least six hours before sampling. Water supply systems may collect the above-mentioned samples in homes or delegate residents to collect samples.

In addition, the water suppliers must collect a sample for lead and copper analyses from a selected number and location of sites, related to the population served:

|  | Number of Sites | |
| --- | --- | --- |
| Population Served | Initial Monitoring | Reduced Monitoring |
| over 100,000 | 100 | 50 |
| 10,001–100,000 | 60 | 30 |
| 3,301–10,000 | 40 | 20 |
| 501–3,300 | 20 | 10 |
| 101–500 | 10 | 5 |
| 100 | 5 | 5 |

## Optimal Corrosion Control Treatment

The first step is to implement monitoring requirements for water quality parameters. All large water supply systems and medium and small systems that exceed the action level for lead and copper shall monitor for: *pH; alkalinity; calcium; conductivity; water temperature; silica* (when an inhibitor containing silicate is used); and *orthophosphate* (when an inhibitor containing phosphate is used).

Samples for the above-mentioned parameters shall be taken at representative points of the distribution system (taking into consideration population, seasons, sources of water supply, and different treatment methods).

## ENHANCED SURFACE WATER TREATMENT RULE (PROPOSED BY USEPA)

On July 29, 1994, USEPA issued a revised Surface Water Treatment Rule, which will remain in effect (since SWTR was a Final Rule) until the Enhanced Surface Water Treatment Rule (ESWTR) will become a final rule, scheduled to be effective around 1999. ESWTR could be called "a post-*Cryptosporidium* rule," because the *Cryptosporidium parvum* has been indicted in several cases of waterborne outbreaks (see Chapter 5) that took place after the publication of SWTR on June 9, 1989.

These outbreaks occurred in cases where disinfection, water treatment, and surveillance of the watershed were questionable. Both SWTR and now ESWTR are applicable to surface water supply systems or groundwater under the direct influence of surface water.*

With ESWTR, USEPA proposed to amend the Surface Water Treatment Rule to provide additional *protection against pathogens* in drinking water. The primary objective is to establish *treatment requirements* for the waterborne pathogens *Giardia, Cryptosporidium,* and viruses.

This proposed rule is applicable to water systems serving 10,000 people or more. For surface water supply systems that wish to avoid filtration, the ESWTR will try to define **stricter watershed control** requirements. ESWTR will also require periodic sanitary surveys. Another goal of ESWTR is the setting of MCLG for *Cryptosporidium* at zero. This 1994 proposal indicates *alternative requirements* for augmenting treatment control of *Giardia, Cryptosporidium,* and viruses.

As regulatory background information, the ESWTR indicates the intention of USEPA to issue the **Groundwater Disinfection Rule (GWDR)** that will regulate further protection for water supply systems using groundwater. GWDR will supplement the **Total Coliform Rule (TCR)** of 1989 issued for control of disease-causing microorganisms (pathogens) in addition to the SWTR. The Enhanced Surface Water Treatment Rule also confirms the intention of USEPA to develop a rule that would limit concentration levels of disinfectants and the chemical disinfection by-products (DBPs) resulting from their use.

USEPA also presents in the regulatory background of ESWTR the procedure used in developing the disinfectants/disinfection by-products (D/DBP) rule which includes NPDWRs for several disinfectants and disinfectant by-products, also published on June 29, 1994, as the ESWTR; both are proposed, not final rules.

The new procedure adopted by USEPA in 1992 is a **formal regulation negotiation process.** A committee is selected to represent water utilities, state and local health and regulatory agencies, environmental groups, and USEPA. The *Negotiating Committee* decided that the SWTR may need to be revised to address health risks from high densities of pathogens in poor-

---

*As defined by USEPA (see the section *Water Works Components—Watersheds or Aquifers* below, page 421) but final judgment on groundwater classification is given by the states. Sometime after the beginning of the new century, USEPA is expected to issue the Groundwater Disinfection Rule (GWDR), under preparation, probably to be issued after the final rule of ESWTR will be issued. In the definition of groundwater under the direct influence of surface water, ESWTR now includes *Cryptosporidium* in addition to *Giardia lamblia*.

quality source waters and from the protozoan, *Cryptosporidium.* If such requirements were deemed necessary, and could be promulgated concurrently with new D/DBP regulations, a water supply system could comply with both regulations and meet the intended public health goals. Due to the necessity of additional field data, the committee agreed to the development of an **Information Collection Rule (ICR).** USEPA proposed the ICR on Feb. 10, 1994 (see *ICR* below, page 502).

In the discussion of the proposed ESWTR rule, USEPA evidenced the following deficiencies of SWTR:

- *Cryptosporidium* was not included.
- Specified pathogen reductions may be inadequate.
- Virus CT values may be greater than assumed by SWTR.
- DBP Rule may undermine pathogen control.

In the evaluation of the post-*Cryptosporidium* experience, USEPA quoted the following statistics in the proposed ESWTR:

It is estimated that over 162 million people are served by public water systems using surface water, most of which are filtered and disinfected. Of these, as of June 1989, an estimated 21 million people were receiving unfiltered surface water that is only disinfected. EPA anticipates that, as a result of the SWTR, more than 80 percent of the unfiltered systems will install filtration. Nevertheless, in spite of filtration and disinfection, *Cryptosporidium* oocysts have been found in filtered drinking water and most waterborne outbreaks of cryptosporidiosis have been associated with filtered surface water systems. Therefore, it appears that surface water systems that filter and disinfect may still be vulnerable to *Cryptosporidium,* depending on source water quality and treatment effectiveness.

In addition, some surface water systems that were able to avoid filtration under the SWTR may need to filter to provide adequate protection against *Cryptosporidium.*

This comment also can be considered an introduction to the decision to revise the annual inspection to evaluate the watershed control program as stated in the WSTR and to propose a more extensive and professional sanitary survey.

## ESWTR Proposed Sanitary Surveys

A periodic sanitary survey is required in the proposed ESWTR for all water supply systems, regardless of whether they filter or not, as long as they use surface water or groundwater under the influence of surface water (see Fig. 11-1).

Sanitary surveys will be conducted by the state or by an agent approved by the state. The water supply system shall report to the state the information on the sanitary survey within 90 days from completion. The state is expected to prepare the outline or minimum requirements for the sanitary surveys. USEPA may or may not specify criteria for the sanitary survey in

Figure 11-1. Represents the Hudson River (New York State) supplying raw water for one of the connected conventional water treatment plants. The pumping station is visible; the intake is located toward the center of the river and near the bottom. The section of the Hudson River, visible above, reveals an extended area upstream.

This area constitutes **the watershed** for these water plants classified as "surface water supplies." The water supply systems are responsible for evaluation, surveillance, and control of their watershed. There are wastewater discharges particularly upstream. They are expected to be under permit and controlled by the New York State Department of Environmental Conservation. Some effects of a tidal wave may reach the water intake during the drought periods.

details in the Final Rule of ESWTR. The sanitary survey will be scheduled every three or five years. Water supply systems that do not filter will be requested to conduct the annual on-site inspection in the years when the sanitary inspection is not conducted.

USEPA "believes that periodic sanitary surveys, along with the appropriate corrective measures, are indispensable for assuring the long-term quality and safety of drinking water." The Negotiating Committee also emphasized the necessity to promote **watershed protection.***

---

*The ESWTR proposal includes the following statement:

"Watershed protection minimizes pathogen contamination in water sources, and hence the amount of physical treatment and/or disinfectant needed to achieve a specified level of microbial risk in a finished water supply. It also may reduce the level of turbidity, pesticides, volatile organic compounds, and other synthetic organic drinking water contaminants found in some water sources. Watershed protection results in benefits for water supply systems by minimizing reservoir sedimentation and eutrophication and by reducing water treatment operation and maintenance costs. Watershed protection also provides other environmental benefits through improvements in fisheries and ecosystem protection."

## ESWTR: Uncovered Finished Water Reservoirs

ESWTR listed under the possible supplemental requirements the intention of USEPA to issue guidelines recommending that all finished water reservoirs and storage tanks be covered. Sanitary/public health engineers, the AWWA, and the American Society for Microbiology agree with USEPA on the necessity to cover reservoirs and tanks containing potable water. Unquestionably, the potential for contamination of the potable water will be reduced in terms of microbiological and chemical contaminants. Potential sources of contamination listed in ESWTR are in uncovered reservoirs: airborne chemicals, surface water runoff, animal carcasses, animal or bird droppings, growth of algae and other aquatic organisms due to sunlight that results in biomass, and violations of reservoir security.

## ESWTR: Cross-Connection Control Program

This program is also included by USEPA among the possible supplemental requirements of ESWTR. Cross-connection problems are listed in this chapter in the *Distribution System* section. USEPA states in the proposed ESWTR that plumbing cross-connections are actual or potential connections between a potable and nonpotable water supply. Citing statistics, 24% of the waterborne disease outbreaks that occurred during 1981–1990 were caused by water contamination in the distribution system, primarily as a result of cross-connections and main repairs. While the vast majority of outbreaks associated with cross-connections are caused by pathogens, a few are caused by chemicals (statistics for the United States). States usually have guidelines for cross-connection control programs and are presently reviewing related programs undertaken by the water supply systems. Also, items like minimum pressure requirements may be or may be not included in the Final Rule of ESWTR.

## State Notification of High Turbidity Level

No decision has been reached in the proposed ESWTR regarding modifications of the rules on turbidity notification as stated in the Surface Water Treatment Rule of 1989; another subject of the "possible supplement requirements" of ESWTR.

## Alternative Treatment Requirements—Proposed

ESWTR presented five alternative treatment requirements. The selection of the proper alternative for the water supply system will be dependent upon the results and interpretation of the data furnished by the Information Collection Rule (ICR). It is expected that the final ESWTR will be one or a combination of the following alternatives:

Alternative A: Enhanced Treatment for *Giardia*
Alternative B: Specific Treatment for *Cryptosporidium*

Alternative C: 99% (2-log) Removal of *Cryptosporidium*
Alternative D: Specific Disinfection Treatment for Viruses
Alternative E: No change in existing SWTR treatment requirements for *Giardia* and viruses.

Table 11-9 has been developed by AWWA.*

## GROUNDWATER DISINFECTION

USEPA has not issued an official disinfection regulation for groundwater that is **not** under the influence of surface water. In 1991 USEPA published a first draft[†] and in 1992 published a second summary.[‡] A legal action against USEPA resulted in a consent decree with a final rule originally scheduled for Aug. 15, 1996, but a court decision has left the new deadline open to negotiations.

The intention of the Groundwater Disinfection Rule (GWDR) is to present regulations for disinfection at the source and for the distribution system, including regulations for qualifications of operators, treatment techniques, MCLGs, "natural" disinfection requirements, and the usual regulatory sections such as monitoring, analytical methods, reporting, variances, and exemptions.

Disinfection is related to waterborne diseases, consequently, the issuance of microbiological parameters is expected. A setting of MCLG of zero is expected for viruses. This draft includes the possibility of an MCLG of zero for *Legionella,* but not for Heterotrophic Plate Count. As per the Surface Water Treatment Rule, a Treatment Technique requirement will represent the MCLs to avoid costly laboratory work for viruses and pathogenic protozoa.

In this type of protected well field, it is not expected to find *Cryptosporidium* and *Giardia,* but viruses are more mobile in groundwater and also more resistant to disinfection.

Disinfection treatment of each well is expected to be regulated to reach the minimum percent inactivation and/or removal of viruses, not yet determined in this draft. Additional studies will be conducted by USEPA and the AWWA Research Foundation. USEPA also intends to provide the states with minimum requirements for design, construction, and operation of each well. The use of the **C × T** concept (see SWTR), i.e., concentration of disinfectant (mg/L) × contact time in minutes, will be adopted unless another effective and economic method is found.

Ultraviolet light (UV) would be acceptable for disinfection; but a sensor and recorder will be required at each well. If minimum disinfection

---

*From the "Enhanced Surface Water Treatment Rule (Proposed)" section of the *SDWA Advisor*, 1995 publication. American Water Works Association, Denver, CO. (By permission. © Copyright 1995.)
†Possible Requirements of the Ground Water Disinfection Rule. USEPA. OGWDW. Washington, DC. (June 20, 1991).
‡Draft Summary. Ground Water Disinfection Rule, Washington, DC. (July 1992).

## Table 11-9. Proposed Alternative ESWTR Treatment Requirements.*

| Alternative | Source Water Cyst or Oocyst Concentration | Removal Required |
|---|---|---|
| Alternative A<br>At least 99.9 percent (3-log) removal/inactivation of *Giardia* is required for systems serving fewer than 10,000 people. | Alternative A<br><1 cyst/100 L<br>1 to 9 cysts/100 L<br><br><br>10 to 99 cysts/100 L | Alternative A<br>99.9 percent (3-log)<br>99.99 percent (4-log) (alternative: 99.9 percent [3-log])<br>99.999 percent (5-log) (alternative: 99.99 percent [4-log]) |
| Systems serving 10,000 people or more must achieve the *Giardia* removal/inactivation as shown in columns 2 and 3 for alternative A. | >99 cysts/100 L | 99.9999 percent (6-log) (alternative: 99.999 percent [5-log]) |
| Alternative B<br>At least 99.9 percent (3-log) removal/inactivation of *Cryptosporidium* oocysts is required for systems serving fewer than 10,000 people. | Alternative B<br><1 oocyst/100 L<br><br><br>1 to 9 oocysts/100 L | Alternative B<br>99.9 percent (3-log) (alternative: 99 percent [2-log];<br>99.99 percent (4-log) (alternative 1: 99.9 percent [3-log]; alternative 2: 99 percent [2-log]) |
| Beginning 18 months after promulgation, systems serving 10,000 people or more must achieve *Cryptosporidium* removal/inactivation as shown in columns 2 and 3 for alternative B. | 10 to 99 oocysts/100 L<br><br><br><br>>99 oocysts/100 L | 99.999 percent (5-log) (alternative 1: 99.99 percent [4-log]; alternative 2: 99.9 percent [3-log])<br><br>99.9999 percent (6-log) (alternative 1: 99.999 percent [5-log]; alternative 2: 99.99 percent [4-log]) |
| Alternative C<br>All systems that filter must achieve at least 99 percent (2-log) removal of *Cryptosporidium* between the source water and the first customer. | | |
| Alternative D<br>Systems serving 10,000 people or more must achieve at least a 0.5-log inactivation of *Giardia* (alternative: 4-log inactivation of viruses) by disinfection alone. | | |
| Alternative E<br>No change in the existing SWTR regarding the level of removal/inactivation requirements. | | |

*Reprinted from *SDWA Advisor,* Enhanced Surface Water Treatment Rule, by permission. ©1995, American Water Works Association.

requirements are not met for at least four continuous hours at the concentration above the minimum, the well must be automatically stopped from supplying water until repairs are completed. The Final Rule will state the minimum concentration required.

## Natural Disinfection Definition

In this Draft, USEPA defined **natural disinfection** as "a source water treatment via virus attenuation by natural subsurface processes such as virus inactivation, dispersion (dilution), and irreversible sorption to aquifer framework solid surfaces." A well or well field that is not vulnerable to viral contamination would be considered to meet the criteria of **natural disinfection** (criteria will be specified in the Final Rule). In addition to meeting the specified criteria, the groundwater supply must also meet the following conditions:

- No waterborne disease outbreaks were identified for that well.
- The well must meet state-approved well construction codes.
- The water supply system must not be in violation of the Total Coliform Rule.

Moreover, to obtain the classification of **natural disinfection**, one of the following rules must be met (according to this GWDR draft):

- The nearest potential source of fecal contamination is at least at a specified minimum distance from the well, and the water must not flow through caves, large fractures, or other similar features.
- The travel time of a groundwater particle is at least at a specified minimum number of days from the nearest potential source of fecal contamination to the receptor well.
- Same as above for the travel time of a microbial pathogen, but including the effects of retardation, dispersion, inactivation, and diffusion.
- A hydrogeological feature, such as a thick unsaturated zone, controls potential contaminant flow to the well, and human activities do not adversely affect the integrity of this feature.

A risk analysis will be developed to determine the amount of treatment by natural disinfection. At this time, USEPA considers an acceptable risk to be less than one gastrointestinal infection per 10,000 people per year. The concern here is to limit gastrointestinal diseases caused by microbial infections in order to protect the sensitive population (the elderly, infants, and the chronically ill) as well as to diminish the risk of more serious diseases such as hepatitis A.

After issuing this draft, USEPA is planning to develop a new contaminant transport model to improve **natural disinfection** criteria.

## GWDR—Distribution System Disinfection

At the entry point to the distribution system, each community water system must maintain at least 0.2 mg/L of disinfectant residual at all times. Concurrently, HPC concentrations should be maintained under 500 CU/mL. Violations will be noted if more than 5% for both parameters, disinfectant residual undetectable and HPC over 500 CFU/mL, are recorded each month for any two consecutive months. Sampling requirements will be at the same frequency and location as for the total coliforms under the Total Coliform Rule.

The **GWDR** is expected to require all water supply systems that use groundwater sources and disinfection to be run by qualified operators (qualifications to be determined by the state).

**GWDR** will determine that 54 months after promulgation of the Final Rule all community water systems not disinfecting at the time of promulgation would be required to install disinfection and meet monitoring and performance requirements, unless the state determines that the water supply system is not required to disinfect because it qualifies for **natural disinfection** or for a variance. All water supply systems that are disinfecting at the time of promulgation would be required to meet monitoring and performance requirements within 18 months (30 months for noncommunity systems), after promulgation of the rule.*

## USEPA—INFORMATION COLLECTION RULE (ICR)—PROPOSED

The Information Collection Rule (ICR) is the more extensive regulatory requirements issued by USEPA regarding the federal intention to monitor treatment and sources of water supplies and will be applicable nationwide. Proposed in the publication of Feb. 10, 1994*, it is dedicated to microbial and disinfectant by-product monitoring, information-gathering in regard to the description of the water treatment plant capacity and infrastructure, and to testing requirements at bench scale, pilot scale, or both.†

**ICR** is intended to be, at the end of this information bank interpretation, a centralized and computerized operation to be conducted at the federal level without the help of the states, as was done with prior regulations. Results will be accomplished through the use of special software (furnished by USEPA) and, implicitly, intelligent cooperation from the water suppliers.

**ICR** lists three major regulatory components:

- Microbiological monitoring for specific parameters.
- DBP monitoring according to the type of disinfection applied for treatment.

---

*Regli, S., Rose; Haas; and Gerba. *Modelling the Risk from Giardia and Viruses in Drinking Water. AWWA Journal.*

†USEPA. Monitoring Requirements for Public Drinking Water Supplies: *Cryptosporidium, Giardia,* Viruses, Disinfection By-Products, Water Treatment Plant Data, and Other Information Requirements. Proposed Rule. *Fed. Reg.,* Vol. 59, No. 28 (Feb. 10, 1994).

- Treatment Plant detailed information to be submitted, including the watershed protection and the characteristics of the raw water used.

The major burden is expected from the systems serving more than 100,000 people, and, with some modification, from the systems serving over 10,000 but less than 100,000 people. USEPA is confident that with the extrapolation of data from the medium and large suppliers it will also be possible to predict regulatory requirements for the small systems.

## Microbial Monitoring

With monitoring of *Giardia, Cryptosporidium, E. coli* or fecal coliforms, and enteroviruses, **ICR** will provide a better understanding of the health risks and the effectiveness of treatment/disinfection applied. **ICR**, then, will assist in the evaluation and revision of the Surface Water Treatment Rule (SWTR) and the Total Coliform Rule.

Microbial monitoring is limited to water supply systems (systems) using surface water, or groundwater under the direct influence of surface water, and serving more than 10,000 people. The use of an approved laboratory is the first step.

For the large systems, **ICR** expected a starting date of source water monitoring not later than October 1995 and the completion of the 18 consecutive months of the monitoring period by March 31, 1997. Samples are to be collected at the intake (or composite samples used if more than one intake); one sample per month to determine total coliforms, *E. coli* or fecal coliforms, *Giardia, Cryptosporidium,* and viruses.

When results are negative for *Giardia, Cryptosporidium,* or enteroviruses in one liter of raw water in the first 12 months, monitoring for an additional 6 months is not required, but data must be reported to USEPA after the fourth month of sampling.

If the system detects pathogens in the one-liter sample, then requirements for sampling of the **finished water** come into effect: one sample per month, starting the month following detection of pathogens. The same parameters used in raw water sampling must be used for finished water. Sampling periods stop after 18 months from the beginning of sampling.

**ICR** does not require monitoring for viruses if the system meets the following conditions:

1. System has tested the raw water for total or fecal coliforms at least five times per week;
2. sampling started at least four months before and two months after publication of this rule in the Federal Register; **and**
3. the results of monitoring indicate either total coliforms in less than 100 colonies/100 mL for at least 90% of the samples, or fecal coliform count is less than 20 colonies/100 mL for at least 90% of the samples.

## Public Water Systems Serving More Than 10,000 But Less Than 100,000 People

The same regulations apply, with the following exceptions:

- Monitoring period: 12 months
- Starting period: not later than April 1996
- Completion by March 31, 1997
- Collect one sample every other month at the intake
- Determine concentrations of same parameters as requested for large systems but **not for viruses**
- Report to USEPA every two months beginning four months after starting

USEPA will at the end extrapolate the data furnished by the large and medium system-monitoring to estimate pathogen occurrence in small systems (those serving less than 10,000 people). USEPA also will waive monitoring requirements for viruses in source water of exceptionally good water quality.

USEPA also has under consideration the possibility of requiring in the final rule the acceptance of previously collected data from systems, and the collection of particle size count data within the treatment plant in lieu of, or in addition to, finished water monitoring for *Giardia* and *Cryptosporidium*. USEPA is also evaluating additional parameters and the possibility of requiring monitoring for *Clostridium perfringens* and/or coliphage, as potential indicators of pathogen occurrence and treatment effectiveness.

## Water Treatment Plant Data

**ICR** is not limited to monitoring and data reporting regulations. To reach the conclusions that ICR needs to provide essential information to complement the regulations contained in the SWTR, ESWTR, and TCR, complete information concerning the plant capacity, type of treatment, theoretical treatment capability, past operating data, etc. is necessary.

USEPA plans to simplify the present draft of the information-gathering schedule prior to issuing the final rule.

The following is a summary list of the subjects of interest that will be reported via electronic media and entry software:

population served
design flow (MGD)
annual water temperature (max. and min.)
type of source (river, stream, reservoir, lake, etc.)
monthly average flow (MGD)
upstream source of microbiological contamination

plant influent data (MGD; peak, average monthly flow)
plant effluent data
sludge treatment (solid production; handling capacity)
general process parameters (units designed or in service)
presedimentation basin (gpm/ft$^2$)
chemical feeder and chemical used (capacity for each unit)
type of mixer (mechanical, hydraulic jump, static, other)
flocculation basin
sedimentation basin (loading at design flow-gpm/ft$^2$)
filtration data (loading at design flow; medium; depth)
contact basin (stable liquid level)
clear well (variable liquid level; minimum volume)
ozone contact basin
clarifier (surface loading; special equipment)
DE filter (surface loading; precoat; body feed; etc.)
Granular activated carbon (design regeneration frequency); membranes; air stripping; adsorption clarifier; dissolved air flotation; low sand filtration; ion exchange; other treatment

## Disinfection By-Products—Monitoring Requirements ICR

Systems that use groundwater not under the direct influence of surface water, and serve over 50,000 and less than 100,000 people, will be required to monitor only for total organic carbon (TOC). Sampling will be scheduled at the entry point to the distribution system. No other monitoring is required for DBP for these systems.

Again, the major burden of monitoring activities is reserved for **community** and **nontransient, noncommunity water systems serving more than 100,000 people**.

The **ICR** requirements specify sampling **monthly** at the sampling points selected at:

- treatment plant influent
- before and after filtration treatment
- each point of disinfection treatment
- entry point to the distribution system

The **analyses** requested are for pH, alkalinity, turbidity, temperature, calcium and total hardness, total organic carbon, and $UV_{254}$.*

In addition, analysis for bromide and ammonia is requested at the plant

---

*$UV_{254}$ analysis is the Ultraviolet Absorption Method standardized at 254 nm, to determine the presence in raw or treated water of some organic compounds and various aromatic compounds that strongly absorb ultraviolet radiation. The Ultraviolet Absorption Method was approved by the Standard Methods Committee in 1994, and reported in the 19th edition of Standard Methods (1995).

influent; ammonia is also necessary after air stripping, and disinfectant residual is requested at the end of each process in which chlorine is applied.

Sampling must also be scheduled **quarterly** at the following points with the stated analyses:

- treatment plant influent: total organic halides
- after filtration (if chlorine is applied prior to filtration); and at the entry point to the distribution systems:

| | |
|---|---|
| trihalomethanes | (THMs) |
| monochloracetic acid | |
| dichloroacetic acid | (HAA6) |
| trichloroacetic acid | |
| monobromoacetic acid | |
| dibromoacetic acid | HAA6 |
| bromochloroacetic acid | |
| dichloroacetonitrile | |
| trichloroacetonitrile | HANs |
| bromochloroacetonitrile | |
| dibromoacetonitrile | |
| chloropicrin | CP |
| 1,1-dichloropropanone | HK |
| 1,1,1-trichloropropanone | |
| chloral hydrate | CH |
| total organic halides | TOX |

In addition, at the entry point to the distribution system the simulated distribution system (SDS) test must be scheduled. Also, in addition for the THM compliance monitoring points,* tests for pH, temperature, alkalinity, total hardness, and disinfectant residual are requested.

### Nonchlorine Disinfectants

If **chloramines** are used, the following additional tests are requested **quarterly**: Cyanogen chloride at the entry point to the distribution system and one of the THM compliance monitoring sample points representing a maximum detention time in the distribution system, same analysis and frequency.

---

*For Trihalomethane monitoring points of the distribution system, four quarterly samples are scheduled at the following points: one point to correspond to the simulated distribution system (SDS) test; one to be selected at the maximum detection time; and the remaining two at representative points of the distribution system.

The SDS test will be analyzed for the same quarterly parameters used for THM testing, but temperature, alkalinity, and total hardness are not required.

When **hypochlorite solutions** are used, frequency remains **quarterly**, and chlorate is analyzed at the treatment plant influent and at the entry point to the distribution system; the hypochlorite stock solution is tested for pH, temperature, free chlorine residual, and chlorate.

When **ozone** is the disinfectant, **monthly** monitoring is requested for the ozone contactor influent with pH, alkalinity, turbidity, temperature, calcium and total hardness, TOC, $UV_{254}$, bromide, and ammonia. Quarterly monitoring is to be scheduled for aldehydes* and for the optional AOC/BDOC† tests.

**Monthly** monitoring is also to be scheduled for the ozone contactor effluent with the analysis of ozone residual, but **quarterly** for aldehides and the optional AOC/BDOC. In addition, **monthly** sampling at points "before filtration" is required for ozone residual. The entry point to the distribution system must be monitored **monthly** for bromate and **quarterly** for aldehydes and the optional tests for AOC/BDOC.

When **chlorine dioxide** is used, the following sampling points are monitored **monthly**:

- Before each "chlorine dioxide application" for pH, alkalinity, turbidity, temperature, total and calcium hardness, TOC, $UV_{254}$, bromide, and ammonia.
- Before the points of application of ferrous salts, sulfur-reducing agents, or GAC, the scheduled parameters are pH, chlorine dioxide residual, chlorite, and chlorate.
- Entry point to the distribution system to be analyzed for chlorite, chlorate, chlorine dioxide residual, and bromate.
- Three distribution system sampling points for chlorite, chlorate, chlorine dioxide residual, pH, and temperature.

The following points are monitored **quarterly**:

- Treatment plant influent: test for chlorate.
- Before first chlorine dioxide application: test for aldehydes and AOC/ BDOC optional.

---

*USEPA listed the following aldehydes to be analyzed: formaldehyde, acetaldehyde, butanal, propanal, pentanal, glyoxal, methyl glyoxal.
†AOC = assimilable organic carbon
 BDOC = biodegradable dissolved organic carbon
 Both AOC and BDOC are part of the group of "Assimilable Organic Carbons" listed as a proposed method when approved by the Standard Methods Committee in 1991. These parameters are connected with the problems or potential problems of bacterial growth in the water distribution and storage systems. This known growth and regrowth is related to viable bacteria that survive the disinfection treatment. To sustain bacterial survival and growth, nutrients are necessary, but contributory factors such as temperature, time (residence in mains, storage tanks, pipes), and disinfectant residual concentration are also necessary. Chemical methods and bioassays are also under consideration. *Standard Methods* (1995) lists the "*Pseudomonas fluorescens* Strain P-17, *Spirillum* Strain NOX Method, the two-species bioassay Method."

- Entry point to the distribution system: test for aldehydes and AOC/BDOC optional.

## Systems Using Less Than 100,000 People—ICR (Proposed)

Monitoring is not required; the only exception remains for groundwater systems **not** under the influence of surface water **and** serving between 50,000 and 100,000 people; in that case analyses are limited to Total Organic Carbon (TOC) and monitoring (limited to 12 months) is expected to begin three months after publication of the final rule. This sampling and testing for TOC is scheduled at the entry point of the distribution system; monitoring data are to be reported to USEPA not later than 17 months after promulgation of the final rule.

## Analytical Methods Used in ICR

All the above mentioned analytes, indicated in the **ICR** proposed requirements, use either approved or proposed analytical methods. All the analytes are listed in the 19th edition of *Standard Methods* (1995) with the exception of the following analytes:

| Analyte | USEPA Method N.* |
| --- | --- |
| chlorate and chlorite | 300.0 |
| bromochloroacetonitrile | 551 |
| chloral hydrate | 5518 |
| dibromoacetonitrile | 551 |
| dichloroacetonitrile | 551 |

Approved laboratories must be used for the analyses of the regulated contaminants. USEPA may reduce the number of the regulated analytes, but monitoring is not expected to be canceled for the following analytes:

trihalomethanes
haloacetic acids
bromate
chlorite
chlorate
total organic halides
total organic carbon
bromide

USEPA expects to delay the starting date of monitoring if adequate laboratory capacity is not available.

---

*USEPA. Methods for the Determination of Organic Compounds in Drinking Water: Supplement I and II, EPA-600/4-90/020 and USEPA EPA-600/R-92/129, respectively, NTIS, Springfield, VA.

## DBP Precursor Removal Studies—ICR (Proposed)

The last section of the monitoring requirements under **ICR** is very limited concerning community and nontransient, noncommunity water systems. First, **groundwater systems** not under the influence of surface water and serving more than 50,000 people, and having a TOC of less than 2.0 mg/L at the entry point of the distribution system are exempt; they are not required to conduct the bench- and/or pilot-scale testing. But if the TOC concentrations are above 2 mg/L, precursor removal studies are also necessary for groundwater systems.

For **surface water systems** and **groundwater systems under the influence of surface water**: When chlorine is the primary disinfectant, TTHM is more than 40 µg/L, HAA6 is more than 39 µg/L, the systems serve more than 100,000 people, and the treatment plant influent is more than 4.0 mg/L of TOC, precursor removal studies are necessary.

**ICR** requires that bench- and/or pilot-scale studies for GAC and/or membranes beginning not later than 18 months following promulgation be conducted. The report of the completed study must be submitted to USEPA no later than Sept. 30, 1997.

These tests have been requested in order to obtain more information on the cost-effectiveness of GAC and membrane technology for removing DBP precursor and DBPs.


## GAC Bench-Scale Tests

This testing is also defined as the **rapid small-scale column tests (RSSCT)**. In order to represent a full-scale empty bed contact times (EBCT) of 10 minutes and 20 minutes respectively, at least two EBCTs shall be tested. The RSSCT shall be run until the effluent TOC is 75% of the running average influent TOC or an RSSCT operation time is reached that represents the equivalent of one year of full-scale operation, whichever is shortest.

For each RSSCT conducted, the following information must be reported:

- pretreatment conditions
- **GAC** type, particle diameter, height and dry weight in the RSSCT column
- RSSCT column inner diameter
- volumetric flow rate and operation time at which each sample is collected

These tests are to be scheduled **quarterly** for one year in order to evaluate seasonal variations. If during the first quarter, RSSCT demonstrates that the effluent TOC reaches 75% of the average influent TOC within 20 full-scale equivalent days for an EBCT of 10 minutes, and within 30 full-scale equivalent days for an EBCT of 20 minutes, then the last three quarterly tests must be conducted using membrane bench-scale testing with one membrane.

**ICR** prescribes that the following parameters be used for evaluation at the **GAC influent**:

- alkalinity, total and calcium hardness, ammonia and bromide (two samples per batch of influent per run)
- pH, turbidity, temperature, TOC, and $UV_{254}$; SDS for THMs, HAA6, TOX, and chlorine demand (three samples per batch, per run)

For the **GAC effluent** at an EBCT of 10 min. (scaled):

- pH, temperature, TOC, and $UV_{254}$; SDS for THMs, HAA6, TOX, and chlorine demand (a minimum of 12 samples; one after one hour and thereafter at 5%–8% increments of the average influent TOC)

For the GAC effluent at an EBCT of 20 min., same parameters above listed for 10 min. and with the same frequency of sampling.

## Membrane Bench-Scale Tests

**ICR** expects a minimum of two membrane types to be tested and designed to have a nominal molecular weight cutoff of less than 1,000. The reactors are expected to be placed in such a configuration in order to yield representative flux loss assessments for membranes.

Testing frequency should be: **quarterly** for one year (four tests for each membrane). For each test the following data are expected to be reported:

- pretreatment conditions
- membrane type, area, configuration, inlet pressure and flow rate, recovery, and operation time at which each sample is taken

For the **bench-scale membrane systems** the sampling points, parameters, and frequency determined by **ICR** are:

- For the **membrane influent**: alkalinity, total dissolved solids, total and calcium hardness, and bromide (two samples per batch, but, if the flow is continuous, a minimum of eight samples is necessary).
- For the **membrane influent**, the additional parameters are: pH, turbidity, temperature, HPC, TOC, and $UV_{254}$; SDS for THMs, HAA6, TOX, and chlorine demand.

The other sampling point is required by **ICR** for the **membrane permeate**, testing for pH, alkalinity, total dissolved solids, turbidity, temperature, total and calcium hardness, bromide, HPC, TOC, and $UV_{254}$; SDS for THMs, HAA6, TOX, and chlorine demand.

## GAC Pilot-Scale Tests

In the pilot-scale testing with **GAC** the following requirements are programmed by **ICR**:

- continuous flow
- minimum inside diameter of the column (2 in.)
- GAC particle size and hydraulic loading rate: to be related to those used in full-scale practice
- EBCTs and RSSCT requirements: same for GAC bench-scale
- Monitoring reports: same for GAC bench-scale

Sampling points, parameters and frequency for pilot-scale GAC systems are:

- For the **GAC influent**: same analyses prescribed for the GAC bench-scale testing (a minimum of 15 samples taken at the same time as the samples for GAC effluent at an EBCT of 20 min).

At the sampling point of the **GAC effluent** prescribed by **IRC** are the same analyses as the GAC influent, both at an EBCT of 10 min. and 20 min. (In both cases the sample frequency is the same: A minimum of 15 samples; one after one day and thereafter at 3%–7% increments of the average influent TOC.)

**Sampling requirements for the pilot-scale membrane systems** are set by **ICR** as follows:

- Sampling at the **membrane influent** for the analyses of pH, alkalinity, total dissolved solids, turbidity, temperature, total and calcium hardness, ammonia, bromide, HPC, TOC, $UV_{25}$; SDS for THMs, HAA6, TOX, and chlorine demand. The frequency is set for a minimum of 15 samples to be taken at the same time as the membrane permeate samples.

The other sampling point is the **membrane permeate**; the analyses are the same scheduled for the membrane influent and the frequency stated is a minimum of 15 samples evenly spaced over the membrane run.

## USEPA—INFORMATION COLLECTION RULE (ICR)—FINAL RULE

Suddenly, on May 14, 1996, after a long delay in congressional budgetary approval, USEPA issued the Final Rule* in order to initiate as soon as possible the sampling and monitoring specified in the ICR (Proposed). The

---

*USEPA. 40 CFR Part 141, "National Primary Drinking Water Regulations: Monitoring Requirements for Public Drinking Water Supplies: *Cryptosporidium, Giardia,* Viruses, Disinfection Byproducts, Water Treatment Plant Data and Other Information Requirements. Final Rule." *Fed. Reg.* Vol. 61, No. 94.

effective date for the Final Rule was June 18, 1996. This rule shall remain effective until December 31, 2000.

Monitoring and data are confined to large public water systems (PWSs). This rule will provide USEPA with sufficient information on the occurrences in drinking water of the following expected parameters:

- disinfection by-products
- disease-causing microorganisms (pathogens) including *Cryptosporidium*
- collection of engineering data (physical descriptions of treatment facilities) necessary to evaluate how PWSs presently control such contaminants

USEPA, with the reporting requirements, stated its intention to notify PWSs that they are subject to this rule. After receiving a Notice of Applicability from USEPA, the PWSs "must respond within 35 days of receipt."

The most important change in the Final Rule is the exemption from any ICR requirements of the PWSs serving from 10,000 to 99,999 people and using surface water.

One important concern developed during the period between the "proposed" and the "final" rule was the reliability of laboratory work in quantification of pathogens like *Cryptosporidium*. USEPA decided in the final rule to "establish stringent laboratory approval criteria to assure adequate quality analysis."

USEPA will conduct microbiological surveys' occurrence data to correlate the data collected from the ICR for large PWSs (over 100,000 people) with the ones collected from the USEPA's survey of medium and small PWSs.

## PWSs REQUIREMENTS

**1. Sampling Plans**  PWSs must submit sampling Plans to USEPA for review and approval no later than **eight weeks** after receiving sampling software and instructions from USEPA. Once USEPA approves the sampling plans, the PWS **must begin monitoring the following month**.

**2. Monitoring**  Even if the sampling plan has not as yet been approved by USEPA, the **PWS must begin treatment study applicability monitoring (i.e., TOC monitoring) no later than three months after the publication of the Final Rule** in the *Federal Register*. The remaining monitoring requirements **must be initiated the month after receiving USEPA's approval of the Sampling Plan**.

**3. Data Reporting**  PWSs must submit **monthly monitoring reports** electronically on diskettes in the format that USEPA has prescribed and will be providing to affected PWSs.

**4. Treatment Studies**   A PWS must begin treatment studies **no later than 23 months** from the date that the ICR was published in the *Federal Register*. PWSs must submit a report of each completed treatment study **no later than 38 months** after the final rule appeared in the *Federal Register*.

## Other Changes of the Final Rule

An additional requirement in sampling microorganisms is the request to collect a full 18 months of virus sampling, with no provisions for reducing monitoring.

USEPA also decided to eliminate *Clostridium perfringens* and *coliphage* as potential microbial indicators. Better results may be obtained by a separate research project, according to USEPA's present opinion.

USEPA will allow **particle counting** in lieu of finished water *Cryptosporidium* and *Giardia* monitoring, and is looking at the possibility of collecting valuable data on particle counting "as a surrogate for *Cryptosporidium* and *Giardia* removal." *Particle Size Count* has been approved by the Standard Methods Committee in 1993 and is reported in the 1995 *Standard Methods* as "Proposed"; see Chapter 2.

# Appendixes

# Appendix A-I

## ABBREVIATIONS AND ACRONYMS

### ABBREVIATIONS

**at. no.**  atomic number

**at. wt.**  atomic weight

**avg.**  average

**BOD**  Biochemical Oxygen Demand

**boil pt.**  boiling point

**Bq**  Becquerel(s)

**Ci**  Curie

**COD**  Chemical Oxygen Demand

**CU**  Color Units (used to quantify determination of color in water)

**DBCP**  dibromochloropropane

**DO**  Dissolved Oxygen

**EDB**  ethylene dibromide

**ETU**  ethylene thiourea

**gal**  gallon = 3.785 liters

**HAA**  haloacetic acid

**hr**  hour(s)

**L**  liter (= 0.2642 gallons or 1.057 quarts)

**LC$_{50}$**  Lethal Concentration—50% kill

**LCLo**  Lowest published lethal concentration

**LD$_{50}$**  Lethal Dose—50% kill

**LMM**  linearized multistage model

**melt. pt.**  melting point

**MFL**  Million Fibers per Liter

**mg**  milligram(s) ($10^{-3}$ grams, $10^{-6}$ kilograms, $10^3$ micrograms)

**mg/L**  milligram per liter (for drinking water with specific density of 1.0) (this unit is equivalent to ppm)

**min.**  minute(s)

**ml**  milliliter(s) = $10^{-3}$ liter

**mq**  square meters

**mREM**  millirem

**μg/L**  micrograms(s) per liter (1/1,000 mg/L)

**μm**  micrometer

**nCi**  Nanocurie(s)

**NTU**  nephelometric turbidity unit

**PAH**  polynuclear aromatic hydrocarbon

**PCB**  polychlorinated biphenyl

**pCi**  Picocurie(s)

**ppb**  parts per billion

**ppm**  parts per million

**PVC**  Polyvinyl Chloride

**RBE**  Relative Biological Effectiveness

**RDL**  reliable detection limit

**REM**   Roentgen Equivalent Man (effective radiation dose)

**RFA**   regulatory flexibility analysis

**RFD**   reference dose

**RIA**   Regulatory Impact Analysis

**s**   second(s)

**sp. gr.**   specific gravity

**SS**   Suspended Solids

**SV**   Sievert

**TCU**   True Color Units (color of water when turbidity is removed)

**TDS**   Total Dissolved Solids

**THM**   Trihalomethane

**TOC**   Total Organic Carbon

**TON**   Threshold Odor Number

**TOX**   Total Organic Halogen

**TSS**   Total Suspended Solids

**UIC**   underground injection control

**URTH**   unreasonable risk to health

**UV**   ultraviolet light

## ACRONYMS

**Used by federal and state agencies and related literature in reference to agencies involved in the drinking water regulations and to parameters used in regulations.**

**AA**   Atomic Absorption

**AADI**   Adjusted ADI (acceptable daily intake) in mg/L

**ACS**   American Chemical Society

**ADI** Acceptable Daily Intake (no significant risk to humans)

**ANPRM** Advance Notice of Proposed Rulemaking (USEPA)

**APHA** American Public Health Association

**ASDWA** Association of State Drinking Water Administrators

**ASTM** American Society for Testing and Materials

**AWWA** American Water Works Association

**BAT** Best Available Technology

**BTGA** Best Technology Generally Available (effective, existing, reasonable water treatment)

**CAG** Carcinogen Assessment Group (research, studies, evaluation, extrapolation, etc., advisory board to USEPA)

**CAS** Chemical Abstracts

**CDC** Center for Disease Control

**CEQ** Council on Environmental Quality

**CFR** Code of Federal Regulations

**CFU** Colony Forming Units

**COE** Corps of Engineers (U.S. Dept. of Army)

**CPSC** Consumer Products Safety Commission

**CWA** Clean Water Act

**CWS** A public water system which serves at least 5 service connections year-round or regularly serves at least 25 year-round residents

**CWSS** Community Water Supply Survey (National survey organized by USEPA prior to issuance of new standards)

**DBP** disinfection by-product

**D/DBP** disinfectant/disinfection by-product

**DWEL**   Drinking Water Equivalent Level*

**DWPL**   Drinking Water Priority List

**DWS**   Drinking Water Standard

**EEC**   European Economic Community (issues drinking water standards to be used as guidelines for the European countries that are members). CEE (French) is the other official denomination.

**EMSL**   USEPA Environmental Monitoring and Support Laboratory (based in Cincinnati, OH)

**ESWTR**   Enhanced Surface Water Treatment Rule

**EU**   European Union = EEC

**FACA**   Federal Advisory Committee Act

**FAO**   United Nations Food and Agricultural Organization (based in Rome, Italy)

**FDA**   U.S. Food and Drug Administration

**FIFRA**   Federal Insecticide, Fungicide, and Rodenticide Act

**FRC**   Federal Radiation Council

**GAC**   Granular Activated Carbon

**GAO**   General Accounting Office

**GC**   Gas Chromatography

**GC/MS**   Gas Chromatography/Mass Spectrometry

**GPO**   Government Printing Office

**GWDR**   Ground Water Disinfection Rule

**GWSS**   Ground Water Supply Survey (organized by USEPA in 1980)

---

*Defined as a minimum concentration of a substance that can be measured and reported with 99% confidence that the true value is greater than zero.

**HPC** Heterotrophic Plate Count

**IARC** International Agency for Research on Cancer

**ICR** Information Collection Request

**ICR** Information Collection Rule

**ICRP** International Commission of Radiological Protection

**IOC** Inorganic Chemical

**IOM** Institute of Medicine

**IRIS** integrated risk information system

**LOAEL** Lowest-Observed-Adverse-Effect-Level

**LOQ** Limit of Quantitation as defined by the American Chemical Society, basically equivalent to PQL, but LOQ is specific to an individual laboratory, while PQL involves available laboratories

**MCL** Maximum Contaminant Level (enforceable standards)

**MCLG** Maximum Contaminant Level Goal; nonenforceable standards to be used as a guide

**MDL** Method Detection Limit

**MF** Membrane Filter Technique for members of the coliform group

**MPN** Multiple-Tube Fermentation Technique to determine the coliform group (*Standard Methods* for microbiological parameters); Most Probable Number

**MRDL** maximum residual disinfectant level

**M/R** monitoring and reporting

**MS** Mass Spectrometry

**MTF** multiple-tube fermentation

**MTP** maximum trihalomethane potential

**MUT**   in vivo Mutagenic Effects

**NAE**   National Academy of Engineering

**NAS**   National Academy of Sciences (Washington, DC.)

**NAS, 1977**   The publication by NAS, *Drinking Water and Health,* prepared by the Safe Drinking Water Committee, issued in 1977, supported by USEPA and NAS

**NAS, 1980**   The previously listed publication, but Volume 3 (also Volume 2 was issued in Sept. 1980)

**NAS, 1982**   Volume 4 of *Drinking Water and Health* prepared for USEPA by the National Research Council, National Academy Press, Washington, DC., 1982

**NAS, 1983**   Volume 5 of *Drinking Water and Health* (see NAS, 1982)

**NBS**   National Bureau of Standards

**NCRP**   National Committee on Radiation Protection and Measurement

**NDWAC**   National Drinking Water Advisory Council

**NEPA**   National Environmental Policy Act

**NGWA**   National Groundwater Association, previously known as the National Water Well Association

**NEO**   Neoplastic Effects

**NIPDWR**   National Interim Primary Drinking Water Regulations

**NOAA**   National Oceanic and Atmospheric Administration

**NOAEL**   No-Observed-Adverse-Effect-Level

**NOMS**   National Organic Monitoring Survey (organized by USEPA in 1976–1977 prior to finalized standards)

**NORS**   National Organics Reconnaissance Survey (conducted by USEPA in 1975 prior to finalized standards)

**NPDWR**   National Primary Drinking Water Regulations (USEPA)

**NPRM**   Notice of Proposed Rule Making (USEPA)

**NRA**  Negotiated Rulemaking Act

**NRC**  The National Research Council (Safe Drinking Water Committee for NAS)

**NRDC**  National Resources Defense Council

**NSDWR**  National Secondary Drinking Water Regulations (USEPA)

**NSP**  National Screening Programs for Organics in Drinking Water (1977–1981, USEPA)

**NTIS**  National Technical Information Service

**NTNCWS**  Non-Transient, Non-Community Water System—a public water system that is not a community water system and that regularly serves at least 25 of the same persons over 6 months per year

**NTP**  National Toxicology Program

**NTU**  Nephelometric Turbidity Unit(s)

**ODW**  USEPA Office of Drinking Water

**OGWDW**  Office of Ground Water and Drinking Water

**OSHA**  U.S. Occupational Health and Safety Administration (or Act)

**P-A**  presence-absence (of coliforms)

**PAC**  powdered activated carbon

**PDWR**  Primary Drinking Water Regulation

**POE**  Point-of-Entry treatment device (a treatment device applied to the drinking water entering a house or building for the purpose of reducing contaminants in the drinking water distributed throughout the house or building)

**POU**  Point-of-Use treatment device (a treatment device applied to a single tap used for the purpose of reducing contaminants in drinking water at that one tap)

**PQL**  Practical Quantification Level for present analytic methodology (lowest level that can be reliably achieved within specified limits of precision and accuracy during routine laboratory operating conditions)

**PTA** Packed-Tower Aeration

**PWS** Public Water System

**PWSS** Public Water System Supervision

**RCRA** Resource Conservation and Recovery Act

**RIA** Regulatory Impact Analysis

**RMCL** Recommended Maximum Contaminant Level (guidelines preparatory to issuance of standards by USEPA)

**RQL** reliable quantitation limit

**RSC** Relative Source Contribution

**SAB** USEPA Science Advisory Board

**SDWA** Safe Drinking Water Act (42 USC 300f), also called "the Act"—the federal law that guides USEPA in setting drinking water standards (1974, and revised in 1977, 1979, 1980, 1986).

**SDWR** Secondary Drinking Water Regulation

**SMCL** Secondary Maximum Contaminant Level

**SG** Surgeon General—the chief medical officer in the U.S. Public Health Service (the highest public health authority at the federal level)

**SNARL** Chronic Suggested No-Adverse-Response-Level (used by NAS in preparation of data advising drinking water standards)

**SNC** significant noncomplier (re violations of USEPA Final Rules)

**SOCs** Synthetic Organic Chemicals (of specific interest for potential contamination of drinking water)

**SRF** state revolving fund

**SWDA** Solid Waste Disposal Act

**SWTR** Surface Water Treatment Rule

**TAW** AWWA Radionuclides Technical Advisory Workgroup

**TCR** Total Coliform Rule

**TDB** Toxicology Data Bank

**TER** Teratogenic Effects

**TLV** Threshold Limit Value (particularly used in determination of ambient air environmental contribution of contaminant)

**TNCWS** transient, noncommunity water system

**TSCA** Toxic Substances Control Act

**TTHMs** total trihalomethanes

**UF** Uncertainty Factor (safety factor) in toxicological evaluation to reach AADI

**UIC** Underground Inspection Control Program

**UNSCEAR** United Nations Scientific Committee on the Effects of Atomic Radiation

**UNS** toxic effects unspecified in source

**URTH** Unreasonable Risk to Health

**USEPA** U.S. Environmental Protection Agency

**USGS** U.S. Geological Survey

**USPHS** U.S. Public Health Service

**VOCs** Volatile Organic Chemicals (of specific interest as potential contaminants of drinking water)

**WEF** Water Environment Federation, previously the WPCF (Water Pollution Control Federation)

**WHO** World Health Organization (with headquarters in Geneva, Switzerland)

**WITAF** Water Industry Technical Action Fund

**WRDA** Water Resources Development Act

**WSRA** Wild and Scenic Rivers Act

# Appendix A-II

---

## WORLD HEALTH ORGANIZATION GUIDELINES

---

## INTRODUCTION

The World Health Organization (WHO) issued the first **guidelines** in 1984, with the intention to provide not only the values for the drinking water guidelines themselves, but also the explanations to justify them. These explanations, which fill three volumes, supply sufficient basic information for all countries to issue national standards. It is expected that the standards will sometimes be different in developing or third world countries.

In 1988 WHO initiated the preparation of a second edition to present modified standards. Over 200 experts from nearly 40 different countries participated in proposals meetings, and review groups. In 1993 WHO published the second edition* with the following tables, reprinted here by courtesy of the WHO.†

---

*WHO publications are available in the United States from the WHO Publications Center, U.S.A., 49 Sheridan Avenue, Albany, NY 12210.

†The World Health Organization is "a specialized agency of the United Nations with primary responsibility for international health matters and public health. Through this organization, which was created in 1948, the health professions of some 185 countries exchange their knowledge and experience with the aim of making possible the attainment by all citizens of the world by the year 2000 of a level of health that will permit them to lead a socially and economically productive life."

## WHO GUIDELINE VALUES

### Table AII-1.  Bacteriological Quality of Drinking Water[a]

| Organisms | Guideline value |
|---|---|
| *All water intended for drinking* | |
| *E. coli* or thermotolerant coliform bacteria[b,c] | Must not be detectable in any 100-ml sample |
| *Treated water entering the distribution system* | |
| *E. coli* or thermotolerant coliform bacteria[b] | Must not be detectable in any 100-ml sample |
| Total coliform bacteria | Must not be detectable in any 100-ml sample |
| *Treated water in the distribution system* | |
| *E. coli* or thermotolerant coliform bacteria[b] | Must not be detectable in any 100-ml sample |
| Total coliform bacteria | Must not be detectable in any 100-ml sample. In the case of large supplies, where sufficient samples are examined, must not be present in 95% of samples taken throughout any 12-month period |

[a]Immediate investigative action must be taken if either *E. coli* or total coliform bacteria is detected. The minimum action in the case of total coliform bacteria is repeat sampling; if these bacteria are detected in the repeat sample, the cause must be determined by immediate further investigation.

[b]Although *E. coli* is the more precise indicator of faecal pollution, the count of thermotolerant coliform bacteria is an acceptable alternative. If necessary, proper confirmatory tests must be carried out. Total coliform bacteria are not acceptable indicators of the sanitary quality of rural water supplies, particularly in tropical areas where many bacteria of no sanitary significance occur in almost all untreated supplies.

[c]It is recognized that, in the great majority of rural water supplies in developing countries, faecal contamination is widespread. Under these conditions, the national surveillance agency should set medium-term targets for the progressive improvement of water supplies, as recommended in Volume 3 of *Guidelines for drinking-water quality*.

## Table AII-2.   Chemicals of Health Significance in Drinking-Water.

### A. INORGANIC CONSTITUENTS

| W.H.O. (1993) | Guideline value (mg/liter) | Remarks |
|---|---|---|
| antimony | 0.005 (P)[a] | |
| arsenic | 0.01[b](P) | for excess skin cancer risk of $6 \times 10^{-4}$ |
| barium | 0.7 | |
| beryllium | | NAD[c] |
| boron | 0.3 | |
| cadmium | 0.003 | |
| chromium | 0.05 (P) | |
| copper | 2 (P) | ATO[d] |
| cyanide | 0.07 | |
| fluoride | 1.5 | Climatic conditions, volume of water consumed, and intake from other sources should be considered when setting national standards |
| lead | 0.01 | It is recognized that not all water will meet the guideline value immediately; meanwhile, all other recommended measures to reduce the total exposure to lead should be implemented |
| manganese | 0.5 (P) | ATO |
| mercury (total) | 0.001 | |
| molybdenum | 0.07 | |
| nickel | 0.02 | |
| nitrate (as $NO_3^-$) | 50 | The sum of the ratio of the concentration of each to its respective guideline value should not exceed 1 |
| nitrite (as $NO_2^-$) | 3 (P) | |
| selenium | 0.01 | |
| uranium | | NAD |

### B. ORGANIC CONSTITUENTS

| W.H.O. (1993) | Guideline value (μg/liter) | Remarks |
|---|---|---|
| *Chlorinated alkanes* | | |
| carbon tetrachloride | 2 | |
| dichloromethane | 20 | |
| 1,1-dichloroethane | | NAD |
| 1,2-dichloroethane | 30[b] | for excess risk of $10^{-5}$ |
| 1,1,1-trichloroethane | 2000 (P) | |
| *Chlorinated ethenes* | | |
| vinyl chloride | 5[b] | for excess risk of $10^{-5}$ |
| 1,1-dichloroethene | 30 | |
| 1,2-dichloroethene | 50 | |
| trichloroethene | 70 (P) | |
| tetrachloroethene | 40 | |
| *Aromatic hydrocarbons* | | |
| benzene | 10[b] | for excess risk of $10^{-5}$ |

## Table AII-2. (*Continued*)

| W.H.O. (1993) | Guideline value (µg/liter) | Remarks |
|---|---|---|
| toluene | 700 | ATO |
| xylenes | 500 | ATO |
| ethylbenzene | 300 | ATO |
| styrene | 20 | ATO |
| benzo[*a*]pyrene | 0.7[b] | for excess risk of $10^{-5}$ |
| *Chlorinated benzenes* | | |
| monochlorobenzene | 300 | ATO |
| 1,2-dichlorobenzene | 1000 | ATO |
| 1,3-dichlorobenzene | | NAD |
| 1,4-dichlorobenzene | 300 | ATO |
| trichlorobenzenes (total) | 20 | ATO |
| *Miscellaneous* | | |
| di(2-ethylhexyl)adipate | 80 | |
| di(2-ethylhexyl)phthalate | 8 | |
| acrylamide | 0.5[b] | for excess risk of $10^{-5}$ |
| epichlorohydrin | 0.4 (P) | |
| hexachlorobutadiene | 0.6 | |
| edetic acid (EDTA) | 200 (P) | |
| nitrilotriacetic acid | 200 | |
| dialkyltins | | NAD |
| tributyltin oxide | 2 | |

## C. PESTICIDES

| W.H.O. (1993) | Guideline value (µg/liter) | Remarks |
|---|---|---|
| alachlor | 20[b] | for excess risk of $10^{-5}$ |
| aldicarb | 10 | |
| aldrin/dieldrin | 0.03 | |
| atrazine | 2 | |
| bentazone | 30 | |
| carbofuran | 5 | |
| chlordane | 0.2 | |
| chlorotoluron | 30 | |
| DDT | 2 | |
| 1,2-dibromo-3-chloropropane | 1[b] | for excess risk of $10^{-5}$ |
| 2,4-D | 30 | |
| 1,2-dichloropropane | 20 (P) | |
| 1,3-dichloropropane | | NAD |
| 1,3-dichloropropene | 20[b] | for excess risk of $10^5$ |
| ethylene dibromide | | NAD |
| heptachlor and heptachlor epoxide | 0.03 | |
| hexachlorobenzene | 1[b] | for excess risk of $10^{-5}$ |
| isoproturon | 9 | |
| lindane | 2 | |

## Table AII-2. (*Continued*)

| W.H.O. (1993) | Guideline value (µg/liter) | Remarks |
|---|---|---|
| MCPA | 2 | |
| methoxychlor | 20 | |
| metolachlor | 10 | |
| molinate | 6 | |
| pendimethalin | 20 | |
| pentachlorophenol | 9 (P) | |
| permethrin | 20 | |
| propanil | 20 | |
| pyridate | 100 | |
| simazine | 2 | |
| trifluralin | 20 | |
| chlorophenoxy herbicides other than 2,4-D and MCPA | | |
| 2,4-DB | 90 | |
| dichlorprop | 100 | |
| fenoprop | 9 | |
| MCPB | | NAD |
| mecoprop | 10 | |
| 2,4,5-T | 9 | |

## D. DISINFECTANTS AND DISINFECTANT BY-PRODUCTS

| Disinfectants | Guideline value (mg/liter) | Remarks W.H.O. (1993) |
|---|---|---|
| monochloramine | 3 | |
| di- and trichloramine | | NAD |
| chlorine | 5 | ATO. For effective disinfection there should be a residual concentration of free chlorine of ≥0.5 mg/litre after at least 30 minutes contact time at pH <8.0 |
| chlorine dioxide | | A guideline value has not been established because of the rapid breakdown of chlorine dioxide and because the chlorite guideline value is adequately protective for potential toxicity from chlorine dioxide |
| iodine | | NAD |

| Disinfectant by-products | Guideline value (µg/liter) | Remarks W.H.O. (1993) |
|---|---|---|
| bromate | 25[b] (P) | for $7 \times 10^{-5}$ excess risk |
| chlorate | | NAD |
| chlorite | 200 (P) | |
| chlorophenols | | |
| 2-chlorophenol | | NAD |

## Table AII-2. (*Continued*)

| Disinfectant by-products | Guideline value (μg/liter) | Remarks |
|---|---|---|
| 2,4-dichlorophenol | | NAD |
| 2,4,6-trichlorophenol | 200[b] | for excess risk of $10^{-5}$, ATO |
| formaldehyde | 900 | |
| MX | | NAD |
| trihalomethanes | | The sum of the ratio of the concentration of each to its respective guideline value should not exceed 1 |
| bromoform | 100 | |
| dibromochloromethane | 100 | |
| bromodichloromethane | 60[b] | for excess risk of $10^{-5}$ |
| chloroform | 200[b] | for excess risk of $10^{-5}$ |
| chlorinated acetic acids | | |
| monochloroacetic acid | | NAD |
| dichloroacetic acid | 50 (P) | |
| trichloroacetic acid | 100 (P) | |
| chloral hydrate (trichloroacetaldehyde) | 10 (P) | |
| chloroacetone | | NAD |
| halogenated acetonitriles | | |
| dichloroacetonitrile | 90 (P) | |
| dibromoacetonitrile | 100 (P) | |
| bromochloroacetonitrile | | NAD |
| trichloroacetonitrile | 1 (P) | |
| cyanogen chloride (as CN) | 70 | |
| chloropicrin | | NAD |

[a](P)—Provisional guideline value. This term is used for constituents for which there is some evidence of a potential hazard but where the available information on health effects is limited; or where an uncertainty factor greater than 1000 has been used in the derivation of the tolerable daily intake (TDI). Provisional guideline values are also recommended: (1) for substances for which the calculated guideline value would be below the practical quantification level, or below the level that can be achieved through practical treatment methods; or (2) where disinfection is likely to result in the guideline value being exceeded.

[b]For substances that are considered to be carcinogenic, the guideline value is the concentration in drinking water associated with an excess lifetime cancer risk of $10^{-5}$ (one additional cancer per 100,000 of the population ingesting drinking water containing the substance at the guideline value for 70 years). Concentrations associated with estimated excess lifetime cancer risks of $10^{-4}$ and $10^{-6}$ can be calculated by multiplying and dividing, respectively, the guideline value by 10.

In cases in which the concentration associated with an excess lifetime cancer risk of $10^{-5}$ is not feasible as a result of inadequate analytical or treatment technology, a provisional guideline value is recommended at a practicable level and the estimated associated excess lifetime cancer risk presented.

It should be emphasized that the guideline values for carcinogenic substances have been computed from hypothetical mathematical models that cannot be verified experimentally and that the values should be interpreted differently than TDI-based values because of the lack of precision of the models. At best, these values must be regarded as rough estimates of cancer risk. However, the models used are conservative and probably err on the side of caution. Moderate short-term exposure to levels exceeding the guideline value for carcinogens does not significantly affect the risk.

[c]NAD—No adequate data to permit recommendation of a health-based guideline value.

[d]ATO—Concentrations of the substance at or below the health-based guideline value may affect the appearance, taste, or odor of the water.

## Table AII-3. Chemicals Not of Health Significance at Concentrations Normally Found in Drinking Water.

| Chemical | Remarks   W.H.O. (1993) |
|---|---|
| asbestos | U |
| silver | U |
| tin | U |

U—It is unnecessary to recommend a health-based guideline value for these compounds because they are not hazardous to human health at concentrations normally found in drinking water.

## Table AII-4. Radioactive Constituents of Drinking Water.

| | Screening value (Bq/litre) | Remarks   W.H.O. (1993) |
|---|---|---|
| gross alpha activity | 0.1 | If a screening value is exceeded, more detailed |
| gross beta activity | 1 | radionuclide analysis is necessary. Higher values do not necessarily imply that the water is unsuitable for human consumption. |

## Table AII-5. Substances and Parameters in Drinking Water That May Give Rise to Complaints from Consumers.

| W.H.O. (1993) | Levels likely to give rise to consumer complaints[a] | Reasons for consumer complaints |
|---|---|---|
| *Physical parameters* | | |
| color | 15 TCU[b] | appearance |
| taste and odor | — | should be acceptable |
| temperature | — | should be acceptable |
| turbidity | 5 NTU[c] | appearance; for effective terminal disinfection, median turbidity ≤1NTU, single sample ≤5NTU |
| *Inorganic constituents* | | |
| aluminium | 0.2 mg/l | depositions, discoloration |
| ammonia | 1.5 mg/l | odor and taste |
| chloride | 250 mg/l | taste, corrosion |
| copper | 1 mg/l | staining of laundry and sanitary ware (health-based provisional guideline value 2 mg/litre) |
| hardness | — | high hardness: scale deposition, scum formation; low hardness: possible corrosion |
| hydrogen sulfide | 0.05 mg/l | odor and taste |

## Table AII-5. (*Continued*)

| W.H.O. (1993) | Levels likely to give rise to consumer complaints[a] | Reasons for consumer complaints |
|---|---|---|
| iron | 0.3 mg/l | staining of laundry and sanitary ware |
| manganese | 0.1 mg/l | staining of laundry and sanitary ware (health-based provisional guideline value 0.5 mg/litre) |
| dissolved oxygen | — | indirect effects |
| pH | — | low pH: corrosion |
| | | high pH: taste, soapy feel preferably <8.0 for effective disinfection with chlorine |
| sodium | 200 mg/l | taste |
| sulfate | 250 mg/l | taste, corrosion |
| total dissolved solids | 1000 mg/l | taste |
| zinc | 3 mg/l | appearance, taste |
| *Organic constituents* | | |
| toluene | 24–170 µg/l | odor, taste (health-based guideline value 700 µg/l) |
| xylene | 20–1800 µg/l | odor, taste (health-based guideline value 500 µg/l) |
| ethylbenzene | 2–200 µg/l | odor, taste (health-based guideline value 300 µg/l) |
| styrene | 4–2600 µg/l | odor, taste (health-based guideline value 20 µg/l) |
| monochlorobenzene | 10–120 µg/l | odor, taste (health-based guideline value 300 µg/l) |
| 1,2-dichlorobenzene | 1–10 µg/l | odor, taste (health-based guideline value 1000 µg/l) |
| 1,4-dichlorobenzene | 0.3–30 µg/l | odor, taste (health-based guideline value 300 µg/l) |
| trichlorobenzenes (total) | 5–50 µg/l | odor, taste (health-based guideline value 20 µg/l) |
| synthetic detergents | — | foaming, taste, odor |
| *Disinfectants and disinfectant by-products* | | |
| chlorine | 600–1000 µg/l | taste and odor (health-based guideline value 5 mg/l) |
| chlorophenols | | |
| 2-chlorophenol | 0.1–10 µg/l | taste, odor |
| 2,4-dichlorophenol | 0.3–40 µg/l | taste, odor |
| 2,4,6-trichlorophenol | 2–300 µg/l | taste, odor (health-based guideline value 200 µg/l) |

[a]The levels indicated are not precise numbers. Problems may occur at lower or higher values according to local circumstances. A range of taste and odor threshold concentrations is given for organic constituents.
[b]TCU, time color unit.
[c]NTU, nephelometric turbidity unit.

# Appendix A-III

## EUROPEAN DRINKING WATER DIRECTIVES

The European Community (also known as EEC or EU) issued these directives in 1980 to be used as a base for each country (12 member nations at that time; now 15) to develop standards that must be as restrictive or stricter than the following standards to be effective as of 1985; the MCL is here called MAC (maximum admissible concentration), and MCLG is presented as GL (guide level).

Table AIII-1. EC Drinking Water Directive-List of Parameters.

| Parameter | Unit of Measurement | Guide Level (GL) | Maximum Admissible Concentration (MAC) | Comments |
|---|---|---|---|---|
| **Organoleptic** | | | | |
| Color | mg/L Pt-Co scale | 1 | 20 | |
| Turbidity | mg $SiO_2$/L | 1 | 10 | |
| Odor | Dilution number | 0 | 2 at 12°C<br>3 at 25°C | To be related to the taste tests |
| Taste | Dilution number | 0 | 2 at 12°C<br>3 at 25°C | To be related to the odor tests |
| | | | | |
| **Physicochemical** | | | | |
| Temperature | °C | 12 | 25 | |
| Hydrogen ion | pH unit | 6.5–8.5 | | Maximum admissible value 9.5 |
| Conductivity | µS/cm at 20°C | 400 | | |
| Chlorides | mg Cl/L | 25 | | |
| Sulfates | mg $SO_4$/L | 25 | 250 | |
| Silica | mg $SiO_2$ | | | |
| Calcium | mg Ca/L | 100 | | |
| Magnesium | mg Mg/L | 30 | 50 | |
| Sodium | mg Na/L | 20 | | |
| Potassium | mg K/L | 10 | 12 | |
| Aluminum | mg Al/L | 0.05 | 0.2 | |
| Total hardness | | | | See lower portion of table |
| Dry residue | mg/L after drying at 180°C | | 1,500 | |
| Dissolved oxygen | Percent $O_2$ saturation | | | |
| Free carbon dioxide | mg $CO_2$/L | | | |
| | | | | |
| **Substances undesirable in excessive amounts** | | | | |
| Nitrates | mg $NO_3$/L | 25 | 50 | |

| Parameter | Unit | | | Observations |
|---|---|---|---|---|
| Nitrites | mg NO$_2$/L | | 0.1 | |
| Ammonium | mg NH$_4$/L | 0.05 | 0.5 | |
| Kjeldahl nitrogen | mg N/L | | 1 | Excluding N in nitrogen NO$_2$ and NO$_3$ |
| Potassium permanganate oxidizability | mg O$_2$/L | 2 | 5 | |
| Total organic carbon (TOC) | mg C/L | | | Measured when heated in acid medium The reason for any increase in the usual concentration must be investigated. |
| Hydrogen sulfide | μgS/L | | Undetectable organoleptically | |
| Substances extractable in chloroform | mg/L dry residue | 0.1 | | |
| Dissolved or emulsified hydrocarbons, mineral oils | μg/L | | 10 | After extraction by petroleum ether |
| Phenols (phenol index) | μg C$_6$H$_5$OH/L | | 0.5 | Excluding natural phenols that do not not react with chlorine |
| Boron | μg B/L | 1,000 | | |
| Surfactants (reacting with methylene blue) | μg/L (lauryl sulfate) | | 200 | |
| Organochlorine compounds not covered by pesticides and related products | μg/L | 1 | | Haloform concentrations must be as low as possible. |
| Iron | μg Fe/L | 50 | 200 | |
| Manganese | μg Mn/L | 20 | 50 | |
| Copper | μg Cu/L | 100 / 3,000 | At plant / At supply point | Above 3,000 μg/L, astringent taste, discoloration, and corrosion may occur. |
| Zinc | μg Zn/L | 100 / 5,000 | At plant / At supply point | Above 5,000 μg/L, astringent taste, discoloration, and corrosion may occur. |
| Phosphorus | μg P$_2$O$_5$/L | 400 | 5,000 | |

*Reprinted from the American Water Works Association *Journal*, Vol. 83, No. 6 (June 1991), by permission.
© Copyright 1991, American Water Works Association.

## Table AIII-1. (Continued)

| Parameter | Unit of Measurement | Guide Level (GL) | Maximum Admissible Concentration (MAC) | Comments |
|---|---|---|---|---|
| Fluoride | µg F/L 8–12°C 25–30°C | | 1,500 700 | MAC varies according to average temperature in geographical area concerned. |
| Cobalt | µg Co/L | None | | |
| Suspended solids | | | | |
| Residual chlorine | µg Cl/L | | | |
| Barium | µg Ba/L | 100 | | |
| Silver | µg Ag/L | 10 | | If silver is used nonsystematically to process the water, a MAC value of 80 µg/L may be authorized. |
| **Toxic substances** | | | | |
| Arsenic | µg As/L | | 50 | |
| Beryllium | µg Be/L | | | |
| Cadmium | µg Cd/L | | 5 | |
| Cyanides | µg CN/L | | 50 | |
| Chromium | µg Cr/L | | 50 | |
| Mercury | µg Hg/L | | 1 | |
| Nickel | µg Ni/L | | 50 | |
| Lead | µg Pb/L | | 50 (in running water) | |
| Antimony | µg Sb/L | | 10 | |
| Selenium | µg Se/L | | 10 | |
| Vanadium | µg V/L | | | |
| Pesticides and related products | µg/L | | | Pesticides and related products means insecticides, herbicides, fungicides, PCBs, and PCTs. |
| Separate | | | 0.1 | |
| Total | | | 0.5 | |
| Polycyclic aromatic hydrocarbons | µg/L | | 0.2 | |

## Table AIII-2.

| Parameter | Volume of Sample mL | Guide Level (GL) | Maximum Admissible Concentration (MAC) | |
|---|---|---|---|---|
| | | | Membrane Filter Method | Multiple Tube Method (MPN) |
| **Microbiological** | | | | |
| Total coliforms* | 100 | | 0 | MPN < 1 |
| Fecal coliforms | 100 | | 0 | MPN < 1 |
| Fecal streptococci | 100 | | 0 | MPN < 1 |
| Sulfite-reducing *Clostridia* | 20 | | | MPN ≤ 1 |
| Total bacteria counts for water supplied for human consumption | | | | |
| 37°C | 1 | | 10 | |
| 22°C | 1 | | 100 | |
| Total bacteria counts for water in closed containers | | | | |
| 37°C | 1 | | 5 | |
| 22°C | 1 | | 20 | |

*Provided a sufficient number of samples is examined (95 percent consistent results).

## Table AIII-3.

| Parameter | Unit of Measurement | Minimum Required Concentration | Comments |
|---|---|---|---|
| **Softened water intended for human consumption*** | | | |
| Total hardness | mg Ca/L | 60 | Calcium or equivalent cations |
| Hydrogen ion | pH | | The water should not be aggressive. |
| Alkalinity | mg HCO3/L | 30 | The water should not be aggressive. |
| Dissolved oxygen | | | The water should not be aggressive. |

*If owing to its excessive natural hardness, the water is softened before being supplied for consumption, its sodium content may, in exceptional cases, be higher than the values given in the MAC column. An effort must be made to keep the sodium content as low as possible, and the essential requirements for the protection of public health may not be disregarded.

# Appendix-A-IV

## LABORATORY REPORTS

City of Poughkeepsie Water Purification Plant Water Quality Data Based on Raw Water, 1992

| Parameter | Units | Freq | NYS STD | Yearly Mean | Yearly Range Max | Yearly Range Min | 1st Q | 2nd Q | 3rd Q | 4th Q |
|---|---|---|---|---|---|---|---|---|---|---|
| TOTAL ALKALINITY | mg/l | 1x/W | *1 | 60.4 | 79.8 | 44.9 | 64.3 | 52.6 | 63.8 | 60.9 |
| ALUMINUM | mg/l | 1x/M | *1 | 0.33 | 0.63 | 0.13 | 0.28 | 0.63 | 0.26 | 0.13 |
| ARSENIC | µg/l | 1x/Y | *1 | <5 | <5 | <5 | <5 | <5 | <5 | <5 |
| ASBESTOS | MF/L | 1x/Y | *1 | *3 | *3 | *3 | *2 | *2 | 0 | *2 |
| BARIUM | mg/l | 1x/Y | *1 | <0.05 | *3 | *3 | *2 | *2 | <0.05 | *2 |
| BERYLLIUM | mg/l | 1x/Y | *1 | <0.005 | *3 | *3 | *2 | *2 | <0.005 | *2 |
| BORON | mg/l | 1x/Y | *1 | <0.05 | *3 | *3 | *2 | *2 | <0.05 | *2 |
| CADMIUM | mg/l | 1x/Y | *1 | <0.005 | *3 | *3 | *2 | *2 | <0.005 | *2 |
| CALCIUM | mg/l | 2x/W | *1 | 27.3 | 34.9 | 17 | 28.4 | 28.5 | 28.4 | 27.6 |
| CHLORIDE | mg/l | 2x/W | *1 | 14.8 | 22.9 | 8.0 | 18.0 | 12.3 | 14.4 | 14.6 |
| CHROMIUM, TOTAL | mg/l | 1x/Y | *1 | <0.01 | *3 | *3 | *2 | *2 | <0.01 | *2 |
| COLOR | CU | 1x/W | *1 | 10.4 | 17.4 | 1.4 | 9.3 | 11.0 | 9.2 | 12.0 |
| COPPER | mg/l | 1x/Y | *1 | 0.02 | *3 | *3 | *2 | *2 | 0.02 | *2 |
| COD | mg/l | 1x/Q | *1 | 11.00 | 24 | <4 | 8 | 12 | 24 | <4 |
| CYANIDE | mg/l | 1x/Q | *1 | <0.01 | <0.01 | <0.01 | <0.01 | <0.01 | <0.01 | <0.01 |
| DISSOLVED OXYGEN | mg/l | 1x/Y | *1 | 6.2 | *3 | *3 | *2 | *2 | 6.2 | *2 |
| TOTAL HARDNESS | mg/l | 2x/W | *1 | 88.5 | 119.5 | 62.6 | 95.1 | 80.5 | 90.0 | 88.6 |
| IRON | mg/l | 1x/Y | *1 | 0.29 | *3 | *3 | *2 | *2 | 0.29 | *2 |
| LEAD | mg/l | 1x/M | *1 | <0.001 | <0.001 | <0.001 | <0.001 | <0.001 | <0.001 | <0.001 |

| Parameter | Unit | Freq | | | | | | | | |
|---|---|---|---|---|---|---|---|---|---|---|
| LITHIUM | mg/l | 1x/Y | *1 | <0.1 | *3 | *3 | *2 | *2 | <0.1 | *2 |
| DETERGENT | mg/l | 1x/Y | *1 | <0.1 | <0.1 | <0.1 | <0.1 | <0.1 | <0.1 | <0.1 |
| MAGNESIUM | mg/l | 1x/Y | *1 | 4.2 | 4.8 | 3.5 | 3.5 | 4 | 4.3 | 4.8 |
| MANGANESE | mg/l | 1x/Y | *1 | 0.07 | *3 | *3 | *2 | *2 | 0.07 | 0.07 |
| MERCURY | µg/L | 1x/Q | *1 | <0.4 | <0.4 | <0.4 | <0.4 | <0.4 | <0.4 | <0.4 |
| NICKEL | mg/l | 1x/Y | *1 | <0.04 | *3 | *3 | *2 | *2 | <0.04 | <0.04 |
| NH3 NITROGEN | mg/l | 1x/Q | *1 | 0.38 | 1.5 | <0.3 | 1.5 | <0.3 | <0.3 | <0.3 |
| NO2 NITROGEN | mg/l | 1x/Q | *1 | 0.02 | 0.032 | <0.01 | 0.032 | 0.013 | <0.01 | <0.01 |
| NO3 NITROGEN | mg/l | 1x/Q | *1 | 0.71 | 0.94 | 0.55 | 0.94 | 0.71 | 0.64 | 0.55 |
| pH | mg/l | 1x/D | *1 | 7.69 | 8.78 | 7.33 | 7.70 | 7.67 | 7.62 | 7.77 |
| PHENOL | mg/l | 1x/Y | *1 | <0.01 | *3 | *3 | *2 | *2 | <0.01 | *2 |
| PHOSPHATE | mg/l | 1x/Y | *1 | <0.1 | *3 | *3 | *2 | *2 | <0.1 | *2 |
| POTASSIUM | mg/l | 1x/Y | *1 | 1.2 | *3 | *3 | *2 | *2 | 1.2 | *2 |
| GROSS a | pci/l | 1x/Y | *1 | <3.8 | *3 | *3 | *2 | *2 | <3.8 | *2 |
| GROSS b | pci/l | 1x/Y | *1 | <4.0 | *3 | *3 | *2 | *2 | <4.0 | *2 |
| RADIUM-228 | pci/l | 1x/Y | *1 | <2.9 | *3 | *3 | *2 | *2 | <2.9 | *2 |
| SELENIUM | µg/l | 1x/Y | *1 | <5 | *3 | *3 | *2 | *2 | <5 | *2 |
| SILICA SiO2 | mg/l | 1x/Y | *1 | <0.5 | *3 | *3 | *2 | *2 | <0.5 | *2 |
| SILVER | mg/l | 1x/Y | *1 | <0.01 | *3 | *3 | *2 | *2 | <0.01 | *2 |
| SODIUM | mg/l | 1x/Y | *1 | 13.8 | 21.3 | 10.5 | 10.5 | 10.8 | 12.9 | 21.3 |
| TOT DISSOLD SOLS | mg/l | 1x/Q | *1 | 129 | 195 | 84 | 108 | 113 | 139 | 125 |
| SPEC CONDUCTANCE | umhos/cm | 2x/W | *1 | 194 | 271 | 119 | 218 | 182 | 189 | 187 |
| STRONTIUM | mg/l | 1x/Y | *1 | 0.15 | *3 | *3 | *2 | *2 | 0.15 | *2 |
| SULFATE | mg/l | 1x/Y | *1 | 19 | *3 | *3 | *2 | *2 | 19 | *2 |
| ODOR | odor uts | 1x/W | *1 | 2 | 5 | 1 | 2 | 2 | 3 | 2 |
| TEMPERATURE | deg C | 1x/D | *1 | 12.1 | 24.6 | 0.7 | 1.6 | 14.2 | 23.2 | 9.3 |
| TURBIDITY | NTU | 1x/D | *1 | 28.7 | 114 | 7 | 23.7 | 34.5 | 29.9 | 26.8 |
| ENDRIN | µg/l | 1x/Y | *1 | <0.1 | *3 | *3 | *2 | *2 | <0.1 | *2 |
| TOXAPHENE | µg/l | 1x/Y | *1 | <1 | *3 | *3 | *2 | *2 | <1 | *2 |
| 2,4 D | µg/l | 1x/Y | *1 | <0.05 | *3 | *3 | *2 | *2 | <0.05 | *2 |
| 2,4,5 -TP SILVEX | µg/l | 1x/Y | *1 | <0.05 | *3 | *3 | *2 | *2 | <0.05 | *2 |
| ALDRIN | µg/l | 1x/Y | *1 | <0.05 | *3 | *3 | *2 | *2 | <0.05 | *2 |

# City of Poughkeepsie Water Purification Plant Water Quality Data Based on Raw Water, 1992 (Continued)

| Parameter | Units | Freq | NYS STD | Yearly Mean | Yearly Range Max | Yearly Range Min | 1st Q | 2nd Q | 3rd Q | 4th Q |
|---|---|---|---|---|---|---|---|---|---|---|
| CCl4 | µg/l | 1x/Y | *1 | <0.5 | *3 | *3 | *2 | *2 | <0.5 | *2 |
| CHLORDANE | µg/l | 1x/Y | *1 | <0.5 | *3 | *3 | *2 | *2 | <0.5 | *2 |
| DDD | µg/l | 1x/Y | *1 | <0.1 | *3 | *3 | *2 | *2 | <0.1 | *2 |
| DDT | µg/l | 1x/Y | *1 | <0.1 | *3 | *3 | *2 | *2 | <0.1 | *2 |
| HEPTACHLOR | µg/l | 1x/Y | *1 | <0.05 | *3 | *3 | *2 | *2 | <0.05 | *2 |
| PCB'S Arochlor 1016 | µg/l | 1x/Q | *1 | <0.5 | <0.5 | <0.5 | <0.5 | <0.5 | <0.5 | <0.5 |
| PCB'S Arochlor 1221 | µg/l | 1x/Q | *1 | <0.5 | <0.5 | <0.5 | <0.5 | <0.5 | <0.5 | <0.5 |
| PCB'S Arochlor 1232 | µg/l | 1x/Q | *1 | <0.5 | <0.5 | <0.5 | <0.5 | <0.5 | <0.5 | <0.5 |
| PCB'S Arochlor 1242 | µg/l | 1x/Q | *1 | <0.5 | <0.5 | <0.5 | <0.5 | <0.5 | <0.5 | <0.5 |
| PCB'S Arochlor 1248 | µg/l | 1x/Q | *1 | <0.5 | <0.5 | <0.5 | <0.5 | <0.5 | <0.5 | <0.5 |
| PCB'S Arochlor 1254 | µg/l | 1x/Q | *1 | <1 | <1 | <1 | <1 | <1 | <1 | <1 |
| PCB'S Arochlor 1260 | µg/l | 1x/Q | *1 | <1 | <1 | <1 | <1 | <1 | <1 | <1 |
| TOLUENE | µg/l | 1x/Y | *1 | <0.5 | *3 | *3 | *2 | *2 | <0.5 | *2 |
| METHYLENE CHLORIDE | µg/l | 1x/Y | *1 | <0.5 | *3 | *3 | *2 | *2 | <0.5 | *2 |
| TRICHLORO ETHYLENE | µg/l | 1x/Y | *1 | <0.5 | *3 | *3 | *2 | *2 | <0.5 | *2 |
| STD PLT COUNT | #/ml | 1x/D | *1 | 1500 | 14000 | 40 | 1500 | 1200 | 1500 | 1700 |
| TOTAL COLIFORM | #/100 ml | 1x/D | *1 | 1400 | 51000 | <1 | 2600 | 970 | 1100 | 970 |

Courtesy of the City of Poughkeepsie, NY
Douglas Fairbanks, Jr.—Chief Operator
Robert A. Ruggiero—Laboratory Director
Key: Y = yearly; Q = quarterly; M = monthly; W = weekly; D = daily
*1 = no state std; 2 = not done; 3 = insuff data for comp; 4 = tested 1xy

City of Poughkeepsie Water Purification Plant Water Quality Data Based on Plant Effluent, 1992.

| Parameter | Units | Freq | NYS STD | Yearly Mean | Yearly Range | | Quarterly Data | | | |
|---|---|---|---|---|---|---|---|---|---|---|
| | | | | | Max | Min | 1st Q | 2nd Q | 3rd Q | 4th Q |
| TOTAL ALKALINITY | mg/l | 1x/W | *1 | 57.7 | 82.3 | 30.9 | 62.6 | 49.9 | 60.6 | 57.8 |
| ARSENIC | μg/l | 1x/Y | <50 | <5 | *3 | *3 | *2 | *2 | <5 | *2 |
| ASBESTOS | MF/L | 1x/Y | 7.1 | 0 | *3 | *3 | *2 | *2 | 0 | *2 |
| STD PLT COUNT | #/ml | 1x/D | *1 | <1 | <1 | <1 | <1 | <1 | <1 | <1 |
| BARIUM | mg/l | 1x/Y | 1 | <0.05 | *3 | *3 | *2 | *2 | <0.05 | *2 |
| BERYLLIUM | mg/l | 1x/Y | *1 | <0.005 | *3 | *3 | *2 | *2 | <0.005 | *2 |
| BORON | mg/l | 1x/Y | *1 | <0.05 | *3 | *3 | *2 | *2 | <0.05 | *2 |
| CADMIUM | μg/L | 1x/Y | <10 | <2 | *3 | *3 | *2 | *2 | <2 | *2 |
| CALCIUM as CaCO3 | mg/l | 2x/W | *1 | 31.9 | 40.3 | 19.5 | 34.5 | 28.1 | 32.4 | 32.9 |
| CHLORINE, FREE | mg/L | 8x/D | | 1.1 | 2.5 | 0.3 | 0.6 | 1.1 | 1.4 | 1.1 |
| CHLORIDE | mg/l | 2x/W | 250 | 18.0 | 27.8 | 11.1 | 21.4 | 15.0 | 18.1 | 17.5 |
| CHROMIUM, TOTAL | mg/l | 1xY | <0.05 | <0.01 | *3 | *3 | *2 | *2 | <0.01 | *2 |
| COLOR | CU | 1x/W | <15 | <0.6 | 1.3 | <0.6 | <0.6 | <0.6 | <0.6 | <0.6 |
| COPPER | mg/l | 1x/Y | <1.3 | <0.01 | *3 | *3 | *2 | *2 | <0.01 | *2 |
| CHEM OXY DEMAND | mg/l | 1x/Q | *1 | 12 | *3 | *3 | *2 | *2 | 12 | *2 |
| CYANIDE | mg/l | 1x/Q | *1 | <0.01 | *3 | *3 | *2 | *2 | <0.01 | *2 |
| DO | mg/l | 1x/Y | *1 | 8.5 | *3 | *3 | *2 | *2 | 8.5 | *2 |
| FLUORIDE | mg/l | 3x/D | 2.2 | 0.85 | 1.51 | 0.18 | 0.85 | 0.84 | 0.90 | 0.80 |
| TOTAL HARDNESS | mg/l | 2x/W | *1 | 106 | 132 | 76 | 112 | 96 | 108 | 107 |
| IRON | mg/l | 1x/Y | 0.3 | <0.03 | *3 | *3 | *2 | *2 | <0.03 | *2 |
| LEAD | mg/l | 1x/Y | <0.015 | <1 | <1 | <1 | <1 | <1 | <1 | <1 |
| LITHIUM | mg/l | 1x/Y | *1 | <0.1 | *3 | *3 | *2 | *2 | <0.1 | *2 |
| DETERGENT | mg/l | 1x/Y | *1 | <0.1 | *3 | *3 | *2 | *2 | <0.1 | *2 |
| MAGNESIUM | mg/l | 1x/Y | *1 | 4.2 | 4.7 | 3.5 | 3.5 | 3.8 | 4.6 | 4.7 |
| MANGANESE | mg/l | 1x/Y | 0.3 | <0.01 | *3 | *3 | *2 | *2 | <0.01 | *2 |
| MERCURY | mg/l | 1x/Q | <0.002 | <0.0004 | *3 | *3 | *2 | *2 | <0.0004 | *2 |

City of Poughkeepsie Water Purification Plant Water Quality Data Based on Plant Effluent, 1992 (Continued).

| Parameter | Units | Freq | NYS STD | Yearly Mean | Yearly Range | | Quarterly Data | | | |
|---|---|---|---|---|---|---|---|---|---|---|
| | | | | | Max | Min | 1st Q | 2nd Q | 3rd Q | 4th Q |
| NICKEL | mg/l | 1x/Y | <0.05 | <0.04 | *3 | *3 | *2 | *2 | <0.04 | *2 |
| NH3 NITROGEN | mg/l | 1x/Q | *1 | <0.3 | *3 | *3 | *2 | *2 | <0.3 | *2 |
| NO2 NITROGEN | mg/l | 1x/Q | 1 | <0.01 | *3 | *3 | *2 | *2 | <0.01 | *2 |
| NO3 NITROGEN | mg/l | 1x/Q | 10 | 0.5 | *3 | *3 | *2 | *2 | 0.5 | *2 |
| pH | mg/l | 1x/D | *1 | 8.05 | 8.92 | 6.87 | 8.07 | 8.05 | 8.02 | 8.05 |
| PHOSPHATE | mg/l | 1x/Y | *1 | <0.1 | *3 | *3 | *2 | *2 | <0.1 | *2 |
| POTASSIUM | mg/l | 1x/Y | *1 | 2 | *3 | *3 | *2 | *2 | 2 | *2 |
| GROSS a | pci/l | 1x/Y | 5 | <3.9 | *3 | *3 | *2 | *2 | <3.9 | *2 |
| GROSS b | pci/l | 1x/Y | 50 | <4.5 | *3 | *3 | *2 | *2 | <4.5 | *2 |
| RADIUM-228 | pci/l | 1x/Y | 5 | <2.9 | *3 | *3 | *2 | *2 | <2.9 | *2 |
| RADIUM-226 | pci/l | 1x/Y | 3 | <2.2 | *3 | *3 | *2 | *2 | <2.2 | *2 |
| SELENIUM | µg/l | 1x/Y | 10 | <2 | *3 | *3 | *2 | *2 | <2 | *2 |
| SILICA Si02 | mg/l | 1x/Y | *1 | 1.3 | *3 | *3 | *2 | *2 | 1.3 | *2 |
| SILVER | mg/l | 1x/Y | <0.05 | <0.01 | *3 | *3 | *2 | *2 | <0.01 | *2 |
| SODIUM | mg/l | 1x/Y | 20/270 | 15.1 | 22.3 | 11.5 | 11.5 | 13.1 | 13.3 | 22.3 |
| TOT DISSOLD SOLS | mg/l | 1x/Q | *1 | 151 | 184 | 116 | 136 | 141 | 159 | 145 |
| SPEC CONDUCTANCE | umhos/cm | 2x/W | *1 | 236 | 307 | 189 | 258 | 216 | 240 | 228 |
| STRONTIUM | mg/l | 1x/Y | *1 | 0.17 | *3 | *3 | *2 | *2 | 0.17 | *2 |
| SULFATE | mg/l | 1x/Y | <250 | 34 | *3 | *3 | *2 | *2 | 34 | *2 |
| ODOR | odor uts | 1x/W | 3 | <1 | <1 | <1 | <1 | <1 | <1 | <1 |
| TEMPERATURE | deg C | 1x/D | *1 | 12.1 | 24.7 | 0.8 | 1.7 | 14.3 | 23.2 | 9.2 |
| TURBIDITY | NTU | 12X/D | <1 | 2.08 | 22 | 0.04 | 0.59 | 1.21 | 5.42 | 1.09 |
| ENDRIN | µg/l | 1x/Y | 0.2 | <0.1 | *3 | *3 | *2 | *2 | <0.1 | *2 |
| TOXAPHENE | µg/l | 1x/Y | 0.5 | <1 | *3 | *3 | *2 | *2 | <1 | *2 |
| 2,4 D | µg/l | 1x/Y | 10 | <0.05 | *3 | *3 | *2 | *2 | <0.05 | *2 |
| 2,4,5-TP SILVEX | µg/l | 1x/Y | 100 | <0.05 | *3 | *3 | *2 | *2 | <0.05 | *2 |

| | | | | | | | | | | |
|---|---|---|---|---|---|---|---|---|---|---|
| ALDRIN | μg/l | 1x/Y | *1 | <0.05 | *3 | *3 | *2 | *2 | <0.05 | *2 |
| CCl4 | μg/l | 1x/Y | 5 | <0.5 | *3 | *3 | *2 | *2 | <0.5 | *2 |
| CHLORDANE | μg/l | 1x/Y | *1 | <0.5 | *3 | *3 | *2 | *2 | <0.5 | *2 |
| DDD | μg/l | 1x/Y | *1 | <0.1 | *3 | *3 | *2 | *2 | <0.1 | *2 |
| DDT | μg/l | 1x/Y | *1 | <0.1 | *3 | *3 | *2 | *2 | <0.1 | *2 |
| HEPTACHLOR | μg/l | 1x/Y | *1 | <0.05 | *3 | *3 | *2 | *2 | <0.05 | *2 |
| TOLUENE | μg/l | 1x/Y | *1 | <0.5 | *3 | *3 | *2 | *2 | <0.5 | *2 |
| METHYLENE CHLORIDE | μg/l | 1x/Y | *1 | <0.5 | *3 | *3 | *2 | *2 | <0.5 | *2 |
| TRICHLORO ETHYLENE | μg/l | 1x/Y | *1 | <0.5 | *3 | *3 | *2 | *2 | <0.5 | *2 |
| BROMOFORM | μg/L | 1x/Q | *1 | <1 | <1 | <1 | <1 | <1 | <1 | <1 |
| BROMODICHLORO-METHANE | μg/L | 1x/Q | *1 | 10.5 | 17.0 | 2.3 | 10.1 | 4.2 | 15.5 | 12.0 |
| CHLOROFORM | μg/L | 1x/Q | *1 | 61.3 | 130.0 | 30.0 | 32.3 | 48.8 | 110.0 | 54.3 |
| DIBROMOCHLORO- | μg/L | 1x/Q | *1 | <1 | 1.1 | <1 | <1 | <1 | <1 | <1 |

Courtesy of the City of Poughkeepsie, NY
Douglas Fairbanks, Jr.—Chief Operator
Robert A. Ruggiero—Laboratory Director
Key: Y = yearly; Q = quarterly; M = monthly; W = weekly; D = daily
*1 = no state std; 2 = not done; 3 = insuff data for comp; 4 = tested 1xy

# Appendix-A-V

## NEW YORK STATE DRINKING WATER STANDARDS

### MINIMUM MONITORING REQUIREMENTS

Table AV-1.  Organic Chemicals: Principal Organic Contaminants, Vinyl Chloride, VOCs (Table 9B).

| Contaminant | Type of Water system | Initial requirement[1] | Continuing requirement where detected [1,2] | Continuing requirement where not detected and vulnerable to contamination[1] | Continuing requirement where not detected and invulnerable to contamination[1] |
|---|---|---|---|---|---|
| Principal organic contaminants listed on Table 9D and vinyl chloride | Community and Nontransient Noncommunity serving 3,300 or more persons | If not sampled between 1/1/88 and 1/1/92, quarterly sample per source for one year | Quarterly | Annually | Not applicable |
| | Community and Nontransient Noncommunity serving fewer than 3,300 persons | If not sampled between 1/1/88 and 1/1/92, quarterly sample per source for one year | Quarterly[3] | Annually[3] | Once every six years[4] for groundwater sources. State discretion for surface water sources. |
| | Noncommunity | State discretion[5] | State discretion[5] | State discretion[5] | State discretion[5] |

| Unspecified organic contaminants and other POCs not listed on Table 9C and 9D. | Community and Noncommunity | State discretion[5] | State discretion[5] | State discretion[5] | State discretion[5] | State discretion[5] |
|---|---|---|---|---|---|---|

[1]The location for sampling of each groundwater source of supply shall be between the individual well and at or before the first service connection and before mixing with other sources, unless otherwise specified by the State to be at the entry point representative of the individual well. Public water systems which rely on a surface water shall sample at points in the distribution system representative of each source or at an entry point or points to the distribution system after any water treatment plant.

[2]The State may decrease the quarterly monitoring requirement to annually, provided that the system is reliably and consistently below the MCL base on a minimum of two quarterly samples from a groundwater source and four quarterly samples from a surface water source. Systems which monitor annually must monitor during the quarter which previously yielded the highest analytical result.

[3]The State may reduce the frequency of monitoring of a groundwater source to once every three years for a public water system which has three consecutive annual samples with no detection of a contaminant.

[4]The State may determine that a public water system is invulnerable to a contaminant or contaminants after evaluating the following factors:

(a) Knowledge of previous use (including transport, storage, or disposal) of the contaminant within the watershed or zone of influence of the system. If a determination by the State reveals no previous use of the contaminant within the watershed or zone of influence, a waiver can be granted.

(b) If previous use of the contaminant is unknown or it has been used previously, then the following factors shall be used to determine whether a waiver can be granted.

1. Previous analytical results.

2. The proximity of the system to a potential point or non-point source of contamination. Point sources include spills and leaks of chemicals at or near a water treatment facility or at manufacturing, distribution, or storage facilities, or from hazardous and municipal waste landfills and other waste handling or treatment facilities.

3. The environmental persistence and transport of the contaminants.

4. The number of persons served by the public water system and the proximity of a smaller system to a larger system.

5. How well the water source is protected against contamination, such as whether it is a surface or groundwater system. Groundwater systems must consider factors such as depth of the well, the type of soil, and wellhead protection. Surface water systems must consider watershed protection.

[5]State discretion shall mean requiring monitoring when the State has reason to believe the MCL has been violated, the potential exists for an MCL violation, or the contaminant may present a risk to public health.

## Table AV-2. Organic Chemicals: Pesticides, Dioxin, PCBs (Table 9C)

Contaminant

| Group 1 Chemicals | Group 2 Chemicals |
|---|---|
| Alachlor | Aldrin |
| Aldicarb | Benzo(a)pyrene |
| Aldicarb sulfoxide | Butachlor |
| Aldicarb sulfone | Carbaryl |
| Atrazine | Dalapon |
| Carbofuran | Di(2-ethylhexyl)adipate |
| Chlordane | Di(2-ethylhexyl)phthalates |
| Dibromochloropropane | Dicamba |
| 2,4-D | Dieldrin |
| Endrin | Dinoseb |
| Ethylene dibromide | Diquat[9] |
| Heptachlor | Endothall[9] |
| Heptachlor epoxide | Glyphosate[9] |
| Lindane | Hexachlorobenzene |
| Methoxychlor | Hexachlorocylclopentadiene |
| Polychlorinated biphenyls | 3-Hydroxycarbofuran |
| Pentachlorophenol | Methomyl |
| Toxaphene | Metolachlor |
| 2,4,5-TP (Silvex) | Metribuzin |
| | Oxamyl (vydate) |
| | Pichloram |
| | Propachlor |
| | Simazine |
| | 2,3,7,8-TCDD (Dioxin)[9] |

| Type of Water System | Initial requirement[1,2] | Continuing requirement where detected[2,3,10] | Continuing requirement where not detected[2] |
|---|---|---|---|
| Community and Nontransient Noncommunity serving 3,300 or more persons[9] | Quarterly sample per source, for one year by 12/31/93[4] | Quarterly | One sample every eighteen months per source |
| Community and Nontransient Noncommunity serving fewer than 3,300 persons and more than 149 service connections | Quarterly samples per entry point, for one year by 12/31/94[5,6,7] | Quarterly | Once per entry point every three years[567] |
| Community and Nontransient Noncommunity serving fewer than 3,300 persons and more than 150 service connections | Quarterly samples per entry point for one year by 12/31/95[5,6,7] | Quarterly | Once per entry point every three years[5,6,7] |
| | State discretion[8] applied to Group 2 | Quarterly | State discretion[8] applied to Group 2 |
| Noncommunity | State discretion[8] | State discretion[8] | State discretion[8] |

[1]If monitoring data collected after January 1, 1990, are consistent with the requirements of Appendix 5-C then the State may allow systems to use that data to satisfy the initial requirements.

[2]The location for sampling of each groundwater source of supply shall be between the individual well and at or before the first service connection and before mixing with other sources, unless otherwise specified by the State to be at the entry point representative of the individual well. Public water systems which take water from a surface body or watercourse shall sample at points in the distribution system representative of each source or at entry point or points to the distribution system after any water treatment plant.

[3]The state may decrease the quarterly monitoring requirement to annually, provided that system is reliably and consistently below the MCL, based on a minimum of two quarterly samples from a groundwater source, and four quarterly samples from a surface water source. Systems which monitor annually must monitor during the quarter that previously yielded the highest analytical result. Systems serving fewer than 3,300 persons and which have three consecutive annual samples without detection may apply to the State for a waiver in accordance with footnote 6.

[4]The State may allow a system to postpone monitoring for a maximum of two years, if an approved laboratory is not reasonably available to do a required analysis within the scheduled monitoring period.

[5]The state may waive the monitoring requirement for a public water system that submits information every three years to demonstrate that a contaminant or contaminants were not used, transported, stored, or disposed within the watershed or zone of influence of the system.

[6]The State may reduce the monitoring requirement for a public water system that submits information every three years to demonstrate that the public water system is invulnerable to contamination. If previous use of the contaminant is unknown or it has been used previously, then the following factors shall be used to determine whether a waiver is granted.

a. Previous analytical results.

b. The proximity of the system to a potential point or non-point source of contamination. Point sources include spills and leaks of chemicals at or near a water treatment facility or at manufacturing, distribution, or storage facilities, or from hazardous and municipal waste landfills and other waste handling or treatment facilities. Non-point sources include the use of pesticides to control insect and weed pests on agricultural areas, forest lands, home, and gardens, and other land application uses.

c. The environmental persistence and transport of the pesticide or PCBs.

d. How well the water source is protected against contamination due to such factors as depth of the well and type of soil and the integrity of the well casing.

e. Elevated nitrate levels at the water supply source.

f. Use of PCBs in equipment used in production, storage, or distribution of water.

[7]The state may allow systems to composite samples in accordance with the conditions in Appendix 5-C.

[8]State discretion shall mean requiring monitoring when the State has reason to believe the MCL has been violated, the potential exists for an MCL violation, or the contaminant may present a risk to public health.

[9]The State may waive monitoring of this contaminant for public water systems serving 3,300 or more persons that meet the conditions of footnote 5.

[10]If a contaminant is detected, repeat analysis must include all analytes contained in the approved analytical method in Appendix 5-C for the detected contaminant.

[11]The Appendix 5-C reports the acceptable methods for the analysis of contaminants in water and related sampling requirements.

## Table AV-3.   Organic Chemicals: Principal Organic Contaminants (Table 9D).

| Contaminant | Specific Contaminants for Analysis | |
| --- | --- | --- |
| POCs | benzene[1] | 1,1-dichloropropene |
| | bromobenzene | cis-1,3-dichloropropene |
| | bromochloromethane | trans-1,3-dichloropropene |
| | bromomethane | ethylbenzene[1] |
| | n-butylbenzene | hexachlorobutadiene |
| | sec-butylbenzene | isopropylbenzene |
| | tert-butylbenzene | p-isopropyltoluene |
| | carbon tetrachloride[1] | methylene chloride |
| | chlorobenzene[1] | n-propylbenzene |
| | chloroethane | styrene[1] |
| | chloromethane | 1,1,1,2-tetrachloroethane |
| | 2-chlorotoluene | 1,1,2,2-tetrachloroethane |
| | 4-chlorotoluene | tetrachloroethene[1] |
| | dibromomethane | toluene[1] |
| | 1,2-dichlorobenzene[1] | 1,2,3-trichlorobenzene |
| | 1,3-dichlorobenzene | 1,2,4-trichlorobenzene |
| | 1,4-dichlorobenzene[1] | 1,1,1-trichloroethane[1] |
| | dichlorodifluoromethane | 1,1,2-trichloroethene |
| | 1,1-dichloroethane | trichloroethane[1] |
| | 1,2-dichloroethane[1] | trichlorofluoromethene |
| | 1,1-dichloroethene[1] | 1,2,3-trichloropropane |
| | cis-1,2-dichloroethene[1] | 1,2,4-trimethylbenzene |
| | trans-1,2-dichloroethene[1] | 1,3,5-trimethylbenzene |
| | 1,2-dichloropropane[1] | m-xylene[1] |
| | 1,3-dichloropropane | o-xylene[1] |
| | 2,2-dichloropropane | p-xylene[1] |

[1]Notification must contain mandatory health effect language.

# Appendix A-VI

## RADIONUCLIDE DATA

Average Relative Source Contribution to the Daily Intake of Natural Radionuclides. (The drinking water contribution assumes that 2 liters per day are ingested.)*

| Radionuclide | Source | pCi/d[†] | Reference |
|---|---|---|---|
| Radium-226 | Air | 0.007 | a |
| | Food | 1.4 | b |
| | | 1.7 | c |
| | | 1.1 | a |
| | Drinking water | Generally small from surface supplies | a |
| | | 0.01–0.2 | c |
| | | 0.6–2 | d |
| Radium-228 | Air | 0.007 | a |
| | Food | 1.1 | b |
| | | 1.1 | a |
| | Drinking water | 0.8–2 | d |
| Uranium-234 | | | |
| Uranium-238 | Air | 0.0007 | a |
| | Food | 0.9 | b |
| | | 0.8 | c |
| | | 0.37 | a |
| | Drinking water | 0.03–1 | c |
| | | Generally negligible | a |
| | | 0.6–4 | d |
| Lend-210 | Air | 0.3 | a |
| | Food | 1.4 | b |
| | | 1.2 | c |
| | | 3.0 | a |
| | Drinking water | 0.02 | c |
| | | <0.22 | d |
| Polonium-210 | Air | 0.06 | a |
| | Food | 1.8 | b |
| | | 1.2 | c |
| | | 3.0 | a |
| | Drinking water | 0.02 | c |
| | | <0.26 | d |
| Thorium-230 | Air | 0.0007 | a |
| | Food | Probably negligible | a |
| | Drinking water | <0.08 | d |

## Appendix A-VI. (*Continued*)

| Radionuclide | Source | pCi/d** | Reference |
|---|---|---|---|
| Thorium-232 | Air | 0.0007 | a |
| | Food | Negligible | a |
| | Drinking water | <0.02 | d |
| Radon-222 | Outdoors (1.8 Bq/m$^3$)$^\ddagger$ | 970 | a,e |
| | Indoors (15 Bq/m$^3$) | 8100 | a |
| | Drinking water | 100–800 | d,e,f, |

a United Nations, 1982
b NCRP report 45
c NCRP report 77
d Cothern, et al., 1986
e Assuming an average adult inhalation rate of 20 cubic meters/day.
f Assuming that a concentration of radon per liter of air is $10^{-4}$ times the concentration per liter of water.
*Environmental Protection Agency—40 CFR Part 141 Water Pollution Control; National Primary Water Regulations; *Radionuclides*; Advance Notice of Proposed Rulemaking (Sept. 30, 1986).
$^\dagger$pCi/d = Picocuries per day
$^\ddagger$Becquerel per cubic meter

# Appendix A-VII

---

## FLUORIDATION DATA

---

**FLUORIDATION***

1. NEW YORK CITY DEPARTMENT OF HEALTH.
   Summary of Recommendations for Fluoridation: "Report to the Mayor on Fluoridation for New York City" by the Board of Health–The City of New York (1955), Leona Baumgartner, M.D., Chairman.
2. NATIONAL ACADEMY OF SCIENCE–NATIONAL RE-SEARCH COUNCIL.
   "Drinking Water and Health" 1977, Summary and Conclusions
3. USEPA PROPOSED NOTIFICATION OF CONSUMERS.
   Connected to a distribution system with high concentration of fluoride.

## NEW YORK CITY DEPARTMENT OF HEALTH

The following summary of recommendations for fluoridation was presented in a "Report to the Mayor on Fluoridation for New York City" by the Board of Health of the City of New York (1955), Leona Baumgartner, M.D., Chairman.

Dental decay constitutes a serious health problem which remains largely neglected due to the undramatic nature of the disease, the cost of full treatment and the lack of sufficient professional personnel. This combination of factors points to the need for a preventive procedure—preferably one that is inexpensive and automatic.

The overwhelming evidence obtained by the application of accepted scientific control standards indicates that water fluoridation offers an ultimate reduction of about 60% in the accumulation of dental defects in New York City's resident population.

Fluoride is a normal component of the body and is distributed widely in natural foods. Most fluoride (at least 75%) is excreted at the recommended

---

*(a) *Fluoride*—As an inorganic chemical parameter, see Chapter 3, *Type A*.
*(b) *Fluoridation*—As public health program and evaluation, see Chapter 9, "Public Health Regulations".

level of one part per million. In general, soft tissues and organs do not store fluoride.

Some fluoride is normally stored in the calcified tissues (bones and teeth) with a self-limiting mechanism helping to control that storage. From the systemic viewpoint, the first observable change is the increase in bone density with continued consumption of water containing more than 5 ppm fluoride.

"Natural" and "artificial" fluoride are identical in their chemical and physiologic action.

As with most other materials introduced into the human body—even the foods we eat—fluoride is a potentially toxic substance. However, the margin of safety at the recommended level is enormous.

The foregoing information derived from many sources reveals that *no harm has been demonstrated or can be anticipated from the consumption of fluoridated water at the recommended level.*

The evaluation of the available scientific data, both theoretical and clinical, fails to reveal any threat of harm to even a small proportion of the population consuming water fluoridated at the 1 ppm level. It is further established that these scientific data are so extensive and broad in scope as to establish the safety of the procedure beyond a reasonable doubt. There is no known measure of health protection which has been subject to as much investigation prior to adoption as has water fluoridation. Water fluoridation presents no industrial problem of practical consequence.

The adjustment of the fluoride content of drinking water to 1 ppm fluoride is in principle and in practice the soundest and most effective approach to caries prevention on a large scale known today.

It is possible to furnish fluoridated water to all of the people of the city for about one-third the price of providing fluoridation for school children only by any other method.

Water fluoridation is endorsed and recommended by all the major national health organizations interested in human well-being. They include: The American Dental Association, American Medical Association, American Public Health Association, National Research Council, Commission on Chronic Illness, American Academy of Pediatrics, and the U.S. Public Health Service. On the local level, the procedure has the endorsement of such groups as the New York State Department of Health, the State and all local dental societies, all the county medical societies in New York City, the New York Academy of Medicine, the Welfare and Health Council, the Commerce and Industry Association of New York City, the United Parents Association, and the Citizens Committee on Children.

There is no doubt that the overwhelming weight of competent scientific opinion in this country and in New York City itself is that water fluoridation is an effective, safe, practical and much needed public health procedure. The New York City Board of Health reaffirms its endorsement on the basis of the exhaustive study outlined in the foregoing report.

## Evaluation of Various Fluoride Vehicles.

| Vehicle | If used as prescribed | | Self-Limitation | Ease of Regulation | Approximate Annual Cost Per Person | Convenience | Consumption by Children | Summary of Drawbacks |
| | Effectiveness | Safety | | | | | | |
|---|---|---|---|---|---|---|---|---|
| Fluoridated Communal Water | Established | Established | Yes | Very good | .9 | No effort required | Yes-established by results | Waste of fluoride but still cheapest |
| Bottled Fluoridated Water | Established | Established | Yes | Fairly good | $18.25 | Special effort | Yes-but related to convenience | Inconvenience, Cost |
| Home-Prepared Fluoridated Water | Established | Established | Yes | Poor | $3.65 | Special effort | Yes-but related to convenience | Inconvenience, Regulation Difficulty |
| Milk | Unknown | Probably safe | Yes | Poor | $2.14 | Selective effort | Variable and selective | Effectiveness Unknown, Cost, Variable Consumption |
| Tablets | Unknown | Probably safe | No | Poor | $3.65 | Special effort | Not normally consumed | Effectiveness Unknown, Inconvenience, Regulation Difficulty, Cost and Accident Hazard |
| Solid Foods (cereals, baby foods, etc) | Unknown | Probably safe | No | Poor | ? | Selective effort | Variable | Effectiveness Unknown, Variable-Consumption, Regulation Difficulty, Cost |
| Salt, Sugar, etc. | Unknown | Probably safe | No | Poor | ? | No effort | Very variable and low | |

## CONCLUSIONS AND RECOMMENDATIONS

This report on the desirability of fluoridation for New York City was prepared by the Board of Health at the request of Mayor Wagner. It is based
on an exhaustive analysis of the scientific literature as well as correspon-
dence and discussion with experts in various related fields. A special effort
was made to solicit documents from opponents of the procedure. Such
extensive study, as is reflected in the body of this report, forms the basis of
the following conclusions:

1. Dental decay is prevalent among all but a small percentage of New
   York City's population. Its extent, widespread neglect, cost of correction and impact on general health make it a true public health problem demanding a preventive approach.
2. Water fluoridation at the level of one part per million has been well
   established as a procedure which offers approximately a 60% reduction in new dental decay not only among children but among adults
   who consume such water throughout life.
3. A substantial body of scientific information attests to the achieve-
   ment of these substantial dental benefits without systemic harm to the
   body. There is no known health measure the safety of which has been
   subjected to as rigorous an investigation prior to adoption as has
   water fluoridation.
4. Evaluation of all known technics for prevention of tooth decay leads
   to the conclusion that water fluoridation combines maximum benefit
   and practicality.
5. The cost of water fluoridation for New York City would involve a
   maximum initial cost of $450,000 and an annual maintenance cost of
   about nine cents per capita.
6. Court decisions and official legal opinions have established that it is
   within the constitutional power of a municipality to fluoridate its
   water supply.

It is therefore urged that appropriate action be taken to improve the
health of the citizens of New York City by providing for the fluoridation of
our municipal water supply.

## NATIONAL ACADEMY OF SCIENCES NATIONAL RESEARCH
## COUNCIL SAFE DRINKING WATER COMMITTEE*

### Summary and Conclusions

In summary, there is no generally accepted evidence that anyone has been
harmed by drinking water with fluoride concentrations considered optimal

---

*National Academy of Sciences, "Drinking Water and Health" (1977) Vol. 1. Washington, DC.

for the annual mean temperatures in the temperate zones. It seems likely, however, that objectionable dental fluorosis occurred in two children with diabetes insipidus. Bone changes, possibly desirable, have been noted in patients being dialyzed against large volumes of fluoridated water. Similar changes can be expected in the rare renal patient with a long history of renal insufficiency and a high fluid intake that includes large amounts of tea. With this particular combination of circumstances, the lowest drinking-water concentration of fluoride associated with symptomatic skeletal fluorosis that has been reported to date is 3 ppm, outside of countries such as India. It should be possible for the medical profession to avoid the possible adverse effects of fluoride under the conditions described above, thereby making it unnecessary to limit the concentrations of fluoride in order to protect these rare patients.

On the basis of studies done more than 15 years ago, occasional objectionable mottling would be expected to occur in communities in the hotter regions of the United States with water that contains fluoride at 1 ppm or higher and in any community with water that contains fluoride at 2 ppm or higher. However, this may not be the case today; more liberal provisional limits seem appropriate while studies are conducted to clarify the subject.

The possibility of fluoride causing other adverse effects (allergic responses, mongolism, and cancer) or beneficial effects other than decreased dental caries has not been adequately documented to carry weight in the practical decision about the desirable levels of fluoride. The questions of mongolism and cancer have been raised on the basis of epidemiological data for which there is contrary evidence and the risk factors involved in any case are too low to establish a causal association. The allergic responses claimed by some reports are based on clinical observations and in some cases double blind tests. The reservation in accepting these at face value is the lack of similar reports in much larger numbers of people who have been exposed to considerably more fluoride than was involved in the original observations. From a scientific point of view none of these effects can be ruled out, but the available data are rather limited or easily improved, so further study is indicated.

## USEPA—PUBLIC NOTIFICATION FOR FLUORIDE*

The notice to be used by systems which exceed the secondary MCL shall contain the following language and no additional language except as necessary to replace the asterisks:

---

*National Primary Drinking Water Regulations—Fluoride—Final Rule, Part II—40 CFR Parts 141, 142, and 143—April 2, 1986, *Fed. Reg.* Vol. 51, No. 63.

Public Notice

Dear User,

The U.S. Environmental Protection Agency requires that we send you this notice on the level of fluoride in your drinking water. The drinking water in your community has a fluoride concentration of[1] _____ milligrams per liter (mg/L).

Federal regulations require that fluoride, which occurs naturally in your water supply, not exceed a concentration of 4.0 mg/L in drinking water. This is an enforceable standard called a Maximum Contaminant Level (MCL), and it has been established to protect the public health. Exposure to drinking water levels above 4.0 mg/L for many years may result in some cases of crippling skeletal fluorosis, which is a serious bone disorder.

Federal law also requires that we notify you when monitoring indicates that the fluoride in your drinking water exceeds 2.0 mg/L. This is intended to alert families about dental problems that might affect children under nine years of age. The fluoride concentration of your water exceeds this federal guideline.

Fluoride in children's drinking water at levels of approximately 1 mg/L reduces the number of dental cavities. However, some children exposed to levels of fluoride greater than about 2.0 mg/L may develop dental fluorosis. Dental fluorosis, in its moderate and severe forms, is a brown staining and/or pitting of the *permanent* teeth.

Because dental fluorosis occurs only when *developing* teeth (before they erupt from the gums) are exposed to elevated fluoride levels, households without children are not expected to be affected by this level of fluoride.

Families with children under the age of nine are encouraged to seek other sources of drinking water for their children to avoid the possibility of staining and pitting.

Your water supplier can lower the concentration of fluoride in your water so that you will still receive the benefits of cavity prevention while the possibility of stained and pitted teeth is minimized.

Removal of fluoride may increase your water costs. Treatment systems are also commercially available for home use. Information on such systems is available at the address given below. Low fluoride bottled drinking water that would meet all standards is also commercially available.

For further information, contact[2] _____ at your water system.

---

[1]PWS shall insert the compliance result which triggered notification under this Part.
[2]PWS shall insert the name, address, and telephone number of a contact person at the PWS.

# Appendix A-VIII

## GENERAL REFERENCES

American Water Works Association. Seminar proceedings on following subjects: **Control of Inorganic Contaminants** (1983 No. 20175). **Chloramination for THM Control: Principles and Practices** (1984 No. 20181). **Ozonation: Recent Advances and Research Needs** (1986 No. 20005). **Water Chlorination—Principles and Practices** (1986 No. 20008). **Water Quality Concerns in the Distribution System** (1986 No. 20009). **Assurance of Adequate Disinfection, or C-T or Not C-T** (1987 No. 20013). **Granular Activated Carbon Installations: Conception to Operation** (1987 No. 20018), Denver, CO: American Water Works Association.

APHA-AWWA-WEF. 1995. **Standard Methods\* for the Examination of Water and Wastewater.** 19th edition. Washington, D.C.: American Public Health Association.

A.S.C.E.—A.W.W.A. 1990. **Water Treatment Plant Design.** 2nd edition. New York, NY: McGraw-Hill.

California Institute of Technology. 1963. **Water Quality Criteria.** Pasadena, CA: NTIS, Springfield, VA.

CRC. 1995. **Handbook of Chemistry and Physics.** 75th edition. Boca Raton, FL: CRC Press, Inc.

Montgomery, J. M. Consulting Engineers, Inc. 1985. **Water Treatment Principles and Design.** New York, NY: John Wiley & Sons.

Dean, J. A. 1985. **Lange's Handbook of Chemistry.** 15th edition. New York, NY: McGraw-Hill.

Lewis, Richard J., Sr. 1993. **Hawley's Condensed Chemical Dictionary** 12th edition. New York, NY: Van Nostrand Reinhold.

National Academy of Sciences. 1977–87. **Drinking Water and Health.** Vols. 1–8. prepared for USEPA by the National Research Council. Washington, D.C.: National Academy Press.

USEPA. 1983. **Methods for Chemical Analysis of Water and Wastes,** EPA-600/4-7S-020, Rev. 3/83. Cincinnati, OH: Env. Monitoring and Support Laboratory.

USEPA. 1985. **Guidance Manual For Compliance With the Filtration and Disinfection Requirements For Public Water Systems Using Surface Water Sources.** Science and Technology Branch-Criteria and Standards Division-Office of Water. Washington, D.C.: USEPA Publications or NTIS, Springfield, VA.

USEPA. **Methods for the Determination of Organic Compounds in Drinking Water.** Supplements I and II, EPA600/4-90/020 (July 1990) and EPA-600/R-92/129 (Aug. 1992). Springfield, VA: NTIS.

---

\*Quoted throughout this book as *Standard Methods.*

# Appendix A-IX

---

## CONVERSION FACTORS

---

### INTERNATIONAL SYSTEM OF UNITS (SI)

SI, also known as MSKA system or MKS system, is the metric system of units based on the meter (m), kilogram (kg), second (s), ampere (A), Kelvin (K) degree, mole (mol), and candela (cd).

The MKS system can also be defined as follows:

| Physical Quantity | Name of SI Unit | Symbol |
|---|---|---|
| Length | meter | m |
| Mass | kilogram | kg |
| Time | second | s |

Standardized multiplying factors from $10^{18}$ to $10^{-18}$ are:

| Prefix | Symbol | Multiplying Factor |
|---|---|---|
| exa | E | $10^{18}$ |
| peta | P | $10^{15}$ |
| tera | T | $10^{12}$ |
| giga | G | $10^9$ |
| mega | M | $10^6$ |
| kilo | k | $10^3$ |
| hecto | h | $10^2$ |
| deka | da | $10^1$ |
| deci | d | $10^{-1}$ |
| centi | c | $10^{-2}$ |
| milli | m | $10^{-3}$ |
| micro | μ | $10^{-6}$ |
| nano | n | $10^{-9}$ |
| pico | p | $10^{-12}$ |
| femto | f | $10^{-15}$ |
| atto | a | $10^{-18}$ |

### GENERAL CONVERSION FACTORS

| Multiply | By | To Obtain |
|---|---|---|
| centigrams | 0.01 | grams |
| centiliters | 0.01 | liters |
| centimeters | 0.01 | meters |
| decigrams | 0.1 | grams |
| deciliters | 0.01 | liters |
| decimeters | 0.1 | meters |
| kilograms | 1000 | grams |

## GENERAL CONVERSION
## FACTORS (*Continued*)

| Multiply | By | To Obtain |
|----------|-----|-----------|
| cubic meters | 1000 | liters |
| kilometers | 1000 | meters |
| milligrams | 0.001 | grams |
| milliliters | 0.001 | liters |
| millimeters | 0.001 | meters |
| tons (metric) | 1000 | kilograms |

## GENERAL CONVERSION FACTORS—U.S. to Metric

| Multiply | By | To Obtain |
|----------|-----|-----------|
| Cubic feet | 1,728 | cubic inches |
| Cubic feet | 7.48 | gallons |
| Cubic yards | 27 | cubic feet |
| Cubic yards | 46,656 | cubic inches |
| Cubic yards | 202 | gallons |
| Feet | 12 | inches |
| Gallons | 231 | cubic inches |
| Miles | 5,280 | feet |
| Miles | 1,760 | yards |
| Miles | 63,360 | inches |
| Yards | 3 | feet |
| Yards | 36 | inches |
| *Length* | | |
| 1 inch | 25.4 | millimeters |
| 1 foot | 304.8 | millimeters |
| 1 yard | 0.9144 | meters |
| 1 mile | 1.60934 | kilometers |
| *Area* | | |
| 1 square inch | 645.16 | square millimeters |
| 1 square foot | 0.092903 | square meters |
| 1 square yard | 0.836127 | square meters |
| 1 square mile | 2.58998 | square kilometers |
| 1 acre | 4,046.86 | square meters |
| *Volume* | | |
| 1 cubic inch | 16,387.1 | cubic millimeters |
| 1 cubic foot | 0.0283168 | cubic meters |
| 1 cubic foot | 28.3168 | liters |
| 1 cubic foot | 6.22884 | U.K. gallon |
| 1 cubic foot | 7.48052 | U.S. gallon |
| 1 cubic yard | 0.764554 | cubic meter |
| 1 U.K. gallon | 0.004546 | cubic meter |
| 1 U.K. gallon | 4.546 | liters |
| 1 U.K. gallon | 1.20095 | U.S. gallon |
| 1 U.K. gallon | 0.160544 | cubic feet |
| 1 U.S. gallon | 0.00378541 | cubic meter |

## GENERAL CONVERSION FACTORS—
### U.S. to Metric (*Continued*)

| Multiply | By | To Obtain |
|---|---|---|
| 1 U.S. gallon | 3.78541 | liters |
| 1 U.S. gallon | 0.832675 | U.K. gallon |
| 1 U.S. gallon | 0.133681 | cubic feet |
| 1 acre-inch | 3,630.0 | cubic feet |
| 1 acre-inch | 22,611 | U.K. gallon |
| 1 acre-inch | 27,154.25 | U.S. gallon |
| *Mass* | | |
| 1 ounce | 28.3495 | grams |
| 1 pound | 0.453592 | kilograms |
| 1 ton (long) | 1,016.05 | kilograms |
| 1 ton | 1.01605 | ton (metric) |
| *Density* | | |
| 1 pound/cubic foot | 16.03 | kilograms/cubic meters or grams/liter |

## CONVERSION FACTORS—METRIC TO U.S.

| Multiply | By | To Obtain |
|---|---|---|
| *Length* | | |
| 1 millimeter | 0.0393701 | inches |
| 1 meter | 3.28084 | feet |
| 1 meter | 1.09361 | yards |
| 1 kilometer | 0.621371 | miles |
| *Area* | | |
| 1 square millimeter | 0.00155 | square inches |
| 1 square meter | 10.7639 | square feet |
| 1 square meter | 1.19598 | square yards |
| 1 square kilometer | 247.105 | acre |
| 1 square kilometer | 0.386102 | square miles |
| *Volume* | | |
| 1 cubic meter | 1,000.0 | liters |
| 1 cubic meter | 35.3146 | cubic feet |
| 1 cubic meter | 1.30795 | cubic yard |
| 1 cubic meter | 219.969 | U.K. gallon |
| 1 cubic meter | 264.172 | U.S. gallon |
| 1 cubic meter | 0.0097285 | acre-in. |
| 1 liter | 0.001 | cubic meter |
| 1 liter | 0.0353147 | cubic feet |
| 1 liter | 0.219969 | U.K. gallon |
| 1 liter | 0.264172 | U.S. gallon |
| *Mass* | | |
| 1 gram | 0.035274 | ounces |
| 1 kilogram | 2.20462 | pounds |
| 1 ton (metric) | 1,000.0 | kilograms |
| 1 ton (metric) | 0.9842 | ton (long) |
| 1 ton (short) | 2,000 | pounds |
| *Density* | | |
| 1 kilogram/cubic meter | 0.0624 | pounds/cubic feet |

## CONVERSION FACTORS (*Continued*)

| Multiply | By | To Obtain |
|---|---|---|
| Acres | 43,560 | square feet |
| Acre-feet | 43,560 | cubic feet |
| Acre-feet | 325,851 | gallons |
| Atmospheres | 29.92 | inches of mercury |
| Atmospheres | 33.90 | feet of water |
| Atmospheres | 14.70 | pounds/square inch |
| Centimeters of mercury | 0.01316 | atmospheres |
| Cubic feet | 0.03704 | cubic yards |
| Cubic feet | 7.48052 | gallons |
| Cubic feet | 28.32 | liters |
| Cubic feet/second | 0.646317 | million gallons/day (MGD) |
| Cubic feet/second | 448.831 | gallons/minute |
| Day | 1,440 | minutes |
| Degrees (angle) | 60 | minutes |
| Degrees (angle) | 0.01745 | radians |
| Degrees (angle) | 3,600 | seconds |
| Feet of water | 0.02950 | atmospheres |
| Feet of water | 0.8826 | inches of mercury |
| Feet of water | 62.43 | pounds/square feet |
| Feet of water | 0.4335 | pounds/square inches |
| Gallons | 0.1337 | cubic feet |
| Gallons | 3.785 | liters |
| Gallons, Imperial | 1.20095 | U.S. gallons |
| Gallons, U.S. | 0.83267 | imperial gallons |
| Gallons, Water | 8.3453 | pounds of water |
| Gallons/minute | $2.228 \times 10^{-3}$ | cubic feet/second |
| Gallons/minute | $1.44 \times 10^{-3}$ | million gallons/day |
| Grains/U.S. gallon | 17,118 | parts/million |
| Grains/U.S. gallon | 142.86 | pounds/million gallons |
| Grams/liter | 1,000 | parts/million |
| Inches of mercury | 0.03342 | atmospheres |
| Inches of mercury | 1.133 | feet of water |
| Inches of mercury | 70.73 | pounds/square feet |
| Inches of mercury | 0.4912 | pounds/square inch |
| Inches of water | 0.002458 | atmospheres |
| Inches of water | 0.07355 | inches of mercury |
| Inches of water | 5.2025 | pounds/square foot |
| Inches of water | 0.03613 | pounds/square inch |
| Kilograms | 2.205 | pounds |
| Liters | 1.057 | quarts (liquid) |
| Liters/second | 0.02283 | million gallons/day |
| Million gallons/day | 1.54723 | cubic feet/second |
| Million gallons/day | 694.444 | gallons/minute |
| Parts/million | 0.0584 | grains/U.S. gallon |
| Parts/million | 8.347 | pounds/million gallons |
| Pounds of water | 0.01602 | cubic feet |
| Pounds of water | 27.68 | cubic inches |
| Pounds of water | 0.1198 | gallons |
| Pounds of water/minute | $2.67 \times 10^{-4}$ | cubic feet/second |
| Pounds/square foot | 0.01602 | feet of water |

## CONVERSION FACTORS (*Continued*)

| Multiply | By | To Obtain |
|---|---|---|
| Pounds/square foot | $6.944 \times 10^{-3}$ | pounds/square inch |
| Pounds/square inch | 0.06803 | atmospheres |
| Pounds/square inch | 2.307 | feet of water |
| Quart (U.S. Liquid) | 0.9461 | liters |
| Square Feet | $2.296 \times 10^{-5}$ | acres |
| Square Miles | 640 | acres |
| | $27.88 \times 10^{6}$ | square feet |

# Index

Boldface pages indicate general discussion.